PRINCIPLES OF DEVELOPMENTAL
BIOLOGY

W . W . NORTON & COMPANY

NEW YORK . LONDON

To our spouses, Diane and Don, our two families,

and mentors Ian Sussex and the late James Ebert

CONTENTS

PART THREE VERTEBRATE ORGANOGENESIS 101

PREFACE

Developmental biology is in the midst of an extraordinary flowering of discovery. For the authors of a new textbook, this was both a blessing and a challenge. The process of choosing from among this cornucopia of material was exhilarating, but the vastness of the field meant we had to make hard choices about what material to include and where to place the critical emphases. In this preface, we shall explain the choices we made in writing *Principles of Developmental Biology*.

The impetus for writing this book was the need we have felt as teachers for an introduction to developmental biology that presents the essentials while not overwhelming students in terms of length and level of difficulty. Our aim was to write a book that can be taught in a one-semester course and that does not presume the student reader will become a professional developmental biologist. Since the backgrounds of students studying developmental biology are so diverse, the only prerequisite we assume is an introductory college-level course in biology that includes the basics of molecular and cell biology as well as standard organismal biology.

Attaining our goals was a tall order. The subject matter is vast and complex, and factual knowledge in this field is increasing at breakneck speed. Hence, we have put the emphasis on the principles, including some descriptive embryology of selected plants and animals, and on providing some historical context. A relatively small number of examples have been chosen. We are painfully aware of many beautiful experiments and developmental systems that have been omitted—the price for presenting an introduction rather than an exhaustive treatment. Nonetheless, it is our experience that most students can tackle and understand more complex and sophisticated experiments once they have truly mastered the basics through a thorough exploration of a limited number of apt examples.

In order to attain some reasonable brevity and to cut through a thicket of vocabulary and examples, we have chosen to focus on the embryos of fruit fly, frog, chick, and mouse and on the development of maize and *Arabidopsis* for most of our examples. Other important developing systems—such as those of *Caenorhabditis elegans*, sea urchins, ascidians, and zebrafish—are briefly introduced where we have judged appropriate. Of course, this results in some loss of breadth, since important experiments have been and are being done using organisms other than those featured here. On the other hand, it has been our experience as teachers that an extensive comparative approach leads most students to acquire more facts but less understanding.

We have followed an order of presentation in which we first focus on describing the actual development of animals and plants, and then in later chapters we emphasize an analysis of the mechanisms of this development. We begin Part One with a brief introduction to the core intellectual issue of developmental biology, namely, how can two cells with the same genetic information become functionally different? This is followed by a description of gametogenesis and fertilization in animals. Part Two describes the early development of *Drosophila*, amphibians, and amniotes. Our teaching experience has shown that the sequence of describing the development of one organism, followed by the development of a second, when coupled with appropriate comments contrasting the two organisms, is easier for beginning students than a stage-by-stage comparative approach. Part Three is a more detailed examination of organogenesis in vertebrates. This topic was formerly the bedrock of beginning embryology courses, and still constitutes an important subject. While the emphasis here, as in Part Two, is on description, these chapters also introduce some analysis of what drives development, emphasizing mechanisms, especially intercellular communication by way of ligand–receptor interactions. Part Four describes plant development in both meristems and embryos. This is an area in which cellular and molecular studies are beginning to show clearly the similarities and differences between plant and animal development, and so in our judgment is now appropriate for inclusion in an introductory course. Part Five presents some necessary cell biology for subsequent chapters. Part Six provides a deeper exploration of morphogenesis by analyzing the molecular basis of differentiation and pattern formation. Included are discussions of the molecular basis of the regulation of transcription, and posttranscriptional events, with special emphasis on inter-

cellular signaling. The last chapter in the book is a brief treatment of the connections between embryo development and evolution, a classical subject area in developmental biology that is being newly revitalized.

Certain important, complex topics are reconsidered throughout the book. For instance, gastrulation is described in Chapters 3, 4, and 5. But details of the cellular and molecular basis of gastrulation are not taken up until Chapters 13 and 16, after the student has acquired sufficient background understanding. Likewise, vertebrate limb development is briefly described in Chapter 7, and the importance of epithelial-mesenchymal interactions in limb development is presented in Chapter 13. Moreover, limb development is revisited in Chapters 16 and 17 because at this point the student can follow complexities in the molecular basis of limb formation. Each of the several discussions of a particular subject area is given either in a different context or at a different level of sophistication. There are substantial pedagogical advantages to revisiting the same topic with a change of focus: the repetition allows the student to learn first about the developmental anatomy, and subsequently about the cellular interactions and changing gene expression underpinning the development.

Textbooks commonly have introductory chapters that cover elementary cell and molecular biology. We have instead chosen to distribute crucial bits of background knowledge and experimental technique throughout the book, where the topic at hand calls for them. Hence, the first chapter contains boxes on nucleic acid hybridization and cloning. Most chapters include various kinds of special information, also in boxes. These may emphasize techniques, as mentioned earlier; or different developmental systems, such as those of the zebrafish or the ascidians; or special topics of interest, such as sex determination, genomics, and stem cells. For instructors who prefer to present introductory and background information at the beginning of the course, two chapters in the middle of our book can be considered at the outset. Chapter 12 is an introduction to cell behavior and the molecules of the extracellular matrix. Chapter 14 is a consideration of the molecular biology of information flow in the cell. (We placed them where they now occur because they contain review information essential for understanding the material immediately following Chapter 13 and Chapters 15 through 17). We hope that teachers using this text will find the CD containing most of the illustrations from the book useful for preparing lectures. The questions posed at the end of each chapter—many without a simple right or wrong answer—are intended to stimulate students to think about the material. The references at each chapter end include some important classics, as well as works providing entrée into what is often a dense literature.

A sampling of the many websites students may find useful includes the following:

Web links and learning tools for developmental biology:
www.ucalgary.ca/~browder/

Excellent tutorials on early development of amphibians, sea urchins, and zebrafish:
worms.zoology.wisc.edu/embryology_main.html

Vertebrate embryo anatomy:
anatomy.med.answ.edu.au/cbl/embryo/

Development in the fruit fly (a database of images):
pbio07.uni-muenster.de/

The development of *C. elegans:*
www.wormbase.org/

Root development:
www.bio.uu.nl/~mcbroots/

There are many other useful websites, and new ones keep appearing. Use a search engine, such as Google or Yahoo, to search on some key words to find them.

We have many people to thank. We are grateful to those who reviewed various portions of the manuscript; errors of fact or opinion that remain are, of course, strictly those of the authors. The reviewers were Anonymous, Steve Benson, Graeme Berlyn, David Epel, John Gerhart, Paul Lasko, Judy Lengyel, Mike Levine, Randy Moon, Lisa Nagy, Steve Oppenheimer, Rodolfo Rivas, Chris Rose, Mark Servetnick, Ian Sussex, and Vic Vacquier.

We want to acknowledge our debt to our own teachers and colleagues, among them the late James Ebert, John Gerhart, Don Kaplan, Ray Keller, the late Dan Mazia, and Ian Sussex. We are grateful to several generations of students, who informed and inspired us. We also wish to thank many colleagues, friends, and family who helped us, with special mention for Jim Fristrom, Rob Grainger, Pat Hamilton, Kristen Shepard, and Diane Wilt. The work of Connie Balek and her colleagues at Precision Graphics saved us. Susan Middleton has been a wonderful copy editor from whom we have learned much, and we are indebted to the team at W. W. Norton, among them Kate Barry, John Byram, Sarah Chamberlin, Tom Gagliano, Chris Granville, Jack Repcheck, Chris Swart, and Joe Wisnovsky.

We are eager to have your comments. Please let us know about errors, idiosyncrasies—in fact, any opinions at all. You can reach us at devbio@wwnorton.com.

FRED WILT
SARAH HAKE

Berkeley, California
March 2003

PRINCIPLES OF DEVELOPMENTAL

BIOLOGY

PART ONE
GETTING STARTED

CHAPTER

OVERVIEW OF DEVELOPMENT

This book presents and analyzes the principles and mechanisms that underlie the development of multicellular animals and plants. The development and maturation of the creatures of the living world invariably assault our sense of wonder and inspire awe. Perhaps this is because we are all keenly interested in our own selves, how we got here, how we became the way we are. Our sense of awe is not completely anthropocentric; the transformations of caterpillar into butterfly or seedling into flowering plant are the kinds of daily reminders we all experience, reminding us that the development and appearance of the new living creature is a poetic, wonderful, and central event of living.

Sometimes development goes awry and is not normal. The young ear of corn has blighted kernels, the colt has a misshapen hoof, or the newborn child possesses some tragic abnormality. Because knowledge of normal development is so necessary for understanding abnormal development, developmental biology has become a

CHAPTER PREVIEW

1. Cell division gives rise to dissimilar cells.

2. Nuclear transfers demonstrate the functional equivalence of the genome in different cells.

3. Cells can respecialize during regeneration.

4. Different cells in the embryo have the same genes.

5. The study of developmental biology requires some knowledge of recombinant DNA, nucleic acid probes, and cell signaling.

standard part of the education of health professionals. And many biologists are also deeply interested in the mechanisms that lead to the great diversity of plants and animals.

Since the late 1800s, biologists have recognized that descriptions of embryonic development, while fascinating, are not enough. As the science of biology matured, it became obvious that what happens in development demands that we ask *how* it happens. What are the mechanisms? And that inquiry leads us smack into some of the more interesting and important problems in the biological sciences.

This book provides two levels of exploration into development. First, we will examine some different species to see the way their embryos develop—step by step, stage by stage. Then, with this vocabulary and background in descriptive embryology in hand, we will dig deeper to ask about the mechanisms for accomplishing this development: How do cells become different from each other? Why are some genes expressed in some cells and not in others? How do groups of cells adopt the appropriate positions in the embryo so that organ systems necessary for the life of the adult are put together properly? These are only some of the questions we will address.

among the simplest eukaryotes, has two different **mating types,** and mating can occur only between unlike types. We know that yeast cells can switch from one mating type to another. However, as shown in Figure 1.2A, this ability is possessed only by a yeast cell that has given rise to a daughter cell. To use the language of yeast genetics, only *mother cells* may undergo **mating-type conversion,** not *daughters.* The two cells formed by mitosis of the budding yeast are not the same; the newly formed daughter cell, even though it has the same genes and is formed by mitosis, lacks an ability that the mother cell now has. The daughter, however, will in turn become a mother cell, after which it acquires the capacity to switch mating type. The mitotic descendants of the mother cell, while they possess the same genes, have different abilities.

Figure 1.2B illustrates an experiment carried out in the early 1900s in Freiburg, Germany, by the famous embryologist Hans Spemann. Using eggs of newts collected from the nearby Black Forest, Spemann was able to separate the first two cells of the fertilized egg by tying a hair loop through the division plane separating them. When this was done, each cell could give rise to a complete but small tadpole. In some cases, however, as shown in Figure 1.2B, only one cell could give rise to a tadpole;

THE BASIC CONUNDRUM OF DEVELOPMENT

The Problem of Development Centers on Nonequivalent Cell Division

Developmental biology is a discipline focused on a paradox. How is it that a single cell can give rise during embryonic development to a creature composed of many different specialized cell types—sometimes as many as 100 different kinds—yet all these different cells have the same genes? These specialized cells are not arranged all higgledy-piggledy, but in functioning organs that are themselves arranged in the familiar body plans of adult plants and animals. Figure 1.1 illustrates a few examples of the extraordinary diversity of specialized cell types that may arise from a single cell.

The way this comes about is that during development a given cell can, and sometimes does, give rise to two *progeny* cells. Even though the two cells may look the same, they have become different from one another and enter different developmental pathways. Or, to put it another way, mitosis can generate progeny cells that are **nonequivalent.** Figure 1.2 shows three different examples of this rather abstract-sounding proposition. The budding yeast (*Saccharomyces cerevisiae*),

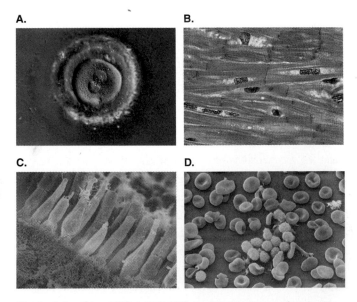

A.   **B.**

C.   **D.**

Figure 1.1   Many Different Cell Types Arising from a Single Egg
(A) Phase-contrast photo of a fertilized human egg, about 100 μm in diameter. It has enough mass to give rise to approximately 400 typical diploid cells (cells with two sets of chromosomes). The remaining cells of the fetus must come from growth, and they are very different both from the egg and from each other. **(B)** Human cardiac muscle tissue in fixed, sectioned, and stained preparation. **(C)** The rod and cone cells of the retina in a scanning electron micrograph. **(D)** A scanning electron micrograph of red blood cells.

A. Yeast Mitosis (Budding)

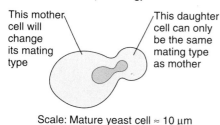

This mother cell will change its mating type

This daughter cell can only be the same mating type as mother

Scale: Mature yeast cell ≈ 10 μm

B. Amphibian Embryo (2-Cell Stage)

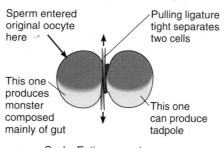

Sperm entered original oocyte here

Pulling ligature tight separates two cells

This one produces monster composed mainly of gut

This one can produce tadpole

Scale: Entire egg ≈ 1 mm

C. Mouse Embryo (16-Cell Stage)

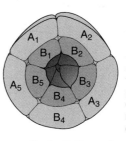

Each AB pair (A_1B_1, A_2B_2, etc.) arose from one mother cell. A cells can only form extra-embryonic tissues like placenta. B cell descendants may form the embryo proper.

Scale: Entire embryo ≈ 100 μm

Figure 1.2   Three Examples of Nonequivalent Cell Divisions   (A) A budding cell of the yeast *Saccharomyces cerevisiae*. Only the mother cell can undergo mating-type switching. **(B)** The two-cell stage of a new embryo. A fine loop of baby's hair is used to separate the two cells that result from the first cell division. Often the plane of this division runs through the site of sperm entrance, but in other instances, as shown here, the site of sperm entrance lies in only one of the two cells. The cell opposite the sperm entrance point has considerable developmental potential and can form a complete embryo, while the other cell is capable only of forming a monster composed primarily of gut tissue. (The developmental potential of the early cells of the amphibian embryo will be discussed in detail in Chapter 4.)

(C) A 16-cell-stage mouse embryo illustrating the positions of "outside" and "inside" cells. For the sake of simplicity, only 10 of the 16 cells are shown; the three pairs not shown lie beyond the plane of the drawing. The outside cells can form placental tissues only, whereas the ancestors of both the inside and the outside cells (from the eight-cell stage) can all form both embryo and placenta.

the other developed into an incomplete embryo—a mass of tissue composed mainly of gut. In the first instance, the first cleavage plane separated the egg into two equivalent cells, each capable of giving rise to a complete embryo. In the other case, the first two cells were not equivalent, only one cell retaining the capability of producing a complete embryo. The cell division in this latter instance is nonequivalent.

Another interesting example is development of the mouse, a mammal whose early development is very similar to that of humans. Very early in the embryo's development, at the eight-cell stage, the cells form a sphere, each cell of which is rounded and only loosely apposed to its neighbors. Then the eight cells develop tight connections with each other, and in the next stage, all the cell divisions occur parallel to the surface of the ball of cells. As a result, eight daughter cells come to lie on the surface of the early embryo, and the other eight progeny lie in the interior (Figure 1.2C). These different cells—or *blastomeres*, as cells in an early animal embryo are usually called—may be isolated, marked, and transplanted to reveal what tissues they develop. All the blastomeres of the eight-cell stage may form either placental or embryonic tissues, but after that crucial fourth division, which generates eight outer and eight inner cells, the outer cells can only form placental tissues, while the inner cells can form embryonic tissues. Some separation of developmental potential has occurred at the fourth division, hence that division is nonequivalent.

We know a considerable amount about how nonequivalent cells are generated. Some substances in the cells of the early embryo may be localized, so that the plane of cell division results in asymmetry in how cytoplasmic materials are distributed. One daughter cell will be endowed with all, or at least more, of some material than the other daughter. The other factor generating nonequivalence is a heterogeneous environment in which mitosis occurs, so that one progeny cell comes to inhabit an environment different from the one inhabited by its sibling. This heterogeneity could result from the presence of another cell or group of cells, some substance lurking in the milieu inhabited by the progenitor cell, or even some physical factor such as light intensity or hydrogen ion concentration.

Cell Lineage and the Cellular Environment Influence Cell Fate

This same issue can also be stated from the standpoints of *lineage* and *position*. These terms describe the mechanisms that regulate nonequivalent cell divisions. The development of a cell or group of cells may depend on their origins; that is, who the parents and grandparents were. This is another way of underlining the importance of materials that are unequally distributed to progeny cells, and this is why the line of descent, the **lineage,** is often so important to know. Lineage is sometimes less important in plant development than in development of animal embryos.

The other consideration, **position,** emphasizes the importance of the environment. From the cellular point of view, who your neighbors are and what the surrounding environs are like are often crucial—certainly the case in both plant and animal development.

Box 1.1 Cell Communication

Sometimes big advances in our scientific understanding occur because of a pivotal idea or discovery by an individual or a small group of pioneers. In recent times in biology, this is exemplified by Watson and Crick's elucidation of the structure of DNA. Just as often, however, big advances are made slowly by hundreds of investigators doing thousands of experiments, and "little facts" accumulate slowly. It is only retrospectively that it becomes apparent some element in the puzzle has been clarified. That is how the relatively recent revolution in our understanding of cell communication has taken place. Though some vertebrate hormones were known to provide signaling between organs by way of the bloodstream, it was not until the 1960s that many new **signaling molecules** began to be discovered. Now we know that there are hundreds, perhaps several thousands of such molecules. Like classical hormones, they come in different chemical forms, though most of them are peptides or proteins. Simultaneously, researchers discovered specific **receptor molecules** for some known hormones and signals, and now many different classes of receptors are known for the signals emitted by cells. See the accompanying table for a list of common signaling and receptor molecules.

As shown in part A of the figure, communication may be seen as a series of individual steps: (1) synthesis and release of the signal by the sending cell, (2) reception of the signal by a receptor on a receiving cell, and (3) modification of the receptor, which leads to (4) activation of a relay, or **second messenger**, which in turn (5) activates some output response by the receiving cell. For example, the receiving cell may alter its cytoskeleton in such a way that its motility is changed. Or, the second messenger, or a product resulting from second-messenger action, may be able to enter the nucleus and affect the **transcription** of one or several genes. The whole process in which a signal evokes a cellular response is called **signal tranduction.**

Part B illustrates the sending system. The signal—often called a **ligand** because it binds to or is tied to a receptor—may be secreted, whereupon it may enter the circulation (*endocrine*), or simply remain in the locality to affect nearby cells (*paracrine*), or even act back on the signaling cell itself (*autocrine*). Some ligands are sequestered by extracellular materials nearby, and have to be released in some way to carry their message. Other ligands are tethered in the membrane of the signaling cell of origin, and can communicate only if the sending and receiving cells are close enough for the ligand and receptor to interact.

As shown in part C, the receptors may be located in the lipid bilayer of the plasma membrane of the receiving cell. There are also cytosolic receptors for ligands, such as lipid-soluble steroids or retinoids; these ligands diffuse easily into the cytosol of cells.

When a ligand-receptor interaction occurs, the receptor is changed in some way. There are large families of receptors for various ligand classes, and there are several different modes of receptor modification. Some of the more common modifications involve phosphorylation (or dephosphorylation) of the receptor molecule on its carboxyl portion, which lies on the cytosolic side of the membrane (see part D of the figure). The modification of the receptor may then begin a cascade of changes in the receiving cell. The original signal is now transduced to a chemical change in the receiving cell, which may lead to changes in cell behavior, in transcription of particular genes, or both.

One of the great lessons of the recent past is that cell signaling by means of ligand-receptor interactions occurs in the developing embryo as well as in the adult. Many of the same molecules are used in embryo and adult. Ligand-receptor interactions are at the very heart of how embryos develop.

One of the realizations of the last decade of research in this area is that very different kinds of creatures, both plants and animals, employ similar mechanistic strategies and even use very similar molecular machinery to carry out development. We shall document this rather abstract proposition throughout the book. At this juncture we merely point out that the central players are the cellular machinery for signaling via *ligands* and *receptors*, and the machinery for regulating *transcription*. The basic facts of cell signaling and transduction, and of transcription, are familiar to many of you, unfamiliar to some. Though we shall treat these topics in detail in several subsequent places in this book, a general knowledge of how cells communicate with one another is so

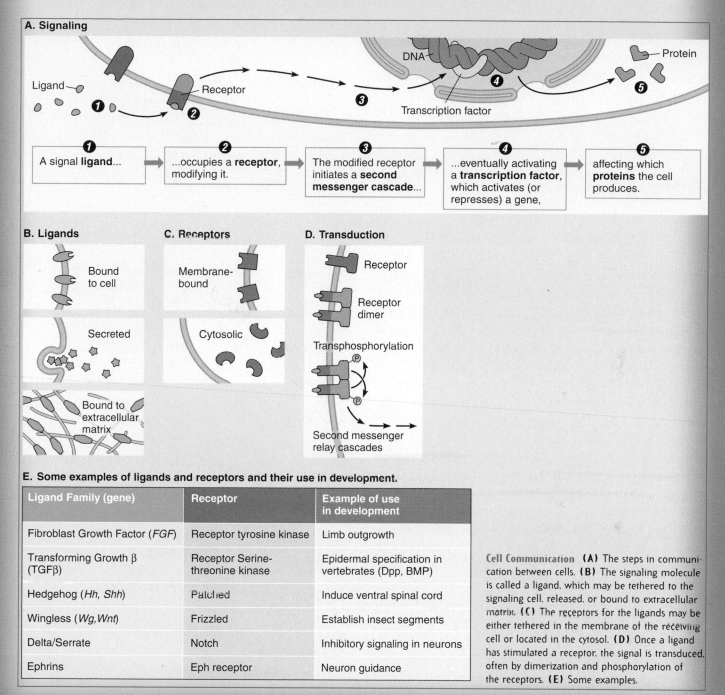

A. Signaling

● 1	● 2	● 3	● 4	● 5
A signal **ligand**...	...occupies a **receptor**, modifying it.	The modified receptor initiates a **second messenger cascade**...	...eventually activating a **transcription factor**, which activates (or represses) a gene,	affecting which **proteins** the cell produces.

B. Ligands

Bound to cell

Secreted

Bound to extracellular matrix

C. Receptors

Membrane-bound

Cytosolic

D. Transduction

Receptor

Receptor dimer

Transphosphorylation

Second messenger relay cascades

E. Some examples of ligands and receptors and their use in development.

Ligand Family (gene)	Receptor	Example of use in development
Fibroblast Growth Factor (*FGF*)	Receptor tyrosine kinase	Limb outgrowth
Transforming Growth β (*TGFβ*)	Receptor Serine-threonine kinase	Epidermal specification in vertebrates (Dpp, BMP)
Hedgehog (*Hh, Shh*)	Patched	Induce ventral spinal cord
Wingless (*Wg, Wnt*)	Frizzled	Establish insect segments
Delta/Serrate	Notch	Inhibitory signaling in neurons
Ephrins	Eph receptor	Neuron guidance

Cell Communication **(A)** The steps in communication between cells. **(B)** The signaling molecule is called a ligand, which may be tethered to the signaling cell, released, or bound to extracellular matrix **(C)** The receptors for the ligands may be either tethered in the membrane of the receiving cell or located in the cytosol. **(D)** Once a ligand has stimulated a receptor, the signal is transduced, often by dimerization and phosphorylation of the receptors. **(E)** Some examples.

central to our discussion that a brief digression on this subject is presented here (see Box 1.1, on cell communication).

If, as we implied, there are real similarities in the ways all creatures undergo embryonic development, then we can ask, Are there principles of development that are widespread, perhaps even universal? Recent work indicates that there

may be positive answers to that question. We shall attempt in this book to elucidate and underline those principles.

We shall reach some places, however, where the principles and mechanisms are not at all clear; these constitute the present and future challenges for research. For instance, it is difficult to understand how the different biochemical paths and mechanisms that are employed by developing cells are all stitched together into a "program" that produces the results. How is the "whole thing" regulated and organized? Or does one event just inexorably entrain the next, and the next, and so on?

Assuming that All Cells in an Embryo Are Genetically Identical Is Justified

We have assumed from the outset that the genes of all cells generated by mitosis from a single cell are identical. That is, all the nuclear genes in the cells of the embryo are generated by mitosis, which guarantees an equal partitioning of the genome to the two progeny. There could possibly be differences in localized substances in the cytoplasm, or in the environments the progeny cells come to inhabit, but the accepted dogma of modern biology is that each cell generated by mitosis has the same chromosomes (two homologs of each kind), and each chromosome contains its invariant repertoire of genes and their regulatory regions. But is this dogma justified?

An experimental approach to this question was invented by Robert Briggs and Tom King, working in Philadelphia in the late 1940s and '50s, and was later refined by John Gurdon and his students in Oxford, England. These investigators asked whether a nucleus could be plucked from a diploid cell (a cell with two sets of chromosomes), transplanted into an egg from which the nucleus has been removed (*enucleated*), and then serve as the founder of all subsequent nuclei in a normally developing embryo. While this may seem an impossible experimental task—indeed, it was originally daunting—different versions of this experiment have been carried out hundreds of times using the large eggs of amphibians.

At first, experimenters encountered difficulty with the "age" of the nuclei used for transplantation. Nuclei taken from early embryos proved very capable of sustaining normal development. But as nuclei were taken from increasingly older embryos, the success rate went down. And nuclei from the tissues of adult frogs had very low success. Did the success rate drop off because the nuclei somehow became genetically "different" in older cells, or was it for technical reasons? Older cells are smaller, so perhaps their nuclei can be damaged more easily, or are less able to become entrained in the very rapid cell cycles characteristic of the early cell divisions of the frog embryo. Eventually, Gurdon and his colleagues, using the South African clawed frog, *Xenopus laevis*, were able to obtain normal tadpoles using nuclei from several different kinds of adult tissues; Figure 1.3 outlines the experimental procedure they used.

Thus the issue was resolved: nuclei from specialized tissues of the adult *can* serve as precursors for nuclei of all the

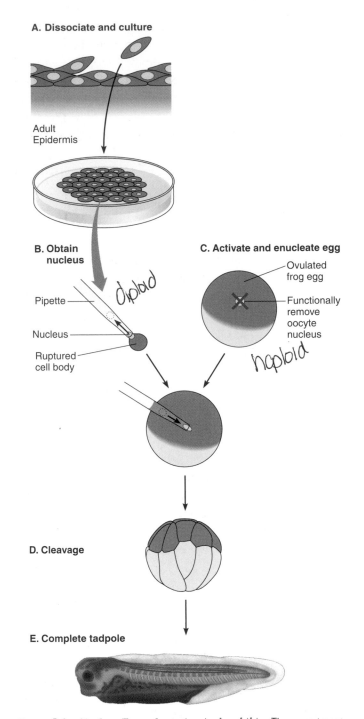

A. Dissociate and culture

Adult Epidermis

B. Obtain nucleus

Pipette

Nucleus

Ruptured cell body

diploid

C. Activate and enucleate egg

Ovulated frog egg

Functionally remove oocyte nucleus

haploid

D. Cleavage

E. Complete tadpole

Figure 1.3 Nuclear Transplantation in Amphibia The experiment Gurdon and his colleagues conducted using nuclei from an adult frog tissue (skin) as a source of nuclei to inject into an activated frog egg. **(A)** Adult skin is excised and placed in tissue culture. **(B)** A nucleus is obtained from a tissue culture cell by drawing a cell up into a narrow pipet, which causes the cell membrane to rupture. **(C)** An ovulated frog egg is activated by physical manipulation, and the egg nucleus is functionally removed by killing it with ultraviolet (UV) irradiation, after which the cultured cell nucleus is microinjected into the activated, enucleated egg. The resultant recipient may then **(D)** undergo cleavage and **(E)** form a complete normal tadpole.

tissues of the developing embryo. However, for reasons still not understood, the embryos derived from these nuclear transplants, while they produce healthy tadpoles with all tissues fully formed, rarely undergo metamorphosis to form an adult. The result clearly rules out loss of or some systematic change to the genetic material as a general explanation for the nonequivalence of cells formed during embryonic development.

Nuclear transplantation has also been carried out with eggs from many other species, including insects and mammals. A recent spectacular case involves the transplantation of nuclei from tissue culture cells derived from the udder of an adult ewe. A few cases of successful development, including the famous lamb "Dolly," are testament to these techniques pioneered in amphibians. While spectacular, we should emphasize that the success rates are very low and the technology incredibly expensive.

The Regeneration of Organs Often Involves the Respecialization of Cells

Other compelling examples also make the case that alterations to the genetic material cannot be a general explanation for cell specialization during development. Many different animals can regenerate organs or parts of organs lost by trauma or experiment. If the lens of an adult newt eye is removed (an operation similar to a cataract removal in humans), the pigmented iris tissue just dorsal to the pupil loses its pigment granules, proliferates, and differentiates into functional lens tissue, thereby replacing the lost lens. Pigmented iris tissue has changed from a pigmented cell to the highly specialized cells of the lens. Figure 1.4 illustrates another famous example of regeneration. In this case, the limb of a salamander can regenerate a complete and perfect new limb and digits after surgical amputation at the forearm level. It has been argued that perhaps the new limb parts arise from a reservoir of unspecialized cells lurking in the limb tissue. While this is difficult to completely exclude, a considerable body of evidence indicates that at least some of the new limb tissues arise from *despecialization* of the limb dermis (the connective tissue underlying the epidermis) and muscle cells, followed by their *respecialization* into many different tissues.

Regeneration of new tissues is, of course, commonplace in plant development. New branches may arise when the tip of a shoot is removed. Once again, the question is whether the source of new shoots and terminal organs such as leaves and flowers is specialized, differentiated cells or some reserve of undifferentiated precursor cells. An experimental answer was provided in unambiguous form in the 1950s by Frederick

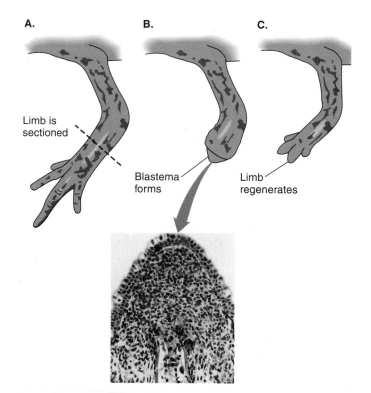

Figure 1.4   The Regeneration of a Salamander Limb (A) When the forelimb of a salamander is severed at the level of the elbow, **(B)** the stump heals, and a mound of cells (the blastema) accumulates under the epithelium over the stump. **(C)** Cell division, cell movements, and subsequent formation of all the elements of the lost limb (muscle, bone, nerve, blood vessels, and so on) which eventually redifferentiate in an appropriate pattern.

Steward, working at Cornell University, who was able to isolate single differentiated vascular cells (phloem) from carrots, culture them in carefully constituted tissue culture medium, and eventually raise a complete new carrot plant from the descendants of this single diploid cell. Figure 1.5 outlines this experimental procedure.

The DNA Sequences in Different Tissues Are the Same

A powerful new way to examine the issue of genomic constancy exists using the tools of molecular biology. It is now commonplace to isolate and clone the DNA sequences that encode various proteins, and these cloned sequences may be used as probes to detect the presence of a given gene in a sample of DNA. The procedures for cloning a gene or a mRNA encoded by a gene, and the ways these cloned fragments may be used as probes, are outlined in Boxes 1.2 and 1.3. Let's imagine, for example, that we have isolated, cloned, and prepared a probe for β-globin (the β chain of hemoglobin) of

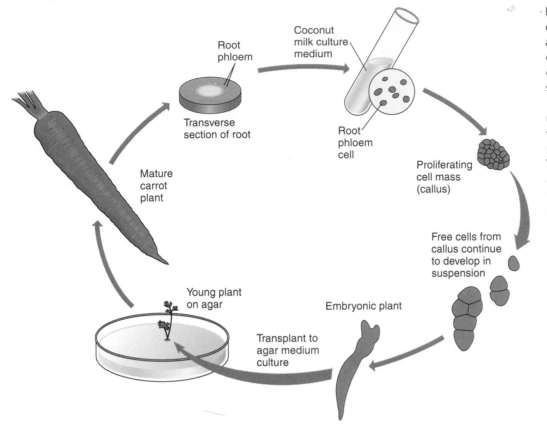

Coconut milk culture medium

Root phloem

Transverse section of root

Root phloem cell

Mature carrot plant

Proliferating cell mass (callus)

Free cells from callus continue to develop in suspension

Young plant on agar

Embryonic plant

Transplant to agar medium culture

Figure 1.5   The Formation of a Carrot from Differentiated Adult Plant Tissue The diagram shows the steps in cloning an entire carrot from a single differentiated phloem (conducting tissue) cell. A mature carrot is sliced transversely, and small pieces of phloem are surgically excised. When the phloem fragments are cultured in medium containing coconut milk, some cells become detached from the mass of callus growing in the medium. They proliferate and form structures resembling plant embryos. When these "plant embryos" are transplanted to a solid culture medium, they form fully developed carrots. The scale used in different portions of the diagram is not uniform.

BOX 1.2 RECOMBINANT DNA TECHNOLOGY (CLONING)

The techniques of recombinant DNA technology, sometimes called **DNA cloning,** are simple and powerful. They allow scientists to synthesize large amounts of any DNA sequence of great length. This is a kind of engineering that cannot yet be done in a chemistry laboratory; indeed, the reproductive power of bacteria is harnessed. Though this technology has many practical applications, the three that concern us here—creating libraries, creating probes, and expression cloning—accomplish important technological feats that were not possible only a few years ago.

Recombinant DNA technology requires two elements: a vector and an insert. A **vector** is any DNA used to introduce, into a host organism, genes that can be replicated by the host cell. The vector is derived from naturally occurring bits of DNA that most bacteria harbor in addition to their own genetic DNA. This extra DNA may be a bacterial virus, called a *bacteriophage*, or it may be a quasi-genetic element called a *plasmid*. Plasmids harbor only a few genes but do have the DNA sequences necessary for the bacterial host DNA to replicate the plasmid's DNA. Plasmids can possess genes that af-

fect bacterial sexual conjugation. Often they harbor genes that encode proteins that confer resistance in the host bacterium to a particular antibiotic. Thus, a bacterium containing a plasmid with a gene for resistance to ampicillin will replicate that plasmid and so pass on to its progeny the ability to grow in the presence of ampicillin, whereas a bacterium of the same species but without this plasmid will not grow under the same conditions. In the example shown in the figure, we use a plasmid as our vector. The vector literally carries the insert along.

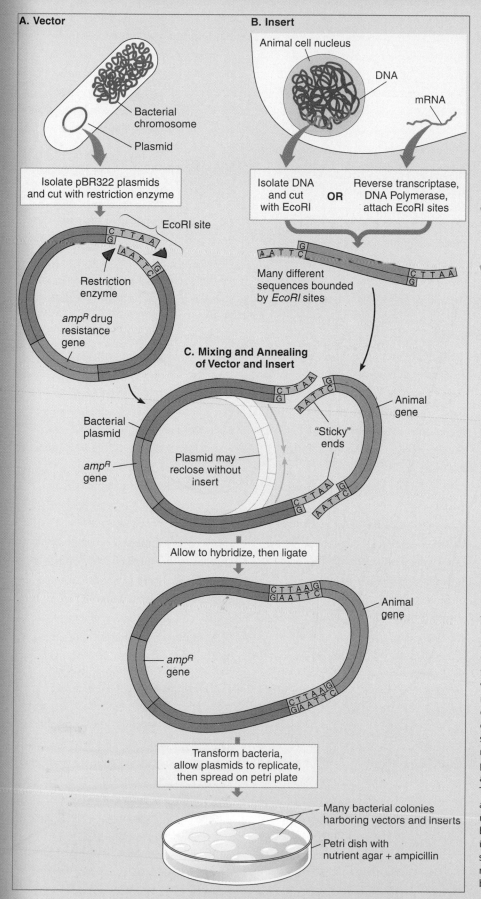

A. Vector

B. Insert

C. Mixing and Annealing of Vector and Insert

The second element—the **insert**—is a piece of double-stranded DNA, taken from any source, that we will stitch into the vector. It can be synthesized in the chemistry laboratory, produced by cutting natural DNA from some source with nucleases that produce pieces of DNA of some convenient length: the DNA may even be a copy of one or many messenger RNAs from a tissue. In this last approach, we can use the enzyme **reverse transcriptase** to make a copy of the RNA molecules using single-stranded **complementary DNA (cDNA)**. (Reverse transcriptase is an enzyme that catalyzes the formation of cDNA from an RNA template.) Then, by appropriate use of the enzyme **DNA polymerase**, we can synthesize a second strand of DNA that is complementary to the first strand and remains hydrogen-bonded to it.

Once the insert has been manufactured, it must be stitched into the vector. It is essential that this stitching be precise. The discovery of **restriction enzymes** (also called **restriction endonucleases**) enables one to do this in a routine way. These

DNA Cloning Several steps are involved in cloning DNA. The vector in this example is the plasmid pBR322, containing the gene *amp^R*, which confers resistance to the antibiotic ampicillin. **(A)** Preparing the vector: The plasmid is cut with the restriction enzyme Eco RI, whose recognition sequence is shown here. **(B)** Making the insert: In this case, the insert is nuclear DNA, which is also cut with Eco RI. The six-base recognition sequence will occur, on average, about every 4,000 base pairs in the DNA, and almost each piece thereby liberated will have a unique sequence bounded by the cut sequence. Alternatively, mRNA may be copied with enzymes to produce double-stranded DNA that can then serve as an insert. **(C)** Putting the insert into the vector: The vector and the insert are then mixed and allowed to form hybrid (recombinant) DNA molecules, which may then be used to transform bacteria. Or the plasmid may reclose without an insert, in which case the resulting plasmid would still be able to confer antibiotic resistance—but not able to serve as a vector for transforming the bacterial DNA.

enzymes, which are found in a variety of microorganisms, cut double-stranded DNA with a high specificity: usually the sequence it attacks is six bases long (sometimes less, occasionally more), and usually the order of bases in one strand is the reverse of the bases on the apposing strand. In our example in the figure, the site cut by the restriction enzyme Eco RI is the DNA sequence GAATTC.

If the vector—in this case, the plasmid DNA—happens to have a sequence recognized by a restriction enzyme, then its own DNA will be severed at that site. The two strands are often cut asymmetrically, producing overhangs (often called *sticky ends*). The bases of the two overhanging ends are exactly complementary, which follows from the fact that they were originally paired in a double helix before digestion. If we cut the plasmid DNA using a restriction endonuclease that recognizes only one site in the plasmid, then the plasmid—which, like bacterial chromosomes, is normally circular—will be "linearized" and the two ends will have overhangs. Also, the nature of the protrusions will be precisely known because we know the specificity of the restriction enzyme.

The accompanying figure shows how this works, using as the vector the plasmid pBR322, which has 4,362 base pairs and is digested with the enzyme Eco RI. If our insert has sticky ends that are complementary to those of the cut plasmid vector, then, when vector and insert are mixed together, there is a good probability that the overhangs of the insert will pair with the overhangs of the "opened" vector. The enzyme **DNA ligase** will ligate (tie together) the ends of such apposed, base-paired pieces.

As shown in the figure, mixing digested plasmid with an insert possessing Eco RI-digested ends, followed by treatment with DNA ligase, may produce concatenated inserts, re-formed vectors, and sometimes recombinant plasmids that harbor the insert. The ligation mixture is then used to transform bacteria. During this routine procedure, the host bacteria not harboring a plasmid are exposed to the foreign DNA under carefully controlled conditions. Some bacteria take up the foreign DNA, which then replicates as the bacteria grow.

Transformation is a very inefficient process: only a tiny percentage of the bacteria take up any of the foreign DNA. The use of the antibiotic resistance marker means that only bacteria that have taken up plasmids, with or without an insert, will be able to grow in the presence of the antibiotic. In order to distinguish the bacterial colonies possessing vectors with inserts from those harboring only recircularized plasmids, the DNA from several colonies has to be extracted and characterized to see which colonies have replicated the inserts.

How does the insert come to possess ends that are compatible with Eco RI-digested DNA overhangs? This is not as daunting as you might imagine. If we are using natural DNA as the source of the insert, we can digest it with the same enzyme, say Eco RI. So all the DNA pieces will have compatible sticky ends. We could, for example, prepare DNA from the embryos of the fruit fly *Drosophila melanogaster*, digest that DNA with Eco RI, ligate it to Eco RI-digested plasmid, transform the bacterium *E. coli*, and grow the *E. coli* in the appropriate antibiotic. The fully grown bacterial culture (about 10^9 bacteria/ml) would contain different pieces of Eco RI-digested fruit fly DNA.

A colony grown from a single bacterium of such a mixture would harbor one, and only one, particular piece of the genetic material of the fruit fly. If we grow the bacteria under conditions in which a large number of colonies are produced, each appearing as a separate colony on a surface of nutrient agar, one of these colonies will likely harbor any given sequence of double-stranded DNA that came from the *D. melanogaster* genome.

When a very large number of different sequences is cloned together, the result is called a **library.** If we knew how to recognize a particular bit of DNA, we could select that particular colony, culture it, and use it as a source for pure DNA of a given kind that we desired. That indeed can be done, as we shall see in Box 1.3. The use of the term *library* is really very apt; it is as if the DNA had all been broken up and packaged neatly between covers. The person who wants that book simply has to have a way to search through all the titles to find the one desired.

It is possible on occasion to have the host bacteria actually process the information in the cloned insert DNA and thereby produce the protein that may be encoded in that piece of DNA. Suppose we have cloned a piece of human DNA that contains the information for the protein growth hormone. If the recombinant DNA molecule is constructed appropriately, a human DNA may be transcribed into a mRNA molecule for growth hormone, and the bacterial protein-synthesizing machinery will then translate this mRNA into human growth hormone. This feat is called **expression cloning,** because not only is the DNA cloned, but it is expressed in the host cell and translated into the protein encoded by the foreign DNA. It is this technology that is behind much of the current biotech industry.

adult frogs. We can now prepare DNA from different stages of the frog embryo or from different tissues of the adult frog. By using the cloned probe, we can show that the β-globin DNA sequence is present in all these DNA samples, and at the same level as in red blood cells. The inescapable conclusion is that the reason hemoglobin is synthesized by erythroblasts and not by brain cells is *not* because the hemoglobin gene is present in one and absent in the other—all cells have the gene—but because expression of the gene is regulated differently in the two types of cells. This conclusion is one of the great accomplishments of modern developmental biology and one of its dominant principles. Simply stated, the differences that arise between cells during development are the result of differences in the expression of a constant genome in the different cell types.

These different developmental pathways that lead to the different cell types of the juvenile and adult are often called pathways of **differentiation.** Cells that have the morphological characteristics traditionally associated with specialized functions of adult animals (for instance, nerve cells and muscle cells) are called *differentiated cells.* In the embryology literature, you may often encounter discussion about when differentiation begins and ends and what its nature is. These discussions have little or no meaning: differentiation is a process (or a series of processes), much of which cannot be observed under the microscope, and which leads to the eventual formation of typical, differentiated cells.

We shall encounter some exceptions to the proposition that the genome is constant and the same in all cells, especially during egg formation in certain animals, and in the formation of the cells of the immune system. Notwithstanding, these exceptions are infrequent, accomplish special and particular purposes, and do not stand as a serious challenge to the dictum that an important part of understanding development is to study regulation of gene expression. The other part of our study, we shall come to see, is the regulation of cell behavior.

THE STUDY OF DEVELOPMENTAL BIOLOGY

This Textbook Is Organized into Three Parts

It is difficult to analyze underlying principles and understand abstractions meaningfully without some acquaintance with real developmental phenomena, and without some working vocabulary and lexicon of experimental techniques. Hence, we will describe the basis of the actual development of a few selected organisms: the fruit fly, the frog, birds and mammals, and vascular plants. Each of these creatures is organized very differently, and seems to develop in very different ways. These very differences help one to discover the wonderful variety of invention that serves development.

However, our main emphasis is on those aspects that are, if not universal, extremely widespread and important underlying principles. For instance, we shall see that all of these creatures activate rapid mitotic division as a consequence of fertilization, and little or no growth will occur during this active first phase of development. This phase of rapid division is called **cleavage.** In many animal embryos, the different cells generated by cleavage, called **blastomeres,** surround a hollow, fluid-filled cavity called a **blastula.** In animal embryos, there follows a period of extensive cell and tissue rearrangement, called **gastrulation,** and then the early differentiation of the various organs (**organogenesis**).

The terms used for plant development are quite different. While active cell movement is crucial for the development of form in animals, we shall see that this is almost completely absent in plants, whose final form is a consequence of changes in the number and shapes of cells.

Following our descriptions of the actual raw stuff of development, we shall concentrate on **morphogenesis,** the mechanisms by which an organism attains its form. Finally, we shall focus on the mechanisms of differential gene expression that we have alluded to at the beginning of this chapter. How is it that some proteins are found in some cells, other proteins in other cells?

The Study of Development Is Circular

In some sense, separating the treatment of morphogenesis from gene expression is artificial, erected mainly for pedagogical reasons. Obviously, gene expression is important for organizing morphogenesis, and the converse is also true. In fact, there is a nagging pedagogical problem in the study of development that troubles both the student and the teacher: Development is complicated. There is so much involved that coming to grips with the underlying principles requires knowing a considerable amount of modern biology. We are currently in the midst of an explosion of information about the cellular and molecular basis of development. Each week sees the publication of experiments implicating another gene or protein in a developmental pathway. Recent advances in

Box 1.3 Nucleic Acid Probes

An important use of cloned DNA is to construct probes. This is simply a kind of lab slang: a **probe** is nothing but a particular sequence of nucleic acid that is somehow tagged, making it easy to identify when it reacts with its "target." The tag may be a radioactive label, like ^{32}P or ^{3}H, or an organic molecule attached to a pyrimidine or purine base in the nucleic acid that can be identified by chemical or immunochemical reactions. And the target is nothing more or less than the DNA or RNA from the cells being examined.

Using nucleic acids as probes to detect a particular gene or mRNA is based on the fact that the DNA of genetic material has a double-stranded structure. Each strand is bonded to its partner by G–C and A–T specific base pairing. Messenger RNA is transcribed from only one of the two complementary strands, and the DNA strand from which mRNA is transcribed is called the *sense* strand. The *antisense* strand, which is complementary to the sense strand, contains only redundant information and is not copied during transcription of mRNA. This second strand is, of course, essential for transmission of the information in the DNA to each of the two progeny cells during mitosis.

When the two strands of DNA are separated, which is usually done by heating (which provides the energy to break all the hydrogen bonds), the DNA is denatured. If the temperature is lowered, under certain conditions a sense and its complementary antisense strand may collide and a double-stranded DNA molecule may re-

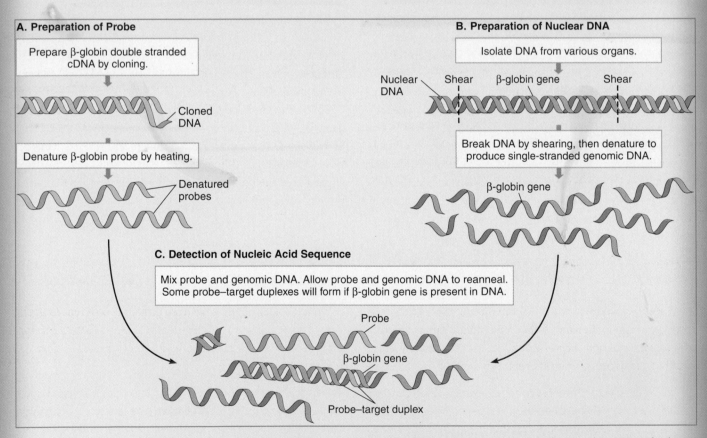

A. Preparation of Probe

Prepare β-globin double stranded cDNA by cloning.

Cloned DNA

Denature β-globin probe by heating.

Denatured probes

B. Preparation of Nuclear DNA

Isolate DNA from various organs.

Nuclear DNA   Shear   β-globin gene   Shear

Break DNA by shearing, then denature to produce single-stranded genomic DNA.

β-globin gene

C. Detection of Nucleic Acid Sequence

Mix probe and genomic DNA. Allow probe and genomic DNA to reanneal. Some probe–target duplexes will form if β-globin gene is present in DNA.

Probe

β-globin gene

Probe–target duplex

The Use of Probes for Gene and mRNA Detection The steps for detecting a given nucleic acid sequence, either DNA (a gene) or RNA (mRNA). **(A)** A particular DNA probe is prepared by cloning; in this instance we shall use a probe for the β chain of hemoglobin, prepared by making a cDNA from β-globin mRNA. **(B)** The double-stranded DNA probe has been labeled with the radioisotope ^{32}P and denatured to separate its two strands. **(C)** Nuclear DNA from different tissues is prepared, broken into shorter pieces by shearing, and also denatured. The probe and genomic DNA are then mixed, and incubated under conditions in which double-stranded DNA may reform. The probe can only hybridize with pieces of sheared genomic DNA that contain sense or antisense sequences corresponding to the coding portions of the β-globin gene.

form. Or the experimentalist may "dope" the denatured DNA with a sequence of single-stranded DNA or even with RNA. If a complement to the introduced singled-stranded DNA or RNA, which we shall designate as probe, is present in the mixture of denatured molecules, and if conditions are appropriate, the introduced probe molecule may re-form a double-stranded form, either a DNA-DNA molecule or possibly an RNA-DNA molecule. Even RNA-RNA hybrids may form if both a particular sequence and its complementary sequence are present as RNA. The formation of double-stranded probe-target duplexes may occur either in solution, using extracted and soluble nucleic acids, or in histological preparations of fixed cells if the nucleic acids in these cells have not been too badly damaged during histological preparation (see the accompanying figure).

So, if we can tag a particular piece of single-stranded DNA or RNA that is either the sense or antisense sequence of a given gene (or portion of a gene, or even the mRNA made from that gene) and carry out this molecular hybridization procedure, then it is possible to evaluate whether the target is present. We can indeed find out if a particular gene is present in a DNA extract, or if the mRNA of that gene is present in RNA extracted from a tissue. And we can do this using preserved cells, indicating to us visually whether a particular gene or mRNA is present or has been expressed.

cell and molecular biology make it hard for even specialists to keep up. Yet, if we wait until we "know enough," it could be a long wait. So it is necessary to jump in.

Often we shall have occasion to revisit the same material, looking at it from different points of view, and, when needed, introducing ancillary but essential information from cell and molecular biology that may be unfamiliar to some of you. Development is a circle, the zygote giving rise to a juvenile, which then develops into a sexually mature adult, in which gametes arise, which then may participate in fertilization, and the cycle begins all over again. So we shall revisit the areas of morphogenesis and differential gene expression many times as we grapple with the vocabulary and the experimental methodology. We shall try to sort out what we really do know at a fundamental level, and what more we need to know, in order to understand how plants and animals develop.

KEY CONCEPTS

1. With a few exceptions, the information content of all cells in an organism is the same. What makes cells vary is the portion of information expressed in different cells. In other words, gene expression is differentially regulated.

2. During development there are cell divisions which give rise to daughter cells that are not the same. Even though the content of genetic information in their nuclei is identical, they are not equivalent. They may come to express different genes. They may display different cellular behaviors.

3. Nonequivalent cells arise from either (1) differences in the cytoplasmic contents distributed to the two daughters during mitosis, or (2) differences in the environment that the two daughters come to occupy.

4. All cells communicate by means of specific ligand molecules interacting with specific receptors on the receiving cell.

STUDY QUESTIONS

1. When Gurdon and his students transplanted nuclei from adult frog tissues into enucleated, unfertilized amphibian eggs, they found that the proportion of injected eggs which developed past the blastula stage was influenced by the original status of the adult cells from which the transplanted nucleus was plucked. For example, nuclei from cells in the skin were poor donors; on the other hand, if the skin cells were placed in tissue culture and allowed to multiply in vitro, the nuclei from cultured cells gave a much higher proportion of developed embryos. Can you speculate why this might be so?

2. The probe for detecting a particular DNA sequence may be RNA or DNA from either the sense or antisense strand. On the other hand, a probe used to detect a particular mRNA, while it still may be either RNA or DNA, has to be the sequence from the sense strand. Can you explain why this is so?

3. There has been considerable controversy about the nature of cells involved in regeneration of an organ or part of an organ. It has often been argued that nonspecialized "reserve" cells (those not very far along on a given differentiation pathway) are the source of cells for new tissues, and that overtly differentiated cells cannot change to form another cell type. Using the example of limb regeneration in salamanders, propose an experiment or approach that might help evaluate this hypothesis.

SELECTED REFERENCES

Alberts, B., Bray, D., Johnson, A., Lewis, J., Raff, M., Roberts, K., and Walter, P. 1998. *Essential Cell Biology*. Garland Publishing, New York.

An excellent introduction to modern cell biology. Chapter 10 discusses cloning technology, and Chapter 15 discusses intercellular communication by means of ligands and receptors.

Briggs, R., and King, T. 1952. Transplantation of living nuclei from blastula cells into enucleated frogs' eggs. *Proc. Natl. Acad. Sci. USA* 38:455–463.

The original report of nuclear transplantation, giving details of how the experiments were carried out and evaluated.

Diberadino, M. A. 1987. Genomic potential of differentiated cells analyzed by nuclear transplantation. *Am. Zool.* 27:623–644.

A scholarly, detailed review of nuclear transplantation from differentiated cells.

Gurdon, J., Laskey, R. A., and Reeves, O. R. 1975. The developmental capacity of nuclei transplanted from keratinized cells of adult frogs. *J. Embryol. Exp. Morphol.* 34:93–112.

A classic report on the transplantation of nuclei from adult cells.

Steward, F. C., Mapes, M. O., Kent, A. E., and Hosten, R. D. 1964. Growth and development of cultured plant cells. *Science* 143:20–27.

A review of the formation of complete plants from cultured differentiated plant cells.

Wilmut, I., Schnieke, A. E., McWhir, J. K., Kind, A. J., and Campbell, K. H. S. 1997. Viable offspring from fetal and adult mammalian cells. *Nature* 385:810–813.

A recent report documenting the cloning of the lamb "Dolly."

CHAPTER

2

GAMETOGENESIS, FERTILIZATION, AND LINEAGE TRACING

Development begins long before the embryo is formed. First, there is **gametogenesis:** the parents must produce the female and male **gametes** (the eggs and sperm). Then comes **fertilization:** the gametes must react with one another to form the union that initiates development of the embryo. Although we will examine actual examples of particular embryos, many of the features of gametogenesis and fertilization are so widespread that it is simpler and more economical to treat these subjects in a general form at the outset. Then, as we go on to discuss the fruit fly, frog, bird, and mammal, we will note special features peculiar to one or another of them.

CHAPTER PREVIEW

1. Eggs are made in ovaries and are necessary to make an embryo.

2. Egg formation requires both meiosis and tremendous growth during oogenesis.

3. Sperm formation takes place in testes; it, too, requires meiosis.

4. The fertilization of eggs by sperm initiates development by causing a temporary rise in intracellular Ca^{2+}.

5. Several mechanisms ensure that only one sperm nucleus fuses with the egg nucleus.

6. Marking individual cells with a "tag" helps embryologists analyze development.

7. Fertilization unleashes a sequence of rapid mitotic divisions.

It is also appropriate to start with gametogenesis because the processes involved in the formation of these single cells illustrate almost all the basic issues that we shall encounter over and over again in examining development. Gametogenesis consists of an orderly program of changes in cellular shape and form, the activation of certain sets of genes, and the suppression of others. Moreover, as we move more deeply into our description of development to try to understand the mechanisms that bring the changes about, we shall see that many of the answers are locked in the structure of the egg itself.

All plants and animals have special organs for reproduction in which the gametes are formed. In animals, these special organs are called **gonads;** the female gonad is the **ovary;** the male gonad is the **testis.**

OOGENESIS

Female Gametes Are Formed in the Ovary

Although there is a bewildering array of special morphologies in the ovaries of different creatures, some basic design features, shown in Figure 2.1, are widespread and important. In animals, the cells that will undergo **meiosis** to form the egg are a special population, called **primordial germ cells,** that arise during embryonic development and come to reside in the developing female gonad (see Chapter 7). Once resident in the gonad, the primordial germ cells are called **oogonia.** The oogonia may form a limited population that does not undergo mitosis in the adult, and hence forms a finite reservoir of eggs. This is the case in humans, where the ovary of a newborn female contains only a few thousand oogonia, and it is the sequential development of them, one at a time, a month at a time, that constitutes the repertoire of eggs for the human female. In contrast, the female frog has oogonia that can undergo mitosis, thereby replenishing themselves. A female frog might, for instance, produce several thousand eggs during the annual breeding season, and the next season accomplish the same feat because her population of oogonia is sustained by mitotic renewal.

It is worth noting that, in this text, we shall often encounter cell populations capable of giving rise to highly differentiated progeny and also sustaining their numbers by mitosis. Such populations are called **stem cells.** The frog's primordial germ cells constitute a true stem cell population. In the case of humans, however, the primordial germ cells are a finite reservoir from which cells are "withdrawn" to become eggs; thus these primordial germ cells are not stem cells.

Oogonia are always surrounded and accompanied by closely associated accessory cells. While these investing cellular layers may have many different names in different creatures, in every well-analyzed case they are known to make important contributions to the developing oogonium as it undergoes differentiation into an egg. As soon as the oogonium enters prophase of meiosis I, it is designated an **oocyte.** (There is a distinction between primary and secondary oocytes that we shall make later.) Often associated with the developing egg and its accessory cells are other cells that we shall designate *supporting cells.* They may produce hormones, help make layers that surround the egg (like shells), or aid in release of the mature egg from the ovary.

It is not unusual for many of the substances stockpiled by the oocyte during its development to be synthesized in organs located elsewhere in the body. These materials are then transported by the circulatory system to the gonad and enter the developing oocyte. In some species, the accessory and supporting cells are involved in the selective import of molecules from the circulation into the oocyte. It is also obvious that the developing egg itself must possess machinery for the import and storage of these molecules. We shall discuss a prime example of this import—formation of yolk proteins in the liver of vertebrates—in Chapter 4.

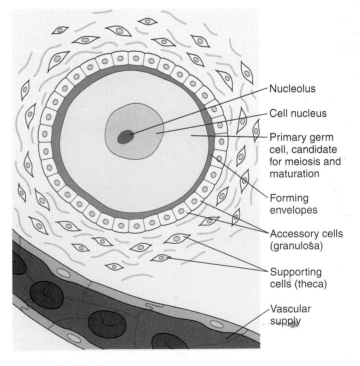

Nucleolus

Cell nucleus

Primary germ cell, candidate for meiosis and maturation

Forming envelopes

Accessory cells (granulosa)

Supporting cells (theca)

Vascular supply

Figure 2.1   The Ovary A generalized vertebrate ovary, showing the relationships between the primordial germ cell (oocyte) and its accessory cells, surrounding membranes, and supporting cells.

Oogenesis Is Characterized by Extensive Growth

The hallmark of oogenesis is the incredible increase in size that occurs during oogenesis. In most embryos there is no growth—in other words, no increase in dry weight—during development until feeding can occur. The growth is done ahead of time, during oogenesis. **Oogenesis**—the orderly stockpiling of all the materials needed for development—is the most impressive example of biosynthesis encountered in the living world. The eggs of birds are a common example; the enormous amount of yolk, all of which is synthesized in the liver of the bird, serves as both the "fuel" and the "mortar and bricks" for development of the bird embryo. Almost all animal eggs employ a similar stockpiling. Even though the egg of the fruit fly *Drosophila melanogaster* is small compared with that of a chicken, it possesses its own store of yolk, enough to provide all the materials needed for the some 20,000 cells that will be generated as the egg develops into a small larva.

Egg Formation Involves Meiosis

The other characteristic feature of oogenesis is meiosis. Scientists have only recently begun to explore experimentally how the cells' choice between mitosis and meiosis is regulated. This decision is obviously of critical importance: Only one of the four meiotic descendants of the primordial germ cells becomes the actual egg; the other three obtain little substance during meiosis because the division plane is asymmetric.

Figure 2.2 outlines the essential features of oogenesis. The primordial germ cells (or their mitotic descendants) that enter the path of egg formation begin a period of exceptional growth. This growth results from the germ cells' own biosynthetic capacities stretched to the utmost, and from the oocyte importing yolk and other materials from the maternal circulation. DNA replication also occurs during this time. By the time the developing **primary oocyte** (as this growing cell is now called) enters the extended prophase of the first meiotic division, it is exceptionally large, with a very prominent nucleus often called the **germinal vesicle.**

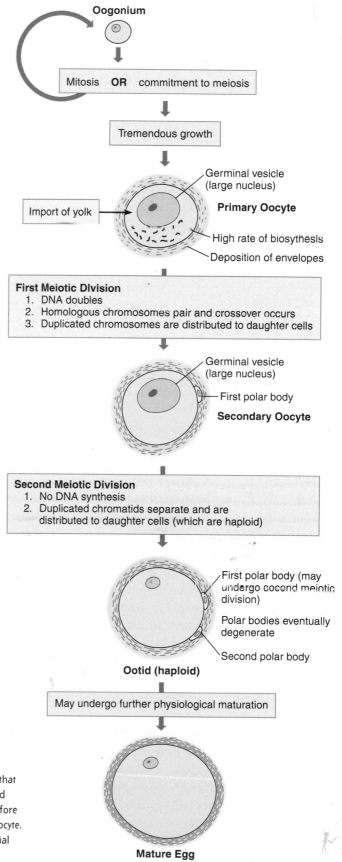

Figure 2.2 **Meiosis During Oogenesis** The sequence of steps that occurs when a female primordial germ cell undergoes meiosis and becomes a fertilizable egg. Fertilization in many species occurs before meiosis is completed. There is enormous growth of the primary oocyte, and meiosis produces only one functional egg from each primordial germ cell.

Recall the essential features of the **first meiotic division (meiosis I):** during the first prophase, the duplicated homologous chromosomes become intimately associated, forming bundles of four chromatids, and portions of the paired chromatids may, and usually do, cross over and physically exchange double-stranded DNA. Only during the later prophase stages (zygotene, diplotene, and pachytene) are the assembled tetrads of paired, duplicated chromosomes visible. When metaphase ensues, the tetrads are juggled into a plane of the cytoplasm approximately equidistant from the two centrosomal poles of the cell. Then in anaphase the duplicated homologous chromosomes are separated from one another, so that each of the two daughter cells receives one of each duplicated chromosome type. (If the processes of pairing, crossing over, and distribution of paired homologs to the two daughter cells is not familiar, consult an introductory biology textbook for review.)

The centrosomes for this first meiotic division are placed not in the "middle" of this gigantic primary oocyte, but close to its perimeter. Hence, when *cytokinesis* (the pinching off of the mother cell's cytoplasm to produce the two daughters) occurs following the first telophase, the two daughter cells differ greatly in size. Most of the cytoplasm—often more than 95% of it, including all the stockpiled materials from the growth phase—is retained by the cell; this cell is now called a **secondary oocyte** (see Figure 2.2). The other cell, which is very small, becomes the **polar body.** Sometimes the polar body undergoes a second meiotic division to produce two very small daughter cells. In any event, the first polar body and any of its progeny soon degenerate, playing no further role in development.

The growth phase of the primary oocyte prior to the first meiotic division may last a very long time. It takes about four months for a primary oocyte of the frog *Xenopus laevis* to complete its growth. The growth of a human primary oocyte takes several weeks. In some animals the primary oocyte can remain in that enlarged state without further progress through meiosis for a very long time, a kind of developmental suspended animation. Some signal, often an environmentally cued increase in hormone levels, can facilitate the next part of its development.

Or, as we shall presently discuss, the primary oocyte may even serve as the developed gamete, able to participate in fertilization reactions and become activated by a sperm. In this instance, the primary oocyte completes the first meiotic division, producing a polar body; then it rapidly completes the **second meiotic division (meiosis II),** producing yet another polar body plus the egg, sometimes called an **ootid.** The nucleus of the egg resulting from the two divisions of meiosis is called the female **pronucleus;** it contains only one of each chromosome type and hence is **haploid.** The completion of meiosis in this instance takes place while the sperm nucleus is in the cytoplasm of the egg. It occurs at a rapid pace, and then the two haploid nuclei fuse to reestablish the **diploid** condition (that is, with two sets of chromosomes per nucleus).

More often the large primary oocyte will complete its first division and then enter an extended prophase condition during the second meiotic division. Once again the nucleus of the secondary oocyte is large and called a germinal vesicle. The egg may remain in this early stage of meiosis II prophase for a long time, and as before, some cued signal—either a hormone or the act of fertilization itself—propels the secondary oocyte to complete meiosis II.

Remember that between meiosis I and II, there has been no replication of DNA. Each secondary oocyte resulting from meiosis I has a representative of each chromosome of its haploid set, and this representative is composed of two chromatids, each of which is a complete double-stranded DNA representative of that particular chromosome. Meiosis II is simply a mechanical distribution of those chromatids, which as the centromeres double and separate, are now simply called chromosomes. They are distributed to the two daughter cells as meiosis II is completed. Once again, the cytokinesis is highly asymmetric, the resultant egg or ootid receiving almost all the cytoplasm; the second polar body forms on the perimeter and soon degenerates.

In animals the fertilization reaction occurs in either the arrested primary or secondary oocyte, depending upon the species. The oocyte's reaction to the sperm is the "hurry up" signal for completing meiosis. Some animals—for example the sea urchin—complete meiosis in the ovary prior to fertilization.

The egg formed during oogenesis is a truly remarkable cell. It is huge, usually thousands or tens of thousands times the volume of the usual diploid cell. It contains a large reservoir of materials for the developmental program soon to commence. Its membranes and metabolic machinery are "on hold," ready to become activated by fertilization. In some creatures, it remains in this metabolically quiescent but physiologically alert state for extended periods of time.

The Egg Is Highly Organized

It is impossible to overstate the importance of the proposition that eggs are organized. The large storehouse of materials possessed by the egg are not just stuffed into a phospholipid membrane bag. In many eggs, evidence of this organization is easily visible. Figure 2.3 gives a generic representation of this organization as well as photos of actual eggs. Yolk, which is assembled in organelles called *yolk platelets*, is often unevenly distributed. The yolk-rich side of eggs is often called the **vegetal** side, while the less yolky hemisphere is designated the **animal** side. There are often pigment granules of various hues that may be more or less localized. The egg nucleus is often located near the periphery of the egg. There are often special, abundant vesicles, called *cortical granules*, near the surface of the egg. The peripheral cytoplasm of the egg often has a somewhat different *ultrastructure* (structure visible only by electron microscopy), especially its cytoskeleton, than does the interior, and the surface cytoplasm is often given a special name, the **cortex.**

The importance of this organization of constituents of the egg will become apparent as we proceed. We may anticipate, however, and point out that, since regions of the egg cytoplasm are different and only limited mixing of different regions occurs during mitosis, the mitotic descendants of this egg (after fertilization) must possess cytoplasms that are different from one another.

SPERMATOGENESIS

Male Gametes Are Formed in the Testis

The male also contains primordial germ cells, which develop in the embryo and come to inhabit the differentiating male gonad (testis). These primordial germ cells also are intimately associated with accessory cells, called *Sertoli cells* in mammals. Germ cells and their associated accessory cells line the perimeter of connective tissue tubes called **seminiferous tubules.** Between the tubules is vascularized connective tissue containing supporting cells. In mammals these connective tissue cells (*Leydig cells*) are responsible for secretion of testosterone. The seminiferous tubules are blind-ended at one end, the other ends joining with other tubules to form a common duct. This common duct, which in vertebrates is comprised of the *vas efferens* and *vas deferens*, may receive secretions from other glands, like the prostate, that help form the seminal fluid, and the duct wends its way through the genitalia in order to deliver sperm and seminal fluid for the fertilization reaction.

Spermatogenesis Involves Formation of a Streamlined Cell

Spermatogenesis, in which meiosis is again a central event, is nevertheless very different from oogenesis. Instead of a stockpiling of materials, many of the contents of the cells are discarded; indeed, little of the primordial germ cell's cytoplasm remains in the mature sperm. The sperm is highly specialized for delivery of the haploid nucleus to the egg. The primordial

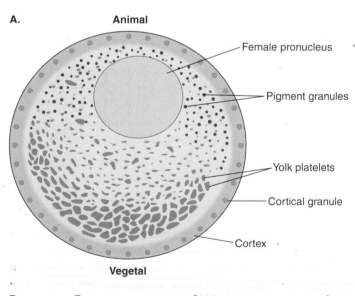

A.

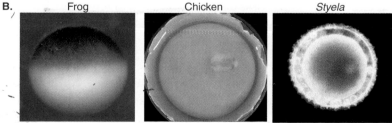

B.

Figure 2.3   The Structure of Eggs   (A) Eggs have various kinds of visible cytoplasmic inclusions, especially yolk platelets and pigment granules. These inclusions are often present in the egg in patterned, stereotyped locations that are characteristic of the species. The less "yolky" side is often called the animal pole, the more yolky, the vegetal. Sometimes the yolk and pigment are evenly distributed and no polarity in the egg is visible. **(B)** Three examples of different kinds of eggs. The frog egg is obviously yolky, with the larger inclusions (called yolk platelets) concentrated on the vegetal side and dark melanin granules on the animal side. The bird's egg is quite different, having huge amounts of yolk, a small island of cytoplasm, complex surrounding membranes, and a shell. The small egg of the marine ascidian *Styela plicata* (sea squirt, an urochordate) shows pigment granules that are localized to a particular region of the egg's cytoplasm.

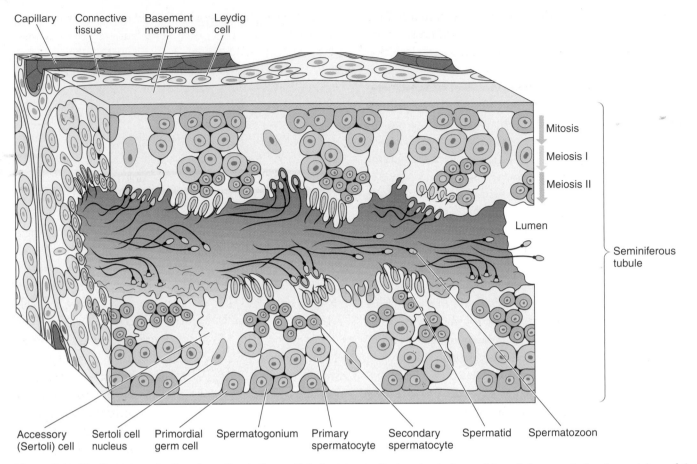

Capillary   Connective tissue   Basement membrane   Leydig cell

Mitosis

Meiosis I

Meiosis II

Lumen

Seminiferous tubule

Accessory (Sertoli) cell   Sertoli cell nucleus   Primordial germ cell   Spermatogonium   Primary spermatocyte   Secondary spermatocyte   Spermatid   Spermatozoon

Figure 2.4   The Structure of a Vertebrate Seminiferous Tubule   A longitudinal section is shown. Primordial germ cells line the periphery of the tubule; sperm formation occurs as they undergo meiosis and move toward the central open lumen of the tubule. Note the presence of supporting Sertoli cells within the tubule. Blood vessels, connective tissue, and special testosterone-secreting cells (Leydig cells) are present in tissue between individual tubules.

germ cell may continuously undergo mitosis in its location near the walls of the seminiferous tubules (see Figure 2.4); these mitotic descendants are called **spermatogonia.** In mammals the spermatogonia adhere closely to the Sertoli cells and remain close to the periphery in the seminiferous tubules. Some of the spermatogonia will enter the meiotic pathway. Hence, the primordial germ cells are a true stem cell population. Those spermatogonia that undergo meiosis, called **primary spermatocytes,** have grown somewhat, replicated their DNA, and are moved away slightly from the tubule wall. Their first meiotic division produces two equal-sized **secondary spermatocytes,** which then undergo the second meiotic division to produce four **spermatids,** all of which may remain joined by thin cytoplasmic bridges and are closely associated with the Sertoli cell.

These small haploid spermatids will undergo a dramatic reorganization. The transition from spermatid to **spermatozoon** (plural: *spermatozoa*; commonly called sperm) occurs near

the center of the tubule and is called **spermiogenesis.** The resultant spermatozoon may be released free into the lumen (the interior space) of the tubule. Most of the cytoplasm and its organelles are sloughed off, including most cytoskeletal elements and vesicular and membranous systems like the Golgi apparatus and the endoplasmic reticulum. The nucleus becomes smaller and more compact, and the chromatin changes its staining characteristics and chemical constitution. The surviving centrioles, mitochondria, and microtubules are reorganized; before its disappearance the Golgi apparatus participates in constructing a new type of vesicle called the *acrosome.*

Figure 2.5 compares the original spermatogonium with the differentiated spermatozoon. While the spermatogonium is a regular "textbook"-type cell, the mature sperm cell is smaller and has discarded almost all of its cytoplasm. The spermatozoon is typically described as having a head, a midpiece, and a tail. The head contains the compact nucleus. The nucleolus is no longer apparent. The histone proteins of the chromatin have been partially or completely replaced by other basic proteins called protamines in most species, which presumably facilitate the more dense packing of the chromatin. At the tip of the head is the **acrosome.** Sperm of invertebrates contain unpolymerized actin adjacent to the acrosome. In the midpiece are the surviving mitochondria; they may coalesce to form a single mitochondrion or a few very large mitochondria, or numerous smaller mitochondria may be present. The two centrioles of the spermatid also reside here, one at the posterior end of the nucleus, the other near the base of the midpiece. From this latter centriole emanates the highly organized microtubule array of the sperm tail, a flagellum specialized for the wavelike motions that propel the sperm.

Sperm extracted from the seminiferous tubules may not be able to participate in fertilization. In many species the sperm must be exposed to substances released by other glands in the male or female reproductive tracts that somehow enable the sperm to reach maturity and become competent in fertilization. In mammals, the maturation of sperm occurs in the *epididymis*, a tubular structure on the outside surface of the testis; when further physiological maturation of sperm occurs in the female reproductive tracts, it is is called **capacitation.**

FERTILIZATION

Fertilization Produces Two Distinct and Important Results

Fertilization occurs when the sperm and egg unite, forming a single **zygote.** As a result of this union, the diploid chromosomal constitution is once again established. This is the *sine qua non* of sexual reproduction: each parent contributes genes to the offspring. The fusion of haploid nuclei contributed by the two parents to form a diploid nucleus is called **syngamy.** Though occasionally haploid zygotes may be produced, generally they do not fare well and usually arrest during development; this may be, in part at least, because when there is only one allele of each gene present, recessive and less well adapted versions of a particular gene may exert their effects unimpeded by a better-adapted allele.

The second crucial result of fertilization is activation of the egg. Prior to fertilization the egg is not able to carry out cell division—neither completion of meiosis nor resumption of mitosis. The interaction between egg and sperm releases the egg from its resting state, and mitosis quickly ensues. The eggs of some species may be experimentally activated through **parthenogenesis,** that is without a contribution from the sperm pronucleus. Parthenogenesis occasionally occurs in nature, and surviving progeny are usually homozygous diploids that result from a chromosome doubling without cytokinesis prior to the first normal mitosis. Examples are flocks of turkeys, in which parthenogenesis sometimes occurs, though rarely, and bees, in which the activation of eggs without sperm produces haploid, sterile drones. For the most part, however, parthenogenesis is a curiosity in nature, but it is a useful experimental tool for the physiologist studying the mechanism(s) of egg activation.

Before we proceed to analyze the phenomenon of egg activation, it is important to underline that the interaction of sperm and egg commonly occurs when eggs are still blocked in metaphase II, or even during meiosis I. In these instances the activation results in the completion of meiosis to produce a haploid egg pronucleus, which then fuses with the newly delivered male pronucleus. Only after that has been accomplished do DNA replication and mitosis ensue.

Both Gametes Are Activated at Fertilization

The interaction of sperm and egg activates not only the egg; the spermatozoa are also activated. The release of sperm from male reproductive tracts is usually accompanied by changes in sperm activity: The mitochondria of the sperm's midpiece activate respiration and ATP synthesis. Sinusoidal beating of the flagellum, powered by this ATP, propels the sperm, moving them through the medium in which fertilization takes place.

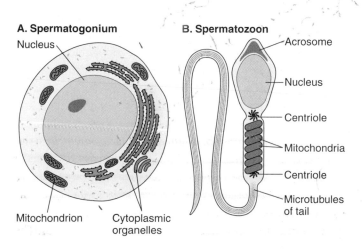

A. Spermatogonium
Nucleus

Mitochondrion   Cytoplasmic organelles

B. Spermatozoon
Acrosome
Nucleus
Centriole
Mitochondria
Centriole
Microtubules of tail

Figure 2.5   Comparing a Spermatogonium and a Spermatozoon
(A) A spermatogonium is a typical cell with the usual organelles.
(B) Its meiotic descendant, the spermatozoon, or sperm cell, is much smaller and has almost no cytoplasm.

Figure 2.6 Sperm Activation The head of a sperm undergoing eversion of its acrosomal filament. Note the formation of ATP by the midpiece mitochondrion and the flagellar motion of the tail.

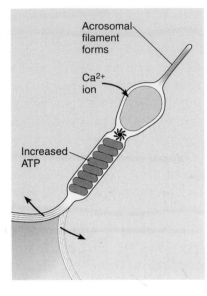

Acrosomal filament forms

Ca^{2+} ion

Increased ATP

Substances released from nearby eggs may enhance this activation of metabolism and motility. In some species, substances released by the egg may direct sperm motility so that the sperm are attracted to the vicinity of the egg (*chemoattraction*).

Furthermore, in most marine invertebrates, a major structural rearrangement occurs, in which the acrosome fuses with the plasma membrane at the anterior tip of the sperm, and this newly added membrane then everts to form a long slender filament (Figure 2.6). Visualization of the **acrosomal filament** requires the resolution of an electron microscope, so it was not until the 1950s that Arthur and Laura Colwin working at the Marine Biological Laboratory in Woods Hole, Massachusetts, and Jean Dan working in Japan were able to see it. The acrosomal filament of invertebrate sperm contains microfilaments (long fibrils of polymerized actin). The reservoir of unpolymerized actin from which the microfilaments form lies just posterior to the acrosomal granule of the unactivated sperm. There is also an acrosome in mammalian sperm, but it does not contain actin. Instead, when the sperm is activated by contact with the *zona pellucida* (the outer layer of the egg), the acrosome fuses with the sperm surface, releasing different enzymes and incorporating the acrosomal membrane into the egg surface.

The metabolic activation and structural rearrangement of the sperm is known to require Ca^{2+} ions; it will not occur in media from which calcium is removed. Furthermore, it is known that in many instances, macromolecules that emanate from the egg, or from the epithelial linings of the female reproductive tract in the case of internal fertilization in mammals, play a part in sperm activation. In the best-studied

example, the sea urchin, a slow solubilization of a sulfated polysaccharide (fucose) in the jelly coat surrounding the egg provides an essential component for sperm activation.

The surface of the acrosomal filament of sea urchin sperm contains a specific protein, bindin, which interacts specifically with a sperm receptor protein of the egg. It is presumed that in all species there is a specific binding between a sperm surface protein and a receptor on the egg membrane. (Sperm–egg binding is a rapidly moving field, and one of special interest regarding mammals because of the human desire to control the reproductive capacity of domesticated animals and of themselves. We shall revisit this topic in Chapter 5, where we discuss mammalian development.)

The species specificity of sperm–egg interaction is often considerable. The occurrence of hybrids between closely related species is rare; usually the sperm fails to activate the egg. (And when hybrids do occur, they often result in sterile offspring.) In the case of sea urchins, bindins of the sperm of one species usually fail to interact with the receptor of eggs of another species, so we know that the bindin–receptor interaction is at the heart of reproductive isolation of different echinoderm species. While bindin is found only in echinoderms, it is likely (though not known for sure) that some kind of specific protein–protein interaction plays a role throughout the animal kingdom, thereby helping to ensure reproductive isolation.

Activation of the Egg Involves the Cortex

The initial interaction between the sperm's acrosome and the egg's receptors sets in motion a dramatic and complex train of events. The egg reacts with a multitude of complicated changes; one of the experimental difficulties in analyzing what is going on is to distinguish the important mainline events from subsidiary consequences. The end game to fertilization is DNA synthesis and mitosis. How does the interaction of the gametes produce this result?

First, let's describe what happens morphologically, as shown in Figure 2.7. Much of what we know comes from work on sea urchin eggs, the prime material for this kind of study; where it has been possible to study the reaction in eggs of other animals, the main outlines are very similar. Interaction of the acrosomal filament with the egg receptor causes profound rearrangement in the egg's cortex. The cortical granules undergo *exocytosis*, releasing their contents to the narrow space between the *vitelline envelope* and the egg's plasma membrane. The macromolecules released into this space raise the osmotic pressure, causing the vitelline envelope to

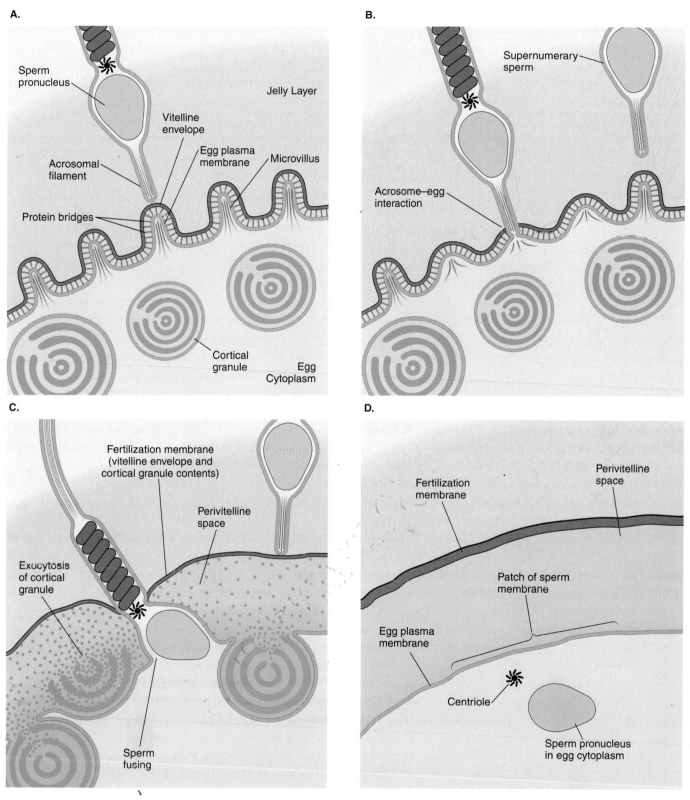

Figure 2.7   The Cortical Reaction of Fertilization The different steps in the interaction of sperm and egg, based on work in sea urchins. (A) The sperm has been activated and the acrosome approaches the egg surface. (B) The acrosome and egg plasma membrane have come into contact, the ligand-receptor interaction has occurred, and now the structure of the egg cortex is beginning to change. (C) The sperm head is fusing with the egg plasma membrane, admitting the sperm pronucleus. The cortical granules have undergone exocytosis, releasing materials into the perivitelline space. The fertilization envelope is forming from cortical granule material and the vitelline envelope. (D) The sperm nucleus has entered the egg cytoplasm, and the fertilization envelope has become elevated.

lift off the egg surface. Materials released from the cortical granules occupy the space between the vitelline envelope and the egg surface (the *perivitelline space*), and some of the macromolecules interact with and become integrated into the former vitelline envelope. This thicker, tougher membrane, a combination of the original vitelline envelope and some of the constituents of the cortical granules, is called the **fertilization envelope,** or membrane. It prevents other sperm from contacting the egg surface with their acrosomal filaments, and as such constitutes a formidable barrier to **polyspermy** (more than one sperm fusing with the egg). While all this is happening, the sperm head is entrapped within the fertilization envelope, and the sperm head's membrane fuses with the egg's plasma membrane. The result is a single cell with two (pro)nuclei, one each from the sperm and egg. The sperm's plasma membrane becomes incorporated into the egg's plasma membrane.

In sea urchins, meiosis is completed before fertilization; in almost all other forms, meiosis continues after fertilization. In either case, the resultant haploid egg pronucleus and sperm pronucleus move closer together within the cytoplasm, a movement that requires the integrity of microtubules. In most cases the sperm centrioles also become integrated into the egg cell. The two pronuclei fuse their nuclear membranes, reestablishing diploidy, followed by a round of DNA synthesis—the **S period** of the first mitotic cleavage division.

These activation events happen rapidly. The cortical reaction in sea urchins, as indicated by the formation of the fertilization envelope, begins within 20 to 30 seconds of sperm–egg mixing. The lifting of the fertilization envelope spreads as a wave originating from the point of sperm contact over the surface of the egg and is complete in 30 seconds. Pronuclear fusion and DNA synthesis may occur by 30 minutes.

Fertilization reactions are very rapid in other species as well. Mammalian eggs are surrounded by several layers of different materials, so the penetration, activation, and binding of sperm are more complex than in marine invertebrates. We shall discuss it in more detail in Chapter 5.

Changes in Function of Egg Membrane Proteins Drive Egg Activation

What are the molecular underpinnings of this dramatic change in the egg? In one sense, we know a lot. Protein-protein interactions at the egg cell surface activate second messenger pathways to cause dramatic changes in egg metabolism and structure, leading to reentry into the mitotic cell cycle. But many steps are only poorly understood; in particu-

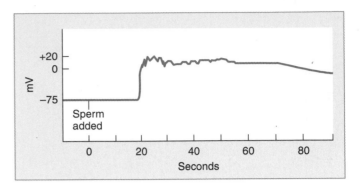

Figure 2.8 Egg Activation Caused by Changes in Membrane Potential A recording microelectrode has been inserted into a sea urchin egg, and the membrane potential measured before and after fertilization. In this instance, the sperm only contacted the egg after approximately 17 seconds. This change is caused by an inrush of sodium ions (and calcium ions) through transiently opened ion channels.

lar, how the very earliest events at the membrane cause the important changes later in time is still an enigma.

One of the earliest changes that occurs after gamete fusion is an increase in the activity of egg membrane ion channels, with a resultant change in membrane potential. For example, in sea urchin eggs the membrane potential is about −70 millivolts (the inside of the cell is negative), as measured by inserting very fine microelectrodes into the egg. Within the first few seconds after fusion, there is an inrush of Na^+ ions into the egg (and Ca^{2+} also enters) and the potential changes to about +20 mV. The membrane potential gradually returns to a negative value in a few minutes (Figure 2.8). Though the exact details of ion conductance changes may differ in different animals, it seems to be a general phenomenon that membrane potential changes occur very rapidly. Soon we shall discuss the importance of these changes for blocking polyspermy.

Calcium Ion Release Is Essential for Egg Activation

The central event of the egg reaction is a transient release of Ca^{2+} ions from vesicles sequestering Ca^{2+} in the egg cytoplasm. The released calcium ion is soon resequestered, so that Ca^{2+} release passes like a wave across the surface of the cell. Again, although this discovery was made in sea urchin eggs, it seems to be a widespread feature of the fertilization reaction. This early calcium ion release is essential for the egg reaction. If we sequester internal calcium in the cell experimentally (by injecting chelating agents or drugs that prevent calcium release), then we can prevent the fertilization reaction. On the other hand, experimental release of internal

A.

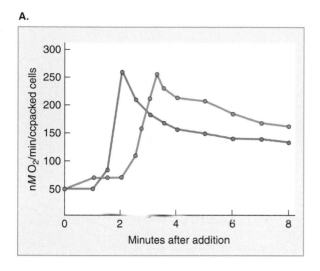

B.

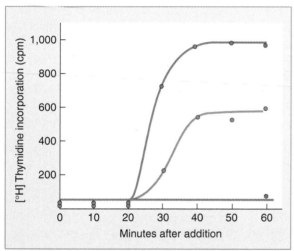

KEY

```
●——● Fertilized
●——● 7.5 μm A23187
```

Figure 2.9 Egg Activation by a Calcium Ionophore Compared with Fertilization When sea urchin eggs are exposed to the calcium ionophore A23187, a fertilization reaction occurs, even without sperm (*triangles*), that is almost as rapid and robust as fertilization with sperm (*circles*). For instance, the fertilization envelope elevates, just as in fertilized eggs. **(A)** There is also a rapid increase in respiration after exposure to A23187, a characteristic response of sea urchin eggs to fertilization. **(B)** Even more important, DNA synthesis in the female pronucleus is activated after exposure to ionophore. Since the parthenogenetically activated egg has only half the DNA of a fertilized egg, the rate of DNA synthesis in artifically activated eggs is only half that in fertilized eggs.

calcium ions from the reservoir is sufficient to mount a complete fertilization reaction. This can be accomplished by exposing eggs to substances called *calcium ionophores*, organic lipid-soluble molecules that carry calcium ions piggyback through phospholipid membranes. Exposure of eggs to calcium ionophores, even in calcium-free seawater, initiates a sustained fertilization response (Figure 2.9).

What is the machinery that causes the transient release of calcium? There is some evidence, based on work in starfish, that the interaction of sperm surface protein(s) with a receptor on the egg surface could activate a second-messenger pathway that utilizes *G proteins*. This leads to the hypothesis that sperm–egg interactions might stimulate inositol phosphate production in a G protein–mediated event, which in turn could lead to calcium release. The transient calcium release could then act as another messenger to initiate a host of important metabolic changes. The likelihood that this hypothesis applies widely is controversial. The background needed to evaluate this model involves some familiarity with G protein–mediated signal transduction, which is outlined in Box 2.1. Some recent experiments that use specific inhibitors of the G protein–linked phospholipase C suggest that there may be multiple pathways; phospholipase C may be involved in Ca^{2+} release but not in the change in membrane potential.

It has been difficult to define precisely the receptor of the sperm stimulus. There have been many attempts to isolate and characterize a sperm "receptor" located on the egg surface. The existence of such a receptor is consistent with our current understanding of intercellular signaling. Definite proof for such a receptor in the well-studied case of sea urchin egg activation has proved elusive. Furthermore, it is necessary to show that any "receptor" is actually the crucial ignition element for the egg response.

There is also some evidence for an alternative view in which the sperm actually delivers to the egg cytoplasm a protein, presumably located in the acrosome, that signals egg activation. New experiments carried out by David Epel and his colleagues at the Hopkins Marine Station of Stanford University have shown that sea urchin sperm contain the enzyme nitric oxide (NO) synthase, and that the synthase is active after activation of the sperm acrosome reaction. An increase in the NO levels in eggs is evident within seconds after fertilization, and this increase precedes the rise in intracellular free Ca^{2+}. Microinjection of NO donors into an unfertilized egg effects a fertilization reaction in these eggs. These researchers speculate that NO synthesis in the eggs catalyzed by the enzyme delivered by the sperm may be the missing link between the sperm–egg interaction and the rise

Box 2.1 G Protein Signaling

As we mentioned in Chapter 1, the study of how cells receive and interpret signals from the environment is a dynamic and important area of modern cell biology. All signaling molecules (ligands) interact with receptor proteins of the receiving cell. This interaction may activate proteins that initiate second-messenger cascades in the responding cell. One set of membrane proteins, called **G proteins**, is an important link in the second-messenger system of cells.

G proteins bind GTP (guanosine triphosphate), which may be converted to GDP (guanosine diphosphate) and then rephosphorylated to GTP, and so on. The G protein is a trimer of three subunit proteins: α, β, and γ. When the G protein trimer is in the active state, the βγ dimer dissociates from the α subunit, which contains a GTP molecule. When GTP is hydrolyzed to form GDP, the inactive αβγ trimer again forms.

The active form of the G protein may then activate other players in the signal transduction pathway, which are neighboring membrane-associated proteins. One pathway may activate enzymes that synthesize cAMP (cyclic adenosine monophos-

phate), an important second messenger. Shown in the accompanying figure is another pathway, which is our immediate concern—the activation of a hydrolytic enzyme that specifically cleaves a phosphorylated hexose called inositol from a membrane phospholipid. Active G protein activates phospholipase C, which hydrolyzes a phosphoinositol-containing membrane

lipid to form diacylglycerol, which remains in the lipid membrane, and the released phosphosugar. The phosphorylated inositol then proceeds through a cycle of sequential removal of phosphates by phosphatases, generating free inositol, which is then enzymatically coupled to the available diacylglycerol to resynthesize the original phospholipid. Thus the signaling event has activated a cycle of hydrolysis and resynthesis of these membrane phospholipid constituents, as is shown in the figure.

The initial products of the hydrolysis—inositol phosphates and diacylglycerol—are potent second messengers. Inositol-*tris*phosphate may cause the release of calcium ions from membrane vesicle reservoirs of calcium, and calcium has profound effects on many different enzymes, including certain protein kinases, which can markedly affect the metabolism of the cell. Likewise, diacylglycerol activates a membrane-associated protein kinase called protein kinase C (PKC), and PKC may also phosphorylate several proteins that lead to a cascade of events affecting metabolism and the transcription of genes.

G Protein Signaling Through Phospholipid Hydrolysis The main steps in the G protein-mediated transduction of signaling. Active protein stimulates phospholipase C to hydrolyze membrane-bound phosphatidylinositol into inositol trisphosphate (IP3) and diacylglycerol, which in turn mobilize intracellular calcium ion (by IP3) and activate protein kinase C (PKC).

in intracellular calcium. The exact sequence linking surface membrane–associated events with the propagated calcium transient is unknown; the implication of a role for NO could provide a way to delineate such a sequence. It also is not clear how the calcium transient produces the many events, shown in Figure 2.10, which are entrained by egg activation. For instance, in many eggs, following the early events that occur within the first minute or two, there is an extrusion of protons from the egg into the medium, causing a rise in the internal pH of the egg, sometimes by as much as 0.5 pH units. Just how the proton pump is activated is not known.

Fertilization May Activate Protein Synthesis on Stored mRNA

Many eggs, though not all, display a marked increase in the rate of protein synthesis shortly after the early events of fertilization. In the sea urchin egg this increased synthesis will begin about 5 to 7 minutes after the addition of sperm. A considerable body of work shows that the rise in pH caused by proton extrusion helps support an increased rate of protein synthesis. There is also direct evidence that many mRNAs not being translated into protein in the unfertilized egg become associated with ribosomes and begin translation after fertilization. This increase in translatable mRNA also helps to sustain an increased rate of protein synthesis. We still need to know how the rise in pH causes an increase in protein synthesis, and how the calcium transient or other associated events increase mRNA access to the translational machinery.

There is a considerable store of mRNA in the unfertilized eggs of most species, which should come as no surprise after our discussion of oogenesis. Much of this mRNA is not translated until after fertilization. Furthermore, the poly A tail, which is characteristic of the mRNA of eukaryotes and is often rather short in the mRNA found in unfertilized eggs, becomes lengthened after fertilization by a poly A polymerase in the cytoplasm. It would be interesting to know whether the calcium transient activates the poly A polymerase. What is the importance of this lengthening of the poly A tail? While this lengthening accompanies an increase in mRNA translation after fertilization, the kinetics of polyadenylation are different from the kinetics of an increase in bulk protein synthesis, at least in sea urchins, so there may not be a simple relationship between polyadenylation and increased translation. It is known, however, that the presence of a poly A tail confers greater stability on a translated mRNA, which may be the function of this cytoplasmic poly A tail lengthening.

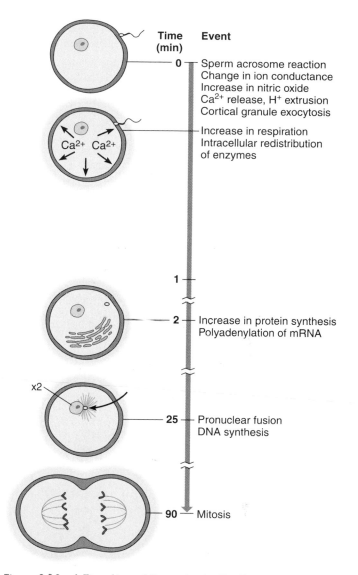

Figure 2.10   A Time Line of Events in the Fertilization Reaction in Sea Urchins Notice that the time scale changes from top to bottom.

The payoff of the fertilization reaction is the reengagement of the cell division cycle. Recent advances in cell biology are unraveling the molecular basis of the control of the cell cycle, but as of this writing, there is very little experimental work that indicates how the early events, especially the crucial calcium transients, bring about pronuclear movements, replication of DNA, and cell division. We may speculate, however, that since early events probably do entrain second-messenger pathways, events downstream, like the phosphorylation of different proteins, are probably the links between fertilization and the cell cycle machinery. A graphical summary of some of the events of fertilization is shown on a timeline in Figure

2.10. It is obvious that activation is very complex; many events that are probably important have been identified and studied. Understanding how they are linked to one another, and how one event drives another, are the present challenges.

Several Mechanisms Prevent Polyspermy

It is important that only one sperm pronucleus join with the egg pronucleus; otherwise not only would a polyploid nucleus result, but the additional sperm would contribute additional centrioles, which would cause multipolar mitotic divisions, leading to unequal distributions of chromosomes to progeny cells and to disturbed patterns of cytokinesis. In other words, polyspermy (Figure 2.11) is a disaster. There are some rare instances where many sperm are admitted to the egg, but then all pronuclei but one degenerate. The usual instance, however, is to admit only one sperm pronucleus to the egg.

There are three principal mechanisms by which this is accomplished (two of which we have already mentioned):

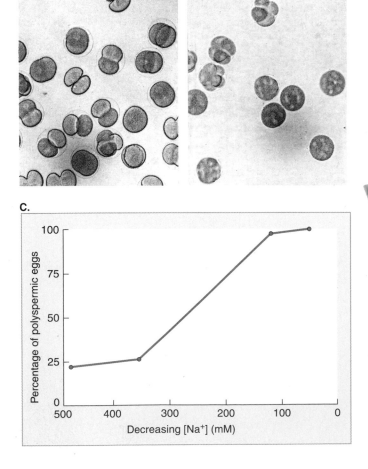

First, there is a very rapid change in the membrane potential of the egg, within the first few seconds after gamete fusion, and this prevents additional sperm from fusing with the egg (Figure 2.11). Second, when the cortical granules discharge, some of the materials released into the perivitelline space are proteases that could degrade egg receptors for sperm. And third, the fertilization membrane that forms in some species constitutes a formidable mechanical barrier to the admission of sperm. Variations of these different kinds of mechanisms are found throughout the animal world, and constitute a suite of fail-safe devices for preventing polyspermy.

TRACING CELL LINEAGES

Lineage Tracing Is Essential to Understanding Development

Fertilization leads to mitosis, and the huge egg is literally quickly carved into a large number of smaller cells that will constitute the tissue fabrics of the early embryo. We need to know if there is any regularity to early development. Does the same part of the egg cytoplasm, when it becomes cellularized, always give rise to the same structures in the adult? Who does what? How do cell groups get to their final locations? If one is going to do experiments, it is obviously essential to have a map. Most embryos are in fact fairly regular in their development, the same part of the egg giving rise to the same structures. Some embryos are absolutely *stereotyped*, giving rise to the same number of cells, each of which is derived from precisely the same volume of the original egg.

A map that tells us what region of an egg, or of an early embryonic stage, becomes what organ or tissue of a juvenile or adult is called a **fate map.** Originally these maps were made by simply watching the living embryos divide, to see

Figure 2.11 The Electrical Block to Polyspermy When the increase in membrane potential (shown in Figure 2.8) is prevented, eggs can no longer prevent the admittance of additional sperm. In this experiment, the rise in membrane potential is prevented by reducing the sodium ion content of the seawater. (Here choline was substituted for sodium in order to maintain the osmotic strength of the artificial seawater.) As the potential changes become weaker and weaker, the degree of polyspermy increases. **(A)** Control eggs fertilized in seawater divide normally. **(B)** Eggs fertilized in low-Na$^+$ (120 mM) seawater are polyspermic. **(C)** The degree of polyspermy increases as Na$^+$ is reduced, causing membrane potential to drop.

A. Mark one cell with fluorescent dye.

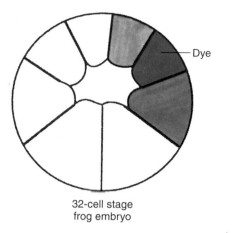

32-cell stage
frog embryo

B. Injected embryo.

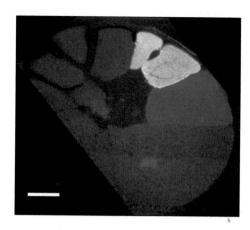

C. Look for fluorescence in sections.

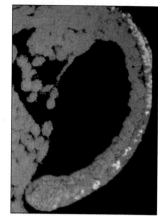

Stage 12 gastrula

Figure 2.12   Lineage Tracing   (A) Diagram of a parasagittal section through a 32-cell-stage embryo of *Xenophus laevis*. Three blastomeres have been colorized to show that each was injected with a different colored fluorescent dye. **(B)** Immediately following injection, the embryo was fixed and sectioned. The blastomeres appear somewhat compressed due to the manipulations of the histological processing; nonetheless, the appropriate blastomeres have been labeled with the different dyes.

(C) In this parasagittal section of a late gastrula embryo (stage 12) that developed after injection of the three blastomeres, the fates of the blue-, red-, and green-marked blastomeres are clearly visible in their descendant cells; where different colored cells have intermixed, other colors, such as yellow or orange, can result. Notice how much smaller the cells are in this gastrula than in the 32-cell embryo. More detail about embryonic development of *X. laevis* will be presented in Chapter 4.

where differently colored granules already present in the cytoplasm in some embryos turned up. Also, very early on, embryologists learned to use *vital dyes* (dyes that stain groups of cells without toxic side effects) in order to follow the fate of such stained cell groups.

The ultimate in tracing is to mark a single cell and examine its mitotic descendants. This is useful in embryos that follow a fairly regular pattern of *mitoses* (mitotic divisions) and lines of descent, from which one can construct a **lineage diagram.** This kind of map not only gives spatial information but also allows us to trace the exact parentage by mitosis of any cell at any given time. Not all embryos will yield this kind of detailed information, but it is extremely valuable when it can be obtained. Figure 2.12 illustrates how a fate map can be obtained for an amphibian egg. A particular **blastomere** is marked with a dye, and later the embryo is observed to find out what tissues are derived from the marked cell. There are many new and powerful ways to mark cells that were not available to early embryologists, some of which are listed in Table 2.1. Our information about what cells give rise to what tissues during the development of many embryos has become very precise, and is now used even to follow the development of complex organ systems in which there is extensive migration of cells, such as the central nervous system.

A famous example of lineage tracing, an instance where it has served as a powerful tool in experimental analysis, is the development of *Caenorhabditis elegans*. The embryos of this soil nematode undergo a completely regular and stereotyped pattern of cell division, generating 558 cells, and after four molts, contain 959 somatic cells and a large number of germ cells. All individuals have that number of cells, and all the cells are generated in the same way. A complete fate map and lineage

TABLE 2.1 AGENTS FOR MARKING AN EMBRYO

Dye	Example
Membrane lipophilic dyes	diI
Inert fluorescent dyes	Rhodamine-dextran
Nontoxic enzymes	Horseradish peroxidase
Vital stains	Nile Blue Sulfate
Nuclear morphology	Quail–chick chimeras
Radioactivity	[³H]thymidine in DNA

diagram of this creature has been assembled. Figure 2.13 shows the early divisions of the embryo and a portion of its lineage diagram. Given such information, we can surgically remove cells or kill them with a laser or even isolate a known particular cell, and then study what happens. Suppose, for example, that the E cell (shaded in Figure 2.13A) is killed at the eight-cell stage. Since we know this cell gives rise to the gut, we would expect the resulting embryo to be missing the gut, and this is what researchers have found. If the E cell is isolated and cultured, it will differentiate some features of a gut. However, if the E cell is isolated at the early four-cell stage, it does not develop into a gut, but it will do so if allowed to remain in contact with cell P_2. The experiment suggests that not only does E give rise to the gut, but also P_2 is necessary to allow the EMS mother cell to give rise to a competent E cell. Fate maps and lineage diagrams are not just road maps; they help us devise and interpret experiments on how cells interact with one another.

Cleavage Is a Period of Rapid Mitosis

We are going to consider some different examples of embryo development in the following chapters. Each of them has different features, yet there are many common tactics. The fertilization reaction leads to rapid mitosis, and in most eggs there is no increase in mass of the zygote during these early stages. Because the zygote is literally cut up into smaller cells,

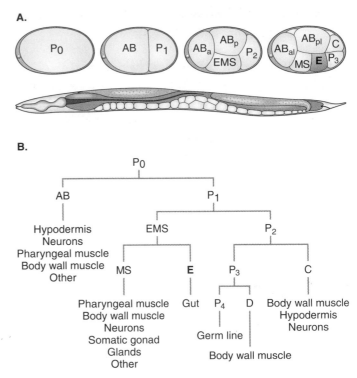

Figure 2.13   Early Development and a Lineage Diagram of the Nematode *Caenorhabditis elegans*   (A) The names of the blastomeres of the first four cleavage embryonic divisions, shown above an adult worm with its different tissues. **(B)** A lineage diagram constructed from careful observation of the patterns of these divisions. Researchers also followed the later divisions so they could accurately assess the tissues to which each blastomere contributed.

TABLE 2.2 DISTRIBUTION AND AMOUNT OF YOLK IN EGGS

Amount	Distribution	Occurs In	Cleavage Type
Isolecithal (sparse amount)	Even	Many invertebrates; mammals	Variable but complete; radial, spiral, bilateral
Mesolecithal (moderate amount)	Predominates on one side	Amphibia	Radial
Telolecithal (large amount)	Very dense, exludes cytoplasm to one side	Birds, reptiles, fish	Incomplete, meroblastic
Centrolecithal (large amount)	Yolk in center	Insects	Surface cleavage

Source: Adapted from Scott F. Gilbert. 2000. *Developmental biology.* Sinauer Associates, Sunderland, Mass. fig. 8.5 (p. 227).

this period is called **cleavage.** Cleavage obviously involves rapid DNA replication and chromosome assembly, and an extensive increase in cell membrane as total cell surface area increases. Patterns of cytokinesis may be considerably influenced by the presence of large internal reserves of yolk, which by impeding cytokinesis affect the progress of cleavage. In some very yolky eggs, such as those found in birds and reptiles, the cytoplasm occupies a small portion of the egg volume, and cytokinesis proceeds only in this island of cytoplasm, cleaving the yolky portion not at all. This is called **meroblastic** cleavage (*mero* means "partial" or "part"). Traditional embryologists classify eggs of different species depending on how much yolk they contain, where the mass of yolk is concentrated, and whether the yolk becomes allocated to different cells by cytokinesis (Table 2.2). Furthermore, the exact pattern of cell divisions that take place in cleavage is determined by where the centrosomal poles of the mitotic apparatus are placed. We know that centrosome placement is governed by cellular ultrastructure, but the mechanisms involved are still unknown. It is clear that the fertilization reaction is a key that unlocks the beginnings of a complex developmental program.

KEY CONCEPTS

1. Gametes arise from a stem cell population, a group of specialized cells that can reproduce themselves and also give rise to descendants that enter a pathway of differentiation.

2. Forming an oocyte places huge demands on the organism, and the oocyte is stocked with materials necessary for early development.

3. Fertilization restores the chromosome number that was halved by meiosis and activates the egg. This activation involves profound metabolic changes brought about by liberation of the second messenger, calcium ion. As a result, the activated gamete reenters the cell cycle, thereby beginning cleavage.

STUDY QUESTIONS

1. Suppose an egg of some creature is brought to your laboratory, and you need to evaluate what stage of meiosis the egg is in. You also have sperm from the same species. The egg is opaque, so you can't peer into its interior. Can you devise a way to judge the meiotic state of the egg?

2. The eggs of some species may survive without being fertilized for extended periods of time. For example, in places where the weather can be very arid, followed by rains, the eggs of some species can be fertilized and develop only when water is present. Can you think of some of the adaptations that eggs from a given species might possess in order to survive for extended periods?

3. The cortex of eggs of many species changes its physical properties after fertilization, becoming considerably more rigid, so that it resists physical deformation when pressed with a needle. From your knowledge of the ultrastructure and organization of cytoplasm, what is the likely molecular change or changes that might be responsible for such changes in the cortex?

4. It is possible to modify the internal pH of cells by adding a dilute base that is able to pass from the medium in which the eggs are suspended through the plasma membrane. Ammonia is one such base. Design an experiment using ammonia to alter intracellular pH in order to evaluate whether pH is a factor in the regulation of cytoplasmic polyadenylation of mRNA in fertilized eggs.

SELECTED REFERENCES

Colwin, A. L., and Colwin, L. H. 1963. Role of the gamete membranes in fertilization in *Saccoglossus kowalevskii* I. The acrosome reaction and its changes in early stages of fertilization. *J. Cell Biol.* 119:477–500.

An early research paper describing the acrosome reaction.

Epel, D. 1997. Activation of sperm and egg during fertilization. In *Handbook of Physiology*, sect. 14: Cell physiology, ed. J. F. Hoffman and J. J. K. Jamieson, pp. 859–884. Oxford University Press, New York.

An important review of the sequence of events in the egg fertilization reaction.

Glabe, C. G., and Vacquier, V. D. 1978. Egg surface glycoprotein receptor for sea urchin sperm bindin. *Proc. Natl. Acad. Sci. USA* 75:881–885.

A description of the interaction of sperm binding with eggs.

Humphreys, T. 1971. Measurements of mRNA entering polysomes upon fertilization in sea urchins. *Dev. Biol.* 26:201–208.

An elegant experimental analysis of the mobilization of stored mRNA following fertilization.

Jaffe, L. A., and Gould, M. 1985. Polyspermy preventing mechanisms. *Biol. Fert.* 3:223–250.

A review of the various devices for preventing polyspermy by one of the leading figures in the field.

Kuo, R. C., Baxter, G. T., Thompson, S. H., Stricker, S. A., Patton, C., Bonaventura, J., and Epel, D. 2000. NO is necessary and sufficient for egg activation at fertilization. *Nature* 406:633–636.

An incisive report of the importance of delivery, by the sperm, of NO synthase to the egg.

Metz, C., and Palumbi, S. 1996. Positive selection and sequence rearrangements generate extensive polymorphisms in the gamete recognition protein, bindin. *Mol. Biol. Evol.* 13:397–406.

A research paper on how bindin may influence species-specific fertilization.

Shen, S. S., and Steinhardt, R. A. 1978. Direct measurement of intracellular pH during metabolic depression of the sea urchin egg. *Nature* 272:253–254.

Another classic describing the changes in pH after fertilization.

Snell, W. J., and White, J. M. 1996. The molecules of mammalian fertilization. *Cell* 85:629–637.

This is an excellent recent review of the molecules involved in sperm–egg interaction in mammals.

Steinhardt, R. A., and Epel, D. 1974. Activation of sea urchin eggs by a calcium ionophore. *Proc. Natl. Acad. Sci. USA* 71:1915–1919.

An important milestone in demonstrating the importance of calcium mobilization in the egg response.

Sulston, J. E., Schierenberg, E., White, J., and Thomson, N. 1983. The embryonic cell lineage of the nematode *Caenorhabditis elegans*. *Dev. Biol.* 100:64–119.

A definitive study of lineage.

Vacquier, V. D. 1998. The evolution of gamete recognition proteins. *Science* 281:1995–1998.

A review of cell surface proteins mediating the fertilization reaction.

Vacquier, V. D., Swanson, W. J., and Hellberg, M. E. 1995. What we have learned about sea urchin bindin. *Dev. Growth Differ.* 37:1–10.

A general and extensive review of the discovery and importance of bindin.

Whitaker, M. 1993. Lighting the fuse at fertilization. *Development* 117:1–12.

An exploration of the wave of calcium release in the fertilization reaction.

Part Two

Early Development of Animals

CHAPTER

3

OOGENESIS AND EARLY DEVELOPMENT OF *DROSOPHILA*

The fruit fly *Drosophila melanogaster* has become an object of intense study of development during the past two decades. The 1995 Nobel Prize was awarded to Eric Wieschaus and Christiane Nusslein-Volhard for their seminal studies, begun in the late 1970s, which have resulted in a revolution in the study of development. The small fruit fly is not simple, nor is it so different from us vertebrates as to be unworthy of our attention. Even though the ancestors of flies and humans may have become distinct from one another over 600 million years ago, we shall come to see that at the molecular and cellular level many of the tactics of development are similar in such disparate creatures.

CHAPTER PREVIEW

1. The fruit fly *Drosophila melanogaster* is an important model organism for studying development because sophisticated genetic manipulations are possible.

2. Growth of the egg depends on input from nurse cells.

3. Cleavage proceeds without cytokinesis, which results in a syncytium.

4. The syncytial blastoderm becomes cellularized, and gastrulation produces germ layers (ectoderm, mesoderm, and endoderm).

5. Factors contributed by nurse cells are localized in the egg, where they establish its anteroposterior axis.

6. Molecular signals originating in the follicle cells and present in membranes around the egg establish the dorsoventral axis of the egg.

One reason investigations using *Drosophila* have been so productive is that genetic studies of animals really began with the study of *Drosophila*, and there is more accumulated knowledge here than with any other multicellular animal. Geneticists have devised elegant ways to produce, identify, and study mutations that interfere with many of the basic processes of development (see Box 3.1). The coupling of classical genetic techniques with modern cell and molecular biology and embryology is producing a torrent of information about development, and recently it has become possible to see the outlines of general principles.

EMBRYOGENESIS

The Formation of an Egg Is a Prelude to Embryo Formation

The general "recipe" for animal development was set forth in Chapter 1: make an egg, cut it up into many diploid cells, move cell groups from place to place to produce a three layered embryo, and selectively activate gene expression in different forming tissues and organs. The formation of the egg in *Drosophila* takes approximately eight days; embryonic and larval development and pupation together occupy about 13 days.

The materials of the oocyte, called *maternal components*, come from a number of sources. The oocyte itself does make some kinds of mRNA and proteins, but most of the mRNAs,

proteins, ribosomes, and organelles are made in the adjacent *nurse* cells. Nurse cells and the oocyte are enclosed within a layer of somatic cells called *follicle cells*. (A *somatic cell* is any cell in the body that is not a primordial germ cell, or one of its descendants.) It is the follicle cells that synthesize the tough covering surrounding the oocyte called the *chorion*. The yolk, composed primarily of a phosphorylated lipoprotein called *vitellogenin*, is synthesized in the *fat body*, which has a function in insects somewhat analogous to liver function in vertebrates. These many biosynthetic activities are regulated by hormones synthesized by secretory cells associated with the nervous system. Oocyte development thus results from the contributions of many organ systems in the adult female.

Figure 3.1 shows the position of the fruit fly's two ovaries connected to the oviducts. Each ovary is a collection of long tubules, called **ovarioles**, in which oogenesis takes place, with younger stages at the distal tip of the ovariole (the end opposite the ducts). Oocytes are generated in the tip; as oocyte growth and maturation take place, the oocyte, associated nurse cells, and enclosing follicle cells move down the ovariole during an approximate eight-day journey (Figure 3.2); the exact timing of events is, of course, dependent on temperature. During the first two days a special set of four cell divisions occurs in the oocyte stem cell, the **oogonium.** As shown in Figure 3.3, the progeny of these mitotic divisions remain connected by cytoplasmic bridges. The bridges assume a special structure called *ring canals*. Only two of the

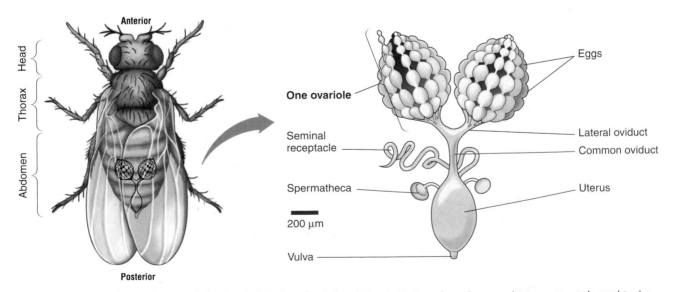

Figure 3.1 The Reproductive System of the *Drosophila* Female A dorsal view indicating where the reproductive system is located in the abdomen. The enlargement on the right shows the basic parts of the gonad, the tracts, and some of the accessory organs, including the seminal receptacle and spermatheca, where sperm are stored.

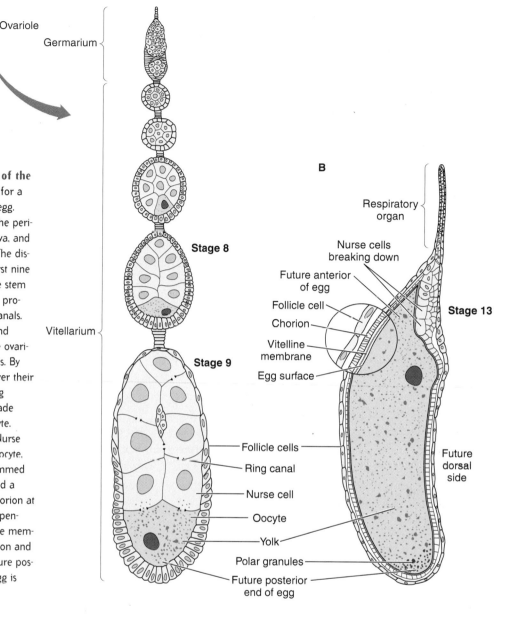

Figure 3.2 Progressive Development of the Oocyte It takes eight days and 14 stages for a germ-line stem cell to become a finished egg. (Recall that only one day is required for the period from fertilization to formation of a larva, and 12 days to go from larva to adult.) **(A)** The distal portion of the ovarioles showing the first nine stages and six days of oogenesis. Germ-line stem cells at the tip (the germarium) divide to produce nests of 16 cells connected by ring canals. As the nests mature, forming nurse cells and oocyte, they move down the portion of the ovariole called the vitellarium toward the uterus. By stage 9 the nurse cells have begun to deliver their contents, via the ring canals, to the growing oocyte. Between days 6 and 8, the yolk made in the fat body is transported into the oocyte. **(B)** A nearly mature oocyte at stage 13. Nurse cells have delivered their contents to the oocyte, are shrinking, and are undergoing programmed cell death. The follicle cells have synthesized a chorion, and specialized portions of the chorion at a dorsoanterior location are respiratory appendages. The oocyte has synthesized a vitelline membrane that is interposed between the chorion and the oocyte. Polar granules marking the future posterior portion of the egg are visible. The egg is about 400 μm long and 150 μm thick.

16 cells are connected to four of these canals, the others having one, two, or three. One of the two cells with four canals enters meiosis to become an oocyte, the other 15 remaining connected to it. These 15 become the nurse cells, and they undergo duplications of chromosomes without cytokinesis, becoming highly polyploid (256 × diploid) and very large. All 15 of the nurse cells are connected to the oocyte by ring canals (Fig 3.4). Then the nurse cells pump mRNA, proteins, and organelles into the oocyte, after which vitellogenin, circulating in the hemolymph, is taken up by the oocyte. Almost half of the oocyte is yolk. During the last day of oogenesis, the oocyte becomes covered by a vitelline membrane made by the oocyte and follicle cells, and the follicle cells synthesize and deposit the chorion (Figure 3.5).

The oocyte is asymmetric; two projections of somatic cells from one end of the oocyte are respiratory organs, marking the future dorsal and anterior end of the embryo. The future ventral side of the embryo has a more rounded appearance. The yolk spheres are present mainly in the center of the egg, while most of the cytoplasm is located peripherally, next to the plasma membrane. As we shall see, these superficial indications of egg organization are underlain by a complex molecular organization. In summary, oogenesis is a period of meiosis and accompanying genetic recombination, a stock

Oogonium/stem cell division

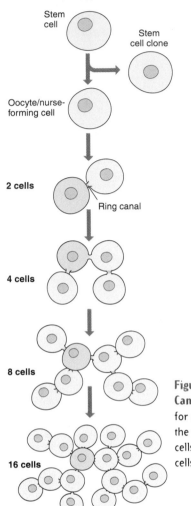

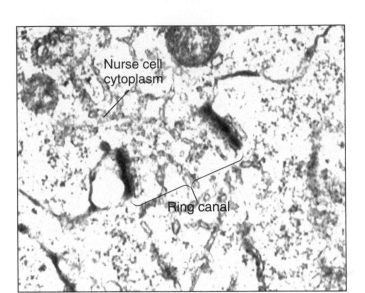

Figure 3.4 The Ring Canals An electron micrograph showing the ring canal structures that connect nurse cells to one another and to the oocyte.

Figure 3.3 The Formation of Nurse Cells, Oocyte, and Ring Canals in *Drosophila* Oogenesis On days 1 and 2, the stem cell for oogenesis, the oogonium, gives rise to a nest of 16 cells and the ring canals, shown here schematically. Notice that the nurse cells and oocyte come from the same precursor germ cell—nurse cells are *not* somatic cells.

A.

KEY

Follicle cells:	Germline cells:
Main body	Nurse
Polar	Oocyte cytoplasm
Terminal	Oocyte nucleus

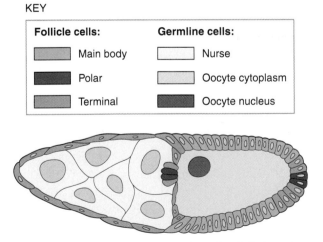

Figure 3.5 The *Drosophila* Egg Chamber After Yolk Deposition Has Started, Stage 10A (A) Follicle cells are subdivided into three cell types: main body cells (shown in blue), terminal follicle cells

B.

(pink), and polar follicle cells (red). The oocyte nucleus is shown in green. **(B)** The accompanying photograph shows how Gurken mRNA (a signaling molecule important in establishing a difference between dorsal and ventral follicle cells, and discussed later in the chapter) has been localized to the dorsoanterior corner of the oocyte. The image was obtained using whole-mount, in situ hybridization procedures and a tagged probe against Gurken mRNA. Notice the tag with an antibody against it. The antibody has the alkaline phosphatase enzyme attached to it, and the actual color obtained in the hybridization procedure was produced by the local action of the phosphatase.

piling of materials from the outside (including nurse cells, fat body, and follicle cells), and a spatial ordering of these materials in the oocyte.

Fertilization and Cleavage Initiate Embryo Formation

Fertilization occurs when sperm (stored in female reproductive tracts) enters an opening in the chorion called the *micropyle*; the egg is activated, meiosis is completed, syngamy takes place, and a series of extraordinarily rapid nuclear divisions occurs. The first eight cycles take place about every 9 minutes. Figure 3.6 and Table 3.1 illustrate the early stages. At first the zygote is a **syncytium:** no cell membranes are formed because there is no cytokinesis, just nuclear duplica-

tion. From cycles 7 to 10, most of the nuclei begin to migrate to the superficial cortical cytoplasmic layer, and very low levels of transcription are initiated. The embryo is called a **syncytial blastoderm** at this stage. Between cycles 9 and 10, some nuclei that lie at the future posterior end become enclosed by cell membranes; they are now the first true cells, called **pole cells.** Their cytoplasm contains unique characteristic granules, and they will form the primordial germ cells of the developing embryo (Figure 3.7).

Nuclear divisions then slow somewhat in tempo. Nuclei continue to move to the periphery, and cell membranes form over the entire surface of the oocyte beginning early in cycle 14. The approximate 6,000 cells will form a continuous cellular layer on the surface surrounding a central yolky area. The nuclei are now enclosed by invaginations of the oocyte surface accompanied by the assembly of new membrane at the

BOX 3.1 DEVELOPMENTAL GENETICS

Genetic analysis has been a powerful tool for discovering the underlying principles of development. The basis of classical genetics is the mating of two parents to produce offspring. When geneticists began to take certain matings and study the early stages of development of their offspring, rather than wait for the appearance of viable juveniles or adults, many new possibilities appeared. Researchers could now observe mutations affecting development so profoundly that formation of a juvenile was severely compromised, even lethal.

We now know a lot about the chemical and cellular basis of how mutations are expressed. A change in the nucleotide sequence in the DNA results in changes in the amount of, or the amino acid sequence

of, a particular protein, and that change may have phenotypic effects: that is, some function or structure is detectably changed. If the function or structure is compromised or does not work so well, the mutation is called a **loss of function.** Occasionally a mutation bringing about the change in the protein results in greater activity or function, and may then be called a **gain of function** mutation. And when two different mutations affect the same structure, activity, or function (even though they may not affect exactly the same protein), the researcher may hypothesize that the two genes (and hence, the two proteins) are linked in some functional pathway. Double mutants in which both of the suspect genes are in the mutant form can allow a geneticist, after

careful analysis of the phenotype of the affected creatures, to deduce which gene is acting prior to another in such a common pathway. In lab jargon, genes are said to be acting *upstream* or *downstream* from one another; in the language of genetics, this is called **epistasis.** A short course in genetics would be out of place here, but you may find it useful to consult the chapters devoted to genetics in an introductory textbook.

The power and importance of genetic analysis will become increasingly apparent as we proceed with our study of development. The reason why the fruit fly, the mouse, the zebra fish, and the nematode *Caenorhabditis elegans* are so widely used is precisely because they lend themselves to detailed genetic manipulation.

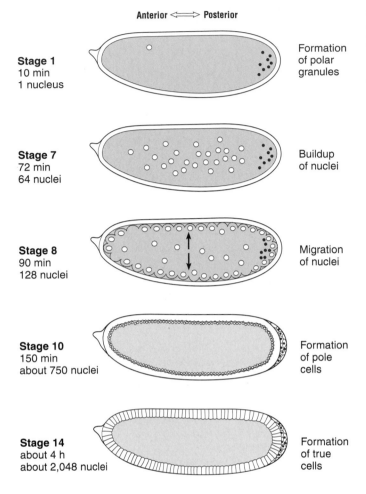

Anterior ⟺ Posterior

Stage 1
10 min
1 nucleus
— Formation of polar granules

Stage 7
72 min
64 nuclei
— Buildup of nuclei

Stage 8
90 min
128 nuclei
— Migration of nuclei

Stage 10
150 min
about 750 nuclei
— Formation of pole cells

Stage 14
about 4 h
about 2,048 nuclei
— Formation of true cells

tip of the invagination (Figure 3.8). Within approximately 4 hours the oocyte has become transformed into the **cellular blastoderm.**

The Cellular Blastoderm Is Organized

During cleavage and before the cellular blastoderm stage, communication between different regions of the embryo is unimpeded because there are no cell membranes; signals can pass between nuclei, and the yolky central regions and the periphery are not segregated. Moreover, all nuclei are developmentally equivalent. Nuclear transplantation experiments have been carried out with nuclei from the syncytial blastoderm stage, and the transferred nuclei can inhabit any

Figure 3.6   Cleavage and Cellular Blastoderm Formation   In the early cleavage stages in *Drosophila*, the nuclei multiply without cytokinesis, after which most migrate to the cortical periphery. Nuclei at the posterior end first form pole cells after the ninth round of replication. The blastoderm becomes completely cellularized during cycle 14. Small white circles indicate nuclei (except in stage 14, from which the nuclei have been omitted). The approximate elapsed time since fertilization to reach the various stages is shown. Table 3.1 gives further information about the stages of cleavage and cellular blastoderm formation.

TABLE 3.1 EARLY DEVELOPMENT OF *DROSOPHILA*

Activity	Cycles
Nuclear proliferation occurs very rapidly (9 min)	1–8
Nuclei migrate to the cortex (egg surface layer)	7–10
Pole cells bud off from the rest of the egg; a low level of zygotic gene expression occurs	9
Cell membranes form between cortical nuclei, and zygotic gene expression increases	Early in 14
Morphogenesis begins (about 5.5 h after fertilization)	Late in 14
Extensive anteroposterior pattern (16 stripes of different cells) and dorsoventral pattern (4 zones of cells) are complete before morphogenesis begins	By middle of 14
Larva hatches (at 22 h) and starts feeding	

Source: Adapted from data in M. Zalokar and I. Erk. 1976. Division and migration of nuclei during early embryogenesis of *Drosophila melanogaster.* J. Microscopic Biol. Cell. 28:97–106.

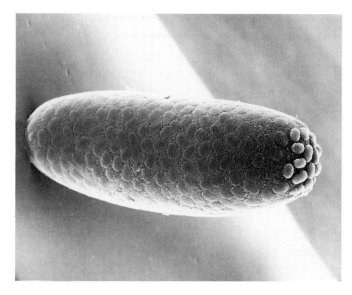

Figure 3.7 Pole Cells A scanning electron micrograph of the surface of a *Drosophila* embryo near the end of cleavage shows the distinctive group of pole cells at the posterior end of the embryo.

A. Chromosomes separate on the mitotic spindle.

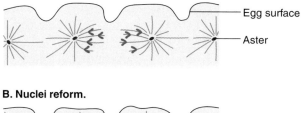

Egg surface

Aster

B. Nuclei reform.

Nucleus

C. Nuclei enlarge. Furrow canals form, added at the tips by membrane vesicles.

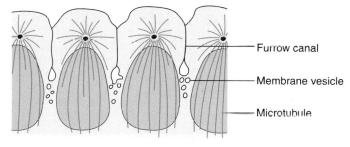

Furrow canal

Membrane vesicle

Microtubule

D. Nuclei continue to grow and the furrow canals progress.

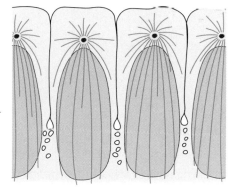

E. Cellularization is completed and yolk membrane is in place.

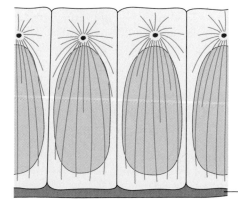

Yolk membrane

region of the cellular blastoderm of the recipient egg and can form any part of the embryo. Transplantation of whole cells, possible only after the 14th cycle, gives a very different kind of result: usually the cellular fate of the transplanted cell has become more or less restricted along particular pathways.

One way to examine the developmental potential of a cell or group of cells in an embryo is to move them from their normal residence to a different location either in the same or another embryo. Transplantation to a different embryo results in a **chimera** (from the Latin for "monster"), an organ or tissue composed of cells from two genetically distinct sources that in the adult are easy to distinguish phenotypically. Many embryos are remarkably resilient and will tolerate these transplantations, which have been a favorite and powerful tool of experimental embryologists for over 100 years.

The earliest true cells to form in *Drosophila*, the pole cells, are accessible and relatively easy to remove from the posterior end of the egg. Furthermore, they can be tagged with genetic markers that allow easy identification of transplanted cells in

Figure 3.8 Cellularization of the Syncytial Blastoderm Shown here are some of the events that occur as the nuclei move to the surface of the oocyte and become gradually enclosed by new cell membranes. to form the cellular blastoderm.

a genetically different host. When pole cells are removed, gonads form, but they are sterile because no primary germ cells are present (Figure 3.9A). When pole cells are transferred from a marked donor to the pole region of a different host, they populate the gonad of the host and form germ cells (Figure 3.9B). But only pole cells form primary germ cells; the

nuclei that come to reside in this polar region during late cleavage have become determined somehow.

We can glimpse how this determination occurs by carrying out the transplantation experiment shown in Figure 3.9C. From the posterior of a donor egg, we withdraw some cytoplasm (called **pole plasm**) with a pipette early in cleav-

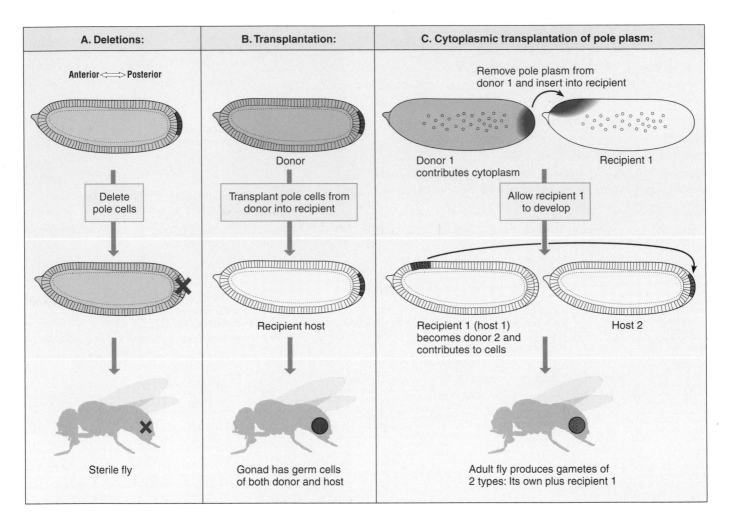

Figure 3.9   Transplantation Analysis of Pole Cells and Polar Plasm   The three different microsurgical manipulations shown here were designed to analyze the role of intact pole cells and of cytoplasm (polar plasm) from the posterior pole. **(A)** Pole cells once formed are deleted, and the embryo is allowed to develop. Gonads form, but no germ cells are present. From this we can conclude that pole cells are necessary and sufficient to form all the germ cells. **(B)** Pole cells from an early embryo (colored red) are removed and transplanted into the pole region of a recipient host. The germ cells derived from transplant donor (red) and host (green)

can be distinguished because they each bear distinctive dominant genetic markers. If we mate this host fly, the progeny will display the markers of *both* the transplant donor and the recipient host. The host is a germ cell chimera. **(C)** Transplantations of cytoplasm from the posterior pole are carried out before pole cells form, that is, there are no nuclei at the cortex yet. Pole plasm is removed from donor 1 and placed in a distinct anterior location in host 1. (Donor 1 and host 1 are genetically marked so that if in error a nucleus were plucked from the donor, it would be subsequently recognized as an artefact in host 1.) Host 1 is allowed to become a cellularized blastoderm.

at which point cells from the area that had received the pole plasm transplant from donor 1 are grafted to the pole cell area of a genetically marked second host. Host 2 is allowed to develop to adulthood, and the appropriate matings are carried out to determine what kinds of germ cells it forms. Host 2 is a chimera composed of germ cells of both the first and second hosts. Host 1 must have contributed germ cell precursors. These precursors could only have arisen if a pole plasm transplant had endowed the cells of host 1 with this potential. A cytoplasmic transfer has caused determination of cells that arise from that cytoplasm.

age, before it has become inhabited with nuclei and "cellularized." We deposit this pole plasm into the anterior region of a genetically marked recipient, thereby providing a way to detect inadvertent transplantation of a nucleus from the donor. The recipient embryo is now allowed to continue development—to form a cellular blastoderm. The cellular region in this recipient that contains the original donor pole cytoplasm is transplanted once again, this time to the primary germ cell region of a second genetically marked host. This second recipient now has some primary germ cells containing pole cytoplasm from the original donor and nuclei from the future head region of the first host. Once the second host develops into an adult, we can easily demonstrate—by mating it with genetically marked partners—that the transplanted cells it received from the first host have formed gametes. But this happens only if pole cytoplasm was transplanted! Polar cytoplasm somehow imposes a primary germ cell fate on the nuclei that reside in it. In other words, there is a cytoplasmic–nuclear interaction that directs the subsequent course of gene expression in the nucleated cells descending from it.

Cells from all over the cellular blastoderm have been transplanted, so we know that different cellular regions reliably give rise to different kinds of particular structures when transplanted to the same (**orthotopic**) or different (**heterotopic**) regions of the recipient. Using the results of these experiments, we can project a map onto a diagram of the surface of the cellular blastoderm to indicate what cells from different regions actually do form; this is the *fate map*, as introduced in Chapter 2 (Figure 3.10). We can also construct a map showing what cells will form if transplanted to a different site; the result is a *map of developmental potential*. Since developmental potential is changing, and after cellularization is almost the same as the fate map, we do not show it here. Nuclei taken from a syncytial blastoderm embryo are **totipotent,** which means they are not fixed and may form any tissue, depending on the cytoplasm they inhabit and their location. This is not so after cellularization; then developmental potential becomes restricted. From these observations, we can make the following generalization: developmentally equivalent nuclei generated during cleavage interact with different cytoplasmic areas of the egg. This begins the early restrictions of developmental potential that eventually produce the different cell types of the embryo.

When the developmental potential of a group of cells becomes restricted, embryologists often say that **determination** has taken place. Determination as a concept goes back to the very beginning of experiments on embryos in the late 19th cen-

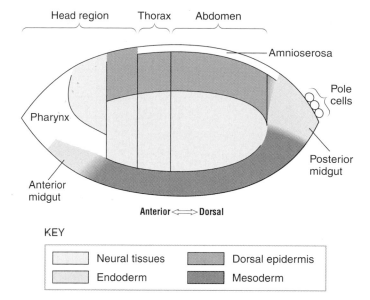

Figure 3.10 A Fate Map of *Drosophila* A stylized diagram of the cellular blastoderm of *D. melanogaster*. Transplanting small numbers of marked cells allowed Gerhard Technau and his colleagues to map the fate of cells from the blastoderm. The positions of various tissue types and germ layers are indicated.

tury. Notice that it is an operationally defined term: manipulation, usually a transplantation experiment, is what informs the experimenter whether developmental options have been narrowed or not. Like differentiation, determination is not a static state; it is a process.

Gastrulation Is a Translocation of Surface Cells to the Interior

The organized cellular blastoderm now undergoes a series of cell and tissue movements that profoundly rearrange its form. Details of *gastrulation movements* differ widely in different animals; they never occur in plant development. (The meaning of "gastrulation" will become clearer by the end of this section.) In *Drosophila* a subset of cells along the future ventral midline invaginates. The invagination results in a *ventral furrow* appearing on the ventral surface of the embryo. The invaginated cells form a transient, internalized tube, which pinches off as shown in Figure 3.11. The invaginated epithelial layer gradually becomes a loosely constituted, motile mesenchymal layer. This segregated group of cells will form the *mesoderm* of the developing embryo. (See Box 3.2 for a brief explanation of mesoderm and other germ layers.) Mesoderm is the precursor of muscles and parts of all other internal organs.

A. Sagittal sections

Dorsal

Anterior ◁⊣⊢▷ Posterior

Ventral

B. Cross sections

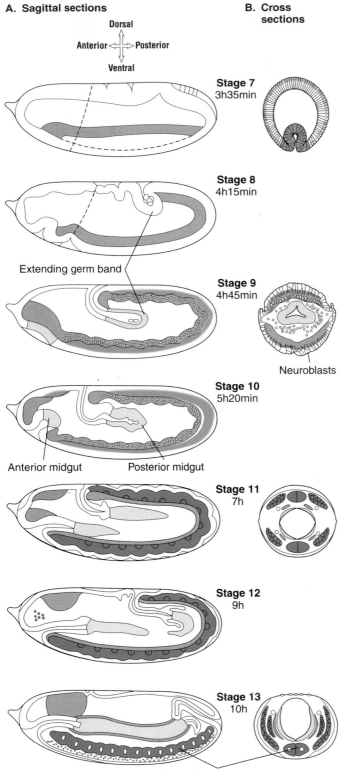

Stage 7
3h35min

Stage 8
4h15min

Extending germ band

Stage 9
4h45min

Neuroblasts

Stage 10
5h20min

Anterior midgut   Posterior midgut

Stage 11
7h

Stage 12
9h

Stage 13
10h

Central nervous system

Figure 3.11 Gastrulation in *Drosophila* (A) Sagittal sections through the embryo are shown, from the end of cleavage until the gut is formed (approximately 10 hours after fertilization). The invagination of the mesoderm, the anterior and posterior invaginations to form the midgut, and the germ band extension are indicated. After 7 hours, the germ band begins to retract, and by stage 13 the anterior and posterior invaginations have met to form the complete gut. **(B)** Cross sections through the middle of the embryo show more clearly the formation of mesoderm by invagination at stage 7. The developing endodermal gut begins to appear in sections at stage 8. The mesoderm transforms from epithelium to mesenchyme, and prospective neuroblasts begin to move inward at later stages.

Please note the color scheme used in Figure 3.11. These are the colors conventionally used to represent the germ layers and their derivatives: ectoderm is usually blue, mesoderm is usually red, and endoderm usually yellow.

The terms *epithelium* and *mesenchymal*, used above, are useful general descriptions of tissue structure. In **epithelium,** a tissue that lines surfaces and cavities, the cell layers are closely packed and the cells are bound together with cell junctions. The epidermis of skin and the internal lining of the stomach are both epithelial. On the other hand, in **mesenchyme,** the cells are not closely packed; considerable extracellular material may lie around and between mesenchymal cells. Cartilage and the dermis of the skin are examples of mesenchymal tissues.

After the ventral furrow has formed prospective mesoderm, there is a second round of invagination at the anterior and posterior ends of the furrow. (Embryological literature sometimes uses *presumptive* to mean *prospective.*) These anterior and posterior groups of invaginated cells will form pockets, which later elongate, extend within the embryo, meet and fuse, and form a long internal tube of endoderm that will become the midgut of the larva. Cells on the surface attached to each end of the forming midgut will later also become interiorized and form the foregut and hindgut.

The ventral cells of the embryo that remain on the surface, together with the internalized mesoderm, undergo **convergent extension,** a general term for certain kinds of intercellular arrangements (see Chapter 12). In the case of *Drosophila* and other insects, this "shuffling" of cells produces an elongation and folding back of the entire posterior two-thirds of the embryo. Constrained within the chorion, the elongating embryo folds back upon itself, rather like a scorpion in attack posture, as shown in Figure 3.11A, stage 9. This elongation, called **germ band extension,** is peculiar to insects, and though it makes following the anatomy of the embryo difficult at these stages, it need not concern us unduly, as the extended germ band later retracts and the posterior cells once again come to lie at the posterior terminus.

While anterior and posterior invaginations and germ band extension are occurring, some ventral cells still on the surface detach from their epithelial neighbors and **ingress** (migrate inward) to a new location in the yolky interior (Figure 3.12). Later these cells will divide to form neurons and supporting cells of the nervous system.

Thus, cell reshuffling (or germ band extension), invagination (of the mesoderm and midgut), and ingression of individual cells (to form neural tissue) have produced a radically reorganized embryo. There are different opinions on what exactly to designate as gastrulation, and when exactly it begins and ends; for our purposes, **gastrulation** is the entire suite of movements that begins when cells move interiorly and produce the distinctive endoderm, mesoderm, and ectoderm germ layers.

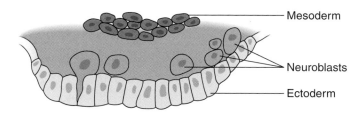

Figure 3.12   The Formation of Neuroblasts Cells from the ventral surface move into the interior. Here the progenitor cell divides. giving rise to both neuroblasts and glial cells.

The Embryo Becomes Segmented

About the time the germ band is fully extended (Figure 3.11), the surface of the embryo displays grooves running from side to side, and these demarcate 14 evenly spaced bands called *parasegments.* These correspond, more or less, with three of their segments forming the head and mouth parts, three forming the thorax, and eight forming the abdomen

BOX 3.2 GERM LAYERS

The term **germ layers** has been used for a long time by biologists. It is a complicated term because it refers to both a description and an idea. Many animal embryos at the conclusion of gastrulation movements clearly have a three-layered morphology. There are cells still on the surface. called **ectoderm,** and those lining an internal tube. This internal tubular structure. which you recall is formed in *Drosophila* by anterior and posterior invaginations. is often called an *archenteron* and is a precursor to some or all of the digestive system. It is the **endoderm** germ layer. The cells between

the surface ectoderm and endoderm are called the **mesoderm.** and much of the internal organs originate from these cells. When the changing developmental anatomy of a creature presents such an appearance. as it always does in vertebrates and in many invertebrates too. the descriptive terms are useful.

The second aspect of germ layers is the idea that cells in germ layers are more restricted in their developmental potential than they were before germ-layer formation. This is because cell determination often progresses during gastrulation. It is important to

keep the use of these terms as anatomical regions distinct from the idea of changes in developmental potential.

A third aspect of germ layers—its use in discussions of evolution—is the most complex. Sometimes it is assumed. while at other times it is explicitly stated, that the germ layer is more than a descriptive term. that it implies some homology and common evolutionary history. Only recently. however. has it been possible to subject this kind of argument to experimental analysis. We shall examine this subject briefly in Chapter 17.

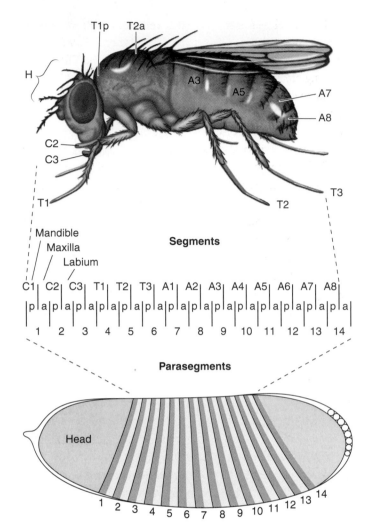

Figure 3.13 The Registration of Embryonic Parasegments with Adult Body Segments In the embryo the "head" region and 14 parasegments become transformed into the head, thorax, and abdomen of the adult. The anterior (a) and posterior (p) boundaries of the adult segments are uniformly slightly anterior of the respective portions (brown and beige) of the embryonic parasegments.

(Figure 3.13). (Some recent evidence suggests that the three morphological parasegments contributing to the adult head may be comprised of seven segments of gene expression.) But the parasegments seen in the embryo are out of register, by about half a parasegment, with the body segmentation seen in the thorax and abdomen of an adult, and the precise details of registration of adult segments and embryonic parasegments is a topic for specialists in insect embryology. Refer to Figure 3.13 when you need to check the registration.

Most of the cells of the larval ectoderm divide only a few more times, and they secrete the cuticle of the larva and various larval surface structures. Ventrally each para-

segment comes to possess curious little projections called *denticles*. The pattern of denticles is somewhat different in each parasegment, thereby allowing an expert to identify a parasegment solely on the basis of the arrangement of its denticles. Each parasegment also possesses small nests of about 10 cells that retain the capacity for extensive cell division— and these nests of *histoblasts* will replace the dying larval epithelium in the pupa. There are also small groups of about 40 cells in the newly hatched larva called **imaginal discs.** They form anatomically complex infoldings of the surface epithelium, and during the transition from larva to adult life, which happens during *metamorphosis*, these discs will evert and form many body parts and organs of the adult. The larval epidermis will die at metamorphosis, and the imaginal discs will form organs of the surface and some of the internal organs of the adult fly. Figure 3.14 shows the scattered nests of imaginal disc cells in a first instar larva. (**Instars** are the successive stages of larval development punctuated by molts that allow growth in the larva.) We shall examine this process of larval development in Chapter 8.

PATTERNING OF THE EMBRYO

Localized Determinants in the Egg Specify a Cell's Fate

Now that early development has been described, we may ask what is really happening at the level of cellular fate assignment and differentiation. We already mentioned that transplanta-

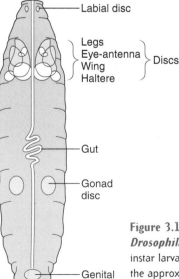

Figure 3.14 Imaginal Discs in the *Drosophila* Larva Diagram of a third instar larva, viewed from above, showing the approximate location of some of the more important imaginal discs.

tions and removal of nuclei, cytoplasm, and blastoderm cells show that regions of the cellular blastoderm have become restricted in their fate in a predictable way. Cells at the anterior of the egg form head parts, those in the middle form thorax, and so on. In fact, we may project onto the cellular blastoderm a fate map, shown in Figure 3.10. Along the anteroposterior axis (that is, from front to back), seven different regions may be distinguished, forming the head, thorax, and abdominal parts. And along the circumference of the egg, four different areas may be distinguished; moving from most ventral to most dorsal, they are mesoderm, neurogenic ectoderm, dorsolateral ectoderm, and an area called the *amnioserosa*, which will form an extraembryonic membrane. Combining the seven anteroposterior regions and the four dorsoventral ones makes for 28 separate areas, all shown experimentally to have fixed and different identities (Figure 3.15).

Classical embryological experiments, carried out particularly by Klaus Sander in Germany in the 1960s, showed that the anterior portion of the insect egg probably contains some substance or substances that favor development of anterior body parts. (Sander actually used eggs of the leaf hopper *Eucelis*, but similar results have been obtained with *Drosophila*.) Likewise, a different, "posterior-favoring" substance was located at the posterior end of the egg. And importantly, different graded concentrations of the anterior and posterior substances merged in the middle of the egg, counteracting each other and resulting in middle body parts.

This model was attractive and well supported by experimental evidence. But what are the substances? How can we get a hold on them to understand how they work? One of the exciting stories of recent biological research is how many of the answers to this question were unraveled through the methods of genetics and molecular biology. The substances are indeed real, as we shall soon discuss in detail. These are molecules whose local concentration determines a particular kind of development. More of the anterior substance favors headlike development, while less favors thoracic development. In other words, the concentration gradient of the substance determines the pattern of differentiations. Such molecules whose gradients of concentration produce patterning in the embryo are called **morphogens.** Researchers invoked them as a causal agent for a long time, but until recently, very few morphogens had been discovered. Now we know they are one of the great strategic inventions of development; we shall encounter them over and over again.

Maternal Effect Mutants of **Drosophila** Encode Morphogens

The breakthrough in the study of morphogens came in the 1970s as a result of an arduous series of genetic experiments by Eric Wieschaus and Christiane Nusslein-Volhard. They were interested in genes that orchestrated the earliest stages of development. Such genes, if they existed, were probably so

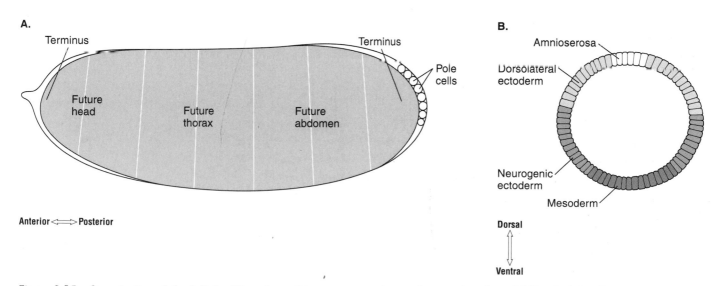

Figure 3.15 Organization of the Cellular Blastoderm Schematic representations of the embryo at the cellular blastoderm stage. **(A)** A sagittal section, running through the midbody from anterior to posterior, shows the seven main regions of anteroposterior organization. The embryo at this stage has about 6,000 cells (about 80 cells long by 72 cells in circumference). **(B)** A cross section through the midbody region, identifying the approximate locations of prospective tissues.

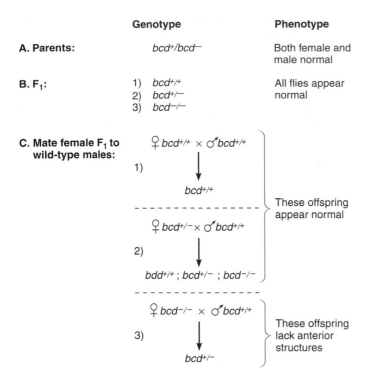

	Genotype	Phenotype
A. Parents:	*bcd⁺/bcd⁻*	Both female and male normal
B. F₁:	1) *bcd⁺/⁺* 2) *bcd⁺/⁻* 3) *bcd⁻/⁻*	All flies appear normal

C. Mate female F₁ to wild-type males:

1) ♀ *bcd⁺/⁺* × ♂ *bcd⁺/⁺*
   ↓
   bcd⁺/⁺

2) ♀ *bcd⁺/⁻* × ♂ *bcd⁺/⁺*
   ↓
   bdd⁺/⁺ ; bcd⁺/⁻ ; bcd⁻/⁻

These offspring appear normal

3) ♀ *bcd⁻/⁻* × ♂ *bcd⁺/⁺*
   ↓
   bcd⁺/⁻

These offspring lack anterior structures

D. The vitellarium of a *bcd⁻/⁻* female:

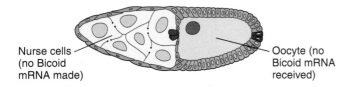

Nurse cells (no Bicoid mRNA made)

Oocyte (no Bicoid mRNA received)

Figure 3.16 The Inheritance of a Maternal Effect Mutation An illustration of the inheritance of a maternal effect mutation is shown, using the *bicoid* gene (*bcd*) as an example. (**A**) The parents are both heterozygous for the mutation, and since it is recessive, they both appear normal. (**B**) When the offspring of the first generation (F₁) are analyzed, they are also all phenotypically normal. However, ¼ of these apparently normal flies are homozygous for the *bcd⁻* gene. They appear normal, however, since their parents were heterozygous, and in particular, the mother had a copy of the normal gene in the diploid nurse cells. (**C**) When these first-generation females are mated to any male, the ¼ of the females that are *bcd⁻/bcd⁻* homozygotes (type 3 in part B) give rise to embryos that fail to develop anterior structures. This is because the homozygous mutant nurse cells failed to deliver Bicoid mRNA to the oocyte. (**D**) A diagram of the vitellarium of a *bcd⁻* female underscoring the reason for the maternal effect: it is the genetic constitution of the nurse cells that counts.

Nusslein-Volhard and Wieschaus, later joined in this work by their colleagues and students, found 34 such genes, a very small number considering that the *Drosophila* genome has over 12,000 genes. Of these 34 maternal effect mutants, 12 affected the pattern along the anteroposterior axis; six affected only structures at the extreme ends of the embryo (the most anterior portion of the head, called the acron, and the posterior tip of the tail, called the telson), and 16 affected the pattern along the dorsoventral axis. Astonishingly, each of these systems seemed independent of the others; we now know with a fair degree of certainty that the different morphogens work in different ways, though the very earliest establishment of asymmetries in the developing embryo does involve utilization of components affecting more than one system. (Later in the chapter, we shall see more of this, in our discussion of the Gurken morphogen.) There are really four systems: one each for anterior and posterior ends, one for terminal regions, and one for dorsoventral organization.

The Anterior Morphogen Is Bicoid

One recessive maternal effect mutant that gave a clear phenotype was *bicoid*; a wild-type (*bcd⁺/bcd⁺*) and mutant (*bcd⁻/bcd⁻* mother) larva are shown in Figure 3.16. Mutants have a normal posterior but no head; instead an extra telson and spiracles (both structures found posteriorly) are found where the head ought to be. If some anterior cytoplasm from eggs of a normal-looking heterozygous mother is injected into the anterior end of an egg from a homozygous mutant mother, then there will be some rescue of anterior structures.

vital that mutations to them would be lethal; so the two researchers had to invent clever ways of producing these mutants. They did so by crossing two suspected heterozygotes carrying recessive mutations and then looking at the offspring very early in development. In this way they could identify abnormalities that would shortly lead to death.

Wieschaus and Nusslein-Volhard were also interested in the class of mutations that geneticists call *maternal effect mutants*. These mutants are seen only in the offspring of a homozygous recessive mother, since these genes affect egg formation during oogenesis, before the egg becomes haploid. We now know that nurse and follicle cells may also be affected by these genes. Thus, while the mother may appear perfectly normal, she will produce eggs that are defective in some way. If there are morphogens laid down in the egg that influence the very first events of determination, then mutations in the genes that encode the morphogens should produce maternal effect mutants (sometimes called grandchildless mutants; see Figure 3.16).

If the donor cytoplasm comes from a wild-type (+/+) egg, then the degree of rescue is even greater. The *bcd* gene seems to encode a morphogen; the more morphogen present, the greater the development of more anterior structures. Bicoid itself seems to produce anterior structures; if the transplanted *bcd*⁺ cytoplasm is deposited into the middle of an egg from a *bcd*⁻/*bcd*⁻ mother, then anterior structures develop in the *middle* of the mutant egg.

We shall have more to say about how Bicoid does all this when we come to discuss transcription in detail. Bicoid protein is a transcription factor containing the homeobox motif (we shall discuss homeoboxes in detail in Chapter 15); when Bicoid is present at certain concentrations, some responsive genes are turned on and begin to synthesize their cognate mRNAs. Thus an agent that turns on specific genes, leading to the formation of anterior structures, is located in the cytoplasm. When cleavage nuclei inhabit cytoplasm containing Bicoid, they import Bicoid into the nucleus, where Bicoid can interact with responsive genes. The *bcd* gene has been cloned, and antibodies directed against the Bicoid protein are available. Hence, it is relatively straightforward to identify the location of Bicoid mRNA and Bicoid protein, as is shown in Figure 3.17.

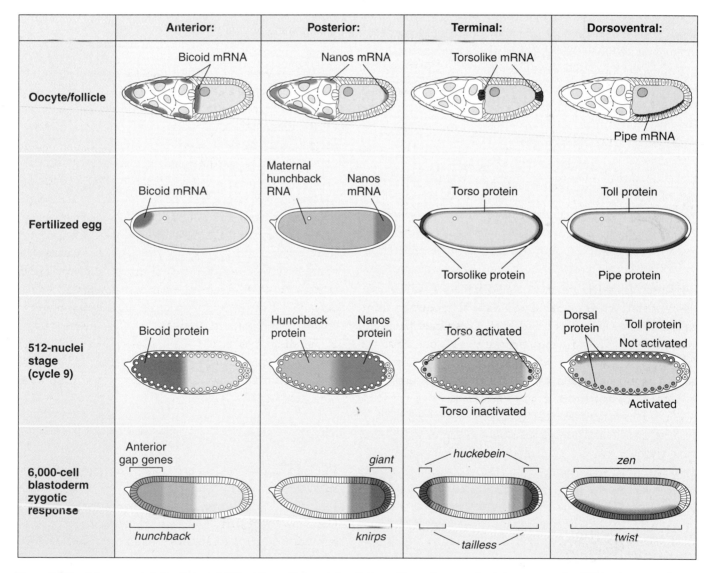

Figure 3.17   A Summary of the Maternal Effect Genes Determining Pattern These diagrams show the position of mRNA and protein of the systems involved in anteroposterior, terminal, and dorsoventral specification in *Drosophila*. The situation is depicted for the oogenesis, fertilization, syncytial blastoderm, and cellular blastoderm stages.

Use of these tools shows not only that Bicoid mRNA is located in the anterior (and dorsal) cytoplasm of the unfertilized egg, but that during oogensis it is the nurse cells that synthesize the Bicoid mRNA. The Bicoid mRNA is then transported to the egg via the ring canals. The Bicoid mRNA is trapped in some way as it enters the prospective anterior portion of the egg, and it is tethered to the cytoskeleton in the anterior cytoplasm. During early cleavages, Bicoid protein is synthesized and begins to diffuse some slight distance from the anterior pole, producing an anteroposterior gradient of Bicoid protein with a maximum concentration at the anterior end of the egg.

The other genes identified in the screen for maternal effect anterior-pattern mutants have also been studied. They are part of the apparatus that functions to localize and tether the Bicoid mRNA in the proper anterior location. Mutations in the *swallow*, *exuperantia*, and *staufen* genes all affect proper localization of the Bicoid mRNA. Sequences near the 3′ untranslated end of Bicoid mRNA are necessary for its proper localization.

The Posterior Pattern Is Determined by the Concentration of Nanos

Nine maternal effect genes were discovered that affect the posterior of the embryo; in these mutants the abdomen is essentially missing and an expanded thoracic region is present. The gene encoding a posterior morphogen is *nanos*. Once again, the Nanos mRNA is synthesized in nurse cells. Then the mRNA is transported into the egg. In this instance the mRNA is not trapped anteriorly but becomes localized in the extreme posterior cytoplasm of the egg, where it associates with polar granules. And, once again, the other genes affect-

ing posterior development and pole cell formation encode proteins that help transport and tether Nanos in the appropriate location; Nanos also requires a sequence in the 3′ untranslated region of the mRNA for its proper localization. Nanos mRNA is translated locally to produce a gradient of Nanos protein with a high concentration at the posterior (Figure 3.17). But there the similarity to Bicoid stops.

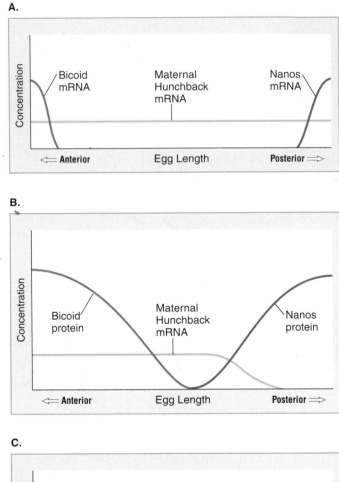

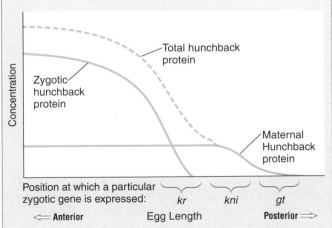

Figure 3.18   The Suppression by Nanos of Posterior Hunchback Translation The concentration of Bicoid, Hunchback, and Nanos mRNA and protein at various positions in the egg/embryo and at different times is shown. **(A)** The concentration of Bicoid, Hunchback, and Nanos mRNA at the time of fertilization. **(B)** Once the mRNAs have been translated: levels are high anteriorly for Bicoid, high posteriorly for Nanos, and largely absent posteriorly for Hunchback because of the suppressive effect of Nanos. **(C)** The concentration of Hunchback at a later time in development: Hunchback is now very high anteriorly due to stimulation of the *hunchback* gene by Bicoid, but Nanos continues to suppress Hunchback mRNA translation posteriorly. The approximate positions at which *krüppel* (*Kr*), *knirps* (*kni*), and *giant* (*gt*) are expressed. The expression of *Kr* requires higher levels of Hunchback than does *kni*, and *gt* is repressed by Hunchback. Hence a gradient of Hunchback dictates the spatial expression along the egg axis of *Kr*, *kni*, and *gt*.

Nanos protein inhibits the translation (and thus protein synthesis) of a few particular mRNAs; most importantly, it inhibits protein synthesis encoded by the Hunchback mRNA already stored in the oocyte ("maternal" Hunchback). *Hunchback* encodes a transcription factor that turns on many genes favoring anterior development. In fact, *hunchback* is the primary target of Bicoid. But there is a complication; some Hunchback protein has been synthesized during oogenesis and is present in the egg, though it is not localized. When Nanos is present, no Hunchback protein can be made posteriorly, and so while Hunchback protein can act anteriorly, and is stimulated by Bicoid, Hunchback cannot act posteriorly (Figure 3.18). Hunchback is a complicated transcription factor; not only does it stimulate the transcription of certain genes, especially those involved in head development; it also *inhibits* the transcription of other genes, among them *giant* and *knirps*, which are needed for the development of posterior structures. We shall revisit these different genes stimulated or inhibited by Nanos in Chapters 14 and 15. Nanos works, then, by inhibiting *hunchback*, and Hunchback inhibits posterior development—a "double negative"! One would predict that the double mutant of *nanos⁻/nanos⁻*, *hunchback⁻/hunchback⁻* would have normal posterior development, and that is the case!

The Morphogen for the Termini of the Egg Is Torso

The genes of the terminal class (there are six) produce embryos that lack the acron and telson, the anterior and posterior ends of the embryo. Instead, a somewhat expanded head, thorax, and abdomen are present. Torso mRNA is made in nurse cells and transported throughout the oocyte, where Torso protein is synthesized and adopts its proper location as a membrane protein throughout the entire egg. The function of Torso in the membrane is to act as a receptor of signals; it is a member of the class tyrosine kinase transmembrane receptors, which we shall discuss later in Chapter 12.

The pattern of Torso activity is determined by another gene, *torsolike*. The protein encoded by this gene is synthesized by special follicle cells located at the anterior and posterior ends. Thus, the Torso receptor is activated only at the anterior and posterior ends by the localized Torsolike protein. Activation of Torso sets in motion a complex chain of chemical commands in the localized cytoplasmic environments that eventually results in activation of *tailless* in the embryo, and *tailless* is the essential first gene needed for devel-

opment of the acron and telson (see Figure 3.17). The action of Bicoid and Nanos involves the localization of a mRNA that produces the morphogen and organizes pattern. In contrast, terminal specification by Torso depends on localized activation of a ubiquitous receptor. We shall see that localized activation of a receptor is also involved in dorsoventral pattern organization.

The Morphogen for Dorsoventral Organization Is Dorsal Protein

Sixteen genes are involved in the organization of the dorsoventral pattern; mutations in 15 of these (13 strictly maternal, two expressed after fertilization in the zygote) lead to a "dorsalized" phenotype, that is, lacking in *ventral* structures such as mesoderm and the ventral nerve cord. One of the maternal genes, *cactus*, when present as a homozygous mutant, leads to a ventralized phenotype, in which the amnioserosa and dorsolateral ectoderm may be absent. Mutations in two other maternally acting genes and five zygotically active genes sometimes produce partially ventralized phenotypes.

The entire set of dorsal genes, including *cactus*, are part of a pathway needed for ventral development; the localized morphogen is the product of the *dorsal* gene, and the Dorsal protein is present in all the cells around the circumference of the embryo. It is the localized activity of Dorsal, not its mere presence, that is crucial here. Figure 3.17 (bottom row) presents some of the key steps involved in the localized activation of Dorsal.

The protein Dorsal is a transcription factor. When antibodies against Dorsal are used to determine where Dorsal is in the embryo, the protein is found in the cytoplasm around the entire circumference of the embryo. After cellularization, Dorsal adopts a nuclear location in ventral cells, and as one's view shifts along the circumference, less and less Dorsal protein is visible in the nucleus, being essentially absent in dorsal cells. The Dorsal transcription factor stimulates transcription of the genes *twist* and *snail*, which are essential for the development of ventral mesoderm. Dorsal also inhibits transcription of the gene *zen*, which is needed for dorsal development (Figure 3.17, bottom row).

How is the ventral localization of Dorsal accomplished? Again, the game plan is established during oogenesis; a localized signaling mechanism activates a ubiquitously present receptor only on the future ventral side of the oocyte. Ventral follicle cells are responsible for synthesis and assembly of a

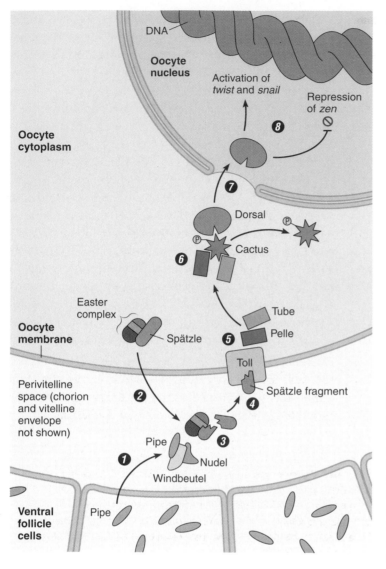

❶ Only ventral follicle cells make Pipe. Complex of Pipe, Nudel, and Windbeutel proteins is secreted (only on ventral side). Pipe sulfates glycosaminoglycans (see Chapter 12) on the ventral side.

❷ Nurse cells have deposited mRNA for Easter complex in the egg, and the Easter protein complex, which includes Snake and Gastrulation defective (Gdp), is secreted around the egg.

❸ Easter complex and Pipe complex together cleave Spätzle.

❹ Spätzle fragment activates the Toll receptor.

❺ Toll activates Tube and Pelle.

❻ Tube and Pelle phosphorylate Cactus so that it dissociates from Dorsal.

❼ Free Dorsal enters the nucleus. It is a transcription factor.

❽ Dorsal stimulates *twist* and *snail* (in mesoderm) and represses *zen* (in dorsal epithelium).

Figure 3.19   The Pathway of Dorsoventral Pattern Specification   This highly schematic diagram of the ventral follicle cell layer, the egg cell plasma membrane, and the cytoplasm illustrates the steps in regulating Dorsal protein's entry into the nuclei on the ventral side. Only follicle cells on the ventral side have synthesized a complex of Pipe, Nudel, and Windbeutel proteins. When a multiprotein complex of Spätzle, Easter, Snake, and Gdp encounters the Pipe complex, Spätzle is cleaved. This fragment is the ligand for the Toll receptor in the egg plasma membrane, which then sets in motion, using Tube and Pelle proteins, a reaction removing the inhibitor Cactus from Dorsal. Dorsal may now enter the nucleus, where it can activate some genes (*twist* and *snail*) and repress others (*zen*).

multiprotein complex, designated X in Figure 3.19; the proteins comprising X, with the colorful names Pipe, Nudel, and Windbeutel, act together with a protein synthesized and secreted by the oocyte called Easter. Ventrally localized Pipe adds sulfate to extracellular glycosaminoglyans during oogenesis. This ventral, localized modification somehow works together with Easter to cause a proteolytic cleavage of a signaling molecule called Spätzle. Because of the localized ventral presence of Pipe, the only place an active form of Spätzle is found is in the perivitelline space adjacent to the prospective ventral side of the egg. The Spätzle signal activates a membrane receptor in the egg called Toll, which acts by way of intermediate steps to remove an inhibitor protein, Cactus, from Dorsal. Once Cactus has been removed, Dorsal is free to enter the nucleus and act on its various target genes as either a positively acting or a negatively acting transcription factor.

The Patterning of the Follicle Cells Drives the Patterning of the Oocyte

The four elegant patterning systems of the *Drosophila* oocyte—anterior, posterior, terminal, and dorsoventral—operate independently of one another. However, antecedent events must be operating very early in oogenesis to help set up these systems. How is "anterior" (or "posterior," for that matter) identified in the oocyte so that Bicoid and Nanos can be localized? What identifies the places in the follicle that contain the Torsolike molecule? What specifies the ventral follicular position of X?

A. Gurken induces polar follicle cells to adopt a posterior fate

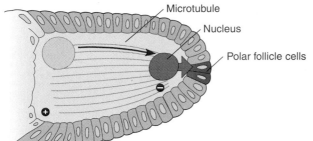

- Microtubule
- Nucleus
- Polar follicle cells

B. Unidentified signal induces repolarization of oocyte microtubules

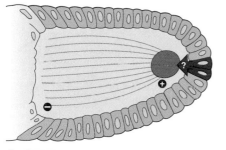

C. Nucleus moves up, to anterior pole, via microtubules where Gurken induces follicle cells to adopt dorsal fate

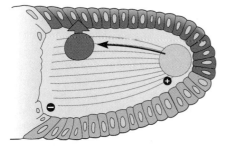

Recent work has shown that a reciprocal signaling system between the developing oocyte and follicle cells plays a crucial role in organizing the cytoplasm of the developing oocyte, probably mainly by way of its microtubule-based cytoskeleton. As shown in Figure 3.20, a signaling molecule called Gurken, a member of the transforming growth factor–alpha (TGF-α) family of molecules, is released from the oocyte nucleus when it is near the posterior portion of the oocyte; this signal is received by a follicle cell receptor called Torpedo, which then causes the follicle cells to signal back to the oocyte and organize the oocyte cytoskeleton so that an anteroposterior axis is laid down. Later the oocyte nucleus moves along this polarized cytoskeleton to a more anterior location, and the Gurken signal is again sent and received. This second signaling induces nearby follicle cells to adopt a dorsal type of existence. This prevents the formation dorsally of the X complex crucial for ventral development, and induces activity of a complicated circuit that leads to development of the bilaterally situated dorsal respiratory organs of the oocyte. A reciprocal signaling between developing follicle cells and the oocyte, together with contributions from nurse cells, results in an elegantly patterned egg surrounded by a heterogeneous, organized follicular epithelium.

Figure 3.20 Interactions Between the Early Oocyte and Follicle Cells (A) The *gurken* gene encodes a signal (Gurken) that causes follicle cells at the posterior end of the egg chamber to adopt a posterior, polar fate. **(B)** Subsequently, the oocyte nucleus moves via the microtubule cytoskeleton to a dorsoanterior location. **(C)** Once again there is Gurken signaling, this time specifying a dorsoanterior fate to the adjacent follicle cells.

KEY CONCEPTS

1. The cytoplasm in which a nucleus comes to reside may, and sometimes does, affect which genes are expressed, as shown in the pole plasm transplantation experiments.

2. A cell or group of cells may have a developmental potential at early times in embryonic development that far exceeds what it actually becomes, its fate.

3. Gastrulation in animal embryos involves a suite of coordinated cell behaviors that results in translocating cells from the surface, or near to the surface, to an interior location. The exact suite of behaviors may differ in different species.

4. *Drosophila* and almost all other embryos (mammals excepted) have localized materials in the cytoplasm of the egg that are heterogeneously distributed. These act as "determinants," signals that initiate gene expression in cells in some areas of the embryo and not in others. These determinants are either transcription factors; molecules that tether, activate, modify, or somehow regulate the activity and placement of them; or localized members of signal reception and transduction pathways. Arriving at this generalization is one of the triumphs of modern developmental biology.

STUDY QUESTIONS

1. Suppose a recessive mutation has been found that impedes the proper development of ring canals so that the canals are never open for the flow of material through them. What do you think the phenotype might be? Would the observed phenotypes classify this as a maternal effect mutation or a zygotic mutation when homozygous?

2. Suppose it were possible to remove, by laser irradiation, the area of cortical cytoplasm in the syncytial blastoderm from which the cells forming the wing disc arise. What would be the effect of the laser ablation if carried out in cycle 4 of cleavage?

3. It is possible to artificially increase the number of copies of *bcd* genes from one per haploid chromosome set to two or three. Can you predict what would be the result of increasing the number of *bcd* genes present in a female fly?

4. Can you speculate what the effect of loss-of-function mutations of *tube* and *pelle* might be? (Refer to Figure 3.19.)

SELECTED REFERENCES

Anderson, K. V. 1998. Pinning down positional information: Dorsal-ventral polarity in the *Drosophila* embryo. *Cell* 95:439–442.

A review of dorsoventral patterning by one of the main contributors in this field.

Knust, E., and Muller, H.-A. J. 1998. *Drosophila* morphogenesis: Orchestrating cell rearrangements. *Curr. Biol.* 8:R853–R855.

A review of various genes involved in gastrulation in *Drosophila*.

Lawrence, P. 1992. *The making of a fly.* Blackwell Scientific Publications, Oxford.

A brilliant, short book reviewing and explaining the analysis of early *Drosophila* development as it occurred in the years following Nusslein-Volhard's and Wieschaus's discoveries.

Misra, S., Hecht, P., Maeda, R., and Anderson, K. V. 1998. Positive and negative regulation of Easter, a member of the serine protease family that controls dorsal-ventral patterning in the *Drosophila* embryo. *Development* 125:1261–1267.

A research paper on the localized proteolytic activation of the ligand leading to the dorsoventral axis. An excellent list of references to this entire subject may be found here.

Nusslein-Volhard, C., Frohnhofer, H. G., and Lehmann, R. 1987. Determination of anteroposterior polarity in *Drosophila*. *Science* 238:1675–1681.

A review of the work on maternal effect mutants, especially *bicoid*.

Perrimon, N., and Duffy, J. B. 1998. Sending all the right signals. *Nature* 396:18–19.

A brief discussion of the complexities of Gurken signaling in the oocyte. For another view of the same subject, see L. Stevens, 1998. Twin peaks: Spitz and Argos star in patterning of the *Drosophila* egg. *Cell* 95:291–294.

Technau, G. M. 1987. A single cell approach to problems of cell lineage and commitment during embryogenesis of *Drosophila melanogaster*. *Development* 100:1–12.

A masterful review of work on the fate and specification maps of *Drosophila*.

Theurkauf, W. E., and Hazelrigg, T. I. 1998. In vivo analyses of cytoplasmic transport and cytoskeletal organization during *Drosophila* oogenesis: Characterization of a multi-step anterior localization pathway. *Development* 125:3655–3666.

Research on how molecules get localized in *Drosophila* eggs.

AMPHIBIAN DEVELOPMENT

In the late 19th and early 20th centuries, biologists interested in problems of how creatures develop turned to the available local fauna, which still form much of the backbone of our knowledge of development. Frogs and salamanders were often used, especially by German biologists, who were on the cutting edge of research at this time. One advantage of using amphibian embryos was (and still is) the relative ease of performing many of the experiments. Fertilization occurs externally, the embryos are relatively large, and development can be observed with the naked eye or low-power dissecting microscopes. Simple methods of transplanting cell groups were devised. The study of many different kinds of vertebrate embryos made it clear that the formation of the different organ systems is similar in the different vertebrate orders. For example, neural tube formation in salamander and mouse embryos shows great similarities.

CHAPTER PREVIEW

1. The frog *Xenopus laevis* is an important model organism for studying vertebrate development.

2. Yolk proteins are synthesized in the liver and transported to the developing oocyte.

3. During oogenesis the number of ribosomal RNA genes is temporarily increased.

4. Complete maturation of the egg requires stimulation by progesterone.

5. The superficial cortex of the fertilized egg rotates, which establishes the anteroposterior and dorsoventral axes of the embryo.

6. Cleavage produces a hollow blastula.

7. Gastrulation occurs by the movement of surface cells and their neighbors into the interior.

8. The neural tube forms in the ectoderm, which is produced by gastrulation.

It was supposed, then, that study of processes of development in amphibians could illuminate similar processes in mammals, and that expectation has been fulfilled. A large body of accumulated information now provides a foundation for current research, and new techniques and discoveries act as a kind of positive feedback loop to sustain the use of amphibians. *Drosophila* has assumed special importance because of its suitability for genetics. The frog (and, as we shall see, the mouse) has become important for several reasons: because it is a vertebrate, because we have a large amount of information about its development on both morphological and molecular levels, and because there is one especially suitable species. The South African clawed frog, *Xenopus laevis*, was used by a few embryologists in the 1950s and became the favorite of experimentalists using nuclear transplantation techniques, mentioned in Chapter 1. Natural populations of readily available North American frogs, such as *Rana pipiens*, have diminished, for reasons probably ranging from environmental insults to overcollecting. *Xenopus* is easily maintained in the laboratory and is now the predominant species used in developmental biology research on amphibians.

GAMETOGENESIS

Oogenesis Involves Cyclic, Progressive Maturation of Oogonia

It should be no surprise that, just as with *Drosophila*, oogenesis involves meiosis, a tremendous stockpiling of materials needed for early development, and an elegant organization of the intracellular materials in the oocyte. Figure 4.1 shows several different stages of oogenesis in a fragment of *Xenopus* ovary. Early stages of oogenesis have not yet begun to accumulate yolk (a process called vitellogenesis). However, amplification of nucleoli, which we will discuss shortly, does occur in these young, translucent oocytes.

The oogonia are stem cells in the ovary that undergo repeated mitotic divisions, increasing cell numbers as much as 10,000 times. Many of these oogonia will enter a state in which they will divide four more times, forming nests of 16 diploid cells connected by cytoplasmic bridges. After these nests are formed, the oogonial descendants are committed to meiosis and are now called primary oocytes. While this pattern may seem reminiscent of oogenesis in *Drosophila*, here all 16 of the oogonial descendants (rather than only one) will enter meiosis, and each oogonium will form a mature egg.

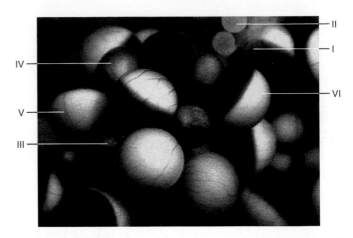

Figure 4.1 *Xenopus laevis* **Oocytes** A photograph of a fragment of ovary displays different stages of oogenesis. The early, translucent cells (stages I and II) have not yet begun to accumulate yolk produced in the liver. Nucleolar amplification is occurring during these early pre-vitellogenic stages. Subsequently, pigment accumulates in the animal hemisphere, and yolk is imported and is concentrated in the vegetal hemisphere.

The oogonium begins to grow and undergo genetic recombination as it enters the extended prophase of meiosis I. The recombination process takes about two weeks. As soon as the oogonium begins to grow, it is called a primary oocyte. The primary oocyte then forms huge extended chromosomes during the diplotene stage of meiotic prophase I. These chromosomes are huge because the chromatin has adopted an uncoiled configuration along much of its length. Because of their brushlike morphology, they were called *lampbrush chromosomes* by early cytologists (Figure 4.2). We now know that they are the site of extremely active transcription of genes. Each primary oocyte in the "nest" becomes surrounded by an envelope of somatic cells called **follicle cells.** The lampbrush chromosomes gradually recompact, the large oocyte nucleus (which, as with *Drosophila*, is called a *germinal vesicle*) is seen to contain about 1,500 separate nucleoli, and the oocyte itself, now some 0.3 mm in diameter, begins to accumulate yolk. Just as in *Drosophila*, the yolk is made elsewhere, in this case in the liver, and the yolk precursor, vitellogenin, is transported in the bloodstream. It leaves the capillary bed in the surrounding follicle and is imported into the growing oocyte. The oocyte with its large germinal vesicle processes the vitellogenin into different-sized "platelets" of yolk, and accumulates them to a greater or lesser extent, in different regions of the egg cytoplasm. Pigment granules are synthesized by the oocyte and localized in the cortex of the egg. The succession of these steps is shown in Figure 4.3.

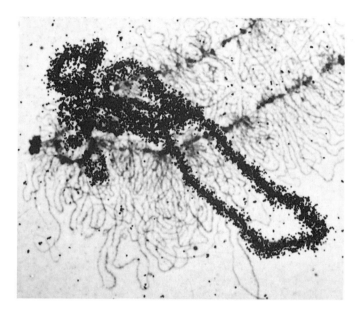

Figure 4.2 A Lampbrush Chromosome A female newt was injected with radioactive uridine in order to label the newly synthesized RNA. Then an isolated lampbrush chromosome was dissected from the germinal vesicle of an amphibian oocyte. The loops of the chromosome are clearly visible. One very large loop is actively synthesizing RNA, as indicated by the dense cluster of black dots caused by the radioactivity.

The oocyte has a clear polarity: the germinal vesicle is located in the hemisphere with more cortical pigment granules, and the yolk platelets are larger and more numerous in the opposite hemisphere. It is usual for developmental biologists to refer to the pigment-rich but yolk-poor side as the **animal pole,** the other as the **vegetal pole.**

Clearly the egg is organized morphologically along an animal-vegetal axis, indicated by gradations of pigment and yolk and the placement of mitochondria and various granules. We shall also come to see that various RNA molecules and proteins are localized along this axis. The egg is radially symmetric, with no indication of anterior or posterior directions. Though the causal links are not definitely established, the vegetal pole corresponds to the former location of the cytoplasmic bridges that linked the 16 sister

Figure 4.3 Oogenesis in *Xenopus laevis* The various stages of oogenesis are shown in linear sequence. The oocytes at different stages become much larger as they mature, so the relative sizes of oocytes here are not quite to scale.

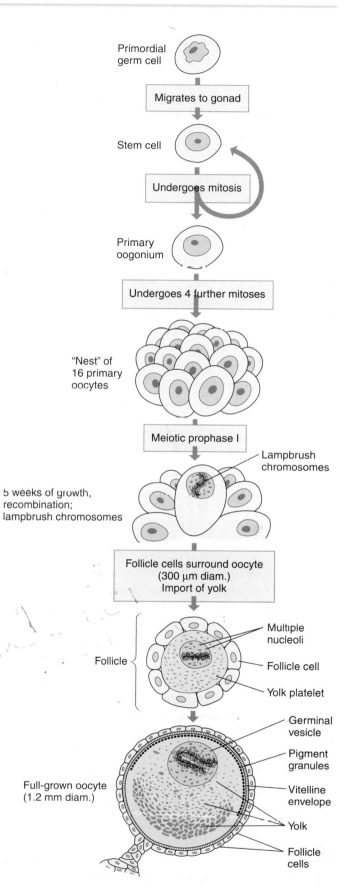

oogonia at the beginnings of oogenesis (see Figure 4.1). Perhaps animal-vegetal polarity is first preordained in connection with formation of the cytoplasmic bridges.

Genes for Ribosomal RNA Are Transiently Amplified

During this growth phase of oogenesis, there is a tremendous accumulation of materials for early development. The mature oocyte of *Xenopus* will attain about 1.3 mm in diameter, and it is about 20,000 times larger than a typical somatic cell. We have mentioned previously that the vitellogenin that forms the yolk is synthesized in the liver and imported. There are no nurse cells here as in *Drosophila*. Instead, the amphibian egg for some of its needs pulls itself up by its own bootstraps, so to speak. Ribosomes, those small organelles that preside over protein synthesis, are needed in large amounts. In the fly they are made by nurse cells. In the frog, some of the genes for the structural RNAs (see Box 4.1) are *transiently amplified* (temporarily duplicated many times over). The genes for the 28S and 18S units of ribosomal RNA are made from a common precursor. This rRNA gene set, which is normally present in about 450 copies per haploid chromosome set, is located in the nucleolus; the rRNA genes are duplicated many times,

and packaged into about 1,500 extra nucleoli (Figure 4.3). This repetition breaks one of the cardinal rules of biology, namely, that the content of the genes for all cells is identical. The transiently amplified rRNA genes serve to provide templates for the synthesis of an excess of enough 28S and 18S rRNA to form 200,000 times the normal level of ribosomes found in the cells of the adult. This selective gene amplification is temporary. The amplified level of nucleoli and the genes they contain returns to "normal" during the later phases of oogenesis and in early development.

The other structural unit of the ribosome, 5S rRNA, also has to be stockpiled. The genes for 5S rRNA are not found in the nucleolar organizer region of the chromatin, however, but on a different chromosome altogether (see Box 4.1). The requirement for large amounts of 5S rRNA has been solved in a completely different, yet elegant way. There is a huge family of 5S genes, all closely linked together on the chromosome. In fact, there are 20,000 copies per haploid genome. But only 400 of these are used for synthesis of 5S rRNA in diploid somatic cells under ordinary circumstances. The other 19,600 are used strictly during oogenesis, to provide the huge stockpile needed. The differences between the "oogenetic" (the 19,600) and the "zygotic" (the other 400) forms of 5S RNA are not great,

BOX 4.1 RIBOSOME FORMATION AND STRUCTURE

Ribosomes are small (about 30 nm) cytoplasmic organelles that serve as "translation factories," that is, sites of protein synthesis. Each ribosome is composed of two ribonucleoprotein subunits, containing both large, "structural" RNA molecules and several dozen different proteins (see accompanying figure). The two subunits are absolutely dependent on magnesium ion (Mg^{2+}) to hold them together, and this association in turn is absolutely essential for protein synthesis; without it, messenger RNAs can-

not bind to the ribosome, and peptide bonds in turn cannot get formed. It is convenient to think of the fully formed ribosome as a template, or scaffolding, for holding the enzymes used in protein synthesis.

Before protein synthesis can occur, however, ribosomes themselves must be manufactured, which occurs both in the nucleolus and in the cytoplasm. To understand how this is done, it helps to know how ribosome parts are named. Ribo-

somes, their subunits, and the structural RNAs that comprise them are all expressed in an unusual (dimensionless) size unit called the svedberg, or S unit, named after Theodor Svedberg, the Swedish scientist who invented the ultracentrifuge. (The S value of a particle indicates how rapidly it moves through a liquid under the influence of the force generated by a centrifuge. Centrifuges used in biological laboratories may generate forces 100,000 times the force of gravity or more. Generally, the bigger the S value of a particle, the more rapidly it sediments when centrifuged. This sedimentation rate depends on the density, shape, and mass of the particle and the medium through which it moves.) The ribosomes of eukaryotes sediment about 80S. The two subunits have S values of 60 and 40. (Note that S units are not linearly additive: the 60S and 40S subunits come together to form an 80S particle.) The 60S subunit is composed of three structural RNA molecules (28S, 5.8S, and 5S), as well as 49 different proteins. The 40S subunit is composed of an 18S RNA molecule and another set of 33 proteins.

The genes encoding the proteins are not in the nucleolus but are located on several different chromosomes, and the mRNAs for these rRNA proteins are translated on already-formed ribosomes in the cytoplasm. The new ribosomal proteins are then shuttled back into the nucleus, in particular, the nucleolus, where new ribosomes are assembled. Here the genes encoding the 28S, 18S, and 5.8S structural RNAs are located. Several identical, tandemly arranged copies of the gene encode the 45S RNA precursors of the 28S, 18S, and 5.8S RNAs, as shown in the figure. The 5S RNA molecules are made from another family of repeated genes, located on another chromosome. The 45S precursor RNA, still in the nucleolus, is then converted into the 28S and 18S rRNA structural rRNAs. The nucleolus is also where the 60S and 40S subunits are assembled, together with the 5S RNA and imported, new ribosomal proteins. The completed subunits are then exported to the cytoplasm, where they combine to form functional ribosomes, ready for synthesizing proteins.

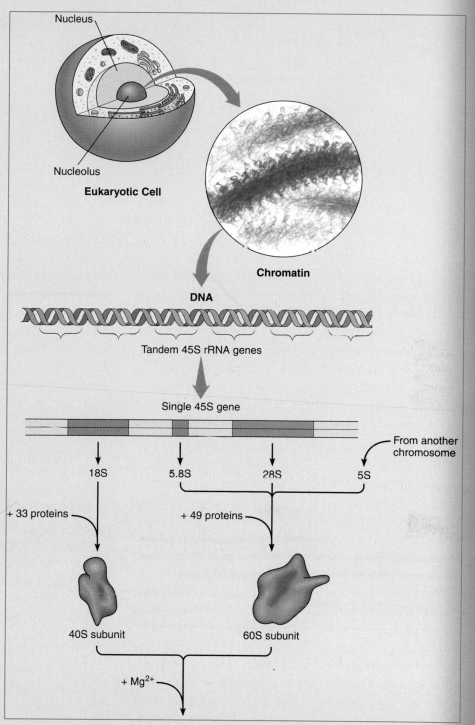

The Eukaryotic Ribosome A diagram of the formation and organization of the ribosome. The genes for the 45S precursor of the 18S, 28S, and 5.8S structural ribosomal RNAs are tandemly arranged in the chromatin of the nucleolus. The 45S precursor gives rise to these three rRNAs, and 5S rRNA is imported into the nucleolus from its site of transcription. The rRNAs combine with the many different ribosomal proteins, which have been translated in the cytoplasm and imported into the nucleolar region. The combinations of appropriate proteins and rRNAs produce two ribosomal subunits, which are joined to form the complete ribosome and then exported to the cytoplasm, where they subsequently function together in protein synthesis.

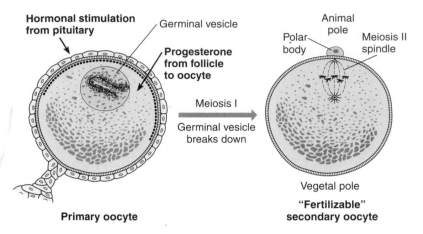
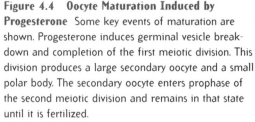

Figure 4.4 Oocyte Maturation Induced by Progesterone Some key events of maturation are shown. Progesterone induces germinal vesicle breakdown and completion of the first meiotic division. This division produces a large secondary oocyte and a small polar body. The secondary oocyte enters prophase of the second meiotic division and remains in that state until it is fertilized.

differing in only a few bases. The control over which group of 5S genes (oogenetic or zygotic) gets transcribed was the subject of a lovely bit of detective work that gave us some of our first insights into the control of transcription. Control of transcription is described in more detail in Chapter 14.

The Completion of Meiosis I Is Regulated by Progesterone

The fully grown oocyte, with its large germinal vesicle, resides in the ovary until hormonal signals impel it to continue maturation. Environmental signals in the natural habitat that influence pituitary gonadotropin secretions, or injected pituitary hormones in the laboratory, stimulate follicle cells to secrete progesterone. Progesterone then reacts with steroid hormone receptors of the oocyte, setting in motion a complex developmental sequence called **oocyte maturation.**

Many biochemical and structural changes occur after progesterone secretion. The most obvious is the breakdown of the germinal vesicle, which occurs at the completion of prophase I. The membrane of the large nucleus disassembles, the contents of the germinal vesicle mix with oocyte cytoplasm, and chromosomes coil to their metaphase length. As meiosis I proceeds, a very small polar body and a large secondary oocyte are generated (Figure 4.4). Furthermore, the follicle surrounding the oocyte loses its integrity, allowing the matured oocyte to enter the enlarged opening of the oviduct and move down to the cloaca, where it can be spawned. Movement of the eggs through the oviduct is accomplished either naturally, during amplexus (where the male frog grasps the female from behind and kneads her abdomen), or artificially, through gentle massage in the laboratory. At the time of mating, meiosis I has completed, and the nucleus has re-formed and is in the early stages of prophase II of meiosis, where it will remain until fertilization.

Xenopus is an example of an organism where fertilization takes place when the egg is in the prophase of meiosis II. Oocyte maturation is under control of an initial chemical signal, progesterone, which acts through complex signal transduction pathways to activate a single meiotic division. The analysis of this mechanism contributed greatly to our current understanding of how cell division is controlled (see Box 4.2).

Spermatogenesis in Frogs Produces Four Spermatids

Spermatogenesis in the seminiferous tubules of the adult male follows the conventional sequence, each spermatogonium acting as a stem cell. Some mitotic descendants undergo differentiation and enter meiosis, forming four haploid spermatids, which then are radically reorganized to form the mature spermatozoon. Most of the cytoplasm and many of the organelles are discarded. Though the sperm contribute genes toward fitness of the organism, in the short run their DNA is dispensable. All the information for early development is in the egg, and all the materials necessary to form the organ systems of the tadpole are in place. What is indispensable, though, is the contribution of the centrosome from the sperm; amphibian eggs are relatively easy to activate; even pricking with a pin will accomplish it. If the activated egg acquires a centrosome, development will proceed and a haploid tadpole will result (which, incidentally, does not survive metamorphosis).

FERTILIZATION AND EARLY DEVELOPMENT

Fertilization Establishes Bilateral Organization in the Egg

While the directions of anterior and posterior, dorsal and ventral, are prefigured in the *Drosophila* oocyte, *Xenopus* eggs pos-

Box 4.2 Using Embryos in Cell Biology

Good experimental science is often extremely opportunistic, and biologists are always on the lookout for a favorable organism to use for decisive tests of one hypothesis or another. A lot of interesting and important research into modern cell and molecular biology has made use of eggs and embryos—not to solve problems of development, but for other purposes. Often such discoveries become extremely important for students of development in a very roundabout way.

A good example of this is the study of mitosis and the cell cycle. Since eggs are huge cells, and some are rather transparent, microscopists have been using them for over 100 years to study the details of mitosis in living cells. The mitotic spindle was first isolated from sea urchin eggs by Daniel Mazia and Katsuma Dan in the 1950s. One of the proteins essential for mitosis (cyclin) was discovered in embryos of the surf clam by Tim Hunt, Eric Rosenthal, and Joan Ruderman in 1982. Work by Yoshio Masui and others using frog oocytes showed that a protein factor was necessary for steroid-induced oocyte maturation to occur; one component of this factor is actually cyclin. (The other component is a protein kinase, first discovered in yeast using genetic techniques.)

Eggs and embryos are still being used by cell biologists interested in understanding the biochemistry of transcription control, the function of the centrosome, the regulation of nuclear envelope assembly and other aspects of the cell cycle, molecular aspects of cell motility, and a host of other important contemporary research areas in cell biology.

---

sess only the animal–vegetal axis and a radially symmetric organization. Fertilization is the symmetry-breaking event. Animal cortex is the preferred sperm binding site, and the fertilizing spermatozoon fuses to egg plasma membrane and contributes its nucleus and centrosome at some place in the animal hemisphere. (Figure 4.5 shows the attachment of many sperm to the surface of a sea urchin egg; only one of these sperm will actually fuse with the egg, however.) In *Xenopus* the aster surrounding the sperm centrosome grows by recruiting tubulin (the protein that forms microtubules) from the oocyte reserve (Figure 4.6). The oocyte, meanwhile, is completing meiosis II, after which syngamy occurs (see Chapter 2). There is also a crucial rotation of materials within the egg that establishes the plane of the first cell division, the site of future gastrulation movements, and the prospective dorsal and ventral sides of the embryo. This internal rotation is essential for normal development; if it does not happen, then dorsal development will not occur. If there is no fertilization but parthenogenesis is induced, then cortical rotation will eventually take place along some meridian, thereby breaking radial symmetry. (A *meridian* is one of an infinite number of imaginary great circles inscribed on the surface of the egg that pass through the animal and vegetal poles, hence all meridians are orthogonal to the egg's "equator.")

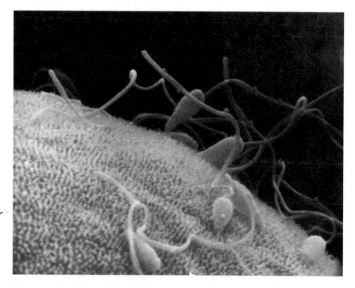

Figure 4.5 Fertilization This scanning electron micrograph shows the surface of a sea urchin egg to which many sperm are attached. However, only one will actually fuse with the egg plasma membrane and enter the egg, since there is a block to polyspermy, as discussed in Chapter 2.

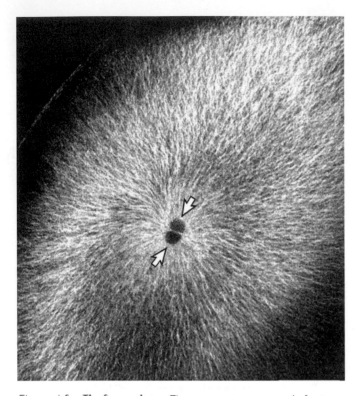

Figure 4.6   The Sperm Aster The sperm aster. composed of micro-tubules. is shown from an amphibian egg that is about halfway through the time period between fertilization and the first cleavage division. The microtubules are stained with an antibody to tubulin. The arrows mark the positions of the two pronuclei.

Cortical Rotation Involves Parallel Vegetal Microtubule Arrays

Recall from Chapter 2 that the surface layers of cytoplasm in the egg have somewhat different properties than the rest of the egg's cytoplasm, and that this "rind" is called the **cortex.** In the amphibian egg, the cortex (about 4 μm thick) is also different from the interior cytoplasm; a network of actin microfilaments causes the cortex to cohere and become stiffer. Dozens of experiments carried out by many workers over the years indicated that there was probably some movement of the cortex relative to the yolky interior, and indeed we now know that cortical rotation does occur (Figure 4.7A). This movement is much like what happens when you fill a plastic beach ball with a viscous fluid; if you rotate the ball slowly in your hands, the internal fluid, because of its inertia, will remain fairly stationary.

This rotation of the cortex was directly demonstrated in the early 1980s by John Gerhart and his colleagues at the University of California at Berkeley. They stained the yolk of an unfertilized egg one color, the surface with a dye of another color, and then embedded the egg in agarose gel just after fertilization. In Figure 4.7B, the internal and external dots, which were in register before fertilization, have visibly moved relative to one another. To show this relative movement experimentally, it proved easier to anchor the cortex than the core, so in the diagram it is the core that appears to move. However, the results are equivalent to what happens in the natural state, where it is the entire spherical cortical shell of the egg that moves relative to the central ball of cytoplasm.

Careful electron microscopic investigation of the interface between the cortex and interior during the time of rotation shows parallel arrays of microtubules associated with the core, all with the same polarity (with the positive end at the prospective dorsal side of the embryo), running along the direction of rotation (Figure 4.7C). There are motor molecules in the cortex, perhaps one or more kinds of kinesin, that move the cortex along these microtubular tracks, thereby producing the cortical rotation. The microtubules take about 40 minutes to become aligned, after which the rotation occurs, the entire cortex moving as shown in Figure 4.7A. Since the surface is moving relative to an interior core, the vegetal surface is moving away from the point of sperm entry and the animal cortex moves toward it. In some amphibian species the relative movement of cortex and core produces a lighter zone, the *gray crescent,* approximately opposite the point of sperm entry. In the shear zone between cortex and microtubules, there is a striking movement of small particles, and possibly signaling molecules as well, toward the dorsally located positive ends of the microtubules.

The region of cytoplasm across the sphere from the sperm's entry point becomes special; this is where rotation has moved vegetal cortical surface to a region of animal core cytoplasm. This area will be bisected by the first cleavage plane and become the location of intense gastrulation activity.

Cortical rotation, which is specified by the point of sperm entry, actually begins 40% of the way into the period between fertilization and first cleavage, and continues until about 80% of the time between fertilization and first cleavage has elapsed. Cortical rotation is essential for establishing bilateral symmetry and developing dorsal structures (Figure 4.8A). If rotation is prevented by, say, UV light (which works by preventing the microtubule polymerization), there is no

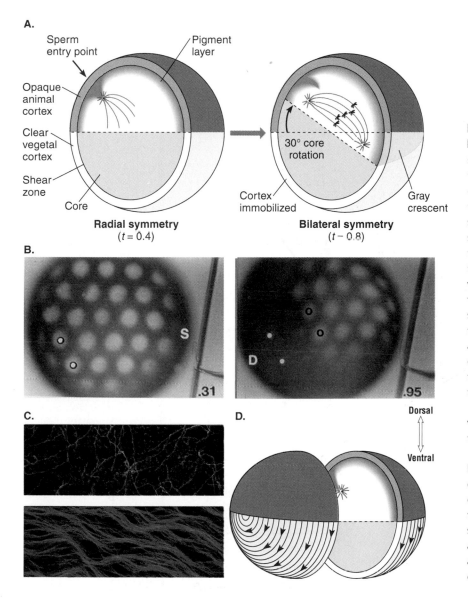

A.

Sperm entry point
Pigment layer
Opaque animal cortex
Clear vegetal cortex
Shear zone
Core

Radial symmetry
($t = 0.4$)

30° core rotation
Cortex immobilized
Gray crescent

Bilateral symmetry
($t - 0.8$)

B.

.31
.95

S
D

C.

D.

Dorsal
Ventral

Figure 4.7 Cortical Rotation in the *Xenopus* Egg After fertilization of the amphibian egg. the sperm aster grows. About four-tenths of the way through the time *t* needed to complete the first cleavage. the cortex rotates with respect to the interior. **(A)** Rotation of the egg shown in sagittal section. The egg is embedded in gelatin so that the cortex and surface are immobilized: the core is free to rotate. **(B)** An experiment in which clusters of yolk platelets are stained with a vital dye. Nile Blue. which has an affinity for lipids. After the yolk is stained. marks on the surface are made with a fluorescent dye that reacts with surface proteins. The egg is fertilized and embedded in gelatin to immobilize the cortex *(left)*. When rotation occurs *(right)*. the relationship of the stained internal yolk to the surface marks has visibly changed. **(C)** Microtubules in the shear zone were stained with fluorescent antibodies directed against tubulin. The microtubules are not well aligned *(upper)* when only four-tenths of the time between fertilization and first mitosis has elapsed: when seven-tenths of the time has elapsed *(lower)*. the orientation of the microtubules is obvious. **(D)** This schematic representation shows that the microtubules are aligned in the direction of the cortical rotation. and the positively charged ends of microtubules correspond to the direction of rotation.

dorsal development (Figure 4.8B). The embryo that forms will still be cylindrical and will develop a gut and blood cells, but no muscles, nervous system, or other axial structure forms. On the other hand, if microtubule formation is stimulated by treatment with heavy water (D_2O), hyperdorsalized embryos with reduced ventral structures can be produced (Figure 4.8C). Finally, if we manipulate the rotation—for example, by centrifugal force—we can impose an extra cortical rotation in another plane, thereby producing Siamese twins with two axes (Figure 4.8D).

This apparently simple cortical rotation cleavage thus dictates several critical aspects of the embryo's future development: where the first cleavage plane and plane of bilateral symmetry will be, and whether dorsal development will occur.

Cleavage Is a Period of Intense Cellular Assembly

You will recall that following the fertilization of the *Drosophila* egg, there is a period of rapid nuclear multiplication. In the frog and almost all other creatures, cytokinesis occurs, so that instead of creating many nuclei in a single "cell," complete, new cells are generated, which happens at breakneck speed. Embryos undergoing cleavage are DNA synthesizing machines; furthermore, because frog embryos package the DNA into chromosomes, they are also busy

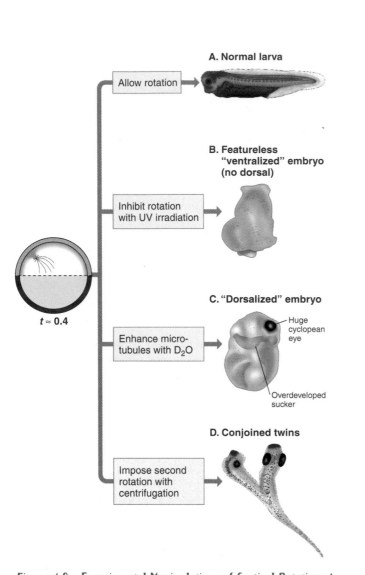

A. Normal larva

Allow rotation

B. Featureless "ventralized" embryo (no dorsal)

Inhibit rotation with UV irradiation

$t \approx 0.4$

C. "Dorsalized" embryo

Enhance micro-tubules with D$_2$O

Huge cyclopean eye

Overdeveloped sucker

D. Conjoined twins

Impose second rotation with centrifugation

Figure 4.8 Experimental Manipulations of Cortical Rotation A fertilized egg undergoing cortical rotation is subjected to various treatments. **(A)** A normal embryo will result if rotation proceeds. **(B)** If the embryo is inhibited by irradiation of the vegetal pole with ultraviolet light, rotation does not occur and there is no dorsal development. **(C)** If the microtubule tracks for rotation are stabilized and enhanced, then more extensive rotation will occur, resulting in embryos in which dorsal development is abnormally robust at the expense of ventral development. **(D)** If an embryo is forced to undergo a second rotation, then it may produce conjoined (Siamese) twins.

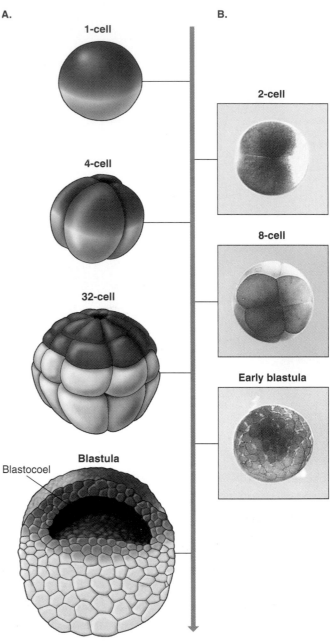

A.

1-cell

4-cell

32-cell

Blastula

Blastocoel

B.

2-cell

8-cell

Early blastula

Figure 4.9 Cleavage (A) Stages in the cleavage of the frog egg, showing the furrows for each mitotic division. Notice how the overall size of the embryo remains the same while the cells become smaller with each division. Also note that the cells are larger in the vegetal hemisphere than the animal hemisphere. This difference is due to the placement of the centrosomes during early divisions. **(B)** Photographs of some living embryos at early cleavage stages.

synthesizing chromosomal proteins. Cytokinesis requires the assembly of huge amounts of new plasma membrane as cell surface area expands. That so much of the machinery for this intense synthesis and assembly is premade during oogenesis should come as no surprise. DNA polymerase, histone mRNA, membrane precursors, and more are stock piled in the egg. In *Xenopus*, mitosis takes place about every 15 minutes, for some 12 divisions, generating over 4,000 cells (Figure 4.9).

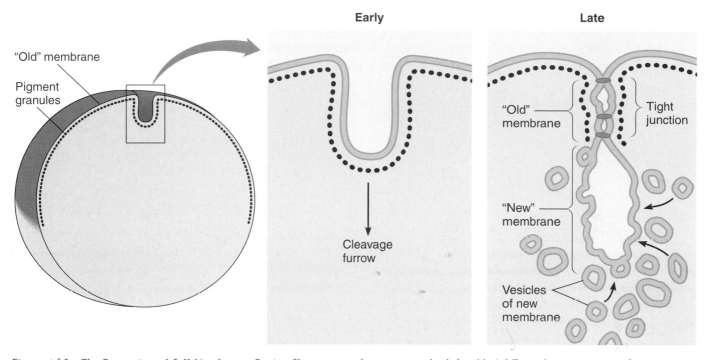

Figure 4.10   The Formation of Cell Membranes During Cleavage Divisions in the Frog The original plasma membrane of the egg is supplemented by assembly of new membrane. The cleavage furrow (*left diagram*) is magnified to show its extent: at first (*middle*), the furrow is completely lined by "old" membrane originating from the egg surface. As the furrow deepens (*right*), "new" membrane, assembled from cytoplasmic vesicles, is added. Tight junctions form at the interface between old and new membrane.

The cleavage furrows that form during these rapid mitoses are partly composed of "old" plasma membrane that has enveloped the egg, and partly from the addition of membrane vesicles that extend the plasma membrane. A *tight junction* forms at the boundary of the "old" and "new" membrane, forming a tight seal (Figure 4.10). The new membrane contains a pump for sodium ions, which causes salt to accumulate inside the embryo. This raises the interior osmotic pressure, eventually resulting in the formation of an internal, fluid-filled cavity, the *blastocoel* (Figure 4.9). (In Chapter 12 we will discuss tight junctions further, along with other cell membrane specializations.) The surface cells of the embryo have become a true epithelium.

The Midblastula Transition Is a Period of Major Changes

After the 12th cell division, three profound changes occur. For the first time, there is active new transcription. (We should add at this juncture that some researchers question whether measurements of early transcription have been sensitive enough; a very low level of transcription has not been completely excluded.) Prior to the **midblastula transition (MBT),** the protein synthesis that occurs has been directed by maternally made mRNAs. Second, some cells in the embryo acquire the potential for motility; we can easily demonstrate this by simply placing small groups of cells in tissue culture and observing their behavior. Third, the cell division cycles—which earlier in the blastula's development were so rapid that they cycled between DNA synthesis and mitosis with no intervening G1 and G2 periods—now begin to occupy much longer time periods, and G1 and G2 periods are present. All of these changes are a prelude to the dramatic morphological reorganization that will soon take place during gastrulation.

Maps of the Late Blastula Reveal Distinct Identities

Before we begin our look at gastrulation, it is useful to analyze some maps of the late blastula (Figure 4.11). As described in earlier chapters, one can obtain a **fate map** of the blastula by staining various of its parts and observing which region develops into which kind of tissue. The fate map of the midblastula would be identical to that of the egg since no cell or tissue movements have yet occurred. A second kind of map, called a **specification map** by amphibian embryologists, is obtained by explantation, that is, by separating parts of the

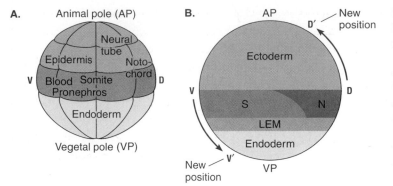

Figure 4.11   Late-Blastula Fate Maps of *Xenopus laevis*
Fate maps are obtained by staining regions, or even single blastomeres, of early embryos, and then observing what they become. **(A)** A composite map of several experimenters projected onto an embryo at the 32-cell stage. **(B)** A more recent revised map, by Ray Keller in 1991, shows that the area from which somites and lateral mesoderm will form is more extensive than previously thought. Recent experiments by Connie Lane and Bill Smith indicate that the prospective dorsal and ventral portions of the embryo are derived from cells closer to the animal and vegetal poles. This assignment of dorsal and ventral is indicated by arrows pointing to D′ and V′. LEM, leading edge of mesoderm; N, notochord; S, somite.

embryo from one another or taking small bits (the *explant*) from different parts of the blastula, putting them in tissue culture medium, and seeing what they become. If bits of a very early cleavage embryo are subjected to such culture in isolation, they may survive in culture. Some ciliated epidermis may form, but little in the way of recognizable structures will differentiate. This particular term, *specification*, is really a version of developmental potential: what can a given cell or group of cells become? Of course, that may depend upon the conditions used by the experimenter, so a complex terminology has grown up around the description of the various ways of examining developmental potential.

By the late-blastula stage, however, different regions have acquired identities. When cultured, cells from the roof of the blastula, overlying the blastocoel, will form ciliated epidermis. The yolky cells of the vegetal pole will form some posterior gut tissue. And the intermediate cells, girdling the equator, will form various mesodermal structures, such as notochord, somites, pronephros, and blood. The development of these mesodermal structures will be considered later in this chapter and in Chapter 7. Those cells from the side opposite the former point of sperm entry will form dorsal mesodermal structures, such as the notochord (a stiff rod that underlies the neural tube) and head mesoderm. Hence this region is called the *dorsal marginal zone (DMZ)*. Cells from the equator on the opposite side of the sphere will form ventral mesodermal structures such as blood, and this area is called the *ventral marginal zone (VMZ)*.

This specification map indicates what cells of the different regions are capable of forming when in isolation. In the embryo, of course, these cells do not develop in isolation, and may develop along somewhat different pathways depending on their exact location. Identification of what a group of cells may form when transplanted to different locations of a host embryo is called a **competence map** by amphibian embryol-

ogists. You will recognize that *competence*, too, is a version of developmental potential, one in which the potential of the cells is tested by transplanting them to different locations. Competence refers to the ability of cells to respond to various developmental signals, and we shall explore this subject in detail in subsequent chapters.

Vegetal Cells of the Blastula Induce Mesoderm from Animal Cells

How does the embryo come to have regions of prospective endoderm, dorsal and ventral mesoderm, and surface ectoderm already specified by the late-blastula stage? The answer comes from some well-known and important experiments carried out in the 1960s by Pieter Nieuwkoop in the Netherlands. As mentioned in our discussion of specification maps, explants of cells from the roof of the blastula—the *animal cap* cells—will form ciliated epidermis in culture, while vegetal explants either do not form recognizable tissue or develop some of the characteristics of posterior endoderm. When the animal cap is combined with vegetal cells and then cultured, however, it will form typical mesodermal structures and anterior pharyngeal endoderm (Figure 4.12). Nieuwkoop concluded that in the normal embryo there is an interaction between the vegetal cells and adjacent animal cells, which together are destined to form the *marginal zone*, and that this interaction leads to mesoderm formation. The molecular basis of this action of vegetal cells that induces mesoderm from the equator of the late blastula will be discussed in Chapter 16.

Additional experiments by Nieuwkoop demonstrated that the vegetal region itself has diverse inductive abilities. When the vegetal cells used in the recombined tissues originate from the prospective dorsal side, the animal cap forms dorsal mesoderm tissues such as notochord and muscle. When the ventral vegetal cells are combined with the animal cap, ventral mesoderm tissues such as blood and mesenchyme form. And vegetal cells isolated from an intermediate position induce intermediate types of mesoderm. These explant experiments are shown diagrammatically in Figure 4.12.

Nieuwkoop and his coworkers concluded that the induction of animal hemisphere cells to form mesoderm occurs in a graded fashion because the vegetal cells emit one or more graded induction signals to the animal hemisphere cells. More recent work on the molecular nature of these signals, to be discussed in Chapter 16, supports the view that the induction process discovered by Nieuwkoop is rather complex: it may consist of a generalized induction of mesoderm and anterior endoderm followed by a subsequent secondary localized signal or signals that induce the development of dorsal-type mesoderm. This latter effect is called **dorsalization.** Some workers consider the vegetal cells on the prospective dorsal side to constitute a dorsal mesoderm signaling center, and have named it a *Nieuwkoop center.* However named, this induction of prospective dorsal mesoderm and anterior endoderm is what brings about the important *Spemann organizer,* which we shall soon discuss.

Nieuwkoop's explantation experiments do not eliminate the possibility that there are autonomous factors, synthesized and localized during oogenesis, that also play a role in predisposing cells in various parts of the blastula to form the three germ layers. In other words, induction does not rule out a contribution from a preexistent tendency (derived from factors localized in the egg) to follow a particular pathway of differentiation. It is exceedingly difficult to remove and explant marginal zone cells from early embryos. If one waits to carry out the surgery until a slightly later stage, when the excision of marginal zone cells is easier, the explant will form mesodermal derivatives in culture. Do the cells do so because they have already been induced in situ by the action of the vegetal cells? Or do they have autonomous potentialities to form mesoderm as well? The final answer is probably "yes," but we shall consider this issue further in Chapter 16.

GASTRULATION, GERM LAYERS, AND ORGANOGENESIS

Gastrulation Involves Massive Movements of Cell Groups

The amphibian embryo is opaque, and several different kinds of movements of cell groups are occurring at nearly the same time. Consequently, it is difficult to visualize just how these rearrangements occur. However, some appreciation of how these movements take place in three dimensions is needed to understand what fundamental processes drive morphogenesis. Over a period of a few hours about a quarter to a third of the surface of the embryo is tucked inside by a rolling process called **involution;** meanwhile the remaining surface area stretches, a process called **epiboly.** Some cells below the

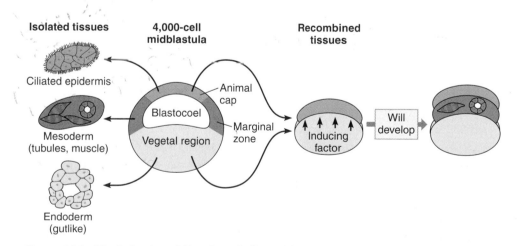

Figure 4.12   The Induction of Mesoderm by Vegetal Signaling Some of the experiments carried out by Nieuwkoop and described in the text. First the animal cap, marginal (equatorial) zone, and vegetal region were surgically isolated and allowed to develop by themselves *(left).* Animal cap formed ciliated epidermis, equatorial cells formed mesodermal derivatives, and the yolky vegetal cells formed gutlike structures. Then the animal cap and vegetal cells were recombined and cultured to see what they would become *(right).* Mesodermal derivatives formed, even though each of the tissues in the recombinant could not form mesoderm by itself.

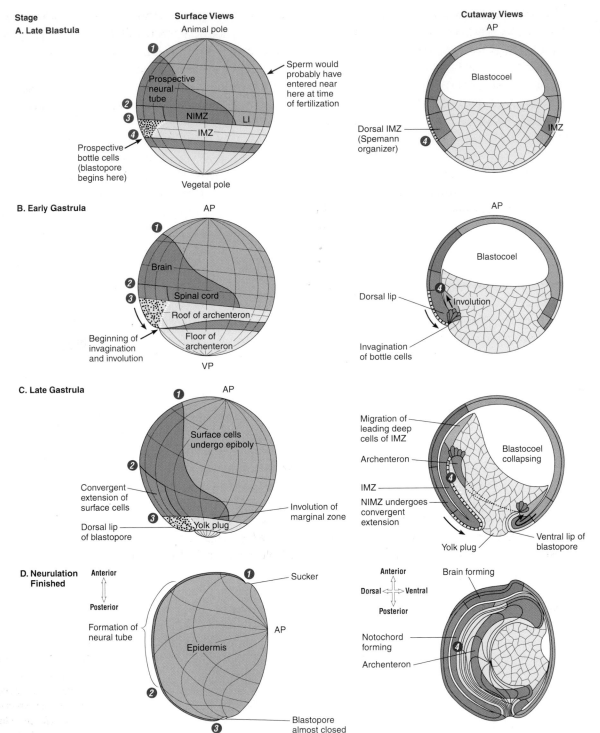

Figure 4.13   Gastrulation Movements in *Xenopus laevis* The movements of various cell groups during progressive stages of gastrulation. The surface views show primarily the morphogenesis of the animal cap and prospective ectoderm, while the cutaways through the embryo primarily reveal the morphogenesis of the involuting marginal zone, which forms mesoderm and endoderm. Arrows on the diagrams indicate the direction of the dominant tissue movements during gastrulation. (**A**) The late blastula, just at the beginning of gastrulation. The blastopore is formed by bottle cells on the vegetal side of the involuting marginal zone (IMZ), shown here on the left side of the blastula. (**B**) The early gastrula.

(**C**) The late gastrula. (**D**) Gastrulation is complete, and the neural tube has formed. The left side of the blastula has become the dorsal side of the embryo. Markers 1 through 4 allow you to trace the movement of a particular group of cells through the entire process. Note that marker 4 becomes internalized early in gastrulation (B), and by late gastrulation (C) the vegetal pole (VP) can no longer be identified, having moved inward as the blastopore closes and the yolk plug is enclosed. Remember that gastrulation is occurring in three dimensions, and the cutaway views show only a single plane cut through the gastrula. Other abbreviations: LI, limit of involution; NIMZ, noninvoluting marginal zone.

surface also roll around the area where involution is occurring, thereby moving even more cells deep into the interior. Some involuted cells become migratory, moving forward, and other involuted and surface cells undergo a mediolateral repacking that causes the entire embryo to lengthen.

As a result of these movements and of cells pushing inward upon it, the blastocoel collapses, and a new interior cavity called the **archenteron** is formed. The archenteron will become the lumen of the gut.

It is virtually impossible to visualize and understand these movements without some reference to diagrams. Let us proceed to examine the various parts of Figure 4.13 to follow the several different kinds of movements.

Bottle Cells Invaginate and Lead the Way for Involution

When gastrulation starts, cells in a specific part on the embryo's surface begin to "duck down" below the epithelial layer, leaving a depression, or "dimple," visible in the surface layer in which they are embedded. This process, called **invagination,** results from the formation of **bottle cells**—surface cells that become long and attenuated, so that their cell bodies sink down into subjacent cells. As they do, they appear to pull on their thin cell "necks" (Figure 4.13A, B). The anatomical structure arising from this invagination is known as the **blastopore.** The beginning dimple is called the **dorsal lip** of the blastopore. It is located about two-thirds of the way down from the animal pole and roughly opposite the original site of sperm entry. In other words, the dorsal lip appears in a region where, due to cortical rotation, cells once part of the

cortex on the vegetal side now appose the cytoplasmic core of the animal side. As bottle cell invagination progresses, the initial dimple spreads in a broader and broader arc around the vegetal circumference, eventually becoming a completely circular blastopore. A close look at the surface views of Figure 4.13 reveals the blastopore's gradual formation.

As the bottle cells invaginate and become interiorized, neighboring surface cells from the animal side move around the lip created by invagination; deeper cells adjacent to the surface also move around the lip (involution, see Figure 4.13B, C). Involution is more easily seen in the cutaway views of Figure 4.13. Involution starts first where bottle cells first invaginate (the future dorsal side of the embryo), and progresses much farther than does the involution on the prospective ventral side. Notice how the involuting surface and deep marginal zone cells actively push the blastocoel to one side and collapse it, creating a cavity lined by involuting surface marginal zone cells. Once involuted, deep marginal zone cells actively migrate, moving the archenteron anteriorly. Perhaps even more important, the stretching surface cells undergoing epiboly become thinner and repack themselves, converging toward the future dorsal midline. This repacking involves *radial intercalation*. The involuted deep marginal zone cells also undergo active *mediolateral intercalation*, a kind of convergent movement, which causes an active extension of the entire length of the embryo. Figure 4.14 shows this convergent extension at the cellular level.

Reexamine Figure 4.13, and you will see that the surface and subjacent marginal zone cells form a torus that has become interiorized, creating a new cavity surrounding the yolk plug, a bit of yolky vegetal pole that did not participate in these movements

A. Radial intercalation

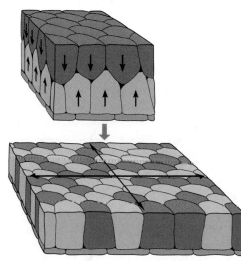

B. Mediolateral intercalation

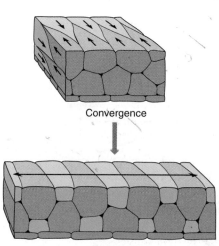

Convergence

Extension

Figure 4.14 Convergent Extension Movements A piece of the forming dorsal side is removed from a gastrula, and the surface epithelium removed. This is called a shaved explant, and the shaving allows us to view movements of subsurface cells. **(A)** In radial intercalation, some cells from deep layers insinuate themselves into more superficial layers. This leads to a tissue with less depth but more surface area, and thus contributes to epiboly. **(B)** Mediolateral intercalation of cells results in cell convergence, causing the cell group to become both narrower and more extended.

(Figure 4.13C). The embryo has become extended. The old animal pole is now over the belly where the heart will form, the vegetal pole has remained over the yolk plug, and the dorsal side of the embryo is derived from a wedge of the animal cap.

Even the anatomical description of this process is complicated. To truly grasp the fundamentals, we shall have to learn more about the cellular basis of invagination, involution, epiboly, motility and migration, and convergent intercalation behaviors. That is one reason we have examined this process is some detail: analysis of this complex case will help illuminate many other morphogenetic movements. And analysis of analogous movements and cell behaviors in other forms can aid us in understanding gastrulation in the frog. We shall continue this analysis in Chapter 13.

Gastrulation Establishes the Germ Layers

Even though the tactics are somewhat different, the end result of gastrulation in the frog is similar to what we saw in the previous chapter in the fruit fly. There is an interior lumen, open in this instance to the exterior through the thin slit of the blastopore, but blind-ended anteriorly. The cells lining this lumen, which of course are the *endoderm*, will form the lining of the alimentary canal (gut) of the tadpole. The cells still remaining on the surface, the *ectoderm*, will form the epidermis, neurons, and a few other cell types. The remaining intermediate layer is the *mesoderm*. An embryo undergoing the gastrulation process is often called a *gastrula*. In Chapters 6 and 7, we will explore how these germ layers give rise to the differentiated organs of the adult. It is helpful at this juncture to preview how this will come about.

Dorsal Ectoderm Is Induced by Dorsal Mesoderm and Will Form a Neural Tube

We know from the fate maps discussed earlier in the chapter that even before gastrulation, some degree of specification has occurred, and the anteroposterior and dorsoventral axes of the nascent tadpole have been set. The cells of the dorsal lip of the blastopore, that region where bottle cells first formed, play a special and crucial role in realizing this organization. The German embryologist Hans Spemann and his student, Hilde Mangold, carried out transplantation experiments with salamanders that revealed the extraordinary role of these dorsal lip cells, which have come to be called the **Spemann organizer.** The cells of the Spemann organizer are destined to form dorsal mesoderm, especially notochord, and some anterior pharyngeal

endoderm. If this group of cells is transplanted, for instance, to an area where only ventral (belly) mesoderm is formed, a remarkable transformation occurs: an entire second axis may form, leading to a second central nervous system and its attendant muscles and other axial organs. Figure 4.15 depicts this dorsal lip transplantation and the conjoined-twin embryos that result.

By use of embryos with distinctive natural pigmentation markers, Spemann and Mangold showed clearly that most of the "new," secondary embryo derived from the host, not the transplant. Hence, they concluded that the organizer region could induce neighboring tissues to change their fate. Since their discovery published in 1924, one often reads of various instances of *embryonic induction*, which simply means that a particular cell or tissue signals an adjacent cell or tissue to adopt a particular pathway of differentiation. Hilde Mangold was tragically killed in a household fire soon after she did these experiments; Spemann went on to receive the Nobel Prize.

It is the dorsal marginal involuting mesoderm that signals the overlying ectoderm to form the central nervous system. We shall explore what is known of the molecular basis of this induction in Chapter 16. The formation of the primordial nervous system is an example of a morphogenetic maneuver that we shall encounter over and over again: a flat plate of cells rolls up into a cylinder (Figure 4.16). The change from a *neural plate* to *neural tube* is called **neurulation,** and an embryo that has completed neurulation is often called a *neurula*. The cells of the surface ectoderm that is induced by the underlying notochordal mesoderm change from roughly square to rectangular in shape when viewed in cross section. This elongation produces a distinctive plate on the dorsal surface of the embryo. The plate is broader anteriorly than posteriorly, so that when you look down on the surface, a pear shape in relief is visible. In cross section, some of the elongated cells of the neural plate then assume a trapezoidal outline, the broad base of each cell lying next to the mesoderm. This shape change requires the microfilament network of the neural plate cells to maintain their integrity. Experiments done in the 1960s using inhibitors that interfered with microfilament function suggested that the plate rolls up, at least in part, because of these shape changes, which are brought about by microfilament functions. The two outer edges of the plate approach one another in the dorsal midline, seemingly dragging the lateral surface epithelium along. The two "lips" of the plate meet and fuse. More recent observations indicate that neurulation is probably much more complex than the process described here, involving not only changes in cell shape, but also changes in the adhesion of cells to one another, protrusive activity, and the mechanical influences of tissues outside the neural plate proper.

The formation of the neural tube does not always come about by plate folding. For instance, in some vertebrates, such as fishes, a chord of cells may form a rod, which then becomes hollow in the center. The posterior portion of the neural tube that forms the lumbar and sacral spinal cord of amphibians does not form from the original neural plate. Instead, a group of cells remaining in the dorsal lip of the blastopore at the end of gastrulation continues to proliferate, and it is these cells that form the solid *medullary cord* and notochord of the tail region. The solid medullary cord hollows out to form a continuous spinal cord. Figure 4.16 also shows that some of the cells of the fusing edges of the plate leave the neural plate–epithelial junction region and move into the adjacent mesoderm. (We will revisit these migratory cells, called **neural crest,** in Chapter 6.)

Mesoderm and Endoderm Will Form Many Organs

The pattern of transforming an epithelial sheet into a tube or sphere is repeated in the formation of many organs of the body. The endoderm of the archenteron, also an epithelium, will bulge out **(evagination)** in various places along its length into the adjacent mesoderm to form salivary glands, lungs, liver, pancreas, and thyroid. The mesoderm layer also is undergoing complex localized changes as neurulation proceeds. The dorsal-most mesoderm just subjacent to the midline of the neural plate undergoes a distinctive intercalation, and the cylindrically disposed stack of cells swell and become surrounded by extracellular matrix. This rodlike structure, the **notochord,** is characteristic of all vertebrate embryos; indeed, the phylum to which vertebrates are assigned is called Chordata for this reason. It is a significant skeletal structure in lower vertebrates but becomes calcified and buried deep within the bony spinal column of terrestrial vertebrates. Figure 4.17A shows this and the other anatomical features characteristic of all chordates.

On each side of the forming notochord, the mesoderm becomes congregated into bilaterally symmetric segmental blocks of mesoderm called **somites.** Much of the skeletal

muscle, all of the dermis, and the cartilage surrounding the spinal chord derive from these structures. Prospective kidneys differentiate from the mesoderm lateral to the somites. The more ventrolateral mesoderm splits into two layers. The more peripheral layer becomes associated with the surface

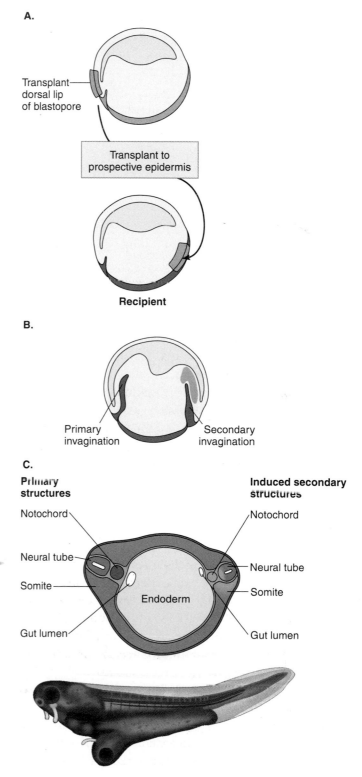

Figure 4.15   The Spemann Organizer The classic experiment by Mangold and Spemann: **(A)** A piece of the dorsal lip of the blastopore of a light-colored newt was surgically removed and grafted into the prospective ventral region of a darkly pigmented newt of the same developmental stage. **(B)** This resulted in two independent sets of gastrulation movements and two axes. **(C)** The two embryos developed into conjoined twins. When Spemann and Mangold fixed, sectioned, and examined such twins histologically, most of the secondary embryo was found to be darkly pigmented. The secondary axial structures (except for some notochord and somites) came from the host, which is explained by an inductive influence of the transplanted "light" dorsal lip.

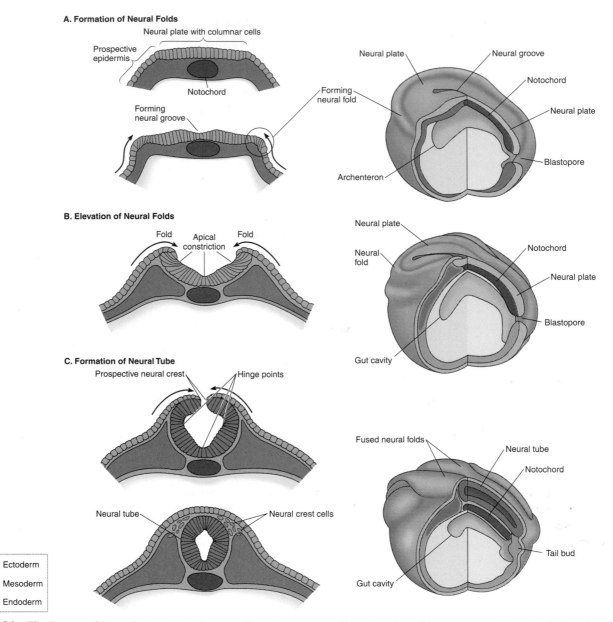

A. Formation of Neural Folds

Neural plate with columnar cells

Prospective epidermis

Forming neural groove

Notochord

Neural plate

Neural groove

Forming neural fold

Notochord

Neural plate

Archenteron

Blastopore

B. Elevation of Neural Folds

Fold

Apical constriction

Fold

Neural plate

Neural fold

Notochord

Neural plate

Blastopore

Gut cavity

C. Formation of Neural Tube

Prospective neural crest

Hinge points

Neural tube

Neural crest cells

Fused neural folds

Neural tube

Notochord

Tail bud

Gut cavity

KEY

Ectoderm

Mesoderm

Endoderm

Figure 4.16 The Process of Neurulation (A) Columnar cells in the dorsal ectoderm overlying the notochord become columnar and form the neural plate. The apices of some of these cells undergo constriction, which in principle could produce a bending. **(B)** The apical constriction and medial movements of prospective epidermis initiate folding, and the flat plate shows raised folds. **(C)** The external forces resulting from medial movements of the ectoderm and cell-shape changes of the neural plate cells produce lateral hinge points, and the plate "rolls up" into a tube. The neural folds fuse in the midline, at which time neural crest cells emigrate from the neural tube. In amphibians, the formation of the neural tube may occur rather quickly along its entire length, but in birds and mammals it takes a considerable length of time, progressing from the brain region posteriorly through the spinal cord region.

ectoderm and is called the **somatic mesoderm** layer. The other layer of mesoderm covers the endoderm of the developing gut and is called the **splanchnic mesoderm.** The space between these two layers is the *coelom.* Figure 4.17B shows a cross-sectional view of the *Xenopus* embryo after neurulation is nearly complete. You should memorize this characteristic arrangement of structures; it is the basic body plan of all vertebrates, the foundation of the huge diversity of anatomical arrangements found in the vertebrate subphylum.

Is All this Detail Really Necessary?

We pause for a moment to ask where we have been and where we are going. What is the big picture, what are details? The

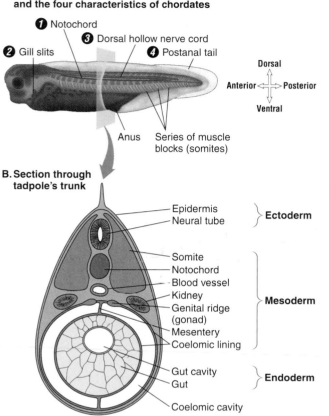

**A. The vertebrate "body axis" (head–trunk–tail)
and the four characteristics of chordates**

❶ Notochord
❸ Dorsal hollow nerve cord
❷ Gill slits
❹ Postanal tail

Dorsal
Anterior ⟷ Posterior
Ventral

Anus   Series of muscle
blocks (somites)

**B. Section through
tadpole's trunk**

Epidermis
Neural tube   } **Ectoderm**

Somite
Notochord
Blood vessel
Kidney
Genital ridge
(gonad)
Mesentery
Coelomic lining   } **Mesoderm**

Gut cavity
Gut   } **Endoderm**

Coelomic cavity

Figure 4.17   The Amphibian Tadpole Body Plan   (A) A side view showing portions of the organization of the body, including the four features characteristic of all chordates. The vertical line indicates where **(B)** was taken from. **(B)** A cross section through the trunk region shows the arrangement of internal organs and their derivation from the three germ layers.

study of development, and of much of biology, is like the study of a foreign language. Some tools, such as acquiring some vocabulary and pronunciation, are necessary in order to have the pleasure of speaking to the waiter when you order a meal in Paris. As you learn more of the language, the deeper character of the country, its people, and the culture becomes apparent. So, we have undertaken to learn some vocabulary, a necessity for any penetrating look at development, and to peel back some of the layers of what happens during development, using concrete examples.

We have chosen two creatures—a fly and a frog—that develop very differently. But hopefully you now realize that their development progresses according to somewhat similar steps. The organized egg is formed with the help of accessory cells and organs. The egg is activated to enter mitosis, thereby generating cells necessary for metazoan (multicellular) existence. During gastrulation, groups of cells, using a program of changes in cellular behavior, move actively or passively to characteristic locations. The localizing of materials during oogenesis, and the sending and receiving of chemical signals between cells, together set in motion characteristic programs of gene expression in various regions.

Before examining the cellular and molecular underpinnings of these tactics in detail (in Part Three), we shall continue our exploration of basic principles by considering two more organisms, each very different from the fly and frog, each telling us something more about the way development proceeds: In Chapter 5 we will look at the development of amniotes (reptiles, birds, and mammals), best studied in the chick and mouse.

KEY CONCEPTS

1. Oocytes may use exotic molecular mechanisms to obtain their stockpile of materials needed for early development. Among these are transient amplification of specific genes, such as those for ribosomal RNA.

2. Hormones are used as signaling ligands to regulate the pace of gamete development and maturation.

3. Though eggs may be organized and have polarity, often their symmetry is different from the symmetry of the resulting organism. Hence, the dominant symmetries must be readjusted. In amphibians this is effected by rotation of the egg cortex, which thereby breaks radial symmetry and converts it

to anteroposterior (front to back) polarity and bilateral (left–right) symmetry.

4. The patterning of the embryo proceeds by establishing potential germ layers. Gastrulation movements—a hallmark of animal development—establish the germ layers. Gastrulation always involves the movement of surface and adjacent cells into the interior of the blastula, thereby creating a multilayered embryo.

5. Tubular structures form in two different ways during amphibian development. The gut forms during gastrulation by involution and invagination. The neural tube forms from a flat plate of cells rolling into a tube.

STUDY QUESTIONS

1. If high-molecular-weight DNA, prepared by standard biochemical procedures, is injected into a frog egg, it may be assembled into chromatin or chromosome-like structures. Yet, if the same experiment is carried out by injecting DNA into somatic cells, the injected DNA does not assemble into chromatin. Can you explain the difference?

2. Three changes occur at the midblastula transition: transcription and motility increase, and the cell cycle lengthens. Perhaps all three events are regulated by some common "switch." Can you propose an experimental way to decide whether these events are linked in some regulatory scheme?

3. Predict what might occur if the animal cap of a living *Xenopus* blastula were microsurgically removed. Would gastrulation occur?

SELECTED REFERENCES

Beck, C. W., and Slack, J. M. W. 1998. Analysis of the developing *Xenopus* tail bud reveals separate phases of gene expression during determination and outgrowth. *Mech. Dev.* 72:41–52.

A current research paper analyzing morphogenesis and molecular details of formation of postanal axial structures.

Elinson, R. P., and Rowning, B. 1988. A transient array of parallel microtubules in frog eggs: potential tracks for a cytoplasmic rotation that specifies the dorso-ventral axis. *Dev. Biol.* 128:185–197.

A research paper documenting the presence of microtubules in the shear zone.

Gerhart, J., Danilchlik, M., Doniach, T., Roberts, S., and Rowning, B. 1989. Cortical rotation of the *Xenopus* egg: consequences for the anteroposterior pattern of embryonic dorsal development. *Development* 107:37–51.

Experiments showing the importance of cortical rotation.

Gimlich, R. L., and Gerhart, J. C. 1984. Early cellular interactions promote embryonic axis formation in *Xenopus laevis*. *Dev. Biol.* 104:117–130.

The role that transplantation of various cells in the early *Xenopus* embryo played in elucidating how signaling centers are placed and how they might work.

Hamburger, V. 1988. The heritage of experimental embryology: Hans Spemann and the organizer. Oxford University Press, New York.

A slender volume about the history of the discovery of the organizer, by one of Spemann's influential students.

Harland, R., and Gerhart, J. 1997. Formation and function of Spemann's organizer. *Annu. Rev. Cell Dev. Biol.* 1997:611–667.

An insightful and detailed review of the Nieuwkoop center and Spemann organizer, by two of the leading researchers in the field.

Holtfreter, J. K., and Hamburger, V. 1955. Embryogenesis: progressive differentiation—amphibians. In *Analysis of development*, ed. B. H. Willer, P. A. Weiss, and V. Hamburger, pp. 230–296. Haffner, New York, reprinted 1971.

A classic chapter of amphibian embryology in a classic textbook.

Jones, E. A., and Woodland, H. R. 1987. The development of animal cap cells in *Xenopus*: A measure of the start of animal cap competence to form mesoderm. *Development* 101:557–564.

Transplantation experiments showing that signaling from the vegetal hemisphere begins very early.

Keller, R. E. 1975. Vital dye mapping of the gastrula and neurula of *Xenopus laevis*, I and II. *Dev. Biol.* 42:222–241 and 51:119–137.

A classic reinvestigation of the movements of gastrulation, which forms the basis for modern work on the subject.

Keller, R. E. 1986. The cellular basis of amphibian gastrulation. In *Developmental biology: A comprehensive synthesis*, vol. 2, ed. L. Browder, pp. 241–327. Plenum, New York.

A detailed review of the cellular basis of gastrulation.

Nieuwkoop, P. D. 1969. The formation of mesoderm in urodelean amphibians, I and II. *Wilh. Roux Arch. Entwick. Org.* 162:341–373 and 163:298–315.

Papers forming the basis for our understanding of vegetal hemisphere signaling. Though tough reading, they are worth the effort.

Rowning, B. A., Wells, J., Wu, M., Gerhart, J. C., and Moon, R. T. 1997. Microtubule mediated transport of organelles and localization of catenin to the future dorsal side of *Xenopus* eggs. *Proc. Natl. Acad. Sci. USA* 94:1224–1229.

A research paper showing that β-catenin is probably moved to the gray crescent area along the tracks of parallel microtubules.

Scharf, S. R., and Gerhart, J. C. 1983. Axis determination in eggs of *Xenopus laevis*: A critical period before first cleavage, identified by the common effects of cold, pressure, and ultraviolet irradiation. *Dev. Biol.* 99:75–87.

Additional careful research on the effects of cortical rotation.

AMNIOTE DEVELOPMENT

All embryos develop in an aqueous environment, and all fish and amphibians have embryos that develop in fresh or salt water. Vertebrates that live and reproduce on land, however, face a special challenge. With a few exceptions, reptiles, birds, and mammals do not spawn their eggs in aquatic habitats, so special adaptations have evolved in these groups. Birds and reptiles have eggs with shells that prevent desiccation of the embryo, and mammals retain the embryo in an internal organ of the female, the *uterus.* All three groups devote a considerable amount of the substance of the embryo to the formation of membranes that enclose the developing embryo in an aquatic environment. The innermost membrane, called the *amnion,* completely encloses the embryo. Hence reptiles, birds, and mammals are called **amniotes.** The amnion and the other membranes that enclose and support these embryos are continuous

CHAPTER PREVIEW

1. Amniotes—reptiles, birds, and mammals—have extraembryonic membranes (special membranes enclosing the embryo), which permit development on land.

2. The eggs of birds and reptiles are large and yolky, with only a small disc of cytoplasm that undergoes cleavage.

3. Mammals possess smaller eggs with little yolk, and cleavage occurs slowly.

4. Gastrulation of amniote embryos occurs by ingression of surface cells.

5. In mammals, parts of the extraembryonic membranes join with tissues of the uterus to form the placenta.

6. Mouse embryos can be used to produce transgenic mice.

with, but distinct from, the cells and germ layers that give rise to the body of the embryo proper. Locations in the membranes are termed **extraembryonic;** for example, the amnion is composed of extraembryonic ectoderm and extraembryonic somatic mesoderm.

Much of what we know about the details of human development comes from studies on other amniotes, in particular, chick and mouse embryos. The presence of a shell in birds and a uterus and *placenta* in mammals creates some substantial differences in the earliest stages of development between these two groups. Nevertheless, there is great similarity between birds and mammals in the stages of gastrulation, germ-layer formation, and organogenesis. This is a great advantage because the physical accessibility of the avian embryo lends itself to transplantation and other microsurgical manipulations, while the extensive body of genetic information available for the mouse renders the mouse particularly suitable for molecular genetic approaches.

OOGENESIS AND THE EARLY DEVELOPMENT OF BIRDS

Gametogenesis in Birds Involves Specialization of the Female Reproductive Tract

Oogenesis in reptiles and birds is spectacular because of the huge amounts of yolk that accumulate during oogenesis. Once again, the liver is the source of yolk proteins, which are deposited in the oocyte developing within the ovary. The oocyte is, of course, undergoing meiosis during this time; it is also synthesizing components for a substantial amount of cytoplasm, which becomes located as a coherent disc on one side of the large yolk mass. The yolk and cytoplasm containing the oocyte nucleus are held within a vitelline membrane laid down by the follicle cells of the ovary.

When the ovarian egg is released from the ovarian follicle under the stimulus of hormonal signals, it travels down a specialized portion of the *oviduct*. As shown in Figure 5.1A, the egg is fertilized soon after it begins its journey through the oviduct. Subsequently, a specialized portion of the oviduct (the *magnum*) synthesizes and secretes the proteins that constitute the "white" of the egg; these proteins are principally ovalbumin and lysozyme, a potent hydrolytic enzyme that can destroy the cell walls of many bacteria, thereby functioning as a natural antibiotic. In another region of the oviduct, the *isthmus*, a complex membrane is elaborated around the egg

white. Finally, in the uterus an elaborate calcium carbonate shell is deposited around the egg.

Since the egg was fertilized soon after release from its follicle, early development occurs while egg white, membranes, and shell are still forming around the egg. The cytoplasmic disc is cleaved by mitosis, generating a multicellular **blastodisc** (another term for blastoderm, used with eggs that undergo meroblastic, or partial, cleavage; Figure 5.1B). A fluid-filled cavity called the *subgerminal cavity* forms between the blastodisc and the underlying yolk. Some cells detach from the blastodisc and pass into the subgerminal cavity, where most of them die. The cells remaining in the original blastodisc layer are now termed **epiblast.** The cells around the periphery of the blastodisc, where the cytoplasm meets the yolk, are syncytial (having many nuclei in a common cytoplasm) and are called the *marginal zone.* We shall have more to say about these various parts of the blastodisc soon, when we discuss early development after egg laying.

At the time of egg laying in chickens, there may be as many as 50,000 to 60,000 cells in the blastodisc perched on one side of the yolk. The egg of a domestic chicken is over 98% yolk. The "spot" of cytoplasm just visible to the naked eye, now cleaved into 60,000 cells in a fertilized egg, carries a huge reservoir of materials for metabolism and biosynthesis during development. It also possesses a shell that resists desiccation yet has pores allowing the passage of oxygen in and carbon dioxide out. The whole package is beautifully adapted for a terrestrial habitat, but in terms of first principles, these "innovations" are more technological than fundamental.

Fertilization occurs within the female reproductive tract after sperm are deposited during mating; the rooster's sperm while resident in the female reproductive tract may remain active and able to fertilize eggs for many days. Fertilization activates mitosis. Because of the huge amounts of yolk, only the disc of cytoplasm undergoes cytokinesis, the mass of yolk remaining uninvolved in cleavage. As we saw in Chapter 2, this kind of early cleavage, where only a disc of cytoplasm is divided by cytokinesis, is called *meroblastic.*

The Newly Laid Hen's Egg Has an Invisible Axis

The disc of 60,000 cells lying over the yolk in a newly laid egg is about 1 mm in diameter. The cleavage divisions in the oviduct have created this inverted saucerlike structure, attached at its edges to the yolk but separated over the central portion by the subgerminal cavity. Figures 5.1B and 5.2 illustrate a section through this disc. Above the subgerminal cavity is a

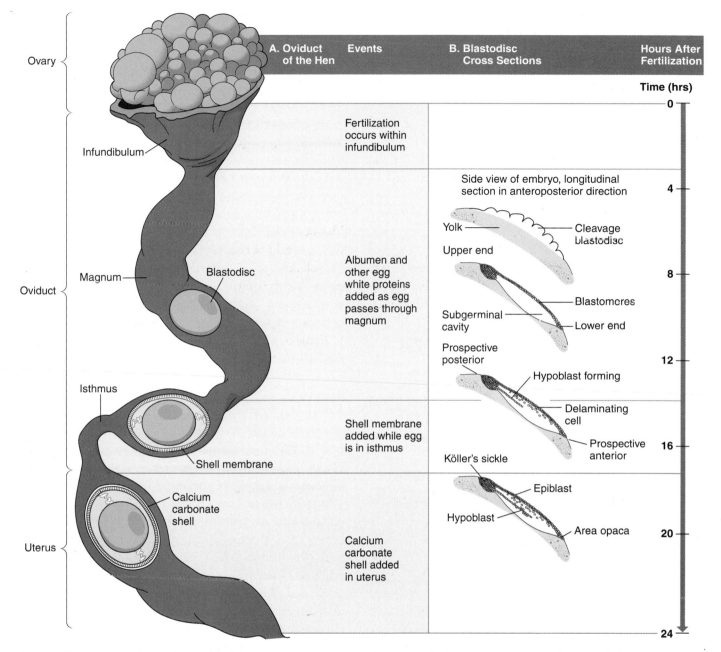

Figure 5.1   The Reproductive Tract of Birds   (A) A diagrammatic view of a hen's oviduct. The ovary is surrounded by the infundibulum of the oviduct, and ovulated eggs enter the oviduct via the infundibulum, where fertilization occurs. As the fertilized egg travels down through the magnum, isthmus, and uterus, the egg white proteins, the shell membrane, and the shell itself are successively added before laying. **(B)** The state of the blastodisc, shown in a longitudinal section. The cleavage is called meroblastic because only the small disc of cytoplasm on one side of the yolk undergoes cell division. By the time the egg is laid, an anteroposterior axis has been established, the hypoblast is forming from the epiblast, and there are about 60,000 cells in the blastodisc. The portion of the blastodisc on the upper end (with respect to the earth's surface) becomes posterior, the lower end anterior. A fluid-filled space below the blastoderm is called the subgerminal cavity.

relatively transparent central region called the **area pelluci-da.** It is more transparent because its constituent cells have less yolk and do not lie directly on the yolk mass. The edges of the blastodisc have yolkier cells; at the periphery they merge into a syncytial area with the yolk. This yolkier periph-eral area is the **area opaca.** The central area pellucida is only a few cell layers thick. As mentioned earlier, during the pas-sage down the oviduct, many cells from the area pellucida are "shed" into the subgerminal cavity (where they die), thereby thinning the area pellucida.

A. Egg before formation of hypoblast

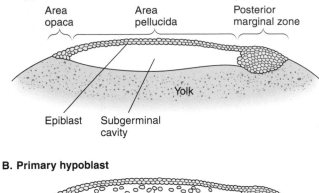

B. Primary hypoblast

Hypoblast cells delaminating from epiblast

C. Secondary hypoblast

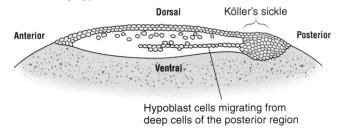

Hypoblast cells migrating from
deep cells of the posterior region

Figure 5.2   The Formation of the Hypoblast   (A) Sagittal sections through the blastodisc of the newly laid egg. The hypoblast forms from cells **(B)** delaminated from the epiblast and **(C)** migrating forward from Köller's sickle.

Though the blastodisc appears radially symmetric, a simple experiment demonstrates that a developmental bias is already present that will determine the future anteroposterior axis. The *blastodisc*—the term used to describe the entire cellular ensemble perched on the yolk—may be dissected from the yolk and thereafter surgically subdivided. Some portions are much better than others at providing a focus for the gastrulation movements necessary for axis formation, as shown in Figure 5.3. The areas with the higher percentage of gastrulation can be shown to form the future posterior midline of the developing embryo. In fact, it is the marginal zone—cells at the future posterior end between the area opaca and the area pellucida—that is the prospective normal center for the initiation of gastrulation movements.

Some ingenious experiments by H. Eyal-Giladi and her collaborators in Israel have shown that the effects of gravity

on the egg during its passage down the oviduct probably influence the location of the posterior marginal zone. During the egg's passage through the oviduct, the muscular peristaltic movements of the oviduct are continuously rotating the egg white around the egg, which remains stationary, with one end "higher" (with respect to the earth's surface) than the other. It is the higher end of the blastodisc that becomes the future posterior end. We know this because when eggs are removed from the oviduct and suspended in various orientations, the future posterior end invariably forms on the high side of each suspended egg. Just how this effect of gravity breaks the radial symmetry of the cleaving blastodisc is not known, but we do know that the cell shedding that participates in forming the area pellucida commences from the future posterior side. At least superficially, this effect of gravitation on the organization of the hen's egg is reminiscent of the rotation of the amphibian cortex after fertilization, a situation in which gravity also plays a role.

The Area Pellucida Becomes a Two-layered Structure

Figure 5.2 shows the appearance of the cellular blastoderm at the time of egg laying. The marginal zone of the prospective posterior end now forms a proliferative ridge termed *Köller's sickle.* A cellular layer advances from this area, forming a loose tissue layer below the surface of the area pellucida called the **hypoblast.** Cells in the anterior and lateral portions of the area pellucida detach from the epiblast, descend-

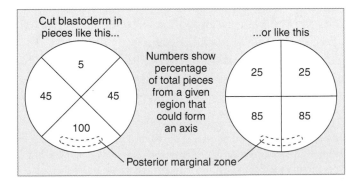

Figure 5.3   Localization of the Ability to Form an Axis in the Early Chick Blastoderm The diagrams help explain the experimental strategy used by Oded Khaner to map the potential to form an axis. First the posterior pole of the early blastoderm was identified as the region of thickening in the marginal zone. Then the blastoderm was cut into quarters, as shown here, and the percentages of pieces in each quarter that form an axis in culture were scored. The results indicate that the posterior marginal zone not only is able to form an axis but also probably suppresses the ability of more lateral regions to do so. This has been confirmed in other transplantation experiments.

A. Preparation of blastoderms

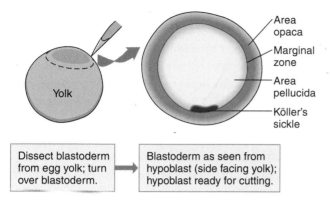

Area opaca
Marginal zone
Area pellucida
Köller's sickle

Yolk

Dissect blastoderm from egg yolk; turn over blastoderm. → Blastoderm as seen from hypoblast (side facing yolk); hypoblast ready for cutting.

Figure 5.4   The Influence of the Hypoblast on Placement of the Axis Khaner's hypoblast rotation experiments shown in diagrammatic form. **(A)** The early blastoderm was removed, and the hypoblast excised and rotated 90° relative to the position of the epiblast. **(B)** The outer limit of the hypoblast as it encroaches upon the area opaca and Köller's sickle was widened in various experiments, as shown. In all instances, rotation of the hypoblast alone failed to change the orientation of the axis that formed in the epiblast, indicating that the epiblast probably has a built-in axial bias.

B. Rotation of hyoblasts by 90°

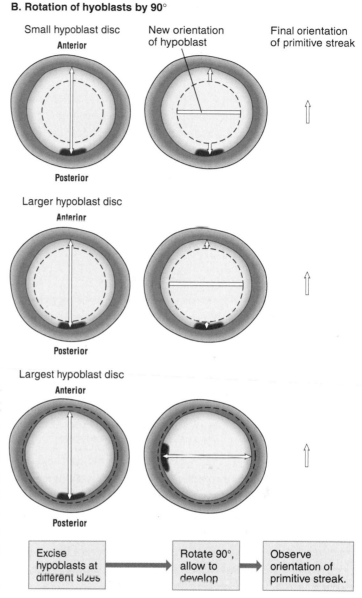

Small hypoblast disc
Anterior / Posterior

New orientation of hypoblast

Final orientation of primitive streak

Larger hypoblast disc
Anterior / Posterior

Largest hypoblast disc
Anterior / Posterior

Excise hypoblasts at different sizes → Rotate 90°, allow to develop → Observe orientation of primitive streak.

ing from the underside of the epiblast to assume a position where they, too, may be incorporated into the developing hypoblast. The detaching cells sometimes detach as a group, rather than singly, and this is called **delamination.** The delaminated and detached cells can form a transient layer called the primary hypoblast, which merges with cells from Köller's sickle to form the secondary hypoblast. (The term *secondary hypoblast* is seldom used, however; the combination of primary hypoblast and cells from Köller's sickle is usually just called the hypoblast, or **endoblast.**) The remaining surface layer of the area pellucida is the epiblast.

After about 12 hours of incubation, the hypoblast is fully formed, and the epiblast begins to show some cellular movement. By 20 hours, the overt morphogenetic movements of gastrulation have begun.

Does the Hypoblast Influence Organization of the Epiblast?

Experiments carried out in the 1930s by the British embryologist Conrad Waddington supported the idea that the anteroposterior axis of the epiblast (and thus, as we shall see, of the whole embryo) results from the hypoblast "inducing" the epiblast. Oded Khaner and others, who reanalyzed and repeated these experiments, have recently challenged this view (Figure 5.4). If the hypoblast is rotated with respect to the epiblast, the axis of the embryo remains unaf-

fected. It is likely that the area pellucida develops an axial bias prior to hypoblast formation and that the epiblast is inherently organized.

In other experiments, when both the marginal zone and the hypoblast were removed from the epiblast, axis formation failed. On the other hand, leaving either the marginal zone or the hypoblast present allows some axis formation to occur. Furthermore, including the growth factor, activin, in a medium in which epiblast alone is cultured allows the formation of axial structures. If Vg1 growth factor is expressed ectopically in the lateral marginal zone, then that portion of the marginal zone becomes competent to induce an axis. Interestingly, activin and Vg1 are also implicated as growth factors involved in the induction of axial structures in amphibians (see Chapter 16). So, while an axial bias in the epiblast apparently cannot be overridden by the hypoblast alone, the hypoblast, the marginal zone, or both are involved in some way in axial formation and mesoderm origination, possibly by way of secreting the same kinds of growth factors that are at work in amphibians. Recent experiments by Bertocchini and Stern indicate that the ligand Nodal (discussed in Chapter 16) and its antagonist, Cerberus, regulate axis formation.

GASTRULATION IN BIRDS

The Epiblast Is the Source of All Embryonic Germ Layers in Amniotes

During gastrulation, cells in the prospective posterior half of the epiblast move in the plane of the epiblast toward the midline. Based on the results of cell-marking experiments, it is probable that only some of these cells—many of them scattered throughout the epiblast as single cells or small clusters—are involved in this migration. This choice (whether or not to move to the midline) implies that individual cells have already acquired an identity prior to these earliest gastrulation movements.

The lateral-to-medial migration advances within the epiblast from posterior to anterior, eventually involving cells located throughout the posterior half of the blastoderm, as shown in Figure 5.5. The result is a marked thickening along the posterior midline. Gradually, this thickening lengthens because cells from the anterolateral area of the epiblast also begin to move toward the midline, and perhaps, too, because the entire blastoderm elongates front to back.

This central line of thickening becomes an area of active invagination and ingression, as cells from surface layers move inward toward the hypoblast. This movement produces a depression in the middle of the area of central congregation. The entire line of ingressing and medially migrating cells is called the **primitive streak.** It is a dynamic structure, just as the dorsal lip of the blastopore in amphibians is dynamic. The primitive streak results from lateral-to-medial movement of cells, like rowboats moving through the water to congregate at a dock, and from the ingression of cells into the interior.

Though the morphogenetic maneuvers are different, gastrulation in amniotes accomplishes the same thing as in other animal groups: cells at or near the surface move to the interior to establish the three germ layers. In amniotes these germ layers develop as follows.

Some epiblast cells move through the middle of the primitive streak (called the *primitive groove*), where they ingress into the interior. This ingression occurs mostly after the primitive streak has reached its maximum length. Once cells leave the surface layers, they take one of two routes: Some cells insert themselves all the way into the hypoblast, literally pushing the former hypoblast to the periphery. Eventually the entire layer beneath the area pellucida, once occupied by the hypoblast, comes to be occupied by cells that originated during gastrulation by ingression from the epiblast. The original hypoblast has been displaced peripherally to positions in the area pellucida and the area opaca.

Other cells migrating through the primitive groove move laterally (and somewhat anteriorly at positions in the anterior

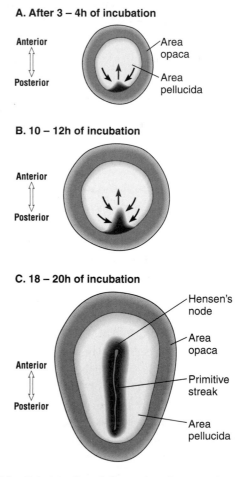

Figure 5.5 Primitive Streak Formation Diagrams showing dorsal views of the chick blastoderm. **(A)** Three to four hours after laying, the movement of epiblast cells toward the center at the posterior margin is marked, and **(B)** continues as more and more cells are recruited anteriorly (stage 3, at 10 to 12 h). The black curved arrows indicate the predominant movement of epiblast cells toward the midline; the red arrow indicates the anterior recruitment of such cells, which causes the primitive streak to lengthen. **(C)** The definitive primitive streak is obvious by 18 to 20 h after laying.

portion of the streak). They do not dive into the hypoblast; nor do they end up in positions contiguous with the hypoblast. Rather, they establish a new, loose mesenchymal layer between the surface epiblast and the remodeled hypoblast.

Thus, we have the origin of the three germ layers: cells remaining in the epiblast become ectoderm, those displacing the original hypoblast become endoderm, and those in the middle, mesenchymal layer become mesoderm.

Two other important morphogenetic activities are occurring during these gastrulation movements. First, cell division continues at the edges of the area opaca, and the blastoderm edges move more and more peripherally to enclose the enormous yolk. This growth and expansion will continue through eight or nine days of development, until the yolk is completely enclosed in extraembryonic tissues (see Figure 5.8). Furthermore, the surface of the marginal zone develops into extraembryonic ectoderm, the old hypoblast next to the yolk becomes extraembryonic endoderm, and a middle layer of cells differentiates as extraembryonic mesoderm. Second, there is an important suite of morphogenetic activities associated with Hensen's node, which we shall now discuss.

The Anterior Border of the Primitive Streak Is Specialized

As shown in Figure 5.6, the primitive streak extends from the posterior margin of the area pellucida to only 50% to 60% of the length of the blastoderm. The anterior terminal end of the streak, called **Hensen's node,** is a hub of complex morphogenetic activities and important cell interactions. Hensen's node has a depressed center, the *anterior primitive pit;* conspicuous ridges partially surround this pit, marking the anterior end of the primitive streak and the position of the node. Cell-marking experiments indicate that cells in the node have been recruited from the epiblast as well as from a subpopulation of cells, which earlier migrated forward from Köller's sickle, that produce the transcription factor xgoosecoid.

The activities in Hensen's node are complex and still a subject of ongoing research. What is known is that the node is a structural way station for passing cells. Some of these cells come from the lateral and anterolateral portions of the epiblast. Other cells ingressing through the anterior primitive pit apparently also move anteriorly and pass through Hensen's node. In addition, the node is a center for cell proliferation, which gradually extends forward from the node to the prospective midbrain region of the forming brain. This strand of cells, called the **head process,** will form the notochord (see Figure 5.6). The node and anterior endoderm probably secrete growth factor antagonists that are essential for development of the head. We shall soon discuss this property when we consider the mouse embryo.

About 24 hours after incubation has begun, the medial movement of epiblast cells to the primitive streak ceases and the ingression of cells through the primitive groove comes to a halt. Hensen's node now moves posteriorly, like a swan gliding on a river, obliterating the primitive streak as it moves through the blastoderm and leaving behind the neural plate, the notochord, and the developing somites (Figure 5.6). Cell-marking experiments have shown that the medial portion of the neural plate, the notochord, and the medial portions of the somites all come from cells that resided in Hensen's node.

The node's migration eventually brings it to an area near the posterior margin of the area pellucida, where its morphogenetic movements cease and it becomes indistinct. In its wake the node leaves an actively forming central nervous system, a notochord, and about 25 pairs of somites. However, this is only about half the total number of somites that will actually form in chickens and other birds. The terminus of Hensen's node, called the *tail bud,* continues to generate yet more axial structures and somites as the embryo lengthens posteriorly during subsequent developmental stages.

Hensen's Node Organizes the Axis and Induces the Central Nervous System

Soon after the discovery of the amphibian organizer by Spemann and Mangold, transplantation experiments in avian embryos by Waddington and others showed that Hensen's node could induce an ectopic (abnormally located) neural tube and somites. Cell-marking experiments have shown that when Hensen's node is transplanted to a host in a position distant from the host's axis, the ectopically formed somites originate from the host's tissues, not the donor's. It is the influence of the transplanted node that produces the somite pattern of bilateral segmentation. The ectopic neural tissue also derives mainly, though not exclusively, from non–Hensen's node tissue of the host. Therefore the node may be thought of as a *primary inducer.* Some genes known to be expressed in the Spemann organizer of amphibians, such as *xgoosecoid,* as well as genes for ligands such as Sonic Hedgehog, are expressed in the node.

Claudio Stern and his colleagues have recently shown that Hensen's node and the adjoining anterior portion of the

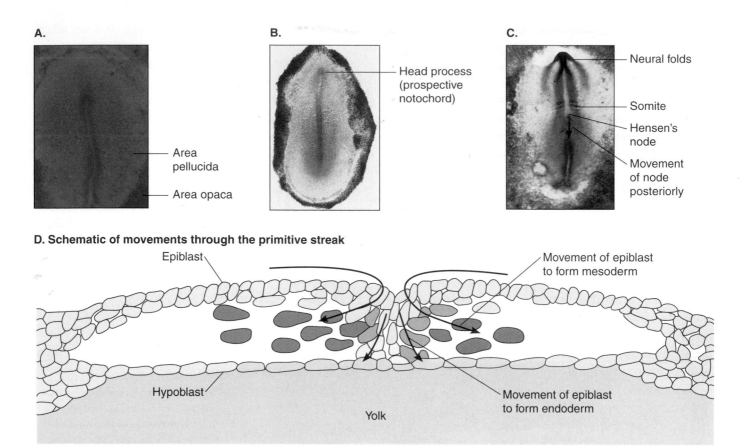

A.

Area pellucida

Area opaca

B.

Head process (prospective notochord)

C.

Neural folds

Somite

Hensen's node

Movement of node posteriorly

D. Schematic of movements through the primitive streak

Epiblast

Movement of epiblast to form mesoderm

Hypoblast

Yolk

Movement of epiblast to form endoderm

Figure 5.6   Hensen's Node and Its Regression (A-C) Photographs showing surface views of the epiblast at various stages plus (D) a schematic of movements through the primitive streak. **(A)** A stage 4 chick embryo (19 to 22 h) shows the streak maximally extended. **(B)** A few hours later. at stage 5. the streak has begun to regress: a noticeable extension. growing forward and below the surface from cells in Hensen's node. indicates the beginnings of the head process. which continues to extend forward during this stage. **(C)** By early stage 8. the embryo already possesses three pairs of somites. and Hensen's node has moved posteriorly. A head fold has delimited the anterior border of the embryonic ectoderm. and neural induction and formation of neural folds have already occurred in the brain region. **(D)** This cross section through the primitive streak shows the movement of epiblast cells (*thick arrows*) to a position between epiblast and hypoblast. thereby forming mesoderm. Other epiblast cells insinuate themselves (*thin arrows*) into the hypoblast. pushing the old hypoblast laterally. The new hypoblast will form endoderm.

primitive streak can be surgically extirpated (removed); as the tissues heal, the area where they knit together regenerates a fully functional Hensen's node, thereby allowing the formation of a more or less normal axis with its central nervous system and segmentally arrayed somites. This outcome implies that node properties may themselves arise as a result of tissue interactions.

In summary, the node is a complex center of morphogenesis and tissue interaction, indispensable for the normal formation of axial structures; the node itself arises as a result of preceding cellular interactions. These are the same properties we associate with the dorsal lip of the blastopore, or Spemann organizer. We should also note that the node of mammalian embryos, insofar as it has been investigated, is like the Spemann organizer and Hensen's node of birds, having the same general properties.

At the end of gastrulation, the very posterior portion of the streak does not have well-defined germ layers. The cells in this posterior portion of the streak will form the tail bud, the source of muscles and spinal cord for the postanal tail. Since all vertebrates have a postanal tail, differentiation of the tail bud is an important, albeit little studied, process.

Gastrulation Results in the Formation of an Archetypal Vertebrate Axis

If we examine a cross section through the chick blastoderm after Hensen's node has regressed, the basic arrangement of germ layers and organ *anlagen* (the technical term for the earliest beginnings of an organ; singular *anlage*) is almost identical to what we encountered in the amphibian neurula. Medially the ectoderm cells are elongating to form a columnar epithelial plate that will be the site of brain and spinal cord formation. Laterally the ectodermal epithelium extends from this plate, across the area pellucida, and into the expanding area opaca as it encloses the yolk. (Compare Figure 4.17 in Chapter 4 with Figure 5.7.)

The middle, or mesodermal, layer forms a central notochordal rod. As we move laterally, the somites become apparent. Even more laterally is a knot called *intermediate mesoderm* (from which the kidneys will arise). Most laterally, the mesoderm forms a thin spreading sheet. This *lateral mesoderm* has actually split into two sheets, one adhering to the overlying ectoderm, called the **somatic mesoderm,** the other adjoining the underlying endoderm, called the **splanchnic mesoderm.** The space between is the **coelom,** which extends out into the area opaca (Figure 5.7).

The endoderm of the area pellucida lies next to the yolk. In its midline portion, the endoderm will undergo morphogenetic movements to form the gut; it will also extend peripherally to enclose the yolk. There are two main differences between this arrangement and the one we saw in amphibians. First, the endoderm has not yet formed a tube; it is as if the gut were cut longitudinally and the embryo splayed out over a yolk mass. Second, a huge amount of each of the three germ layers, which are gradually enclosing the yolk mass, will never become part of the embryo proper but instead will form a series of membranes characteristic of the amniotes.

The Extraembryonic Membranes of Birds Are Comprised of Four Membranous Sacs

Birds and the other amniotes—reptiles and mammals—all form a set of four membranous sacs that are adaptations to reproduction in terrestrial environments. Let us examine the situation in birds by referring

to Figure 5.8. We now view the entire egg, yolk and all, in longitudinal section, where the embryonic portion is dwarfed by the extraembryonic tissues extending over the yolk mass.

The ectoderm–somatic mesoderm layer in the area pellucida begins to undergo tissue movements that undercut the margins of the embryo, demarcating this tissue from the body of the embryo proper (Figure 5.8A). Simultaneously, this same layer grows up and over the embryo, both from the front and later from the back (Figure 5.8B, C), so that the entire embryo is enclosed in a double-layered sac. The double layers eventually separate. The inner ectoderm–somatic mesoderm layer, called the **amnion,** surrounds the embryo and contains amniotic fluid in which the embryo grows and develops. The outer ectoderm–somatic mesoderm layer, which is closer to the shell, eventually extends around the entire mass to become an epithelial layer under the shell

A.

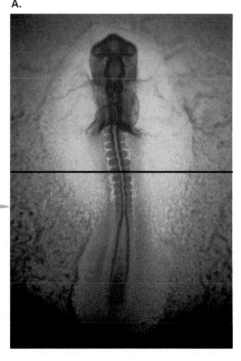

B.

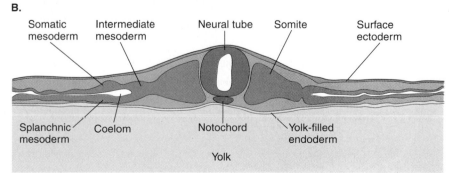

Figure 5.7 An Avian Embryo in Cross Section A stage 10 embryo (33 to 38 h). **(A)** The photograph shows a dorsal view; the line indicates the position of the cross section diagrammed in **(B)**, which shows the arrangement of structures at the midbody level. Notice how similar the basic arrangement is to that of the amphibian embryo, except that the blastoderm is flattened out on the yolk mass.

Somatic mesoderm · Intermediate mesoderm · Neural tube · Somite · Surface ectoderm

Splanchnic mesoderm · Coelom · Notochord · Yolk-filled endoderm

Yolk

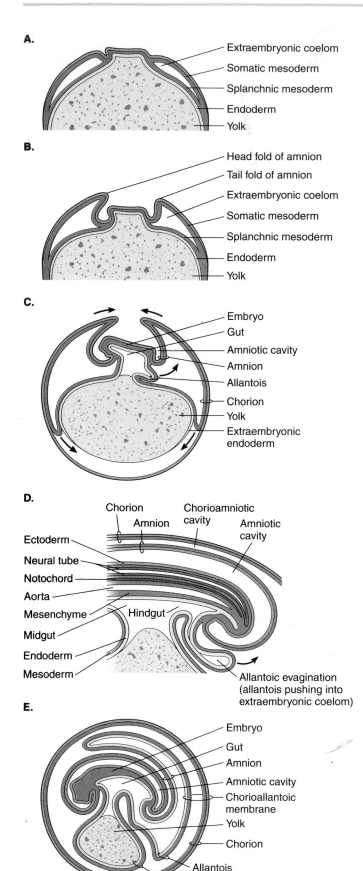

A.
- Extraembryonic coelom
- Somatic mesoderm
- Splanchnic mesoderm
- Endoderm
- Yolk

B.
- Head fold of amnion
- Tail fold of amnion
- Extraembryonic coelom
- Somatic mesoderm
- Splanchnic mesoderm
- Endoderm
- Yolk

C.
- Embryo
- Gut
- Amniotic cavity
- Amnion
- Allantois
- Chorion
- Yolk
- Extraembryonic endoderm

D.
- Chorion
- Amnion
- Chorioamniotic cavity
- Amniotic cavity
- Ectoderm
- Neural tube
- Notochord
- Aorta
- Mesenchyme
- Hindgut
- Midgut
- Endoderm
- Mesoderm
- Allantoic evagination (allantois pushing into extraembryonic coelom)

E.
- Embryo
- Gut
- Amnion
- Amniotic cavity
- Chorioallantoic membrane
- Yolk
- Chorion
- Allantois
- Yolk sac

through which gas exchange will occur. This layer is the **chorion.**

The combination of extraembryonic endoderm and splanchnic mesoderm, lying on top of the yolk mass, eventually encloses the yolk. It becomes richly vascularized and is called the **yolk sac.** Enzymes in the yolk sac mobilize the enclosed yolk, which then is transported by blood vessels to supply nutrients to the embryo.

In addition to the amnion, chorion, and yolk sac, an evagination of the developing hindgut that is composed of endoderm and splanchnic mesoderm will intrude into the space between the yolk sac and chorion. Called the **allantois,** it becomes very large in birds, filling the space between the chorion and yolk sac. It, too, has a rich vascular supply that carries blood to the chorionic layer for gas exchange. In birds the allantois and chorion fuse under the shell into a single, vascularized *chorioallantoic membrane* (Figure 5.8C–E).

EARLY MAMMALIAN DEVELOPMENT

Oogenesis, Fertilization, and Cleavage in Mammals Involve Specialization of the Oviduct

It is interesting to see how different is the strategy used by mammals to survive the threats of desiccation imposed by terrestrial existence. While birds and reptiles enclose huge amounts of yolk in a shell, mammals (with a few exceptions) retain the zygote in a special region of the oviduct called the **uterus;** also unlike birds, mammals do not deposit much yolk in the egg, but provide nutrition to the embryo by allowing it to gain access to the maternal circulation. Mammals invented

Figure 5.8 Extraembryonic Membranes of Avian Embryos Sagittal sections showing the formation and arrangement of the amnion, chorion, yolk sac, and allantois in birds. Note that the embryo and membranes are draped on the yolk with the head of the embryo to the left. **(A)** Tissue anterior to the head begins to undercut the embryo, beginning the formation of the amnion. **(B, C)** This process continues around the entire embryo, until **(D, E)** the folds of the ectoderm and somatic mesoderm meet and fuse. The part of the folds next to the embryo proper is the amnion; the portion next to the shell is the chorion. The extraembryonic endoderm has grown over the yolk, forming the yolk sac. An evagination from the hindgut extends into the space between the chorion and the amnion, forming the allantois. The allantois and chorion fuse to become a fused membranous layer underneath the shell. The yolk sac and chorioallantoic membrane become highly vascularized.

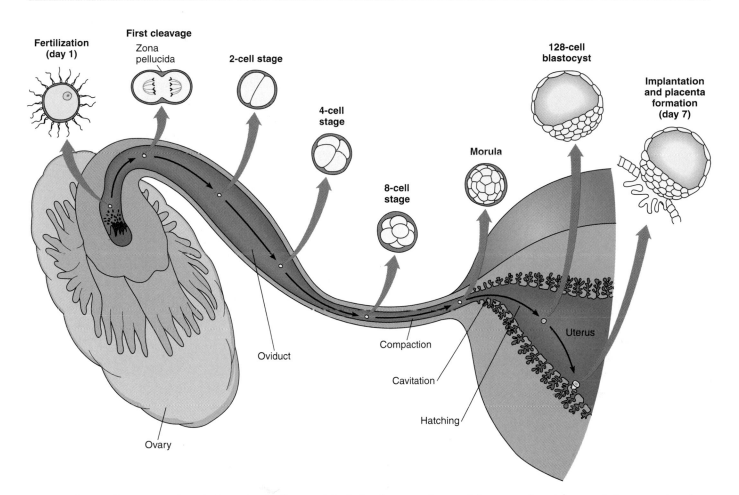

Figure 5.9   The Mammalian Female Reproductive Tract and Early Development Stages of cleavage and implantation of the mammalian (in this case, human) embryo as it travels through the Fallopian tube to the uterus. (Only one side of the uterus and one of the two tubes and ovaries are shown.)

formation of the **placenta.** This organ provides the embryo with several important things: a humid environment, a way to gain nutrition without yolk, and a means for respiration and excretion of wastes through exchange with the maternal circulation. We will consider the placenta in more detail later in the chapter. Figure 5.9 illustrates the gross anatomical features of these specializations of the mammalian reproductive tract.

Oogenesis in mammals involves the following features: growth of the oocyte, meiosis, the formation of membranes by follicle cells surrounding the egg, and regulation of oogenesis and ovulation by hormonal signals initiated by the pituitary. If you are not familiar with the estrous cycle of mammalian reproduction, consult an introductory biology textbook.

The ovulated mammalian egg is fertilized by sperm in the upper reaches of the **Fallopian tubes** (as the two mammalian oviducts are called); the egg is activated, and mitosis ensues. The pace of mitotic division is much slower than in invertebrates and nonamniotes. For instance, in humans or mice, the first mitosis takes about a day to occur, subsequent mitoses may occur at a rate of only about every 10 to 12 hours, and it takes about five to seven days for the zygote to make its journey to the uterus. Thus, when it first enters the uterus, the human egg may consist of little more than 100 cells.

Perhaps because of this slower tempo and the presence of authentic G1 and G2 periods, new gene transcription (called *zygotic gene transcription* since it occurs in the zygote) is detectable in mammals as early as the two-cell stage; by contrast, recall that as many as eight rapid mitotic cycles may occur in *Drosophila* or *Xenopus* before a low level of zygotic transcription is evident. Mammals show other important differences from nonmammals: (1) There may be limited maternal control of organization of the egg or embryo; (2) there is almost no yolk; (3) very early in development the formation of elaborate extracellular membranes begins; and (4) some genes in the male and female may be expressed differentially during embryonic development. This last phenomenon, called *imprinting,* is caused by different extents of methylation of these particular genes in the male and female gametes, as discussed further in Chapter 14.

Cleavage Produces a Blastocyst that Will Embed in the Lining of the Uterus

With few exceptions (such as monotremes), most early mammalian zygotes undergo a similar sequence of events, though the absolute timing may differ. For example, in mice the embryo is ready to attach to the uterine wall four days after fertilization; in humans after about seven. Attachment and implantation are preceded by a slow mitotic cleavage as the egg moves through the Fallopian tubes (Figure 5.9). At the eight-cell stage, a remarkable and important change occurs (Figure 5.10). The loosely attached blastomeres become a true polarized epithelium. The outer cell surfaces come to bear microvilli, and tight junctions develop between adjacent cells. The cells also seem to flatten out, becoming closely adherent—hence the term **compaction** for this change in morphology.

At the next (fourth) division, many of the cell division planes are angled such that the resulting cells remain on the surface. But some divisions take place with the division planes angled parallel to the surface, thus generating some three or four "internal" cells. These internal cells are called the **inner cell mass (ICM),** and they will form the embryo proper and the amnion. The outer cells, which comprise the **trophoblast,** will form the placenta. The early cleaving embryo is sometimes called a *morula.*

Shortly after generation of the ICM, during the next one or two cell divisions, another important transition occurs: the cells of the ICM and trophoblast become irrevocably different from one another. If we were to separate the ICM from the

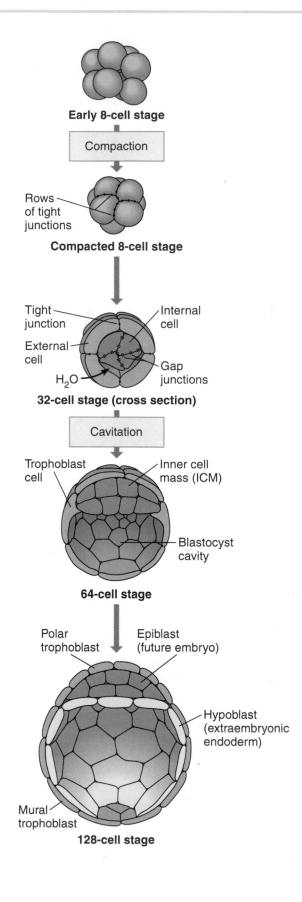

Figure 5.10 Cleavage and Compaction At the early eight-cell stage, the blastomeres are not closely adherent. During the compaction process, epithelial junctions form and ICM cells are generated among surrounding trophoblast cells. A blastocyst cavity, very clear at the 64-cell stage, then forms. By the 128-cell stage, the blastocyst wall begins to secrete a proteolytic "hatching" enzyme, digesting the membranes of the zona pellucida that enclosed the zygote, and preparing the blastocyst to implant in the epithelial wall of the uterus. At this point, the epithelial covering of the embryo, called the trophoblast or trophectoderm, is clearly distinguishable as polar trophoblast, which overlies the epiblast, and mural trophoblast, which encloses the blastocyst cavity.

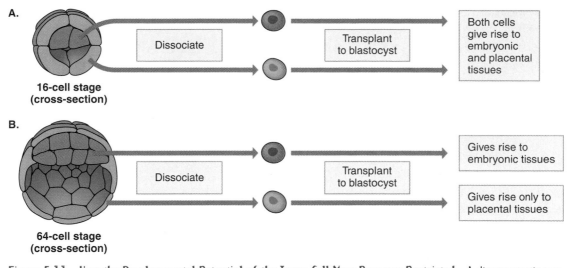

A.

16-cell stage
(cross-section)

Dissociate → Transplant to blastocyst → Both cells give rise to embryonic and placental tissues

B.

64-cell stage
(cross-section)

Dissociate → Transplant to blastocyst → Gives rise to embryonic tissues / Gives rise only to placental tissues

Figure 5.11   How the Developmental Potential of the Inner Cell Mass Becomes Restricted   A diagrammatic representation showing how determination of the ICM cells can be analyzed. **(A)** At the 16-cell stage, the blastocyst is dissociated, and genetically marked ICM and trophoblast cells are implanted in a host blastocyst: both ICM and trophoblast can still give rise to both embryonic and placental tissues. **(B)** But at the 64-cell stage, the same procedure shows that the developmental potential has become restricted: only transplanted ICM cells give rise to embryonic tissues, and transplanted trophoblast cells can form only placenta.

trophoblast at, say, the 32-cell stage, the ICM might regenerate some trophoblast, and the trophoblast some ICM. But by the 64-cell stage (after the sixth cell division), ICM cells are no longer capable of giving rise to trophoblast, nor trophoblast to ICM. These two cell groups have undergone *determination*, or as embryologists commonly say, they are now *determined*. One cell group is now destined to give rise to the embryo, the other to placenta structures. See Figure 5.11.

The polarized epithelium of the trophoblast has asymmetrically distributed sodium pumps on the apical surface. This asymmetry causes Na^+ (and counter anions such as Cl^-) to accumulate in the internal space that is developing alongside the ICM. Drawn by osmotic pressure from the relatively high concentration of Na^+ ions, water flows into and swells the space, a process called **cavitation** (Figure 5.12). The embryo, which in the mouse consists of about 128 cells after four days of development, is now called a **blastocyst** and is ready to implant in the wall of the uterus. (Please note the difference between the two similar-sounding terms: the blastocyst—a term reserved for this stage of mammalian embryos—is *not* a blastula.)

Near the end of the fourth day, the mouse embryo blastocyst will implant in the uterine epithelium. Trophoblast cells opposite the ICM (comprising the *antipodal trophoblast*) secrete a protease that digests the covering around the blastocyst. This covering is the **zona pellucida,**

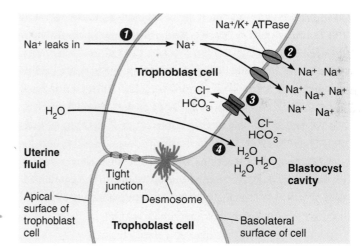

Figure 5.12   The Formation of the Blastocyst Cavity (Cavitation) Na^+ ions leak into trophoblast cells through the apical cell membrane, and the Na^+ is pumped out through the basolateral cell membrane by Na^+/K^+ exchangers. The trophoblast cells are sealed by tight junctions, which prevent passage of Na^+ from basolateral spaces. Trophoblast cells are also held together with desmosomes (see Chapter 12). The active transport out through the basolateral membrane causes Na^+ ions to accumulate between cells, thereby increasing the osmotic pressure in the fluid on the basolateral side of cells. Cl^-/HCO_3^- exchange also occurs, in order to maintain electrical neutrality. The greater osmotic pressure in "internal" spaces (320 milliosmols as opposed to 240 milliosmols "outside") causes inward water movement, which expands the internal space in the embryo and inflates the blastocyst cavity.

A. Early blastocyst at time of implantation (4 days)

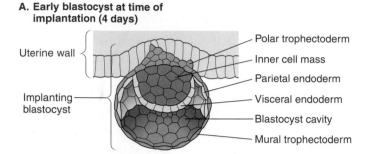

Uterine wall

Implanting blastocyst

Polar trophectoderm
Inner cell mass
Parietal endoderm
Visceral endoderm
Blastocyst cavity
Mural trophectoderm

B. Inner cell mass at 5 days   **C. Inner cell mass at 6 days**

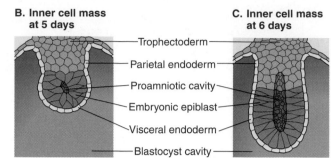

Trophectoderm
Parietal endoderm
Proamniotic cavity
Embryonic epiblast
Visceral endoderm
Blastocyst cavity

Figure 5.13   The Mouse Blastocyst   (A) The early mouse blastocyst shown at the time of implantation (four days after fertilization): the internal cell mass is covered with primitive (visceral) endoderm, and parietal endoderm is beginning to line the mural trophectoderm, which encloses the blastocyst cavity. **(B)** At five days, the ICM portion of the blastocyst is covered with visceral endoderm. Both an embryonic epiblast and a more proximal proamnion are forming from the ICM. **(C)** By day 6, the ICM has formed a cylinder with a proamniotic cavity. Shortly after this, the epiblast undergoes gastrulation.

originally secreted by follicle cells that surrounded the oocyte in the ovary. The trophoblast over the ICM region (called the *polar trophoblast*) adheres to the highly vascularized, mucus-secreting cells of the uterus. This process of **implantation** marks the beginning of placenta formation.

The *blastocyst cavity* is lined with extraembryonic endoderm (*parietal endoderm*), and the ICM is covered with endoderm as well (called *visceral endoderm*). A cavity forms within the ICM that will become the amniotic cavity. Programmed cell death and selective survival of ICM cells are thought to contribute to the formation of this proamniotic cavity. (Programmed cell death will be discussed in more detail in the next chapter.) The visceral endoderm apparently emits crucial signaling ligands (BMP2 and BMP4) that initiate this process. (BMP stands for bone morphogenetic proteins, so named because they were first discovered by their effects on bone formation.) The anatomical terminology can become a source of confusion when following the course of early mammalian development; referring to Figures 5.13 and 5.14 may help you clarify terms.

Signaling Pathways May Be Elucidated by the Use of Dominant Negatives

How do we know that visceral endoderm signals ICM cells to carry out proamniotic cavity formation? Because of its small size, the embryo itself is difficult to use for such studies. However, there are lines of tissue culture cells derived from embryos that form "embryoid bodies" in culture, and these will undergo cavity formation, in a manner similar to the process observed in normal embryos. These virtual embryos are excel-lent material for analyzing how tissues in an embryo communicate with one another. Gail Martin and her colleagues at the University of California at San Francisco have employed this model system. They applied a rather commonly used experimental strategy, borrowed from yeast genetics, called *dominant negatives*. This strategy has been especially useful for identifying the functions of ligands and receptors used by the embryo during interactions between tissues.

The rationale for the experiment goes like this: If a particular protein must interact with other proteins in order to carry out its function, then that function might be compromised by flooding the system with a mutated, counterfeit version of that protein. This protein would need to have been altered so that it could still interact with its normal partners but would not function even though paired with its partner. For example, suppose a receptor serine-threonine kinase protein were mutated so that it lacked its normal kinase domain (essential for signaling) but was still able to insert into the membrane and interact with its ligand, ATP. And suppose the system were exposed to an excess of the mutated protein. Then this mutated protein would dominate the system: So much of the mutated counterfeit would be present that it would, by simple mass action, populate most of the available sites in the membrane. This would compromise the function of that receptor even if the system were making normal versions of the receptor protein. Martin and her associates showed that BMP2/4 could promote proamnion cavity formation. When the tissue culture cells used in the experiment were transformed so that the BMP2/4 receptor was prevented from functioning by a dominant negative version of the receptor, then cavity formation did not occur.

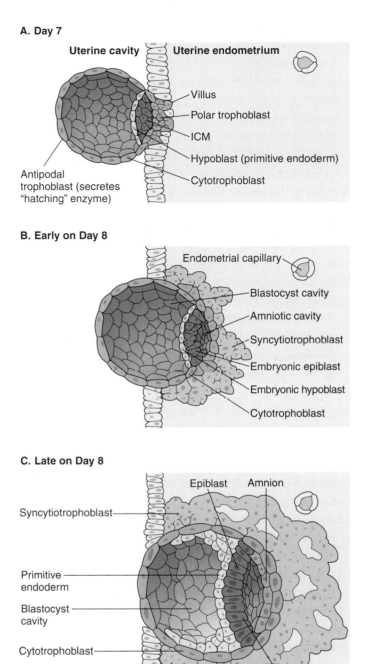

A. Day 7

Uterine cavity   Uterine endometrium

Villus

Polar trophoblast

ICM

Hypoblast (primitive endoderm)

Cytotrophoblast

Antipodal trophoblast (secretes "hatching" enzyme)

B. Early on Day 8

Endometrial capillary

Blastocyst cavity

Amniotic cavity

Syncytiotrophoblast

Embryonic epiblast

Embryonic hypoblast

Cytotrophoblast

C. Late on Day 8

Epiblast   Amnion

Syncytiotrophoblast

Primitive endoderm

Blastocyst cavity

Cytotrophoblast

Blastodisc

Figure 5.14   The Human Blastocyst   (A) A seven-day-old human embryo in the process of implantation. The polar trophoblast cells are invading the uterine endometrium. forming chorionic villi. **(B)** Early on the eighth day. the trophoblast has proliferated even more and is composed of both a cellular and an outer syncytial portion. The amniotic cavity is forming in the ICM. **(C)** Late on the eighth day. the amniotic cavity is well formed. the embryo is completely implanted. and the epiblast and primitive endoderm are apparent.

The Formation of the ICM Is a Strategy Unique to Mammalian Embryos

Before proceeding, let us pause to emphasize the crucial nature of the events that occur during compaction and shortly thereafter, from the fourth to the sixth cleavage division. All the cells of the eight-cell embryo are able in different experimental situations to give rise to any and all cells of the embryo, from placenta to neuron. After this stage, irrevocable changes take place depending on whether a cell comes to lie to the outside (where it will become a polarized epithelial cell, which will form trophoblast), or to the inside (where it will become a nonpolarized stem cell, able to give rise to any tissue of the embryo proper.) Even though each blastomere at the 16-cell stage is able to give rise to all embryonic and extraembryonic tissues, the question still remains whether the early blastomeres are indistinguishable. Recent very careful fate mapping studies by Richard Gardner in Oxford, England, and Karolina Piotrowska and her colleagues in Cambridge, England, have shown that one of the first two cells gives rise to the upper portion of the ICM and polar trophectoderm, while the other cell contributes to mural troophectoderm and primitive endoderm. Hence, the axis of the blastocyst running orthogonal to the plane of the ICM is approximately orthogonal to the plane of the first mitotic spindle.

The precise mechanisms at work in the determination of the ICM are not known, but the driving force underlying this determination is probably not a maternally localized mRNA or protein, a gravitational influence, or a mechanism involving sperm. We do know cellular components associated with cell polarity reflect this crucial inside-outside difference as soon as the asymmetric divisions take place. For example, sodium pumps become localized basolaterally, and ezrin, a microfilament-associated protein present in microvilli, is found apically. Recent work with *Drosophila* and yeast has shown how a nonpolarized cell may distribute some components to only one of its two daughter cells and other components to the other, thereby creating a nonequivalent set of sibling cells. These findings may point the way to similar research with mammals. We shall return to this subject in Chapter 15.

The Mouse Embryo Possesses an Unusual Morphology

Though the mammalian egg displays little or no asymmetry, the blastocyst is highly asymmetric. The ICM marks one side, and the blastocyst cavity is lined with ectodermal polar trophoblast (*polar trophectoderm* for short) adjoining the ICM and mural trophectoderm elsewhere. The primitive endoderm covering the ICM proliferates to form parietal endoderm adjoining mural trophectoderm and visceral endoderm covering the ICM.

Cavitation and proliferation in the embryos of the mouse and a few other mammals give the blastoderm a cylindrical shape. However, in most mammals, including humans, the blastoderm (blastodisc) maintains a more planar arrangement. Hence, in the "typical" mammal, the epiblast and visceral endoderm are much like the flattened chick blastoderm, except that there is no yolk as such in mammals. The so-called yolk sac formed by mammalian visceral endoderm is a fluid-filled bag. In the mouse, the shape of the blastoderm looks as if one had inserted a miniature thumb and pushed the epiblast down in its center, creating a kind of cylinder. The morphology of the human embryo is shown in Figure 5.14. The gastrulation movements that a mouse and human embryo will undergo, as far as we currently know, are similar, and approximate the movements discussed earlier regarding the chick embryo.

The Epiblast Is the Source of Embryonic Germ Layers

While the blastocyst as a whole expands and the trophoblast cells divide and differentiate into placental tissue, the inner cell mass generates a layer of cells over the epiblast layer (see Figure 5.14). This layer becomes the amnion surrounding the developing embryo. The epiblast proper, while undergoing continuous cell division, forms a primitive streak, which possesses an anterior node region analogous to Hensen's node in birds: when a node region is transplanted from a mammalian embryo to the area pellucida of a chick, a secondary axis forms in the chick blastoderm.

Furthermore, transplanting the node region of one mouse embryo to a lateral region of another synchronously developing mouse embryo results in a secondary neural axis in the host. However, anterior brain regions are missing in such a secondary axis. This and other observations support the idea that in the mouse, and perhaps in other mammals, there is a second "head" organizer in the anterior visceral endoderm that influences or induces anterior epiblast to form a head. This is currently a rapidly developing research area; genes that may be involved (*hex, cerberus*) have been identified and are being studied intensively. Another gene, *arkadia*, which is expressed in anterior visceral endoderm, is necessary for the node to form. Furthermore, anterior visceral endoderm is believed to secrete antagonists (Frizzled, Dickkopf) of Wnt ligands, and this antagonism is essential for head formation. In addition, the definitive primitive endoderm, which arises from the move-

ment of cells through the node and primitive streak, also secretes Wnt antagoonists (e.g., *cerberus*), which are necessary for heart development. The chicken homolog of the mouse *hex* gene has been isolated too. Does this mean that birds, or even all the amniotes, have an anterior, head-inducing center in hypoblast or endoderm that is distinct from the node?

Extensive cell-marking experiments by Roger Pedersen, Claudio Stern, and others have shown that the general strategy of gastrulation movements in the mouse and other mammals is similar to that encountered in the chick. Epiblast cells in the posterior epiblast move toward a midline and form a primitive streak. Cells move through the streak and node, establishing the definitive ectoderm, mesoderm, and endoderm in roughly the same way as in the chick. Soon the node regresses, the primitive streak disappears, the definitive germ layers form, and the developing neural tube and somites are evident.

Even though any portion of the zygote may give rise to either the ICM or extraembryonic tissues, this does not mean that zygote organization has no influence whatsoever on the organization of the embryo. Recent cell-marking experiments indicate that cells inheriting cytoplasm from the area near the polar body are much more likely to form the most distal portion of the egg cylinder; this is the region in which the node will eventually appear. Hence, if this recent work on mice is any indication, even the mammalian embryo may be influenced to some extent by the organization of the egg.

MAMMALIAN ADAPTATIONS

At the End of Gastrulation, the Morphology of Mammalian Embryos Is Similar to that of Birds, but the Extraembryonic Membranes Differ

By the end of gastrulation, the embryos of mice (despite their contorted epiblast), humans, and other mammals possess the basic ground plan of the archetypal vertebrate embryo. The epiblast has been sculpted by gastrulation and node regression to generate a midline area where neural tube formation will take place, and lateral areas where epidermis will form. There is somatic and splanchnic mesoderm, separated by a coelom, and the old hypoblast (visceral endoderm) layer has been pushed aside to become extraembryonic endoderm. Cells that ingressed through the primitive streak have formed a coherent layer over the blastocyst cavity, eventually forming the embryonic endoderm, which becomes the gut and associated organs.

The internal cell mass has also generated an umbrella of cells over the epiblast, which form the amniotic membrane enclosing the developing embryo. This membrane secretes amniotic fluid, in which the embryo is bathed. Occasional cells shed from the amnion float in this fluid; a fairly routine obstetrical procedure called *amniocentesis* involves carefully withdrawing amniotic fluid by use of a long syringe needle. The cells in the fluid can be stained to determine whether the chromosomal constitution of the fetus is XX (female) or XY (male).

Some mesoderm near the developing posterior end of the embryo remains connected to the trophoblast, which creates an isthmus through which the blood vessels of the embryo vascularize the embryonic portion of the placenta. This mesodermal isthmus and its blood vessels are homologous to the splanchnic mesoderm of the allantois in birds and rep tiles. Just as in birds, some mesoderm at the posterior end of the primitive streak in the mammalian embryo will become the tail bud, which will generate the tissue of the forming tail. The yolk sac generated by the ICM does not actually enclose any yolk since mammalian eggs do not contain substantial amounts of yolk; the name simply reflects its homology to avian yolk sacs.

The Trophoblast Will Help Form the Placenta

During this early period of growth, as the internal cell mass develops to form both the embryo and the amnion, the trophoblast cells proliferate rapidly and literally invade the mother's richly vascularized mesenchyme that underlies the uterine epithelium (Figures 5.14 and 5.15). The leading "invasive" trophoblast may form a multinucleated syncytium (*syncytiotrophoblast*), while trophoblast nearer the blastocyst forms true cells (*cytotrophoblast*). The developing placenta remains connected to the embryo proper by a mesodermal strand, the *body stalk*, which becomes the *umbilicus*. The formation of the placenta is complex, and the details of this process exceed the bounds of this book. However, we can make two important points here.

First, the degree of intimacy of the anatomical association between the embryonic and maternal portions of the placenta varies in different mammalian species. In the mouse and human, for example, the uterine epithelium eventually erodes, and the maternal capillaries open to form sinuses of blood that directly bathe the endothelial layers of capillaries of the embryo. In contrast, in the guinea pig, both the uterine epithelium and the surface of the trophoblast are retained, albeit very closely apposed, so that in addition to the two endothelial layers, two other epithelial layers are interposed between the two circulatory systems.

A. About 9 Days

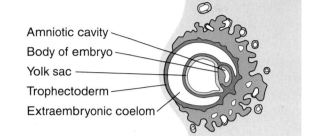

Amniotic cavity
Body of embryo
Yolk sac
Trophectoderm
Extraembryonic coelom

B. About 13 Days

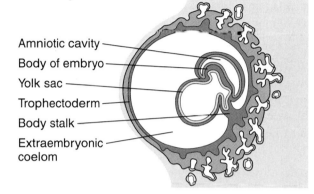

Amniotic cavity
Body of embryo
Yolk sac
Trophectoderm
Body stalk
Extraembryonic coelom

C. About 21 Days

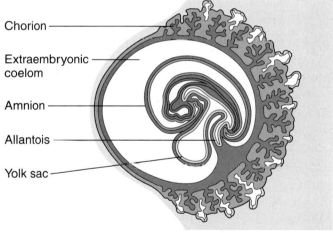

Chorion
Extraembryonic coelom
Amnion
Allantois
Yolk sac

Figure 5.15 The Placenta The developing human embryo and placenta are shown. The fetal portion of the placenta, with its projecting villi, surrounds the extraembryonic coelom (formerly the blastocyst cavity), which in turn contains the fetus, amnion, and yolk sac. **(A)** By eight or nine days after fertilization, the embryo has already implanted and the amniotic cavity is in place. **(B)** By about 13 days, gastrulation is in progress, germ layers have formed, while a stalk of mesoderm—the body stalk—remains connected to the trophoblast. **(C)** By about three weeks, blood vessels have grown through this body stalk to form the fetal blood vessels of the placenta. The body stalk becomes the umbilical cord.

Second, the placenta, which is composed of both fetal and maternal tissues, is a unique and, as mentioned earlier, incredibly important organ. Nutrition, respiration, and excretion of the embryo take place across the epithelia separating the maternal and embryonic circulations. In addition, the maturing placenta serves the same function as did the pituitary and ovary prior to pregnancy—becoming a factory for the production of hormones (chorionic gonadotropins and steroids).

The placenta also performs important immunological functions for both mother and fetus: It prevents the mother's body from mounting an immune reaction against the tissues of the fetus—though exactly how the placenta accomplishes this is still not understood. Since the fetus has a different genetic constitution than the mother (due to the paternal genes that are part of its makeup), the maternal immunological system could initiate an immune reaction against the fetus if this "barrier" did not exist. Furthermore, antibodies present in the maternal circulation are admitted to the embryonic circulation during the latter part of development. Hence, the newborn mammal has some acquired immunity against a variety of common pathogens, which serves it well until its own immune system is more fully developed.

MANIPULATING MOUSE EMBRYOS

Allophenic Mice Provide an Opportunity to Understand the Cellular Dynamics of the Early Embryo

You will recall that at the eight- and 16-cell stages, most of the cells of the embryo remain toward the outside to form the trophoblast, leaving a few inner cells that will eventually generate the embryo. How many cells from these early stages will form the later epiblast? An approach to answering this question was pioneered by Beatriz Mintz working in Philadelphia. The technique involves removing the zona pellucida membrane that encloses these early cleavage stages, a task accomplished by digesting the membrane with proteolytic enzymes. Two cleavage-stage embryos from different mothers, each possessing a different genotype, are then placed next to each other in a culture dish. The embryos are sticky, so they adhere; gradually the cells form a composite embryo called a *chimera* (see Chapter 3). This chimera is then implanted in the uterus of a surrogate mother mouse primed by hormones to allow implantation (a so-called pseudopregnant mouse), in which the chimera will develop and from which it will be born (Figure 5.16).

Suppose many such chimeric embryos are created, each by fusing a blastocyst from a black strain with a blastocyst from a white strain. If only one cell gives rise to the embryo proper, then each of the created embryos will be either all-white or all-black. If two cells in the embryo are required, then, given the laws of chance, we would expect about a quarter of the chimeras to be all-black, a quarter all-white, and half some kind of mixture. And if three cells are required, we can further calculate that the probability of mixed chimeras resulting would be about 75% of the total number—very close to the result that Mintz obtained (73%).

The mice arising from cells derived from more than one embryo are called *allophenic mice*; they have been very useful in probing the cellular dynamics of organogenesis in mammals, a subject we shall explore in chapters to come.

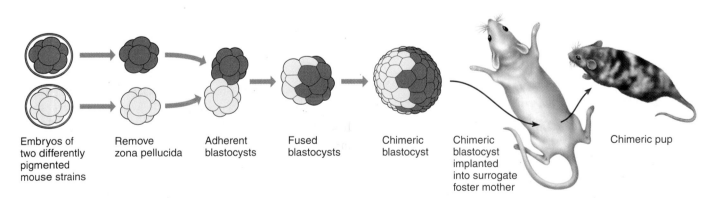

Embryos of two differently pigmented mouse strains → Remove zona pellucida → Adherent blastocysts → Fused blastocysts → Chimeric blastocyst → Chimeric blastocyst implanted into surrogate foster mother → Chimeric pup

Figure 5.16 The Allophenic Mouse A chimera composed of cells from two distinct mouse embryos may be formed by removing the zona pellucida and allowing the sticky embryos to adhere. The resulting chimeric blastocyst is then implanted in a pseudopregnant surrogate mother. When the source of embryos in a chimera is two differently pigmented strains—one white and one black—an allophenic mouse with both black and white coat colors may result.

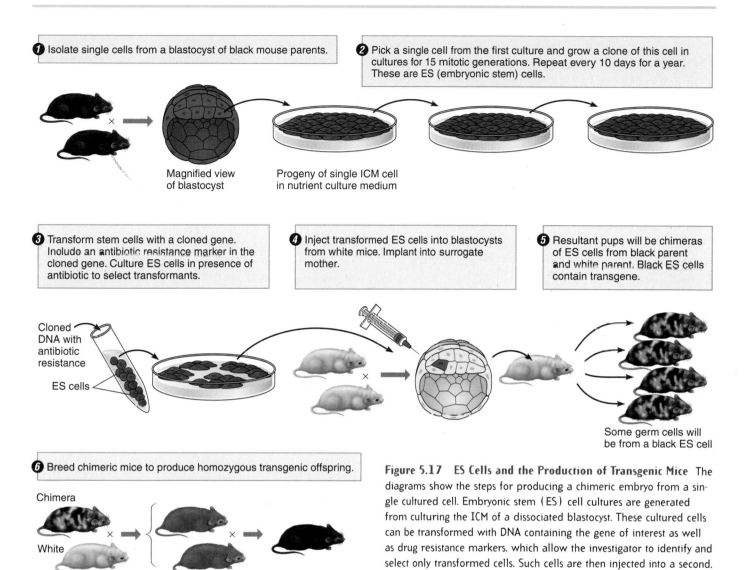

① Isolate single cells from a blastocyst of black mouse parents.

② Pick a single cell from the first culture and grow a clone of this cell in cultures for 15 mitotic generations. Repeat every 10 days for a year. These are ES (embryonic stem) cells.

Magnified view of blastocyst

Progeny of single ICM cell in nutrient culture medium

③ Transform stem cells with a cloned gene. Include an antibiotic resistance marker in the cloned gene. Culture ES cells in presence of antibiotic to select transformants.

④ Inject transformed ES cells into blastocysts from white mice. Implant into surrogate mother.

⑤ Resultant pups will be chimeras of ES cells from black parent and white parent. Black ES cells contain transgene.

Cloned DNA with antibiotic resistance

ES cells

Some germ cells will be from a black ES cell

⑥ Breed chimeric mice to produce homozygous transgenic offspring.

Chimera

White

Parents   **Heterozygous F₁**   **Homozygous strain from ES cell, F₂**

Figure 5.17   ES Cells and the Production of Transgenic Mice The diagrams show the steps for producing a chimeric embryo from a single cultured cell. Embryonic stem (ES) cell cultures are generated from culturing the ICM of a dissociated blastocyst. These cultured cells can be transformed with DNA containing the gene of interest as well as drug resistance markers, which allow the investigator to identify and select only transformed cells. Such cells are then injected into a second, host blastocyst, which is then placed in the uterus of a pseudopregnant surrogate mother. The surviving chimeric embryos will be composed of a mixture of cultured, transformed ES cells and normal embryonic cells. If any germ cells in the chimera are derived from cultured ES cells, the ES genes may be propagated by sexual reproduction.

Cells Injected into the Blastocyst Cavity Can Also Form Chimeras, Which is the Basis for Creating Transgenic Mice

It is not necessary to fuse cleavage-stage embryos to generate a chimera. At later blastocyst stages, groups of cells from the ICM may be removed and dissociated to single cells through mild digestion with proteolytic enzymes such as trypsin. The liberated cells may then be injected into the blastocyst cavity of a recipient host. The injected cells can adhere to the host's ICM, become integrated into it, and give rise to many parts of

the embryo. This has proved to be an incredibly important tool in the analysis of mammalian development; the ability to do this with mouse embryos is one reason the mouse has become the most important embryo for research on mammalian development.

The cell injected into the blastocyst need not be from a freshly dissociated donor embryo. Procedures have been devised for cultivating ICM cells in vitro. These cells are called **ES (embryonic stem) cells.** When ES cells are injected into a blastocyst, they are incorporated into it and give rise to differentiated tissues in the newborn. Figure 5.17 shows the procedure for producing chimeric embryos that are partially composed of tissues derived from ES cells.

There is a "wrinkle" to this technology that makes it incredibly powerful: The ES cells in culture can be tricked into taking up foreign DNA, so that new genes can be inserted into the ES cells, which are now said to harbor *transgenes*. Hence, it is possible to create **transgenic mice,** mice whose tissues contain some foreign genes. If any such ES cell provides progenitors for sperm or egg cells in the chimeric mice (and ES cells do in a reasonable percentage of chimeras), then the transgene may be passed to offspring; brother-sister matings of those offspring may in turn produce mice homozygous for the transgene; in other words, a genetically stable transgenic mouse will have been created.

It is even possible, in a procedure called a *knockout*, to incorporate a transgene that completely eliminates the function of the "normal" gene present in the ES cell before the transgene was introduced. This powerful experiment is rapidly making it possible to diagnose the role of any particular gene in mouse development. For example, the gene *nodal* encodes a protein that is present at high levels in the region of the mouse embryo analogous to Hensen's node. Is this Nodal protein important for the function of the node? This can be tested by establishing a homozygous mouse without the normal *nodal* gene, using the knockout approach outlined above. The result is that all the embryos in which both normal copies of *nodal* have been "knocked out" fail to undergo gastrulation and form almost no mesoderm. Thus *nodal* is an essential gene, and *nodal* mutations are lethal when homozygous. In later chapters, we shall see how important this experimental technique has become for understanding mammalian development.

KEY CONCEPTS

1. Early development in amniotes has some unique features. There is little or no reliance on localized maternal determinants for initiating differential cell behavior and gene expression. Rather, asymmetric divisions generate cells that come to inhabit two very different microenvironments.

2. Amniote species that have adopted terrestrial habits must rely on membranes to enclose the embryo and prevent its desiccation.

3. The development of viviparity (producing live young) in mammals leads to a unique organ, the placenta.

4. Despite the many unique features of amniotes, their development during the stages after gastrulation is similar to what occurs in other vertebrates.

STUDY QUESTIONS

1. The proposition that gastrulation involves movements of cells from the surface to the interior has been made in several places in this book. With respect to *Drosophila, Xenopus,* and the chick embryo, (a) which surface cells are actually interiorized, and (b) what germ layer and/or organ do those cells form?

2. Can you speculate on how the experiments that involve rotation of the chick epiblast with respect to the hypoblast bear on the issue of a possible separate anterior organizing center in birds?

3. Chimeric mouse embryos may be exploited to analyze the role of various signals between the embryonic ectoderm (epiblast) and the extraembryonic ectoderm. Cells from a blastocyst will, of course, form both kinds of ectoderm, but ES cells will contribute only to embryonic ectoderm. There are suggestions in the literature that germ cell development in the mouse may require a signal of BMP4 from extraembryonic ectoderm. Can you devise an experiment using chimeras to evaluate this suggestion?

4. What type of cell junctions would be formed between prospective trophectoderm cells during compaction? What role(s) might these cell junctions play in establishing different microenvironments around forming-ICM cells as distinguished from prospective trophectoderm cells?

SELECTED REFERENCES

Belaoussoff, M., Farrington, S. M., and Baron, M. H. 1998. Hematopoietic induction and respecification of A-P identity by visceral endoderm signaling in the mouse embryo. *Development* 125:5009–5018.

An experimental analysis of how endoderm may induce the formation of blood cells in the mesoderm.

Belo, J. A., Vouwmeester, T., Leyns, L., Kertesz, N., Gallo, M., Follettie, M., and DeRobertis, E. 1997. Cerberus-like is a secreted factor with neuralizing activity expressed in the anterior primitive endoderm of the mouse gastrula. *Mech. Dev.* 68:45–57.

A report of how the mouse homolog of the amphibian *cerberus* gene was isolated, and how this gene may work as a head inducer.

Catala, M., Teillet, M.-A., DeRobertis, E. M., and Le Douarin, N. M. 1996. A spinal cord fate map in the avian embryo. *Development* 122:2599–2610.

Grafts from quail to chick embryos were used to map the fate of cells in the spinal cord.

Coucouvanis, E., and Martin, G. R. 1999. BMP signaling plays a role in visceral endoderm differentiation and cavitation in the early mouse embryo. *Development* 126:535–546.

An experimental analysis involving ES cells to analyze the mechanisms of amniotic cavity formation.

Gardner, R. L. 2001. Specification of embryonic axes begins before cleavage in normal mouse development. *Development* 128:839–847.

Recent research showing that the early mouse egg is organized.

Gosden, R., Krapez, J., and Briggs, D. 1997. Growth and development of the mammalian oocyte. *BioEssays* 19:875–882.

A review of mammalian oogenesis.

Hamilton, H. L. 1965. *Lillie's development of the chick*. Holt, Rinehart & Winston, New York.

A classic description of chick embryology.

Hatada, Y., and Stern, C. D. 1994. A fate map of the epiblast of the early chick embryo. *Development* 120:2879–2889.

Experimental analysis of the fate of different portions of the epiblast of the early chick embryo.

Khaner, O. 1993. Axis determination in the avian embryo. *Curr. Topics Dev. Biol.* 28:155–180.

A review of the experimental embryology showing how the chick axis is determined.

Khaner, O. 1995. The rotated hypoblast of the chick embryo does not initiate an ectopic axis in the epiblast. *Proc. Natl. Acad. Sci. USA* 92:10733–10737.

A research paper that revised our understanding of hypoblast–epiblast interactions.

Kochav, S., and Eyal-Giladi, H. 1971. Bilateral symmetry in chick embryo determination by gravity. *Science* 171:1027–1029.

The demonstration that gravity affects axis formation.

Lawson, K. A., Dunn, N. R., Roelen, B., Zeinstra, L. M., David, A. M., Wright, C., Korving, J., and Hogan, B. L. M. 1999. BMP4 is required for the generation of primordial germ cells in the mouse embryo. *Genes Dev.* 13:424–436.

An experiment, using chimeras with ES cells, that showed the influences of extraembryonic ectoderm on the epiblast of the embryo proper.

Lemaire, L., and Kessel, M. 1997. Gastrulation and homeobox genes in chick embryos. *Mech. Dev.* 67:3–16.

A review of chick gastrulation.

Pera, E., Stein, S., and Kessel, M. 1998. Ectodermal patterning in the avian embryo: Epidermis versus neural plate. *Development* 126:63–73.

A research paper on how Hensen's node interacts with surrounding tissues.

Piotrowska, C., Wianny, F., Pedersen, R. A., and Zernicak-Goetz, M. 2001. Blastomeres arising from the first cleavage division have distinguishable fates in normal mouse development. *Development* 128:3739–3748.

Fate mapping studies showing that the first two blastomeres of the mouse embryo are not equivalent.

Psychoyos, D., and Stern, C. D. 1996a. Fates and migratory routes of primitive streak cells in the chick embryo. *Development* 122:1523–1534.

A cell-marking study that followed the entrance and egress of cells from the streak.

Psychoyos, D., and Stern, C. D. 1996b. Restoration of the organizer after radical ablation of Hensen's node and the anterior primitive streak in the chick embryo. *Development* 122:3263–3273.

A study showing that the node and streak may be regenerated after their removal.

Stern, C. D. 1990. The marginal zone and its contribution to the hypoblast and primitive streak of the chick embryo. *Development* 109:667–682.

Analysis by experimental embryology of the role of the marginal zone in chick embryos.

Tam, P. P. L., and Behringer, R. R. 1997. Mouse gastrulation: The formation of a mammalian body plan. *Mech. Dev.* 68:3–25.

A thorough, recent review of the subject.

Weber, R. J., Pedersen, R. A., Wianny, F., Evans, M. J., and Zernicka-Gopetz, M. 1999. Polarity of the mouse embryo is anticipated before implantation. *Development* 126:5591–5598.

Cell-marking experiments show that there is a bias in the zygote as to which cytoplasmic regions will form the anterior or posterior portion of the epiblast.

Yamaguchi, T. P. 2001. Heads or tails: Wnts and anterior-posterior patterning. *Current Biology* 11:R713–R724.

A review of the role of Wnt antagonists in head and anterior mesoderm formation.

Yuan, S., and Schoenwolf, G. C. 1998. De novo induction of the organizer and formation of the primitive streak in an experimental model of notochord reconstitution in avian embryos. *Development* 125:201–213.

A research paper with new findings on the mechanisms by which the organizer or organizers are formed in the chick embryo.

PART THREE

VERTEBRATE

ORGANOGENESIS

CHAPTER

6

THE DEVELOPMENT OF ECTODERMAL DERIVATIVES IN VERTEBRATES

So far, we have concentrated on how the activated egg is transformed into an organized array of layered primitive tissues, the germ layers. Though we shall have occasion in later chapters to refer to different differentiated tissues and organs in several types of invertebrates, including *Drosophila*, it is cumbersome and confusing at this point to launch into a comparative discussion of how organs are formed in different animal groups. Almost all the important principles in organogenesis are evident when considering the vertebrate case, even if some of the experiments that defined these principles were first done using

CHAPTER PREVIEW

1. In vertebrates, ectoderm forms a neural tube, which gives rise to the central nervous system (CNS).

2. Neuron number in different regions of the CNS is regulated by the number of connections nerves make with their peripheral targets.

3. The neural tube has a complex, segmental organization.

4. The eyes arise from the forebrain region, and eye formation is initiated by the action of "master" genes.

5. Ectoderm also forms the neural crest, a migratory cell population that contributes to many different tissues.

6. Differentiation of the neural crest is controlled by the final location of its migrating cells.

7. Ectoderm also forms various nerve organs and the epidermis.

8. Adult stem cells can differentiate similarly to embryonic "blast" cells.

invertebrates. Therefore, we choose to focus most of the chapter on the story of organ formation in vertebrates.

THE NEURAL PLATE

The Ectoderm Is the Source of the Nervous System and the Skin

You may recall that frogs, chicks, and mammalian embryos adopt a similar "ground plan" as a result of the maneuvers of gastrulation. Cells remaining on the surface of the embryo after gastrulation—in other words, the ectoderm—form an epithelial covering, usually only one or a few cell layers thick. Figure 6.1 presents a summary view of the organs that form from the ectoderm, and also hints at the complexity of interactions between ectoderm and the adjacent mesoderm. The entire nervous system originates from ectoderm, as does the pituitary gland. Many other structures of the integument—including skin, hair, feathers, and sweat and sebaceous glands—have an important ectodermal contribution but are compound organs, composed of both ectoderm and mesoderm. We shall come to see that all these structures—indeed, probably every organ and tissue in the vertebrate body, whatever the germ layer(s) involved—depend on interactions between the different cell types that compose their neighborhoods of origin.

It is also useful to recall the distinction drawn in Chapter 3 between epithelium and mesenchyme, a classification completely separate from that of germ layers. Ectoderm forms nerve tissue, epithelia, and mesenchyme. Mesoderm and endoderm form both epithelial and mesenchymal types of tissues. *Germ layer* refers to embryonic origin, while *tissue type* refers to the type of cells and their architecture within a given organ.

The Neural Plate Results from the Induction of Ectoderm

The Spemann organizer evokes (induces) the formation of the nervous system; this is sometimes called *primary induction* because researchers learned subsequent to the organizer's discovery that a myriad of other inductions are important for the formation of all organs.

What is the organizer, and how does it induce and organize? The answer to the first question is easy. Recall from Chapter 4 that the Spemann organizer is the group of cells that constitutes the dorsal lip of the blastopore. The dorsal lip is a constantly changing population of cells; surface and adjacent cells involute (in amphibians) or ingress (in amniotes) into the interior. The cells of the dorsal lip become the midline of the anterior mesoderm (notochord) and anterior endoderm (the lining of the pharynx).

Just how this transient, moving population of cells induces the ectoderm to adopt a prospective neural fate is an altogether more difficult question. The search for the organizer substance or substances occupied decades of intensive research resulting in many volumes of research papers, and

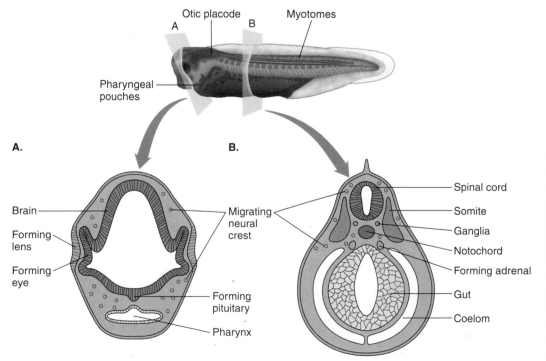

A.

Otic placode
Myotomes
Pharyngeal pouches

Brain
Forming lens
Forming eye
Migrating neural crest
Forming pituitary
Pharynx

B.

Spinal cord
Somite
Ganglia
Notochord
Forming adrenal
Gut
Coelom

Figure 6.1 Derivatives of the Ectoderm A young amphibian larva is shown in side view; the lines indicate the planes of cross sections A and B. **(A)** A section through the head region, slanting downward and posteriorly, showing some ectodermal derivatives and the pharynx. **(B)** A section through the trunk region. Head structures are lacking, but spinal cord, somites, and autonomic ganglia are shown.

although some important characterizations of the inductive process were made, the actual identification of active principles came to naught. Only in the last few years has important progress been made. We have the benefit of hindsight now to appreciate why it was so difficult and took so long. Spemann and Mangold's results were originally taken to mean that the organizer must somehow stimulate cells to become neural; without that stimulation, the cells would just form epidermis. When researchers attempted to isolate the stimulating substance, many different kinds of materials seemed to be candidates. Tests for activity of the putative inducing substances were often carried out by placing an extract of some tissue, such as mammalian liver, into a wound made in the embryo; such embryos were allowed to develop and then examined for the presence of induced tissues. Other tests were carried out by isolating prospective epidermis from a late-blastula or early-gastrula embryo, and then placing this tissue into an in vitro organ culture in a simple saline medium. The candidate material was then added, and days later the explanted tissue was examined for the presence of neurons, which were identified by their morphology and staining characteristics.

The interpretation of the results of such experiments quickly became confusing. So many different substances or events seemed capable of causing induction. And what was considered a real example of nerve tissue often varied in different experiments. Even treating the prospective ectoderm of some creatures, such as the salamanders originally used by Mangold and Spemann, by altering the pH or potassium ion concentration could lead to neuron formation. Experiments in the late 1980s showed that even the process of dissociating the uninduced, prospective ectoderm to single cells and then reaggregating them a few hours later into a coherent tissue could lead to neuron formation. (For a discussion of the dissociation of tissues to single cells and their reaggregation into tissues, see Chapter 12.)

There are two retrospective lessons here. First, biochemical searches for biologically active substances require a simple, "clean" assay. The fact that investigators encountered such a bewildering array of potentially active materials, including changes in pH, is a tip-off that the assay being used was a problem. Second, it is important to assess the results from an appropriate viewpoint. Only recently have researchers come to appreciate that perhaps the active material(s) of the Spemann organizer were not stimulating the formation of neurons, but preventing the cells exposed to it from becoming epidermis. In current jargon, the "default state" of the ectoderm is neural, not epidermal, and what the organizer does is to counteract the tendency to become epidermis. Strong experimental evidence,

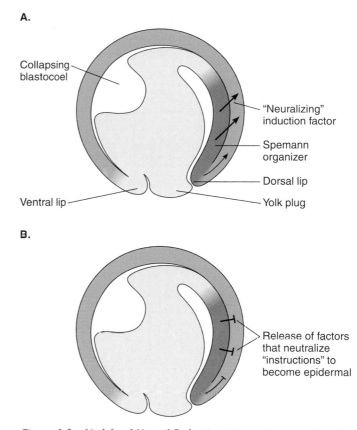

Figure 6.2 Models of Neural Induction A sagittal section through a midgastrula-stage amphibian embryo is shown diagrammatically to illustrate two different models of neural induction. **(A)** The "positive" induction model of the Spemann organizer: a factor or factors, released by the dorsal lip of the blastopore and dorsal mesoderm, induce the overlying ectoderm to form neural tissue of various kinds. **(B)** The "default" model: Signaling molecules from vegetal and lateral tissues would ordinarily cause ectoderm to differentiate along an epidermal pathway. However, a mixture of factors released by the Spemann organizer and dorsal mesoderm counteracts these signals, thereby allowing dorsal ectoderm to differentiate into neural tissue.

which we shall shortly consider, supports this more recent view. It is not uncommon for results of scientific investigations to be interpreted within an accepted framework that leads to a dead end, even though the original facts were mostly correct.

Figure 6.2 illustrates the two models for neural induction just discussed: In one, the organizer stimulates ectoderm, which would otherwise become epidermis, to become neural (Figure 6.2A). In the second model, the organizer counteracts signals that would cause ectoderm to form epidermis, so that its "default," or "ground state," as neural tissue can then be expressed (Figure 6.2.B). Another way to frame the distinction is to ask whether the organizer induces a neural response, or issues commands to stop epidermal development.

Another important aspect in our understanding of the Spemann organizer is that methods of identifying a neuron have advanced greatly. This is mainly due to the power of molecular biology for identifying genes active in, and essential for, neuronal development. Gene activity is often assessed indirectly by using antibodies made against the proteins encoded by these genes, which provides a simple and powerful means of identifying neural phenotypes.

Figure 6.3 outlines a current view of how the Spemann organizer directly influences ectoderm to adopt a neural pathway of development. The organizer is a source of several secreted molecules. The principal ones so far identified are Noggin and Chordin; there are thought to be others as well. The secreted factors as a group are sometimes referred to as a "cocktail." Noggin and Chordin can and do interact with members of the family of growth factors called BMPs (see Chapter 5). BMPs are secreted from cells in the prospective marginal and ventral portion of the embryo. These ligands diffuse from their source, reach the cells in the animal hemisphere, and form a concentration gradient at the gastrula stage. The BMPs—especially BMP4, and to a lesser extent BMP2 and BMP7—interact with

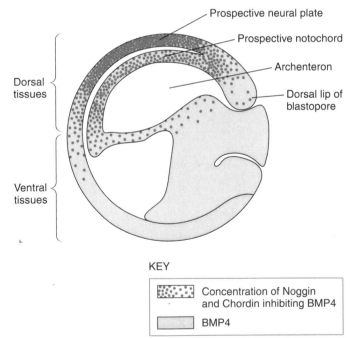

KEY

▒	Concentration of Noggin and Chordin inhibiting BMP4
☐	BMP4

Figure 6.3   Antagonism of BMP4 by Chordin and Noggin from the Spemann Organizer A sagittal section through an amphibian embryo at the late-gastrula stage. While lateral and ventral tissues are releasing BMP4, which drives ectoderm to become epidermis, the notochord is producing and releasing antagonists of BMP4, namely Chordin and Noggin. The antagonism of BMP is most pronounced in the dorsal midline; hence, this is where neural differentiation can proceed.

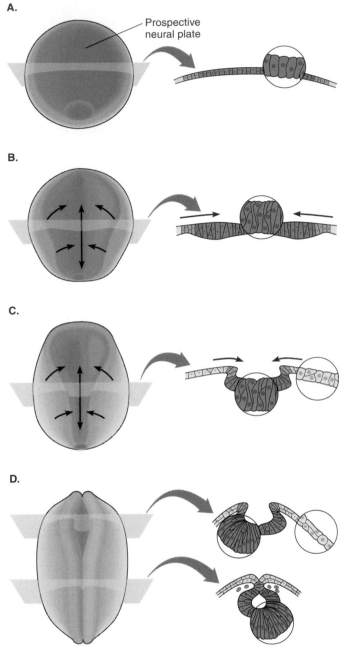

Figure 6.4   The Palisade Reaction and Convergent Extension of the Neural Plate *Left:* Dorsal surface views of a succession of stages of neural plate formation. Arrows indicate the direction of tissue movements as the plate extends in the anteroposterior direction, and the neural folds bend medially to form a tube. *Right:* Cross sections through the forming neural tube show how the cells change from cuboidal to columnar and form bending regions. This cell shape change occurs in the neural plate but not in the adjoining epithelium. **(A)** Late gastrula; **(B–D)** early- to late-neurula stages. Neural crest cells are shown in black.

dimeric receptors in the cell membrane of ectoderm cells, and this interaction leads to the epidermal pathway of differentiation. The organizer tissue in the dorsal midline secretes a cocktail of "anti-BMPs"—proteins that prevent BMPs from acting on prospective **neural plate** cells. Therefore the dorsal midline, where concentrations of BMPs are low and of organizer anti-BMPs are high, becomes neural by default. In Chapters 15 and 16, we shall return for a closer look at how growth factors and receptors produce these and other fundamental changes in development.

At this point, we should raise some questions, since our understanding is not as neat and tidy as the foregoing might imply. For instance, how do the sharp borders between prospective epidermis and neural tube arise? How does the anteroposterior (that is, brain–spinal cord) organization of the neural plate arise from exposure to a single "cocktail" of factors antagonizing BMP signaling? Is the organizer the only source of factors, or are there others? Is all signaling direct, or are some "relay" mechanisms involved? Of these questions, we shall address the second and third in this chapter, and only touch on the first question (about sharp borders) and the last (about "relays") in later chapters, especially 15 and 16.

The Arising Neural Plate Is Organized

An immediate consequence of the action of the Spemann organizer, and of the notochord derived from it, is that the ectoderm cells of the prospective neural plate change shape. In particular, they elongate along the animal–vegetal axis in a *palisade* reaction—forming a raised "pavement" of columnar cells. Furthermore, the entire neural plate region lengthens as the cells in the neural plate rearrange themselves (see Figure 6.4). This rearrangement, called *convergent extension*, is similar to

the convergent extension discussed in earlier chapters, and will be discussed further in Chapter 13. The elongation of the neural plate is also driven in part by lengthening of the notochord. Evidence for this comes from the result that when the notochord is surgically removed, the neural plate does not lengthen, nor does it adopt its characteristic pear-shaped outline (when viewed dorsally). While columnarization of the neural plate is nearing completion, the formation of the neural tube from this elongated, pear-shaped plate is beginning, as discussed in Chapter 4.

Anteroposterior differences are present in the cells of the neural plate even before neurulation is complete. Many experiments have been done in which various portions of the neural plate (with or without underlying mesoderm) were shifted from one position along the anteroposterior axis to another. Transplanted tissues have a marked tendency to differentiate according to the position from which they were taken (in other words, their usual fate) rather than their new surroundings. Recent experiments in which various molecular markers characteristic of a given region are visualized (for example, by antibody staining of homologs of *HOM* genes— genes that specify the identity, or type of differentiation, for *Drosophila* segments; see Chapter 15) lead to the same conclusion: regional specification of portions of the central nervous system along its anteroposterior axis has already begun by the time gastrulation ends.

Experiments carried out with amphibian embryos strongly support the idea that suppressing BMP signaling (primary induction) results in the formation of neuralized ectoderm with an anterior "character." The resulting neurons, their primitive tissue organization—even the molecular markers they express—are characteristic of the brain. A simple experiment (Figure 6.5) illustrates this early onset of

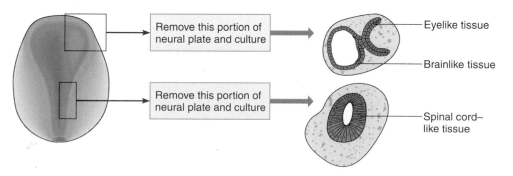

Figure 6.5 The Regional Character of the Neural Plate If anterior neural plate is explanted and cultured, it will form a vesicle of tissue containing neural structures with the form of brain and eye; molecular markers may be used to confirm the "anterior" character of the differentiation. In contrast, explants of the posterior neural plate form spinal cord-like structures and express the appropriate "posterior" molecular markers.

A. The basic situation

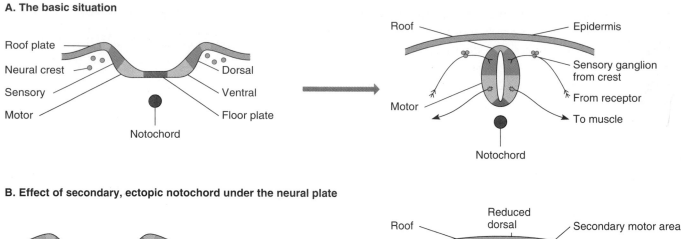

B. Effect of secondary, ectopic notochord under the neural plate

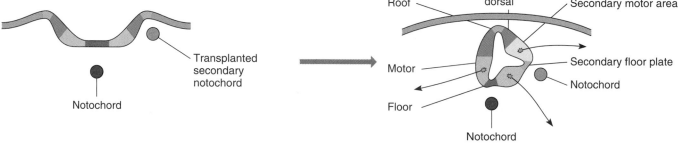

Figure 6.6   Dorsoventral Organization of the Neural Tube by the Notochord **(A)** The open neural plate, identifying the different regions of the spinal cord that will form after it becomes a tube. **(B)** When a piece of notochord from a donor embryo (colored) is inserted into an unusual position under the forming neural tube of the host embryo, it essentially induces an additional "ventral" region to form in the host's spinal cord.

anteroposterior organization. When anterior portions of the neural plate are explanted into organ culture medium, the resulting isolate forms recognizable eye tissue; in contrast, eye tissue never develops when the source of the explant is a more posterior portion of the neural plate.

Most investigators believe that something else besides suppression of BMP signaling needs to happen in order for hindbrain and spinal cord to form. There is debate about what that something is. Some contend that a second group of signaling molecules, originating from subjacent mesoderm and/or the dorsal lip of the blastopore, are necessary to "posteriorize" the result of primary induction. (FGF, working together with members of the Wnt family, and retinoids are the presently favored candidates.) Other workers in the field believe that the length of time and/or the intensity of exposure to the cocktail of substances coming from the developing mesoderm and Spemann organizer are sufficient to produce the hindbrain and spinal cord. It is probable that a variety of secondary intercellular signaling events are involved in these specifications; we shall examine them further in Chapter 16.

Dorsoventral Patterning of the Neural Plate and Tube Is Driven by Localized Release of the Growth Factor Sonic Hedgehog

The organization of the neural plate and neural tube also occurs early in development, at least as soon as the neural plate is evident, and there is some evidence that organization of the plate is under way during gastrulation itself. The key players are the notochord (which in amphibians, as you will recall, derives from the Spemann organizer), and the *floor plate*, the portion of the neural plate immediately above (dorsal to) the notochord.

Figure 6.6 shows an experiment in which a portion of the notochord is transplanted from one embryo into a second embryo, to a position lateral to its newly formed neural tube. This secondary notochord in turn induces a secondary floor plate, and a curious kind of a spinal cord—with two bottoms and a severely disturbed pattern—emerges. Many experiments of this kind have clearly identified the notochord and the adjacent floor plate area of the neural plate and tube as important organizing centers for the dorsoventral organization of the developing spinal cord. Some of the principal signaling molecules involved are even known.

Figure 6.7 outlines the important role of the signaling molecule Sonic Hedgehog (Shh) in establishing the dorsoventral organization of the developing neural tube. The *shh* gene encodes the vertebrate homolog of a signaling molecule important in segment and organ patterning in *Drosophila* (Hedgehog); (see Chapters 14 and 15). Shh is secreted as a precursor molecule, which then cleaves itself in two in an autocatalytic manner. The amino (NH₂) terminal portion of what was the precursor molecule is further modified when cholesterol is attached to its carboxyl (COOH) terminus, at which point it becomes an active ligand. (The amino and carboxyl ends of proteins are often referred to simply as the N-terminus and C-terminus.) It is believed that this cholesterification tends to tether the active Shh to the cell membrane, thus restricting its diffusion to some extent.

The precise way the Shh signal becomes active at a distance from its source is not yet completely understood. What is known is that the notochord and floor plate both secrete Shh at a time and place consistent with an important role for Shh. A substantial body of experimental evidence supports this conclusion: Blocking Shh activity (through the use of antibodies that neutralize Shh) prevents the CNS from developing its proper dorsoventral pattern. When cells that have been engineered to secrete Shh are implanted in locations that cause *shh* to be expressed ectopically, an ectopic dorsoventral pattern is organized, with ventral tissues closest to the additional source of Shh. Exposing cells close to the ventral midline to Shh results in the transcription of genes known to be expressed in the ventral portion of the spinal cord, such as *pax6*. Shh suppresses BMP4 activity in the ventral region of the neural tube. Motor neurons and ventral-type interneurons appear in the ventral neural tube after exposure to Shh. (Interneurons are neurons whose neurites lie entirely within the CNS.) There is good evidence that other factors, notably the BMP antagonist, Chordin, collaborate with Shh to establish the pattern of neural differentiation in the ventral spinal cord.

The influence of Shh extends only so far. It is possible that different concentrations of Shh favor different types of neurons in the ventral spinal cord; in other words, Shh may be acting like a morphogen (see Chapter 3). The dorsal portion of the spinal cord, arising from the lateral portions of the neural plate, which lie next to the surface ectoderm, is exposed to different influences. BMP expression is present in this adjacent nonneural ectoderm, and, as the roof of the neural tube forms, some BMP expression occurs there as well. The ectoderm overlying the dorsal spinal cord is also a

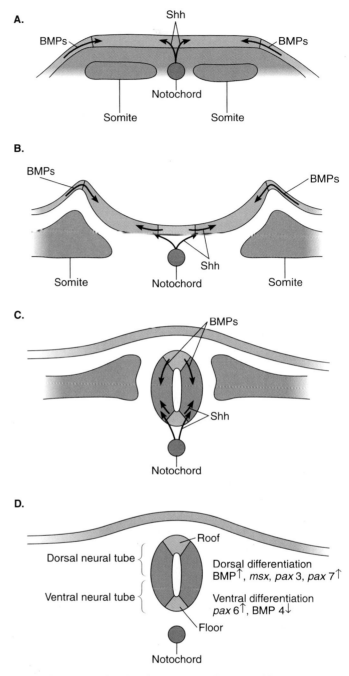

Figure 6.7 Notochord and Floor Plate Signaling Molecules
Diagrams of the developing neural tube show the site(s) of release and action of various signaling molecules during neural tube development. **(A)** The open neural plate is exposed to Shh from the notochord and to BMPs from the lateral tissues. **(B)** As the plate forms a tube, these influences continue; the prospective ventral neural tube (floor plate) also emits Shh. **(C)** The dorsal neural tube emits BMP signals, while the notochord and ventral spinal cord emit Shh. These opposing gradients help organize **(D)** the differentiation shown here.

source of BMP expression. Transcription factor genes such as *msx*, *pax3*, and *pax7*, which are repressed by Shh, are expressed in the dorsal neural tube. Commissural interneurons and dorsal-type interneurons develop from the dorsal neural tube, where Shh influence is low or absent. ("Commissural" describes neurons that send neurites across the midline; see Box 6.1.) Perhaps most important, the lateral neural plate, as it forms the dorsal portion of the neural tube, gives rise to a subpopulation of cells, first encountered in Chapter 4, called the *neural crest*. These cells, which we investigate in the following section, generate the primary sensory neurons that lie just outside the developing spinal cord.

THE NEURAL CREST

The Neural Crest Is a Multipotential Population of Migratory Cells

Neural crest arises from epithelial cells at the border between prospective epidermis and the neural plate. There is some evidence that the relative levels of BMP 4 and its antagonist, Noggin, are important for this initial emigration. At this early stage neural crest can be identified by characteristic expression of genes (*snail* in the mouse; *slug* in the chick) related to the *snail* gene of *Drosophila*; their expression is necessary for neural crest development.

The neural crest forms a very large number of different tissues, including pigment cells, several types of neurons and supporting cells, some endocrine tissues, and a wide array of mesenchymal tissues in the head (Table 6.1). Figure 6.8 illustrates the paths of migration taken by neural crest cells as they emerge from the spinal cord region; some cells migrate under the epidermis, forming pigment cells, while others migrate more ventrally, forming sensory and autonomic ganglia and the adrenal medulla (Figure 6.8A). The neural crest of the brain region contributes to many different sensory ganglia of the cranial nerves; it also forms cartilage, bones, teeth, and other mesenchymal tissues. Muscles in the head form from mesoderm.

In the trunk, neural crest cells migrate along and through the anterior portion of the adjacent somite, avoiding the posterior portion of the somite (Figure 6.8B). The molecular basis for this preference is not yet known, though recent evidence indicates that neural crest cells may be repelled by a molecule (ephrin) that is bound to the surface of cells in the posterior portion but not its anterior portion.

The behavior of the neural crest presents two fundamental and interrelated questions: First, how do neural crest cells come to migrate to their several specific locations? Perhaps they are guided by molecular pathways or attracted by chemical signposts. This kind of question is at the heart of understanding much of morphogenesis; therefore, we shall defer a more detailed exploration of neural crest cell movements until Chapter 12, in which mechanisms of morphogenesis are

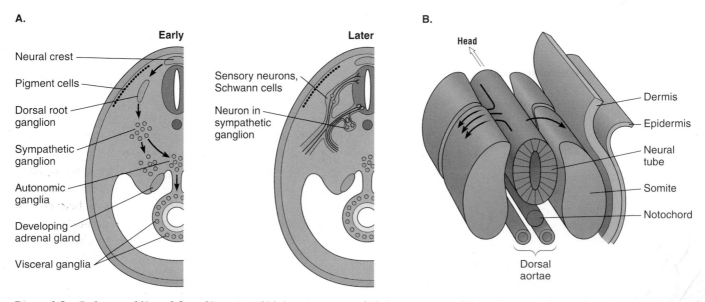

Figure 6.8 Pathways of Neural Crest Migration (A) A cross section through the midbody region of a vertebrate embryo, showing the various types of neural crest differentiation (see also Table 6.1).

(B) A cutaway view of the embryo with the ectoderm removed. Arrows show the pathway of migration of crest cells through the somite area, where migration is restricted to the anterior portion of the somite.

TABLE 6.1 MAJOR NEURAL CREST DERIVATIVES

System or Tissue	Derivatives From:	
	Trunk and Cervical Crest	Cranial Crest
Pigment cells	Melanocytes	Small contribution
Sensory nervous system	Spinal ganglia	Cranial nerves V, VII, IX, X
Autonomic nervous system:		
Sympathetic	Cervical ganglia	
	Vertebral ganglia	
	Visceral and enteric ganglia	
Parasympathetic		Parasympathetic ganglia of head and neck
	Mesenchyme of dorsal fin in fishes and amphibians	Intrinsic ganglia of viscera
Skeletal and connective tissue		Trabecular bone of head
	Walls of aortic arches	Basal plate of skull
	Parathyroid stroma	Parachordal cartilage
		Odontoblasts of teeth
		Head mesenchyme
	Adrenal medulla	Membranous bone of skull
Endocrine	Calcitonin-producing cells	
	Some glia	
Supporting cells	Schwann cells	Some supporting cells
	Contributions to meninges	

discussed. Our understanding of how these cells move to their destinations is far from complete.

The Differentiation of Neural Crest Cells Is Determined Largely by Their Location

The second question is whether neural crest cells, as they emerge from the neural plate during closing of the tube, already know their respective identities, or whether a particular pathway of differentiation is imposed on them by their final location. There is a fairly clear answer: though there are some autonomous limitations, neural crest cells differentiate in accordance with their final location. The final location chosen provides chemical signals that regulate neural crest differentiation.

When individual cells of the newly formed neural crest are labeled with a lineage tracer, the progeny of that single cell are found to contribute to several different structures. Hence many, perhaps all, originating neural crest cells are multipotential.

The importance of final location can be demonstrated by carrying out transplantation experiments. One particularly informative experiment involves transplanting quail neural crest to various locations in chicken embryos and vice versa. Quail and chickens develop very similarly; control grafts of this type demonstrate that neural crest grafts between the two species show normal development. The nuclei of cells from the two species stain rather differently. This difference in staining serves as a natural marker, allowing the researcher to identify quail cells in chicken tissue and vice versa.

Normally, crest cells migrating from the area delimited by somite levels 1 through 7 move into the gut, where they form enteric ganglia and synthesize the neurotransmitter

Box 6.1 Cells in the Nervous System

An extraordinary diversity of cell types can be found in the nervous systems of both invertebrates and vertebrates. Neuron morphology is characterized by very long extensions from the cell body called **neurites**. For example, a neurite originating in a cell body in the human spinal cord may extend a meter to innervate a muscle in the foot! Neurites may conduct nerve impulses either toward the cell body or away from it. One system of terminology designates neurites as dendrites (carrying impulses toward the cell body) and axons (carrying impulses away from the cell body). These days, biologists find it more useful to apply the terms **afferent** to neurites that carry impulses toward the cell body and **efferent** to those that carry impulses away from the cell body.

The morphological diversity of neurons and their neurites is incredible. Afferent neurites may be connected to sensory receptors, such as those in the skin for pain or heat, and in the nasal epithelium for odor. Or afferent neurites may connect via unidirectional synapses to various efferent neurites within the brain or spinal cord. Likewise, efferent neurites may conduct impulses from a center in the CNS (such as the brain or spinal cord), or from an autonomic ganglion, to a muscle or secretory cell. Efferent neurites also form synapses on the afferent neurites and cell bodies of other neurons in the brain and spinal cord. Thus, the functional wiring of the nervous system is composed entirely of synapses between neuron cell bodies and efferent and afferent neurites, together with the junctions of efferent neurites on muscles and secretory cells. Part A of the accompanying figure is a pictorial representation of basic neuron anatomy, while part B indicates the range of neuron morphologies.

The nervous system also contains essential cells that insulate, support, and possibly help nourish neurons. Cells that originate from the neural crest wrap around the afferent (sensory) and efferent (motor) nerves that course outside the brain and spinal cord to the various tissues and organs of the body. These *Schwann cells* form the lipid-rich insulating material called *myelin*, which wraps around the peripheral neurites, giving them a white, glistening appearance (see figure, part A). Within the brain and spinal cord, some cells derived from the neural plate form *neuroglia*, often called glial cells or simply glia, which also form myelin. There are several distinct cellular types of glia, all of them essential for the proper functioning of the brain and spinal cord (see figure, part C). Many brain tumors are gliomas, which form from the misregulated proliferation of glial cells.

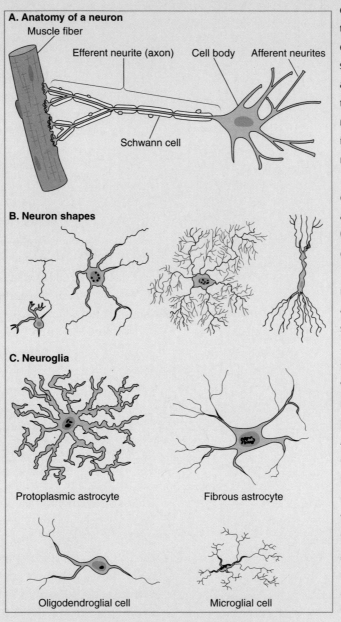

A. Anatomy of a neuron
Muscle fiber
Efferent neurite (axon)   Cell body   Afferent neurites
Schwann cell

B. Neuron shapes

C. Neuroglia
Protoplasmic astrocyte   Fibrous astrocyte
Oligodendroglial cell   Microglial cell

Neuron and Glial Cell Morphology **(A)** A highly formalized diagram of a single neuron identifying its essential elements: cell body; afferent neurites; efferent neurite, here shown connected to a muscle fiber; and Schwann cells, which cover the efferent neurite. **(B)** Tracings of different kinds of neurons found in the CNS; note the wide range in morphology. **(C)** Different kinds of glial cells, showing a variety of morphologies.

acetylcholine. On the other hand, crest cells from somite levels 18 through 24 form the adrenal medulla and secrete norepinephrine. Figure 6.9 illustrates a grafting experiment using exchanges between these two regions. Clearly, the neural crest cells from somite levels 1 through 7 can form a functional adrenal medulla if transplanted to somite levels 18 through 24. Conversely, crest cells from somite levels 18 through 24 will form functional enteric ganglia when moved to a more anterior level of a host embryo.

The multipotentiality of neural crest cells is not, however, absolute. Crest cells from the head region, when transplanted posteriorly, form sensory and autonomic neurons in accord with their new location. However, crest cells from the trunk region implanted into the developing head fail to form skeletal structures, which cranial crest normally does.

CONTROL OF NEURAL SYSTEM GROWTH

The Proliferation of Cells in the Neural Tube Is Strictly Organized

When the neural tube is first formed, it is composed of a single layer of pseudostratified epithelium. Next to the central canal created during neurulation, a cuboidal epithelium called the **ependyma** will form. Portions of the ependyma of the brain will become vascularized, and this ependyma will play an important role in the secretion of cerebrospinal fluid, which is present in the central canal of the brain.

The neural epithelial cells stretch from the ependyma to the surface **marginal zone.** As a cell in the epithelium nears the end of its G2 period, its nucleus migrates to the ependymal side; the cell rounds up, undergoes mitosis, and divides; then the two daughter cells reassume their elongate shape. As development proceeds further, groups of cells undergoing proliferation tend to inhabit the **mantle zone,** between the ependyma and the marginal zone. Cells in the mantle will form neurons and supporting neuroglial glial cells (glia), while cells near the ependyma will continue to generate more cells. The long neurite extensions have a tendency to course along the more superficial layers of the tube; thus the outer marginal zone has fewer cell bodies and more neurites (Figure 6.10). This anatomical arrangement is largely preserved in the spinal cord and hindbrain, but the midbrain and forebrain regions undergo secondary, localized episodes of cell division and cell migration.

Those of you unfamiliar with the various cell types found in the central nervous system may wish to refer to Box 6.1 for a discussion of neurons and glia.

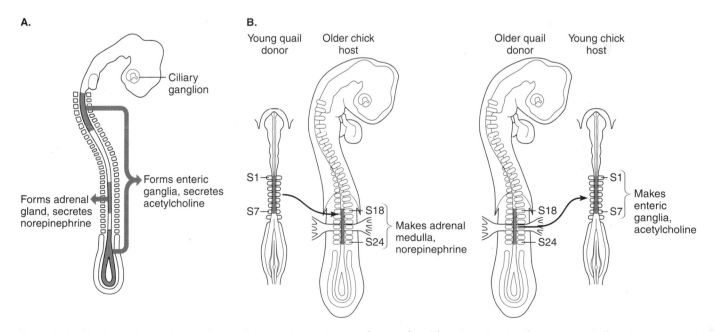

A.

Ciliary ganglion

Forms adrenal gland, secretes norepinephrine

Forms enteric ganglia, secretes acetylcholine

B.

Young quail donor   Older chick host

S1

S7

S18

S24

Makes adrenal medulla, norepinephrine

Older quail donor   Young chick host

S1

S18

S7

S24

Makes enteric ganglia, acetylcholine

Figure 6.9   Final Location and Neural Crest Differentiation An experiment carried out by Nicole Le Douarin and her colleagues in 1982, in which neural crest from different body regions was exchanged. The transplants were made between chick and quail embryos to allow easy identification of host and graft cells. **(A)** In normal development, neural crest from the midbody region (somites 18 to 24) forms the adrenal medulla, which secretes norepinephrine. Crest cells from thoracic region (somites 1 to 7) and sacral region (somites after 28) make sympathetic ganglia, which secrete acetylcholine. **(B)** Crest cells from the thoracic region of a quail donor (somites 1 to 7), when transplanted to the lumbar region (somites 18 to 24) of a chick embryo, now make adrenal medulla and secrete norepinephrine in the host chick embryo. Conversely, lumbar crest cells (somites 18 to 24) of a quail moved to the thoracic region (somites 1 to 7) of a chick host form enteric ganglia and secrete acetylcholine.

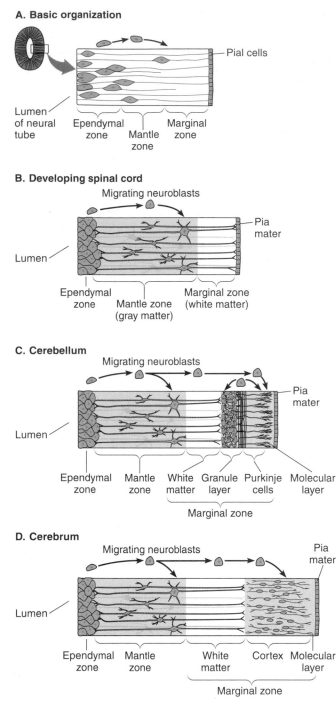

A. Basic organization

Pial cells

Lumen of neural tube — Ependymal zone — Mantle zone — Marginal zone

B. Developing spinal cord

Migrating neuroblasts

Pia mater

Lumen

Ependymal zone — Mantle zone (gray matter) — Marginal zone (white matter)

C. Cerebellum

Migrating neuroblasts

Pia mater

Lumen

Ependymal zone — Mantle zone — White matter — Granule layer — Purkinje cells — Molecular layer

Marginal zone

D. Cerebrum

Migrating neuroblasts

Pia mater

Lumen

Ependymal zone — Mantle zone — White matter — Cortex — Molecular layer

Marginal zone

Figure 6.10 Cell Proliferation in the Neural Tube The patterns of proliferation of neuroblasts in the neural tube of vertebrate embryos are shown diagrammatically. **(A)** The basic organization of the neural tube, with its medial ependymal and intermediate mantle zones (both replete with nucleated cell bodies), and the outermost marginal zone. Overlying the marginal zone is a covering of pial cells, which form a membrane (pia mater) around the neural tube. **(B)** The subsequent stages of differentiation of the spinal cord. **(C)** The developing cerebellum. Migration of neuroblasts in the developing cerebellum creates a much more complex marginal zone. **(D)** The developing cerebrum. Neuroblast migration through the mantle zone and into the marginal zone of the cerebrum establishes new layers of cells and neurites.

The Size of Neuronal Populations Is Sensitive to Peripheral Target Size

The number of cell bodies in the spinal cord varies with location. For example, at the levels of the arms and legs, the spinal cord is larger, primarily due to the larger number of motor neurons in the ventrolateral mantle area. Could this be because these cells innervate a much larger *peripheral target*, that is, a more extensive cohort of muscles, than do cells at other levels of the spinal cord?

Transplantation experiments carried out decades ago showed clearly that this was the case. When a limb was removed on one side in the early embryo, the resultant spinal cord did not display the very high number of motor neurons usually associated with this portion of the cord. Rather, motor neurons on the side of the spinal cord that would have innervated the missing limb died. Sensory ganglia on this side also had fewer neurons than on the normal, unoperated side. On the other hand, grafting a supernumerary (additional) limb adjacent to an existing limb extended the area of increased numbers of neurons and led to enlarged spinal ganglia (Figure 6.11). This and other, similar experiments indicated that the innervation of the peripheral target area and the size or mass of that target influence the number of certain cell types in the central nervous system.

Cell Numbers in the CNS are Likely Controlled Through the Prevention of Apoptosis

Researchers have long recognized that many of the neurons generated by mitosis in the neural tube do not persist into adulthood. This selective elimination of cells during normal development, or *programmed cell death*, occurs extensively in animals throughout their life span, and we shall encounter it in many different instances in development, not just in the nervous system. A recent term for programmed cell death is **apoptosis** (pronounced "ah-po-TOE-sis").

The probable mechanism for regulating cell numbers in the CNS is through *preventing* programmed cell death: When motor or sensory neurons connect with the organs they innervate, apoptosis is prevented. For instance, when neurons in the thoracic and lumbar regions of the spinal cord make connections to the limb muscles, apoptosis is reduced, resulting in a greater number of cells surviving in the regions of the cord that innervate the limbs. Extirpation of a limb prevents the peripheral connections from forming; apoptosis then occurs just as it does in other regions of the spinal cord.

Understanding how programmed cell death is regulated is of considerable clinical importance since improper regula-

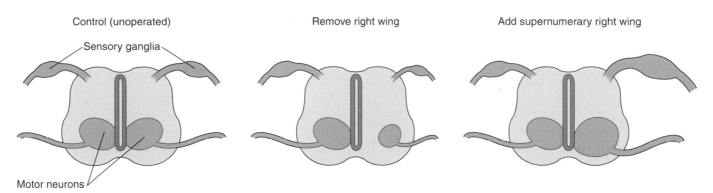

Control (unoperated)   Remove right wing   Add supernumerary right wing

Sensory ganglia

Motor neurons

Figure 6.11   The Effect of Limb Changes on Neuron Number
Experiments showing how the size of the wing field innervated by a region of the spinal cord changes when limb buds are added or removed. In the normal chick embryo, both sensory and motor ganglia at the level of the wing are approximately the same size on either side of the spinal cord. When a limb bud from a normal three-day chick embryo is removed, the collections of motor and sensory neurons innervating the wing field on the side without a wing are much reduced in number. The converse experiment is accomplished by transplanting a supernumerary wing bud to a recipient, thereby creating an animal with a duplicated wing on one side. Here the collections of motor and sensory neurons innervating the wing region are substantially enlarged.

tion of apoptosis is thought to be involved in several different diseases including some kinds of cancers. Consequently, signals that either prevent or permit apoptosis are currently the subject of intense basic and clinical research.

Neurons Are Influenced by Nerve Growth Factors

Proliferation, outgrowth, and survival of neurons are influenced by *neurotrophic* (nerve growth) *factors*. In particular, secreted paracrine factors affect the proliferation, survival, and vigor of neurite outgrowth during development; they may possibly even act as chemoattractants of neurite outgrowth. The list of neurotrophic factors and the receptors they interact with is burgeoning rapidly. These factors, members of the immense collection of growth factors found in metazoans, show specificity for different classes of neurons.

Let us consider the very first nerve growth factor discovered, which helped to establish the important role of growth factors and their receptors, and resulted in a Nobel Prize for Rita Levi-Montalcini and Stanley Cohen. Working at Washington University in St. Louis, Missouri, Levi-Montalcini discovered that certain tumors occurring in chickens produce a factor or factors that stimulate the proliferation and outgrowth of neurons of the developing sympathetic nervous system of chick embryos. She and Cohen went on to show that this factor, a small dimeric glycoprotein now called NGF (for nerve growth factor), stimulates the proliferation and outgrowth of neurites from somatic sensory and sympathetic neurons, both in the embryo and in culture (Figure 6.12). The receptor for NGF is a member of the receptor tyrosine kinase family.

The precise role of NGF in the early development of the nervous system may not be large since the ligand and receptors are

Figure 6.12   The Effect of Nerve Growth Factor on Neuroblasts in Culture
Sensory ganglia from a seven-day chick embryo were cultured in vitro **(A)** in the absence of and **(B)** in the presence of 10 ng/ml of NGF for 24 hours, and then stained to demonstrate the relative abundance of nerve processes.

A.   **B.**

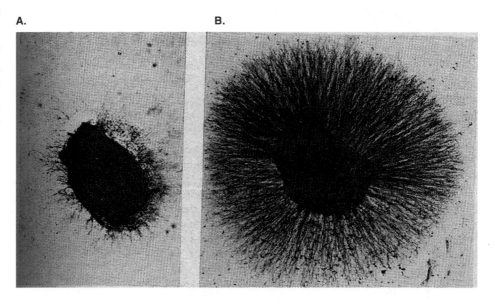

BOX 6.2 STEM CELLS

Many tissues in the body of an adult have the capability for self-renewal. The epidermis of human skin is replaced about every two weeks. This is accomplished by the differentiation of new epidermis from a stem cell population called the basal stem cell layer. Stem cell populations are extraordinarily important for the maintenance of organs in the adult. Not only skin, but hair follicles, the lining of the intestine, and provision of red and white blood cells require the activity of stem cells. We first met stem cells in Chapter 2 when discussing gametogenesis. Oogonia and spermatogonia undergo mitosis in order to form populations of new sperm and eggs as well as to maintain the precursor stem cell population itself. Stem cell populations are capable of self-renewal *and* of giving birth to cells that differentiate along a particular pathway.

The conventional view has been that the nervous system does not possess stem cells, so that once an area of the brain dies, due to disease or vascular accident, the cells will never be replaced. Differentiated neurons, at least in mammals, never do divide again. Hence, there is enormous interest, among researchers and clinicians alike, in recent reports that there may exist stem cell populations in adult mammals that can give rise to neurons. If further study bears this out, then

many clinical applications—ranging from treatments for Alzheimer's and Parkinson's disease to the repair of spinal cord injuries—may be able to exploit these findings (see accompanying figure).

The studies on putative neural stem cells have been carried out in mice and rats;

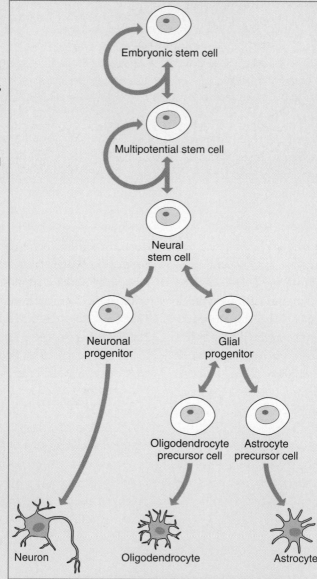

comparable studies in humans will be difficult to carry out. In one study, bone marrow cells tagged with GFP (green fluorescent protein) were injected into mice lacking functional immune systems (the *nude* mutant). Not only did the recipient mice develop immune systems; tagged cells were also found in the brain, and some of these displayed proteins typical of neurons. In addition, there are recent reports of stem cells in the brain that have the appearance of glia and can be reprogrammed to form apparent neurons.

These new studies raise many issues. Some of these, of course, concern the particular experiment: Is the tag used for cells reliable? Are the molecular markers for identifying a particular cell type foolproof? Other matters are more fundamental: Can a cell be "somewhat differentiated," then be nudged into becoming mitotic again to assume the role of a stem cell? And what regulates the formation and function of stem cells? Many of the stem cells used in experimental studies have been cultured and exposed either to various regimens of growth factors or to unknown and mysterious factors present in serum. Furthermore, it has been very difficult to maintain human stem cells in culture; the search for adequate culture media and optimal condi-

Stem Cell Lineages The undifferentiated stem cells of the early embryo can give rise to multipotential stem cells of various kinds, populating bone marrow, muscle, liver, gut, and many other organs. Some workers believe these multipotential stem cells can be nudged back to a state more like that of ES cells. It is also hypothesized that more-specialized types of stem cells might be "reversed" in this manner. For example, one type of more specialized stem cell, the neural stem cell, ordinarily has the potential to form progenitors of either neurons or glia. It may be that glial progenitor cells can be "reversed" to form neural stem cells, which are more widely potent.

tions for growing and maintaining human stem cells in culture remains ongoing.

There is also the matter of the potential of a given stem cell population. The conventional wisdom is that stem cells from early embryos have a much wider developmental potential than do the more specialized stem cell populations found in adult tissues.

Recall from Chapter 5 that cells from the inner cell mass of the mammalian blastocyst can be maintained in tissue culture (ES cells). When the cocktail of growth factors and medium is changed, ES cells can differentiate into a very wide variety of differentiated tissues. Cells taken from adult bone marrow can differentiate into various types of blood cells, but they can also form glial cells called astrocytes. So marrow stem cells are multipotential. But they are probably more restricted in their developmental potential than ES cells are. For example, stem cells for striated muscle, called satellite cells, can proliferate and differentiate to repair damaged muscle but seem not to have the potential to form other tissues.

The range of developmental potential of stem cells could arise from intrinsic, stable conditions of the stem cells themselves. For instance, perhaps certain genes critical for neuron formation have become methylated and are embedded in a region of heterochromatin that cannot be transcribed.

On the other hand, it is possible that the microenvironment—the combination of growth factors and other chemical signals—is critical for the kind and extent of differentiation of a given stem cell population. The role of the microenvironment (sometimes called the niche by stem cell researchers) and the intrinsic capacities of the cells are not mutually exclusive. The biology of stem cells presents a virtual replay of the same issues that concern us when considering the normal development of the embryo.

not present at very early stages. NGF is needed for the maintenance of sensory and sympathetic neurons, however. Injection of antibodies to NGF into young mice leads to a drastic decrease in some sensory neurons and the loss of almost all sympathetic neurons. Curiously, these "sympathectomized" mice carry out life functions well. Though NGF exerts its effect on neurons, many organs, especially the salivary glands and kidneys, secrete it as well.

As mentioned earlier, NGF is only one of many paracrine factors that influence neurons, and there is good evidence that some of them are important in the early development of the nervous system. Fully differentiated neurons in the adult are not able to carry out mitosis, but there is some evidence that some undifferentiated neural stem cells are present in the adult (see Box 6.2).

REGIONAL DIFFERENTIATION OF THE CENTRAL NERVOUS SYSTEM

The Nervous System Is Segmentally Organized

The vertebrate CNS displays a marked anatomical segmentation, which is obvious upon inspection. These segmental dilatations of the neural tube are called *neuromeres*. In the developing brain, noticeable dilatations indicate the future fore-, mid-, and hindbrain regions, technically designated prosencephalon, mesencephalon, and rhombencephalon. In particular, the *prosencephalon* forms the forebrain (cerebral hemispheres) and the diencephalon (hypothalamus); the eyes form as outpocketings from the diencephalon. The *mesencephalon* forms the midbrain. And the *rhombencephalon* forms both the metencephalon (cerebellum and pons area) and the myelencephalon, which in turn forms the hindbrain (brain stem). These principal dilatations of the brain, often called brain vesicles, are indicated in Figure 6.13.

The different brain vesicles show distinctive patterns of gene expression very early in their development. For example, *pax* 6 is characteristic of diencephalon, while *pax* 2 and *pax* 5 are indicative of mesencephalon. *Engrailed* expression in the mesencephalon is necessary for establishing the border between diencephalon and mesencephalon. The strip of tissue between the mesencephalon and rhombencephalon is called the *isthmus*. Embryological and molecular studies have shown that the isthmus has a powerful effect on the development of the anterior rhombencephalon. The growth factor Fgf 8 from the isthmus suppresses *hox* gene expression in the anterior rhombencephalon, while retinoic acid (possibly emitted by the surrounding mesoderm) stimulates *HOX* gene expression; the interplay of these two factors (and possibly others) helps establish the pattern of *HOX* gene expression in the rhombencephalon.

In some vertebrates, the rhombencephalon shows very distinct neuromeres called *rhombomeres* (Figure 6.13). We shall see in Chapter 16, when we discuss the *HOX* genes (which play an important role in patterning the body along the anteroposterior axis), that rhombomeres are borders not only of anatomy, but also, as we have just implied, of specific gene expression. In addition, rhombomeres are developmental boundaries; once formed, cell bodies do not cross from one rhombomere to another.

The spinal cord, too, has a segmental arrangement of neuromeres. Associated with each neuromere is a pair of somites, with one somite lateral to each side of the neuromere. Neural crest cells migrate through the anterior portion of each somite, and some crest cells form dorsal root sensory ganglia, one on each side of the neuromere. The somites form the cartilaginous precursors of vertebrae; these, too, are segmented in register with the ganglia and spinal cord neuromeres. When the somites or their precursor tissues (paraxial mesoderm) are removed, the segmental character of the spinal cord fails to develop. Another way to demonstrate the influence of somites on the patterning of the spinal cord is to transplant additional somites to only one side of the developing cord. When that is done, the number of dorsal root ganglia formed corresponds to the number of somites (see Figure 6.14).

The Eyes Develop from the Diencephalon

A very prominent outpocketing on each side of the diencephalic region of the forebrain occurs early in the development of the brain. These outpocketings develop into the *optic vesicles*, which will form the eyes. As they proliferate and evaginate through the head mesenchyme (much of which,

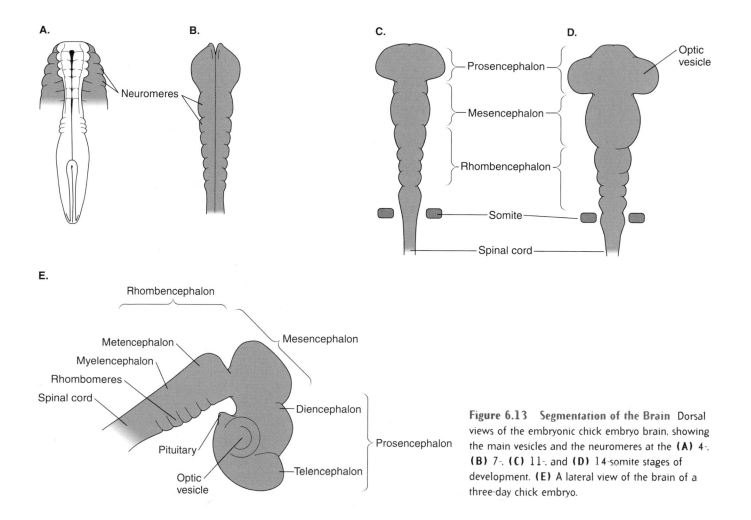

Figure 6.13 Segmentation of the Brain Dorsal views of the embryonic chick embryo brain, showing the main vesicles and the neuromeres at the **(A)** 4-, **(B)** 7-, **(C)** 11-, and **(D)** 14-somite stages of development. **(E)** A lateral view of the brain of a three-day chick embryo.

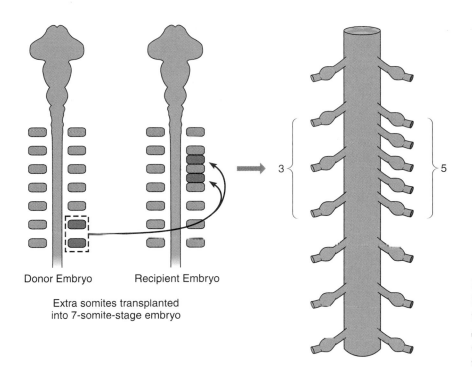

Donor Embryo   Recipient Embryo

Extra somites transplanted
into 7-somite-stage embryo

Extra ganglia induced by extra somites

Figure 6.14 Segmentation of the Spinal Cord Determined by Somites The segmentation of the spinal cord is easily identified by the sensory ganglia that form in each segment: the mature cord has one pair of ganglia per somite. If extra somites are transplanted next to the developing spinal cord, extra ganglia will form on that side.

A.

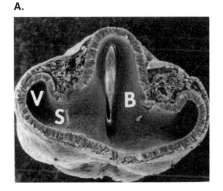

B.

C.

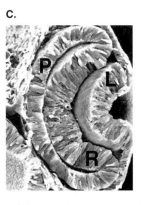

Figure 6.15 Development of the Amniote Eye Scanning electron micrographs through the head region where the eye is formed were prepared at successively later stages of development. **(A)** Evagination of the optic vesicle from the diencephalon. B. brain; V. optic vesicle; S. stalk of optic vesicle. **(B)** Invagination of the lateral portion of the optic vesicle and the formation of the lens placode in overlying ectoderm. **(C)** The formation of the optic cup and beginning of invagination of the lens placode. shown at higher magnification. P. pigmented retina; R. neural retina; L. lens placode.

you will recall, is of neural crest origin), the vesicles abut the surface ectoderm. Then the optic vesicle invaginates on its lateral aspect to form an optic cup (Figure 6.15). As the cup forms, the ectoderm adjacent to it changes from cuboidal to columnar; then this surface ectoderm invaginates and pinches off to lie within the entrance to the optic cup. This raised surface of columnar cells, sometimes called a placode, has now become internalized, develops into the lens. (We shall have more to say about placodes later in the chapter.) The outer surface of the optic cup becomes the pigmented retina, while the inner surface becomes the photoreceptors and neurons of the neural retina and also generates an epithelial iris. The head mesenchyme derived from the mesoderm that surrounds the area becomes the hard scleral covering of the eyeball and attached eye muscles. Figure 6.15 shows many of these developments.

The formation of the eye is so striking that it has fascinated embryologists since the beginning of the 20th century.

Spemann's early experimental work studied the influence of the optic cup on the overlying ectoderm that forms the lens. Many investigators extirpated portions of the anterior neural plate or carried out transplantations, showing that the potential to form eye tissue was present in the prosencephalic region as early as the open neural plate stage. Genetic analysis of the formation of the eye has been an area of intense research activity, especially in *Drosophila*. At this point, there are two especially important sets of experiments to consider, both of which bear directly on our effort to discern underlying principles of organ formation.

Does the Optic Cup Induce the Lens?

First comes the question of whether the optic cup and/or vesicle induces the overlying ectoderm to form the lens. Early research carried out in the early part of the 20th century certainly indicated that this was so; indeed, along with the Spemann organizer, induction of the lens by the optic cup was considered a prime example of embryonic induction. Removal of the optic vesicle prior to lens formation resulted in the absence of a lens. But there were exceptions: in some amphibian species, and in amphibians raised at certain temperatures, the lens formed even when the optic cup was removed. Experiments by Warren Lewis early in the 20th century showed that transplanting an optic cup to an unusual location, such as under the ectoderm of the flank, resulted in a lens forming from the flank ectoderm adjacent to the transplanted optic vesicle. Other transplantation experiments indicated that endoderm and mesoderm of the head region might also be involved in lens induction.

In the early 1990s, Robert Grainger, working at the University of Virginia, analyzed and repeated Lewis's transplantation experiments, using modern cell-marking techniques, to test the source of the ectopic lens. Transplanted optic cups were taken from donors whose tissues had been marked by prior injection of horseradish peroxidase into the zygote. It was then possible to determine whether an ectopic lens originated from donor tissue, which was marked with horseradish peroxidase, or from host tissue, which was not. The results were clear: The optic cup does *not* induce the lens from flank ectoderm; rather this ectopic lens arises from contaminating ectoderm cells from the donor that adhere to the transplanted optic vesicle (Figure 6.16).

We now have a different idea of the interactions between the optic cup and prospective lens ectoderm. Considerable

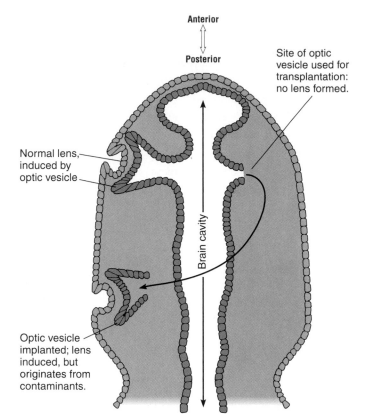

Figure 6.16 Lens Induction A diagram of a generalized early-amniote embryo, showing various transplantation effects: for the sake of simplicity, separate donor and host embryos are shown as a single embryo, and only ectoderm is shown. An optic vesicle was removed from the head region *(top right)* and placed under the ectoderm far posterior to the eye region *(bottom left)*. An ectopic lens developed at the latter site, but subsequent careful investigation showed that the ectopic lens had formed from ectoderm cells carried along with the optic cup transplant, not from host site ectoderm. No lens developed at the site where the optic vesicle was removed.

experimental evidence supports the view that, while the optic cup does provide some of the signals stimulating lens formation, these work in concert with other factors that emanate, earlier in development, from prospective head mesoderm (in particular, from prospective heart mesoderm, which lies very close to the optic vesicle) and from anterior endoderm. The nature of the different ligands and receptors involved remain to be identified. The moral of the story seems to be that, as with the case of the Spemann organizer, a so-called induction may result from multiple signals working in combination; furthermore, the composition of the signal cocktail may change over time.

Executive Genes Are Necessary for Eye Formation

A second issue regarding organ formation is when and how ectoderm first acquires the ability to form an eye. Until the late 1970s, it was common for embryologists to speak of the "eye field," the area of cells capable of giving rise to well-organized eye tissue when put in tissue culture or transplanted to some "neutral" site in another embryo. As mentioned earlier, the development of the eye is one of the earliest consequences of neural induction by the Spemann organizer.

The nature of the signals or combination of signaling molecules that evokes "eyeness" is still poorly known. But research on some mutations in *Drosophila melanogaster* has presented us with a surprise. There are mutations in the fly that result in failure of the eye to form at all, one of which is *eyeless*. In its normal wild-type form, this gene, which encodes a transcription factor, seems to play an important and dominant role in eye formation in *Drosophila*; if the gene can be forced by means of sophisticated genetic manipulations to express itself in tissue outside the prospective eye region, an ectopic eye may form, something that normally never happens (Figure 6.17). Activity of the *eyeless* gene seems to set in motion the whole set of events leading to formation of an eye

in *Drosophila*, no matter where in the embryo the gene is expressed. Some biologists call it a kind of a "master," or "executive," gene. Though other genes closely related to *eyeless* have recently been discovered, we do not yet know the details of how *eyeless* carries out this master control function.

The *Drosophila* eye forms in a radically different way from the vertebrate eye (see Box 6.3). Nevertheless, the homolog of the *Drosophila eyeless* gene exists in vertebrates; it is *pax6*. (We first encountered *pax6*, you may recall, earlier in the chapter in connection with Sonic Hedgehog's role in dorsoventral patterning of the neural plate and tube.) *Pax6* is expressed in anterior neural tissue when the eyes form, and null (complete loss of function) mutations of *pax6* prevent eye formation. *Pax6* is expressed somewhat later in development in other areas of the CNS, so it may serve functions in addition to eye development. When the mouse version of the gene is inserted into an *eyeless Drosophila* mutant, the fly is able to form an eye. The mouse gene works apparently normally in a fly. Does this mean there is a master control gene for eye formation whose existence predates the separation—approximately 500 million years ago—of arthropods from the vertebrate lineage? If so, then a close analysis of how the *eyeless* gene supervises the pathway leading to construction of an eye should be of immense interest.

A.

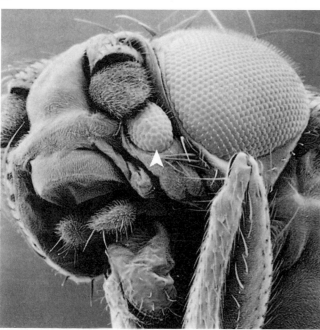

B.

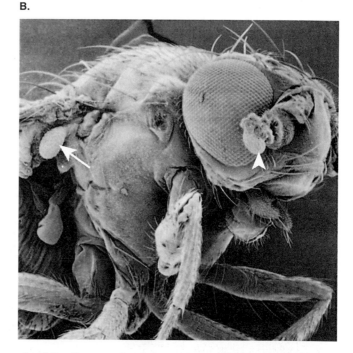

Figure 6.17   Formation of Ectopic Eyes Caused by Misexpression of the *eyeless* Gene Scanning electron micrographs of the fruit fly. *Drosophila*. in which the *eyeless* gene has been experimentally expressed ectopically. resulting in ectopic eyes in various regions. **(A)** The arrowhead points to an ectopic eye in the head region. **(B)** The arrow points to an ectopic eye under the wing. Note the eye tissue on the tip of the antenna (arrowhead).

Ectodermal Placodes Contribute to the Development of the Brain

The lens is not the only surface ectoderm to develop in close concert with the anterior portion of the CNS. As the different brain vesicles start on their particular pathways of differentiation, specific patches of head ectoderm transform from cuboidal to elongated columnar epithelium. These cranial ectodermal *placodes*, together with cranial neural crest cells, form sensory and autonomic neurons in cranial ganglia. The most obvious placodes are the nasal placode, which gives rise to the sensory neurons and nasal epithelium that serve the sense of smell, and the otic placode. The latter contributes to the formation of the inner ear and neurons of the eighth cranial nerve, which serve the senses of hearing and balance. The collaboration of cranial neural crest and head ectoderm involves reciprocal signaling between the two tissues. Differences in the details of these interactions provide variation in the kind of organ that eventually forms. Not only familiar organs of vertebrates (such as eyes and ears), but also special sensory organs, such as the lateral line organ in fishes, arise in this way.

Figure 6.18 diagrams the relationships between the major placodes and the developing brain, while Table 6.2 provides a tabular summary of the names, derivation, and functions of the different cranial nerves.

We should also mention that an inpocketing of the pharyngeal ectoderm called Rathke's pouch (see Figure 6.13E) will form the anterior pituitary. Adjoining Rathke's pouch is an outpocketing of the hypothalamic portion of the diencephalon that will become the posterior pituitary.

The Brain Vesicles Undergo Regional Differentiation

Earlier in the chapter, we mentioned that the migratory behavior of neuron cell bodies originating at the ependymal border of the neural tube varies according to the area of the developing CNS. In different regions of the brain, groups of cell bodies migrate in specific patterns typical of those particular regions. These groups form layers, or congregations of cell bodies, called *nuclei*. Connecting the different nuclei are *tracts* of neurites, which are obvious under the microscope. These tracts constitute the major cables of hard wiring of the brain and are the subject matter of neuroanatomy.

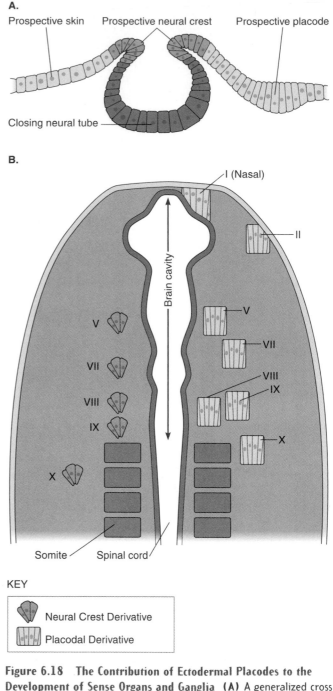

KEY

Neural Crest Derivative

Placodal Derivative

Figure 6.18   The Contribution of Ectodermal Placodes to the Development of Sense Organs and Ganglia   (A) A generalized cross section of the closing neural tube. Arrows indicate the movements of neural crest and incipient placodes. **(B)** A highly diagrammatic dorsal view of the head region of a vertebrate. For simplicity, the contributions from neural crest are shown on the left and those from placodes on the right. The major placodes include the nasal and otic placodes, which form portions of the nose and inner ear; placodes also contribute to many cranial nerve ganglia. Several ganglia receive neural crest as well as placodal contributions. Surface ectoderm also forms the epibranchial placodes (not shown here), which form ganglia associated with the developing pharyngeal arches.

TABLE 6.2 A SUMMARY OF THE FUNCTION AND DERIVATION OF CRANIAL NERVES

Name	Number	Function	Contributions From:	
			Placode	Neural Crest
Olfactory	I	Smell	+	
Optic	II	Sight	+	
Oculomotor	III	Eye muscle (motor)		
Trochlear	IV	Eye muscle (motor)		
Trigeminal	V	Sensory	+	+
Abducens	VI	Eye muscle (motor)		
Facial	VII	Mainly motor		+
Auditory	VIII	Hearing	+	
Glossopharyngeal	IX	Mixed	+	+
Vagus	X	Mixed	+	+
Accessory	XI	Mainly motor		
Hypoglossal	XII	Tongue (motor)		

BOX 6.3 EYE FORMATION IN FLIES AND FROGS

Vertebrate eyes develop from an outpocketing on each side of the prosencephalon, the anterior-most brain vesicle. These outpocketings provide a weak stimulus to overlying ectoderm to form the lens; the optic vesicle also invaginates to form the two-layered optic cup. The inner layer of this cup then differentiates to form the photoreceptors (rods and cones) and neurons of the retina. The efferent neurons of the retina project back to the brain via the optic nerve. Part A of the accompanying figure presents a diagrammatic summary; see also Figure 6.15 for scanning electron micrographs of this sequence.

The eyes of flies originate very differently from those of vertebrates, and are organized in a distinct manner, so it is surprising that executive genes in both vertebrates and flies are so conserved. The fly eye originates from one of the nests of epidermal cells called imaginal discs, mentioned in Chapter 3. Each eye disc arises as a collection of 20 to 40 cells in the anterior epidermis of the first instar larva; later development of the imaginal discs involves tremendous growth and a dramatic differentiation just before and during metamorphosis, a process we will discuss in Chapter 8. The identity of the eye discs is established by the action of *eyeless* and its partners. The fully differentiated eye is composed of about 800 identical facets, each with its own lens and a set of eight photoreceptor cells. Each facet sends an efferent neurite to the ventrally located anterior ganglion of the fly's central nervous system. This type of faceted eye is called a compound eye.

Late in larval life, the eye discs, now composed of tens of thousands of cells each, undergo a progressive differentiation into small groups of cells that will form the individual facets. This wave of differentiation moves from the posterior to the anterior portion of the disc, visible as a furrow that moves across the eye disc as the wave of differentiation proceeds. A complex set of signals is being relayed during this time. (We encounter some of these signaling molecules elsewhere in this chapter, in our discussion of the vertebrate spinal cord; we shall meet some of them again when we explore the development of another imaginal disc, the wing disc, in Chapter 15.) One of these, Dpp—the homolog of BMPs in vertebrates—is secreted in the morphogenetic furrow, while *hedgehog (hh)* is expressed in cells posterior to the furrow. Behind this wave of inductive signaling is a series of interactions among the eight photoreceptor cells that lead to their final differentiation. Cells immediately posterior to the morphogenetic furrow are just beginning their program of differentiation into photoreceptors, while more-posterior cells have progressed further in the program of differentiation. Part B of the accompanying figure outlines the steps in this process.

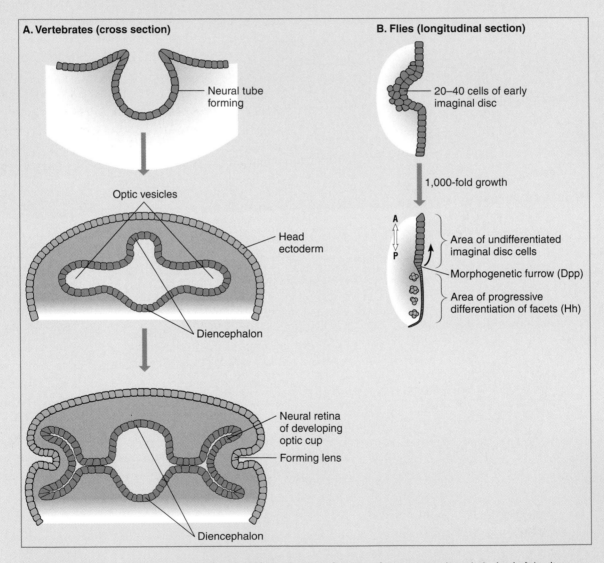

Vertebrate and Dipteran Eye Formation Compared **(A)** Vertebrate eye formation: A cross section through the level of the diencephalon shows the progression from neural tube to optic vesicles, followed by the formation of each lens and optic cup. The inner layer of the optic cup differentiates into three layers of neurons (including photoreceptors), which form the sensory apparatus of the eye. **(B)** Dipteran compound eye formation: The dorsolateral surface epidermis is shown in longitudinal section (single eye only). Some 20 to 40 cells form the early eye imaginal disc, located subjacent to the surface epithelium. These cells divide, and the disc grows manyfold. Late in larval life, the morphogenetic furrow, a site of Dpp secretion, progresses from posterior to anterior end. In its wake, small groups of eight cells form the photoreceptor complex, each group organizing into a separate facet of the compound eye.

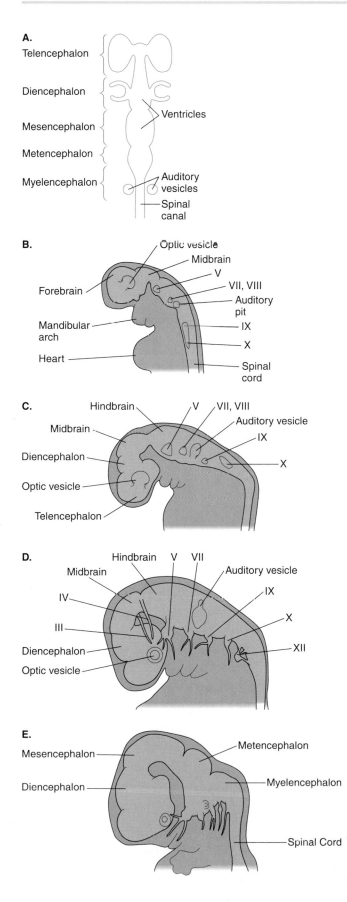

A.
Telencephalon
Diencephalon
Mesencephalon
Metencephalon
Myelencephalon
Ventricles
Auditory vesicles
Spinal canal

B.
Optic vesicle
Midbrain
V
Forebrain
VII, VIII
Auditory pit
Mandibular arch
IX
X
Heart
Spinal cord

C.
Hindbrain
V
VII, VIII
Midbrain
Auditory vesicle
IX
Diencephalon
X
Optic vesicle
Telencephalon

D.
Hindbrain
V
VII
Midbrain
Auditory vesicle
IV
IX
III
X
Diencephalon
XII
Optic vesicle

E.
Mesencephalon
Metencephalon
Diencephalon
Myelencephalon
Spinal Cord

Different rates of cell division, different extents of apoptosis, and cellular interactions leading to morphogenetic movements all sculpt the brain vesicles into their characteristic final form. Throughout development, a central hollow chamber remains in the brain and spinal cord, a remnant of their formation as a neural tube; these hollow, multipocketed chambers are called the ventricles in the brain, and the spinal canal in the cord (Figure 6.19A). Figure 6.19B–E shows the gradual changes in shape and extent of the prosencephalon, mesencephalon, and rhombencephalon of the human embryo.

The fundamental question of how these vesicles develop, both at the gross morphological level and at the level of detailed wiring diagrams, involves many of the basic issues of development: How do the different areas of the developing brain acquire their identities in the first place, and what is the molecular basis of acquiring their identity? What are the signals and programs regulating cell division and morphogenesis? How is specific connectivity between different neurons realized? Is activity of a neuron, or lack of it, a factor in establishing or maintaining connectivity? We shall explore these questions further when we discuss morphogenesis in Chapters 12 and 13.

THE INTEGUMENT

Epidermal Structures Arise from Ectoderm

Ectoderm and mesoderm each make essential contributions to the outer covering of an animal, or **integument**. The ectoderm forms the outermost layer of the skin, the epidermis, which is several cell layers thick. Ongoing mitosis in a basal stem cell layer next to the basal lamina continuously generates new ectoderm cells, which are displaced toward the surface. (See Box 6.2 for more about stem cells.) Here, in a process called *keratinization*, they synthesize large amounts of the protein keratin, which endows the epidermis with the physical properties so useful in the integument:

Figure 6.19 Development of the Brain (A) A dorsal view of the primitive vertebrate brain, showing its main subdivisions. **(B-E)** Lateral views of the developing human brain: **(B)** A 20-somite embryo, at about $3\frac{1}{2}$ weeks of gestation. **(C)** A four-week embryo. **(D)** A $5\frac{1}{2}$-week embryo. **(E)** A seven-week embryo, about 17 mm in overall length. By this stage, the brain has become very complex and the cranial nerve ganglia are evident; the cerebral cortex, however, has not yet developed.

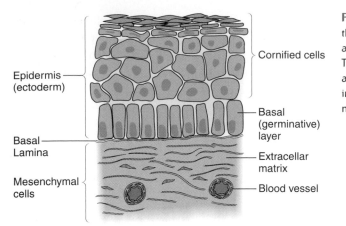

Figure 6.20 The Epidermis and Its Derivatives A stylized diagram of the surface epithelium of a terrestrial vertebrate. The ectoderm gives rise to a germinative layer that forms the stratified. cornified surface of the skin. The mesoderm forms the underlying dermis. comprised of mesenchyme and a vascular supply. The ectoderm also forms portions of specialized structures in the skin. including sweat and sebaceous glands. hair. feathers. gills. and mammary glands.

impermeability, strength, and toughness. Epidermal cells filling with keratin granules undergo apoptosis, resulting in an outer, cornified ("horny") layer that is several cells thick and constantly being sloughed off (Figure 6.20).

In addition to the epidermis, the integument is comprised of numerous specialized structures composed of both ectoderm and mesoderm. These include hairs, feathers, sebaceous glands, and mammary glands in amniote classes; the cement gland, balancers, and portions of the gills in amphibians; and scales in fish.

We shall defer a more detailed description of the skin, including the inner layer (the dermis) and some of its associated glands, to Chapter 7, when we discuss the other essential component of the integument, mesoderm.

KEY CONCEPTS

1. Cell interactions during development, including "inductions," occur by means of ligand–receptor interactions. These may involve multiple ligands and receptors, and the response may be positive or negative. Ligands may be released and sensed at short range, or form gradients over longer distances. The combinations of important ligands may change during time, as do the combinations of receptors.

2. Some multipotential populations of cells, such as the neural crest, are determined by interactions that occur at their final destination.

3. The regulation of cell proliferation and of programmed cell death (apoptosis) by the presence or absence of ligands plays an important role in development.

4. Though the specific tactics used by different embryos to progress from egg through gastrulation may differ greatly in different animals, within the same phylum different species tend to be very similar in their postgastrulation development and organ formation. (We only hint at this concept in this chapter, reserving a detailed discussion of this principle until Chapter 17, when we shall consider evolution and development.)

STUDY QUESTIONS

1. Several examples of tissue interactions (that is, embryonic induction) have been discussed in this chapter. Are there any examples of differentiation of ectoderm that do not involve an induction?

2. What makes the induction of tissues by the Spemann organizer different from any other examples of tissue interactions?

3. We have learned that the emigration of the neural crest cells occurs over an extended period of time. Can you devise a way to determine whether crest cells emerging early from the neural tube have the same fate as those emerging later?

4. A large number of growth factors, some of which act only on nervous tissues (neurotrophic factors), are implicated in neural development. How would you go about finding out whether a particular growth factor, such as FGF8, is active in neural development?

SELECTED REFERENCES

Bally-Cuif, L., and Wassef, M. 1995. Determination events in the nervous system of the vertebrate embryo. *Curr. Biol.* 5:450–458.

A review of patterning events in the brain and the genes and molecules involved.

Begbie, J., Brunet, J.-F., Rubenstein, J. L. R., and Graham, A. 1999. Induction of the epibranchial placodes. *Development* 126:885–902.

A research paper showing that pharyngeal endoderm induces the formation of these placodes by means of BMP7.

Bronner-Fraser, M., and Fraser, S. E. 1997. Differentiation of the vertebrate neural tube. *Curr. Opin. Cell Biol.* 9:885–891.

A review of the molecules involved in the patterning and differentiation of the neural tube.

Crowley, C., and 10 other authors. 1994. Mice lacking nerve growth factor display perinatal loss of sensory and sympathetic neurons yet develop basal forebrain cholinergic neurons. *Cell* 76:1001–1011.

Experiments showing the important role of NGF.

Dodd, J., Jessell, T. M., and Placzek, M. 1998. The when and where of floor plate induction. *Science* 282:1654–1657.

A recent review, by leaders in the field, of the action of the floor plate.

Fuchs, E., 1997. Of mice and men: Genetic disorders of the cytoskeleton. *Mol. Biol. Cell* 8:189–203.

A review of skin development and the role of intermediate filaments.

Fuchs, E., and Segre, J. A. 2000. Stem cells: A new lease on life. *Cell* 100:143–155.

An excellent and detailed review of examples and issues in stem cell biology.

Grainger, R. M. 1992. Embryonic lens induction: Shedding light on vertebrate tissue determination. *Trends Genet.* 8:349–355.

An incisive review of classical and contemporary research on lens formation.

Halder, G., Callaerts, P., and Gehring, W. J. 1995. Induction of ectopic eyes by targeted expression of the *eyeless* gene in *Drosophila. Science* 267:1788–1792.

A research report on the discovery of a master gene for eye formation.

Harland, R., and Gerhart, J. 1997. Formation and function of Spemann's organizer. *Annu. Rev. Cell Dev. Biol.* 13:611–667.

An extensive review of the classical literature and contemporary research on Spemann's organizer.

Jean, D., Ewan, K., and Gruss, P. 1998. Molecular regulators involved in vertebrate eye development. *Mech. Dev.* 76:3–18.

An extensive review of what is known about the genes and signaling systems involved in eye differentiation.

Le Douarin, N. M., and Ziller, C. 1993. Plasticity in neural crest cell differentiation. *Curr. Biol.* 5:1036–1043.

A brief review of neural crest development by one of the leaders in the field.

Lumsden, A., and Krumlauf, R. 1996. Patterning the vertebrate neuraxis. *Science* 274:1109–1114.

A review of how the brain and spinal cord attain their organization.

Martinez, S., Crossley, P. H., Covos, I., Rubenstein, J. L. R., and Martin, G. F. 1999. FGF8 induces formation of an ectopic isthmic organizer and isthmocerebellar development via a repressive effect of *Otx2* expression. *Development* 126:1189–1200.

A research paper showing that FGF8 is probably involved in patterning the midbrain.

McGrew, L. L., Hopler, S., and Moon, R. T. 1997. Wnt and FGF pathways cooperatively pattern anteroposterior neural ectoderm in *Xenopus. Mech. Dev.* 69:105–114.

A research paper demonstrating how the ligands Wnt and FGF play a role in forming the posterior nervous system.

Meier, P., and Evan, G. 1998. Dying like flies. *Cell* 95:295–298.

A short review of contemporary research on apoptosis.

Pituello, F. 1997. Neuronal specification: Generating diversity in the spinal cord. *Curr. Biol.* 7:R701–R704.

A review of how Hedgehog may generate different neuronal types in the spinal cord.

Placzek, M. 1995. The role of the notochord and floor plate in inductive interactions. *Curr. Opin. Genet. Dev.* 5:499–506.

A review of the action of Sonic Hedgehog in the organization of the neural tube.

Tanabe, Y., and Jessell, T. M. 1996. Diversity and pattern in the developing spinal cord. *Science* 274:1115–1122.

The role of Sonic Hedgehog and other ligands in patterning the cord.

CHAPTER

DEVELOPMENT OF MESODERMAL AND ENDODERMAL DERIVATIVES IN VERTEBRATES

In Chapters 4 and 5, we outlined how the middle, mesodermal layer of the early embryo forms during gastrulation: surface and subjacent cells get tucked and stuffed into the middle of the embryo mass by one means or another. This middle layer contributes cells to almost all the developing organs of the embryo.

In addition to contributing cells, mesoderm is sending and receiving signals from neighboring cells in its own and the other germ layers. This signaling is crucial for the proper development of the body's organs. In fact, it is the single most important theme in the development of vertebrate organs. Whether these dialogues between cells are called "cellular interactions," "tissue interactions,"

CHAPTER PREVIEW

1. The mesoderm of vertebrates forms the notochord, somites, kidneys, gonads, blood vessels, and limbs

2. Somites form the muscles, cartilage, and bone surrounding the spinal cord, and the dermis of the skin.

3. The limbs develop from interactions between flank ectoderm and lateral mesoderm.

4. Endoderm forms the gut and portions of associated organs.

5. Many "endodermal" organs, such as lungs, pancreas, and liver, require cellular interactions between mesoderm and endoderm in order to develop.

6. Muscle formation in the nonvertebrate ascidian embryo depends on a localized factor in the egg.

or sometimes "epithelial-mesenchymal interactions," it is hard to overestimate their fundamental importance.

DORSAL MESODERM

Mesoderm Becomes Determined During Gastrulation

As is the case for neural ectoderm, the mesodermal germ layer is regionally organized from its inception. Once gastrulation is complete, and the anteroposterior and mediolateral extent of the mesoderm has been realized, the mesoderm in different areas has been specified to develop along different pathways, thereby forming the various tissues. Mesoderm from the midline, for instance, can form notochord or muscle tissue if placed alone in culture, while mesoderm from more lateral areas can form kidney tubules or blood cells. When does mesoderm acquire this regional ability to differentiate into a specific tissue type in culture? The superficial answer to the question is: just before and during gastrulation.

Determination Is Defined Operationally

The **developmental potentiality** of a cell or group of cells, while an abstraction, is defined experimentally. What does a cell group become if isolated from the embryo and placed in culture? What would this same group of cells become if transplanted to a different location in the recipient embryo than its origin in the donor embryo? Or if placed into a recipient embryo of a different stage from its donor? As for the donor embryo, can it fill in the defect caused by extirpating that cell group? You may remember we briefly discussed how these operations are used to define specification and competence in Chapter 4.

Another aspect of this class of experiments is how to recognize what the transplanted or explanted cell group becomes. It is obviously helpful if the transplanted cell group can be reliably marked, for instance with a lineage tracer. Until recently, recognizing conventional histological hallmarks of different tissue types was the only means of identifying the final differentiated state of a group of cells. New means, now available, provide a powerful way to identify different tissue types: reliable antibodies or cloned cDNAs can be used to diagnose the presence of certain gene expressions, which in turn are indicative of particular kinds of differentiation. Much recent progress in our understanding has come from use of these methods (see Box 7.1).

A huge vocabulary has arisen around the subject of developmental potentiality. One will encounter words such as *determination*, *commitment*, *specification*, *autonomous development*, and *conditional specification*, to mention a few. Some biologists give these terms precise operational definitions, while others do not. It is important to remember that the experiment is what really defines the concept under discussion. Developmental potentiality often changes, gradually and progressively becoming narrower. Developmental fate, on the other hand, simply refers to what a cell or group of cells actually becomes in the normal embryo. Fate is most often ascertained through simple observation or lineage-tracing procedures carried out in the normal embryo.

The challenging experiment defining when mesoderm or endoderm cells are restricted to developing along a fixed pathway is to take a single cell or a few marked cells from different regions of the early embryo, and insert them into an unusual location in a recipient embryo. This has been done extensively with single cells in the *Drosophila* embryo. The results show that the cells of all three germ layers *gradually* acquire their definitive identity during gastrulation. This kind of experiment has also been done, though less extensively, in amphibians and to some extent in the chick embryo. For example, if one transplants some prospective endoderm cells, marked with a cell lineage tracer, from a late-blastula embryo into a region where mesoderm will form, the transplanted cells will sometimes form mesoderm (Figure 7.1). Their endodermal character has not been irrevocably fixed. As gastrulation commences and proceeds, a greater percentage of the transplanted cell groups will conform to their endodermal origin and form endodermal derivatives that express molecular markers characteristic of that germ layer. The story is similar for prospective mesoderm and ectoderm as well.

For vertebrates in general, then, we may say that, as far as the studies go, during the period beginning shortly before and continuing through gastrulation, cells gradually—and eventually irrevocably—adopt the potentiality of the different germ layers. The narrowing of possibilities for development within the confines of a given germ layer is also a gradual and progressive affair in vertebrates: A given cell, even though it may be destined to form only mesoderm tissues, is not necessarily restricted at this time to forming a certain kind of mesoderm. For example, a prospective mesoderm cell may be capable of forming muscle, while its progeny may also be able to form vertebral cartilage or body-wall mesenchyme. Thus, even though we state that the germ layers are determined during gastrulation, the potentialities of a given cell may still

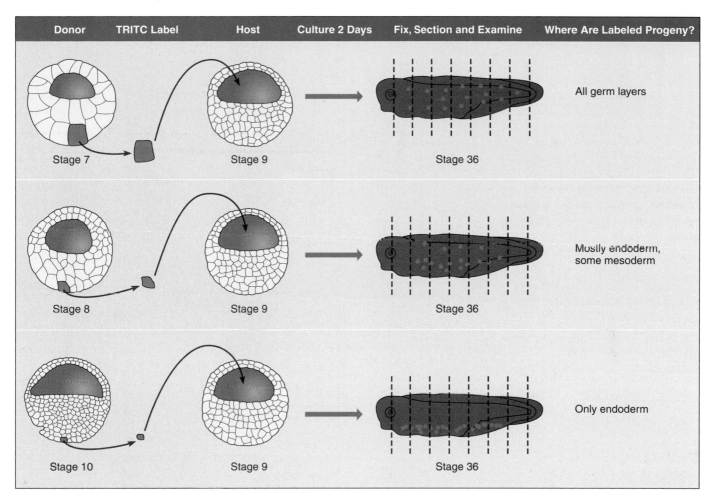

Donor	TRITC Label	Host	Culture 2 Days	Fix, Section and Examine	Where Are Labeled Progeny?

Stage 7 — Stage 9 — Stage 36 — All germ layers

Stage 8 — Stage 9 — Stage 36 — Mostly endoderm, some mesoderm

Stage 10 — Stage 9 — Stage 36 — Only endoderm

Figure 7.1 Testing the Developmental Potential of Single Cells by Transplantation Donor embryos are isolated from embryos of successively older stages and labeled by immersion in tetramethylrhodamine isothyocyanate (TRITC), a dye that stably labels cell surface proteins. A labeled cell from each donor is transplanted into the blastocoel of a late-blastula (stage 9) host, which is then allowed to develop into a stage 36 larva. The embryo is then fixed and sectioned, and the location and state of differentiation of the transplanted single cell are evaluated. In the examples shown here, the donor cell is prospective endoderm, and its commitment to endoderm gradually becomes fixed between stages 7 and 10.

be rather broad. This is a hallmark of vertebrates: the potentiality of a given cell or group of cells may become definitively narrowed and fixed only during the *post*gastrulation stages.

With these observations in mind, let us examine the development of the many tissues and organs formed by mesoderm. Throughout, we will pay particular attention to the cellular interactions that drive the whole process of determination and differentiation.

Dorsal Mesoderm Forms the Notochord and Somites

The mesoderm layer that underlies the surface ectoderm displays a clear organization as gastrulation nears completion.

In amphibians, the dorsalmost mesoderm, arising from the Spemann organizer, is subjacent to the forming neural plate; in amniotes, the corresponding area is the midline mesoderm left in the wake of the regressing Hensen's node or its equivalent in mammals. As shown in Figure 7.2, the dorsal midline cells congregate to form a rod, the *notochord*, only a few cells in diameter, that becomes surrounded by a cellular sheath and a basal lamina. (See Chapter 12 for a more thorough discussion of the basal lamina.)

Progressing from the anterior end of the spinal cord, at the level of the prospective hindbrain, to the posterior end, the notochord becomes clearly distinguishable as the cells form vacuoles into which proteoglycans (see Chapter 12) are secreted. Water enters the vacuoles, and the whole notochord

A. Dorsal View

B. Cross-sectional views

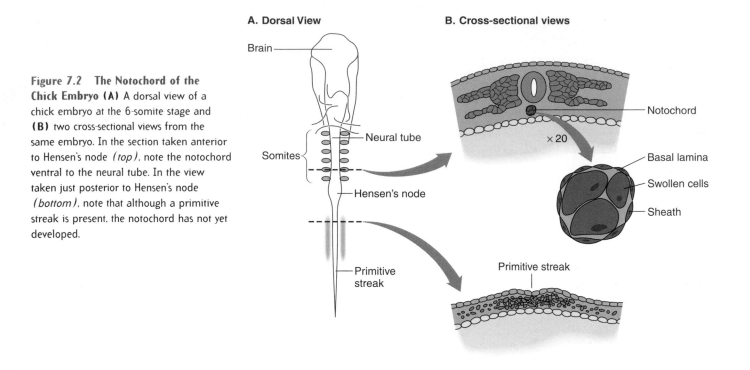

Figure 7.2 The Notochord of the Chick Embryo (A) A dorsal view of a chick embryo at the 6-somite stage and **(B)** two cross-sectional views from the same embryo. In the section taken anterior to Hensen's node *(top)*, note the notochord ventral to the neural tube. In the view taken just posterior to Hensen's node *(bottom)*, note that although a primitive streak is present, the notochord has not yet developed.

BOX 7.1 DETECTING GENE EXPRESSIONS

The tools for detecting expression of a particular gene are now the workhorses of laboratories carrying out research in developmental biology. Detection of gene expression is a vast subject, loaded with jargon and detail, but the principles are simple. Gene expression is usually diagnosed by detecting either a particular mRNA or the protein encoded by that mRNA. Detection of mRNA requires nucleic acid hybridization techniques outlined in Chapter 1 (Box 1.3), while detection of specific proteins often involves immunocytochemical methods that employ specific antibodies. The accompanying figure outlines the basics of these two kinds of techniques, further described below.

Detecting which embryonic cells or tissue is the site for expression of a particular

gene to produce an mRNA is often carried out by **in situ hybridization.** First the embryo or tissue is preserved, or fixed, with chemicals that maintain well the morphology of the tissue and do not chemically alter the mRNAs present. The embryo is then sectioned, just as in traditional histology, or it may be treated with agents (such as alcohols or proteases) that render it more permeable. The sections, or the whole embryo (if it is small enough), is then incubated with a cDNA (or other suitable) probe. As explained in Box 1.3, the probe is complementary to (and thus specific for) the sequences present in the mRNA being detected. The probe diffuses into the fixed, permeabilized tissue and hybridizes to the target sequence. Any

unhybridized copies of the probe are then washed away. If the probe is radioactive, a photographic emulsion placed over the sectioned tissue will detect the radioactivity, which causes visible silver grains to appear in the emulsion.

Another common in situ hybridization approach is to use a probe with a fluorescent reporter attached to it. The term *reporter* is a kind of shorthand used by researchers to designate a detectable molecule (fluorescent in this instance) that "reports on" (indicates the location of) another molecule of interest, such as a protein or mRNA. We shall have more to say about reporters in Chapter 14.

A third alternative is to use a probe that contains a modified purine base that can be detected by antibodies specific for that modification. Then the antibody, which itself is tagged with an appropriate enzyme or fluorescent dye, is used to locate the position of the nucleic acid probe, revealing those places in the embryo where gene expression is detectable.

The second class of techniques detects the protein encoded by the gene, rather than the mRNA. The probe is an antibody directed against the protein in question. If the antibody is specific, then a sensitive way of detecting the presence of the gene product is available. As with in situ hybridization, **immunocytochemical detection** may be carried out on sections or on the entire fixed embryo. After fixation, the tissues are rendered permeable. The sections or embryos are then flooded with nonspecific proteins

(such as serum albumin or milk protein) to reduce the possibility of the antibody binding to the tissues adventitiously. Following this preparation, the tissue is exposed to the antibody, excess antibody is washed away, and the tissue is exposed to a secondary antibody. Because it has been made in another species, this secondary antibody reacts with the species-specific determinants present on the surface of all members of the class of the primary antibody. So it is the secondary antibody that reports the location of the primary antibody.

The process is like making a sandwich: the primary antibody is specific for a given protein; the secondary antibody is specific for the primary antibody; and the secondary antibody possesses the reporter for the detection system, either a fluorescent molecule or

an enzyme. For example, we might expose a section from the tail bud of an amphibian embryo to a primary antibody, made in rabbits, against amphibian myosin. Then we would expose the already treated embryo to a secondary antibody. A secondary antibody would be made in goats against the rabbit globulins (against the blood fraction containing the soluble antibodies). This goat-antirabbit antibody would be given reporting capability by conjugating the enzyme alkaline phosphatase to it. After the embryo is treated with the secondary antibody and excess antibody has been washed away, we could detect the location of the primary antibody by incubating the section or embryo for alkaline phosphatase activity using a substrate that deposits a colored precipitate wherever the enzyme occurs.

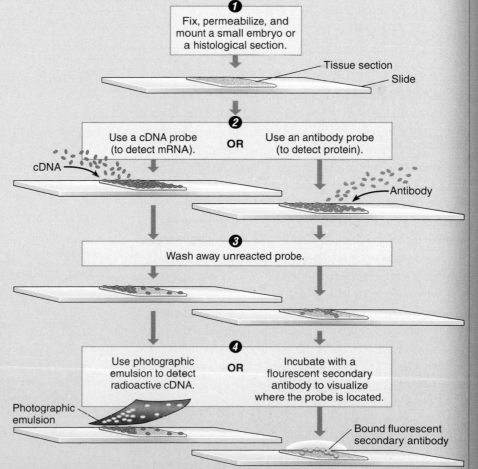

Using Probes to Monitor Gene Expression A small embryo (less than 100 μm in thickness) is fixed, permeabilized with lipid solvents, and exposed to a probe. The sample then can be mounted on a microscope slide, or treated while floating in solutions in a small tube and mounted later. The probe can be radioactive cDNA *(left)*, which will allow interaction with the cognate mRNA; or it can be an antibody directed against the protein encoded by that mRNA *(right)*. Excess, unreacted probe is washed away. *Left:* If the bound cDNA probe is radioactive, the mounted sample is covered with photographic emulsion; radioactive decay will cause formation of silver grains in the emulsion, allowing detection of the probe. An alternate way to detect cDNA probes is to label them with conjugated haptens, which allow detection with antibodies, as discussed in the text. *Right:* Antibody probes are usually detected with a secondary antibody directed against the class-specific antigenic sites of the primary antigen, as discussed in the text. If the secondary antibody has a fluorescent tag conjugated to it, fluorescence microscopy will detect it.

❶ Fix, permeabilize, and mount a small embryo or a histological section.

Tissue section
Slide

❷ Use a cDNA probe (to detect mRNA). **OR** Use an antibody probe (to detect protein).

cDNA
Antibody

❸ Wash away unreacted probe.

❹ Use photographic emulsion to detect radioactive cDNA. **OR** Incubate with a flourescent secondary antibody to visualize where the probe is located.

Photographic emulsion
Bound fluorescent secondary antibody

becomes stiffer, lengthening as the embryo extends. The notochord is thought to provide some skeletal support in the embryo. It is a transient structure, becoming incorporated into the vertebrae surrounding the spinal cord. As we saw in Chapter 6, it is formed by cells descended from the Spemann organizer and has considerable ability to induce neural tissues, probably by blocking of BMP signaling.

Situated on both sides of the developing notochord is the *presomitic (paraxial) mesoderm,* the cells of which passed through the primitive streak near the end of gastrulation in amniotes. Cells in this region will give rise to the complex, paired segmented structures on either side of the midline called *somites.* We should note here that dorsal mesoderm tissues are exposed to higher levels of BMP antagonists, and hence to lower levels of active BMP. As we move laterally and ventrally along the mesoderm layer, the concentration of active BMP increases, thereby creating a gradient. According to one view, this gradient of active BMP plays an important role in the progressive, regional differentiation of the mesoderm—low levels of BMP favoring dorsal mesoderm differentiation, high levels favoring more lateral and ventral mesoderm tissues.

The Somites Arise Progressively

Beginning at the anterior end of the developing spinal cord, the mesoderm cells that lie on either side of the notochord and flank the forming neural tube aggregate into distinct groups. This aggregation process forms the somites, whose development progresses sequentially from front to back, to the terminus of the spinal cord (Figure 7.3). Paraxial mesoderm in the head region does not form somites.

The paraxial mesoderm in which the somites develop shows a periodic expression of several genes, among them *c-hairy,* a homolog of *hairy* in *Drosophila* (see Chapter 15). This periodic gene expression may be responsible for the formation of segments; in *Drosophila,* the timing of *hairy*'s expression coincides well with the timing of segment formation. In the chick, *c-hairy* is expressed at high levels in the paraxial mesoderm about every 90 minutes, which is the time needed for one somite pair to form. When paraxial mesoderm adds a new segment, the high level of *c-hairy* expression is maintained in the posterior portion of that new somite.

Expression of *c-hairy* may be part of a complex molecular clock. Interestingly, when expression of the chick homologs of the *Drosophila* genes *delta, notch,* or *fringe* is altered, *c-hairy* expression cycles and segmentation rates both change concordantly.

The Somites Are Multipotential

The somite is a complex, transient embryonic structure. Soon after it forms, the cells in its medioventral portion become less tightly aggregated, allowing them to abut the notochord and developing spinal cord. This portion of the somite, called the **sclerotome,** gives rise to segmented cartilage that in turn are precursors to the vertebral column and proximal portions of the ribs.

In contrast to the sclerotome, the dorsal portion of the somite retains an epithelial character. It is called the **dermamyotome** because it is comprised of *myotome,* which develops into muscle, and *dermatome,* which develops into dermis (the vascularized layer of mesodermal origin below the epidermis). In particular, the medial portion of the dermamyotome in the anterior portion of each somite forms the segmented epaxial musculature associated with the vertebrae and the spinal cord. The lateral portion of the dermamyotome provides cells that will form the muscles of the body wall, limbs, and distal portions of the ribs. And, as you may remember, some of the ectodermal neural crest cells will migrate into the developing dermis, where they differentiate into the pigment cells of the skin. Figure 7.4 shows the progression from newly formed somite into sclerotome, dermamyotome, and finally separate dermatome and myotome; also shown are the relative positions of portions of the developing kidney—the pronephros and mesonephros. We shall soon consider them.

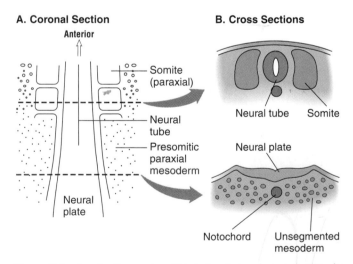

Figure 7.3 Somite Formation (A) A dorsal view (coronal section) through the chick embryo at the level just anterior to Hensen's node. The neural plate is flanked by paraxial mesoderm that is becoming segmented sequentially from anterior to posterior end. **(B)** Cross-sectional views of the same chick embryo, showing the status of the paraxial mesoderm at levels where it is segmented *(top)* and still unsegmented *(bottom).*

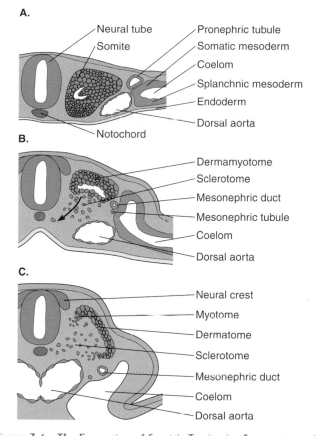

Figure 7.4   The Formation of Somitic Territories Cross sections of a chick embryo, after 36 hours of incubation. **(A)** Early in its development, the somite is an epithelial polygon with some loose mesenchyme at its center. **(B)** As somite development progresses, which happens first in the more anterior somites, some of the cells in the ventromedial portion migrate ventrally *(arrow)*, establishing a loose mesenchyme, the sclerotome, which will later form cartilage. **(C)** At a later developmental stage, the dorsal epithelial portion of the somite forms the dermatome and myotome.

Cellular Interactions Occur in the Somite

Single cells from the prospective somitic mesoderm, when transplanted to ectopic sites, will not form muscle unless transplantation is carried out late in gastrulation. The irrevocable determination to form muscle is finalized only as gastrulation ends, before the actual somite forms. What influences a cell to follow the pathway to becoming cartilage, or muscles of the back or lateral body wall, or dermis? The answer is the immediate environment. Many interactions are occurring between prospective somites and adjacent tissues, including the developing neural tube, the notochord, the overlying ectoderm that will form epidermis, and the nonsomitic mesoderm lying lateral to the prospective somite.

The problem of how cells come to follow one or another alternative pathway has been studied for many decades. In the case of the somite, many researchers have isolated portions of the somites and co-cultured them with various nearby tissues. Figure 7.5 outlines one version of this approach, a variation of the basic isolation experiment discussed previously. If paraxial mesoderm is isolated from a chick or frog embryo before the somites have formed, and placed in a nutrient medium in culture, little or no obvious differentiation will take place (Figure 7.5A). If a fragment of notochord or ventral neural tube is placed next to the mesoderm or newly formed somite, masses of cartilage will differentiate in culture (Figure 7.5B,C). The cartilage is easily identified by its characteristic morphology and staining with various dyes. If neural tube—especially dorsal neural tube—and some overlying ectoderm are cultured with the explanted mesoderm, typical striated muscle will form (Figure 7.5D). If ectoderm or lateral mesoderm (sometimes called lateral plate mesoderm) is included with the paraxial mesoderm, muscle fiber formation may be delayed or stopped (Figure 7.5E).

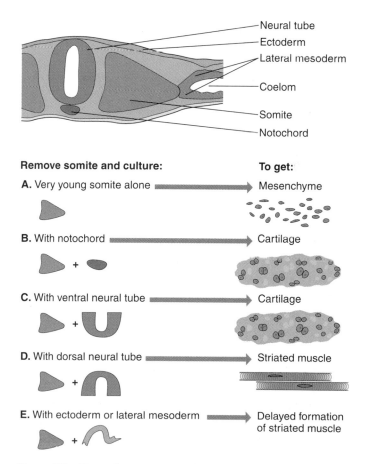

Figure 7.5   Tissue Interactions and Somite Development A diagrammatic representation of various types of somitic explants. To see what kinds of development ensue, the early somite (epithelial stage) is isolated and cultured, either **(A)** alone, or in combination with **(B)** notochord, **(C)** ventral neural tube, **(D)** dorsal neural tube, or **(E)** ectoderm or lateral mesoderm.

While these numerous tissue interactions have been appreciated to some extent for a long time, recent findings have deepened our understanding of them. The recent identification of some of the molecular participants in tissue interactions and differentiation is typical of the progress in most of developmental biology. First, many of the genes necessary for the formation of the tissue-specific proteins characteristic of a given tissue have now been isolated and identified. For instance, in the case of muscle, the genes encoding the proteins that constitute the contractile machinery—actin, myosin, tropomyosin, troponin, among others—have been cloned and characterized. Furthermore, several important members of the transcription factor families that are necessary for these genes to be transcribed have also been isolated and characterized. Two transcription factors, MyoD and Myf5, are specific for muscle cells. If the genes encoding MyoD and Myf5 are inserted into a variety of tissue culture cells, such cells are then converted into muscle cells. So, even if a fully developed muscle cell with contractile abilities does not result in a given experiment, we now have tools to diagnose whether the cells have moved at least partway along the pathway to "muscleness."

Second, and just as important, progress in cell biology has identified many of the proteins serving as signals and signal receptors, so that molecules possibly involved in the tissue interactions are rapidly being identified.

And finally, there are now new means to test, in living cells in live embryos, whether the suspected genes and the protein factors that they encode are actually doing the job suspected for them. We shall discuss many of these approaches and experiments as we proceed. For example, a homolog of the *Drosophila* gene *delta* (which encodes a signaling molecule) has recently been isolated in mice. Transgenic mice that are homozygous for a null mutation in this gene have been created (see Chapter 5). The homozygous mutant mouse dies midway through gestation and has defects in the paraxial mesoderm and the central nervous system. Examination of the somitic regions shows that, although there is some differentiation of cartilage and muscle, there is little or no segmentation, and defined somites with an anteroposterior polarity never form. The protein encoded by *delta* is somehow involved in a signaling pathway necessary for forming the segmental pattern of somites. It is important to note that even though segments do not form, differentiation of muscle and cartilage does occur. Hence, this experiment tells us that the two events—somite border formation, and later differentiation of different tissues from the somite—are not inexorably linked.

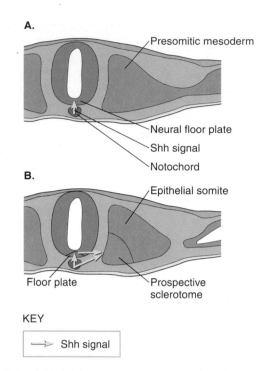

Figure 7.6   A Model for the Role of Shh in Sclerotome Formation (A) The notochord produces Shh, but the paraxial (presomitic) mesoderm is not yet competent to respond. The Shh produced by the notochord does, however, induce the neural floor plate to produce even more Shh. **(B)** It is only when floor plate and notochord together produce high levels of Shh that the nearby ventromedial portion of the somite responds, changing gene expression rates (*pax1* increases, *pax3* decreases) to favor sclerotome differentiation.

In Chapter 6, we met a prominent and important signaling molecule in somites with the curious name Sonic Hedgehog (Shh for short); it is a homolog of the *Drosophila* signaling molecule dubbed Hedgehog (Hh). Shh is synthesized by the notochord and ventral portion of the neural tube (the so-called floor plate), and can be shown to influence nearby cells. When presomitic paraxial mesoderm is removed from the embryo and cultured in vitro with tissue culture cells that secrete Shh, the somite cells show a robust expression of gene markers characteristic of sclerotome, in other words, cartilage differentiation. Thus, Shh is present at the right time and in the correct place to participate in sclerotome induction, and the effects of Shh in vitro lend powerful support to the hypothesis that Shh is important in cartilage formation (Figure 7.6).

The powerful new tools available for studying tissue interactions in molecular terms present a problem, albeit a "good" one. New genes and proteins playing a role in a given kind of embryonic development are being discovered almost daily, making cataloging them and their effects overwhelming, not only for the student but for the specialist as well. Consequently, it becomes increasingly important to identify the salient features of these interactions and to learn how to gain access to available information when needed. It becomes less important, perhaps impossible, to memorize the names and precise roles of all the players. Each tissue interaction considered in this text, insofar as information is currently available, probably involves many different signaling molecules, their cognate receptors, second messengers, and response systems. It is not only the Spemann organizer that emits a cocktail of "factors." Many, perhaps all, of the tissue interactions that underlie organogenesis utilize complex mixtures of signaling factors and activated receptor circuits.

Figure 7.7 depicts in schematic form the tissues present in the dorsal midline, and shows some of the recently identified signaling molecules that influence "decision making" in the

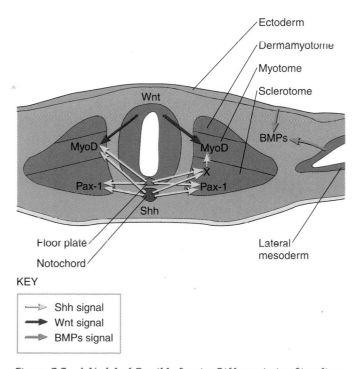

Figure 7.7 A Model of Possible Somite Differentiation Signaling
Shh produced by the notochord and the floor plate signals Pax1 to induce sclerotome formation, as shown in Figure 7.6. In addition, Shh acts on the more dorsal somite cells, either directly (shown on the left) or indirectly through some unidentified intermediate (X, on the right); these dorsal somite cells are also exposed to Wnt ligands from the dorsal neural tube, and the combined action of Shh and Wnt favors myotome differentiation. MyoD is a transcription factor important for muscle differentiation. Ectoderm and lateral mesoderm produce BMPs, which act on the dermamyotome anlage and counteract Wnt-like influences there.

somite. There are three primary effects on the early, naive somite: When influenced by factors emanating from the dorsal neural tube or ectoderm (such as BMPs and Wnt), the somite is encouraged to develop into dermamyotome. Under the influence of factors secreted from the ventrally situated notochord and floor plate (for example, Shh), the somite develops into sclerotome. And when dorsal and ventral influences are both present at appropriate levels, the myotome can form.

Striated Muscle Differentiates as Syncytia

The prospective muscle cells of the myotome are called *myoblasts*. (Remember that *blast* is a general suffix denoting precursor cells of terminally differentiated cell types.) Myoblasts continue to divide, embarking on a remarkable program of muscle formation (Figure 7.8) whose developmental sequence has been studied in detail in tissue culture. The signal for the next step is provided by manipulating the culture medium; shifting the medium from "rich" to "poorer" signals the cells to withdraw from the cell cycle, to remain in a kind of permanent G_1 state, termed G_0. (Differentiated neurons do the same thing.) The signal for myoblasts in the embryo to withdraw from the cell cycle is still unknown, but it is noteworthy that each myoblast population for a given muscle (or muscles) in its own particular location has its own particular timing.

The nondividing myoblasts follow a distinctive pathway resulting in the activation of certain muscle-specific genes and the capability of fusing with one another. The myoblasts become elongated and spindle shaped, and cell extremities fuse, thereby forming long cylinders of many nuclei in a common cytoplasm, each cylinder a syncytium. The terminal program of gene expression directs the synthesis of proteins for the contractile machinery; actin, myosin, and the other contractile proteins are laid down in a repeated pattern along the length of the cell, thus providing a striated appearance to the muscle in histological sections (see Figure 1.1B).

There are many variations on this well-understood theme. Head muscles of the jaw and face are striated and develop similarly but without somite formation. Heart muscle is distinctive, though similar to striated skeletal muscle. Smooth muscle of the digestive tract and other organs also undergoes terminal differentiation, forming contractile assemblies with forms of actin and myosin that are slightly different from those in cardiac and skeletal muscle. Furthermore, smooth muscle cells remain individual and do not fuse to form a syncytium.

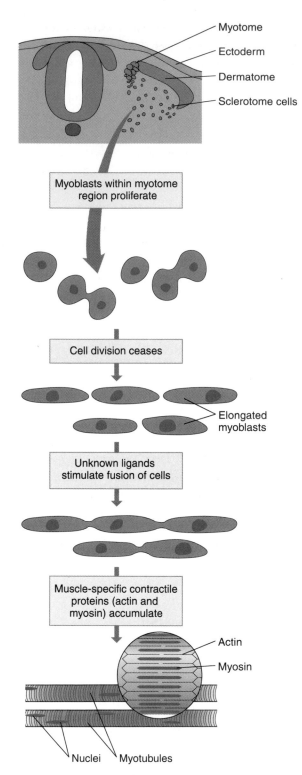

Figure 7.8 Striated Muscle Formation A diagrammatic view of how striated muscle differentiates. The myotome produces proliferating myoblasts, which then stop dividing, elongate, and fuse together to form syncytial myotubes. These myotubes then synthesize and assemble the contractile machinery.

So far, we have seen that the formation of axial musculature from somites is a long, stepwise process in vertebrates. The story is rather different, however, in a group of animals closely related to vertebrates. Because ascidians (commonly called sea squirts) have played such an important role in research, it is worth examining them (see Box 7.2) before continuing our discussion of the organs and structures derived from dorsal mesoderm in vertebrates.

The Sclerotome Forms Axial Cartilage, Which Then Is Converted to Bone

The vertebrate skeleton develops first through the formation of cartilage. The sclerotome forms cartilage of the vertebrae and ribs, while lateral mesoderm will provide cartilage for the long bones of the limbs. The cartilage is a template for bone, which forms within the cartilage by a process called endochondral ossification. (Most bones of the skull form directly in mesenchymal cells, derived from neural crest, without a cartilage template, a process called intramembranous ossification.) When sclerotome cells aggregate, the central cells become chondrocytes (cartilage cells) that secrete copious matrix; other cells form a perichondrial sheath, called the *perichondrium*, which surrounds the central cartilaginous core.

Bone forms within the cartilage by a complex process in which the chondrocytes cease division and swell, when they are referred to as hypertrophic cartilage; in addition, the surrounding matrix undergoes chemical changes (Figure 7.9A,B). Blood vessels invade the hypertrophic cartilage, and bone-forming osteoblasts move into the area. These osteoblasts secrete bone matrix and begin bone formation by depositing hydroxyapatite, a crystalline form of calcium phosphate. Growth and morphogenesis of bone depend on an accurate template and a carefully regulated replacement of the cartilage by bone. Premature ossification results in bones that are small, short, or both; conversely, abnormally slow rates of ossification can lead to bones that are very large and/or long.

Not surprisingly, much of this ossification process is controlled by cellular interactions in which secreted chemical signals and their cognate receptors play a huge role. One of the BMP family (BMP2) is known to stimulate cartilage formation and inhibit muscle formation in mesoderm cells. The chondrocytes secrete a Hedgehog family member called Indian Hedgehog (Ihh); Ihh in turn stimulates the perichondrium to secrete another ligand (parathyroid hormone, or PTH) that inhibits hypertrophic cartilage formation. This negative feedback loop can then inhibit further cartilage differentiation.

Absence of this control loop leads to excessive hypertrophic cartilage and abnormal bone formation (Figure 7.9C).

The Dermatomal Portion of the Somite Forms the Dermis

As we discussed, the medial dermamyotome forms muscle, as do some cells in the somite next to the lateral mesoderm. The remaining portion of the dorsal, epithelial portion of the somite is a population of cells that continue to divide and migrate under the surface ectoderm, forming the dermis of the skin. This process is stimulated by factors released from the overlying ectoderm and the dorsal neural tube.

The development of the skin and its derivatives is a joint venture between the surface ectoderm and the underlying dermal mesenchyme derived from the dermatome. As mentioned in Chapter 6, the ectoderm forms the epidermis, a stratified multilayered epithelium. Surface cells accumulate large quantities of the protein keratin and undergo apoptosis. The surface epidermis cells are constantly being shed and replaced through mitosis of epidermal blast cells lying on the basement membrane. Beneath the basement membrane is the vascularized, mesenchymal dermis. Sweat and sebaceous glands can form in the dermis. Specialized derivatives of the integument (such as scales, feathers, and hair) develop in a complex pathway involving many interactions between the ectoderm and mesoderm tissues.

Tooth formation is also similar to the development of glands in the integument; sequential and reciprocal interactions between the epithelial and mesenchymal tissues lead to the

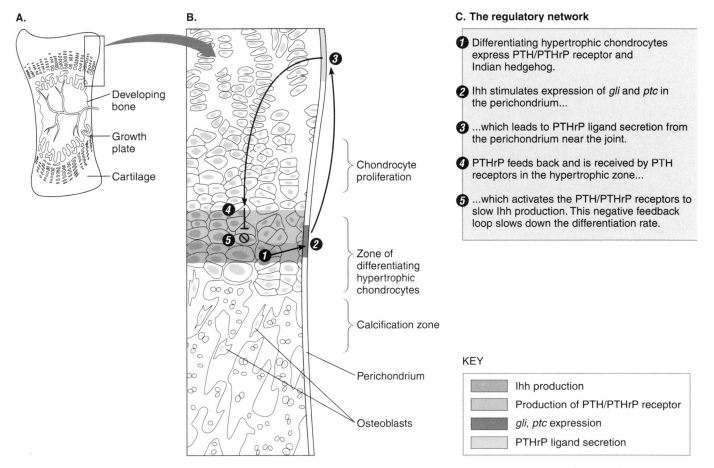

C. The regulatory network

① Differentiating hypertrophic chondrocytes express PTH/PTHrP receptor and Indian hedgehog.

② Ihh stimulates expression of *gli* and *ptc* in the perichondrium...

③ ...which leads to PTHrP ligand secretion from the perichondrium near the joint.

④ PTHrP feeds back and is received by PTH receptors in the hypertrophic zone...

⑤ ...which activates the PTH/PTHrP receptors to slow Ihh production. This negative feedback loop slows down the differentiation rate.

Labels in B:
Developing bone
Growth plate
Cartilage
Chondrocyte proliferation
Zone of differentiating hypertrophic chondrocytes
Calcification zone
Perichondrium
Osteoblasts

KEY

Ihh production
Production of PTH/PTHrP receptor
gli, *ptc* expression
PTHrP ligand secretion

Figure 7.9   The Ossification of Cartilage The conversion of cartilage to bone tissue in a "long" bone. **(A)** The general regions of the developing bone. **(B)** A magnified view of the region of ossification of hypertrophied cartilage. **(C)** The signal pathways regulating cartilage differentiation. Proliferating chondrocytes differentiate so that they express Ihh. The chondrocytes also begin to express the PTH receptor. The perichondrial cells respond to Ihh and upregulate (increase) expression of the genes *gli* and *patched (ptc)*. This leads to the production of parathyroid hormone (PTH). The prehypertrophic chondrocytes, which possess the receptor for both PTH and PTHrP (parathyroid hormone related peptide), are prevented from moving down the differentiation pathway, thereby limiting the extent of differentiation. When this repressive feedback loop is abrogated, chondrocyte differentiation occurs too extensively, disturbing the development of ossified bone.

Box 7.2 Ascidian Development

The ascidians, or sea squirts, are a class consisting of approximately 2,300 species of sessile marine animals in the subphylum Urochordata (tunicates and their allies). Ascidian embryos have played an important historical role in the analysis of mechanisms of development. Like *Caenorhabditis elegans*, ascidians have an invariant, stereotyped pattern of cleavage. The cell lineages of the different tissues of the tadpolelike larvae are well known (see accompanying figure).

Two observations placed ascidians at the forefront of analytical research early in the 20th century. First, it was noted that the eggs of some species contain colored granules that become differentially distributed to different blastomeres. E. G. Conklin, working at the Marine Biological Laboratory at Woods Hole, Massachusetts, in the early 1900s, found that, at the eight-cell stage, yellowish cytoplasm became localized to a single pair of blastomeres known as B4.1 cells; he then

noted that descendants of these cells formed the axial muscles of the larval body. Hence, the localized yellow cytoplasm is often called *myoplasm*. Second, it was observed that removal of these blastomeres causes lack of muscle development. When researchers deleted the B4.1 cells from eight-cell embryos, no muscle developed in otherwise normal embryos. On the other hand, when the isolated B4.1 cells were allowed to continue development in isolation, they formed

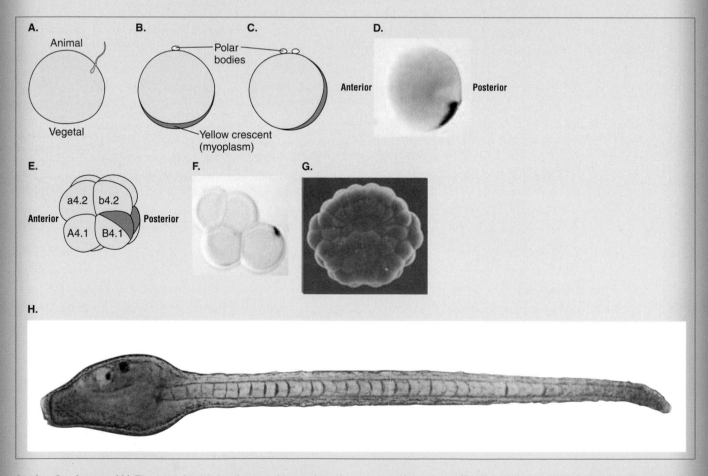

Ascidian Development **(A)** The egg is fertilized in the animal hemisphere. A tough chorion (not shown) surrounds the egg. **(B)** The fertilized egg completes meiosis I, forming a polar body. A cytoplasmic contraction occurs, resulting in a concentration of the myoplasm, called the yellow crescent in some species, at the vegetal pole. **(C)** As meiosis II is completed, the myoplasm shifts to the prospective posterior side. **(D)** A whole-mount in situ hybridization shows that Macho-1 mRNA is localized in the posterior vegetal portion of the zygote. **(E)** After three cell divisions, there are four pairs of cells, as shown in this lateral view. The myoplasm is localized in the B4.1 pair of blastomeres. **(F)** Whole-mount in situ hybridization has shown that Macho-1 is present only in the B4.1 blastomeres. **(G)** A scanning electron micrograph (vegetal view) of a normal embryo at the 76-cell stage, which occurs about 10 hours after fertilization. **(H)** This fully developed tadpole, viewed from the side, reveals the row of muscle cells.

muscle. Later work showed that when cytoplasm was transferred from B4.1 cells to other, non-muscle-forming blastomeres, the recipient cells acquired the ability to form muscle.

These observations lent support to the hypothesis that the yellow cytoplasm localized in B4.1 cells might be a localized maternal determinant for "muscleness" in the egg. Recall that the idea of maternal determinants was discussed in Chapter 1 and exemplified by the discovery of Bicoid in *Drosophila*.

We now know that some of the details of this story need revision. The B4.1 blastomeres give rise to only 28 of the 42 muscle cells of the tadpole. Another 14, the so-called secondary muscle cells, are derived from neighboring blastomere sets b4.2 and A4.1, and the development of these secondary muscle cells is dependent on cellular interactions that occur during gastrulation. Other cellular interactions are also required for the formation of various tissues, including both endodermal and ectodermal derivatives. So the ascidian embryo is not really a mosaic of autonomously developing cells after all.

The development of muscle from B4.1 does, however, seem to exemplify autonomous development, as the elusive determinant present in B4.1 has now been identified. Hiroki Nishida and Kaichiro Sawada, working in Tokyo, have shown that the mRNA for a transcription factor named Macho-1 is localized in B4.1 cells, and that Macho-1 is necessary and sufficient for muscle development. Macho-1 is a member of an important group of transcription factors called *zinc finger* proteins because they contain a protein domain that binds zinc. Transcription factors are discussed further in Chapter 14. If Macho-1 protein is eliminated with an antisense oligonucleotide, primary muscle will not form. (**Antisense oligonucleotides** are short chains—less than about 30 nucleotides long—that are usually composed of some form of deoxynucleotide. Because they are perfectly complementary to some portion of the mRNA, they hybridize to it; the hybrids are susceptible to attack by nucleases specific for RNA-DNA hybrids. In some instances, the hybrid formed between the antisense oligonucleotide and the mRNA cannot be translated by ribosomes, and consequently the protein encoded by the mRNA cannot be synthesized. Whatever the precise mechanism involved in any particular experiment, antisense oligonucleotides can result in loss-of-function phenotypes.) If Macho-1 mRNA is injected into a non-muscle-forming blastomere, that blastomere now has the ability to form muscle. Macho-1 seems to be the identity of the classical localized maternal determinant posited nearly a century ago for primary muscle cell formation in ascidians.

formation of teeth, and many signaling molecules already mentioned—especially the BMPs, members of the Hedgehog family, and fibroblast growth factors—are involved. As we pointed out in Chapter 6, much of the mesenchyme of the head originates from the neural crest. This is true for the jaw mesenchyme; the teeth derive from ectoderm, which forms the enamel, and from neural crest mesenchyme, which forms the dentin.

LATERAL MESODERM

The Kidneys and Gonads Derive from Mesoderm That Is Lateral to the Somite

Moving laterally from the somite, one comes to the region just dorsal to the forming coelom often called the intermediate mesoderm. This area is also called the **nephrotome** because it forms the tubules and ducts that make up the kidney (Figure 7.10). Some of these ducts will be co-opted or shared with the developing reproductive system. Differentiation progresses posteriorly through the nephrotomic mesoderm as development proceeds.

The kidney consists of two parts: tubules, which filter liquid and small molecules from the blood, and ducts, which reabsorb some of these materials and also deliver the processed filtrate to the cloaca. The tubules differentiate progressively from anterior to posterior end, and the more posterior tubular structures are more complex. Some relatively primitive vertebrates, such as the hagfishes (Agnatha), form tubules solely from anterior portions of the nephrotome. These animals possess a primitive and rather inefficient kidney called a **pronephros**. Amphibians and fish develop some pronephric-type anterior tubules that degenerate, after which the nephrotome at midbody levels forms more complex tubules comprising the definitive adult kidney, called a **mesonephros**. Amniotes

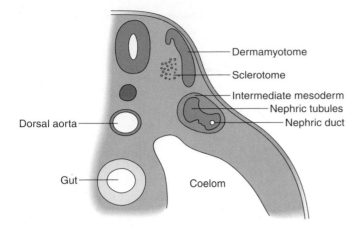

Figure 7.10   The Nephrotome A cross section through the prospective thoracic region of a stereotypical vertebrate. emphasizing the positions of the derivatives of the nephrotome. the nephric tubules. and nephric duct. The mesoderm ventral to the somite. which includes the nephrotome and prospective gonad. is the intermediate mesoderm.

also develop a few transitory pronephric-type tubules and an extensive mesonephric tubule system that functions during part of their embryonic life. The definitive kidney tubules of adult amniotes develop in the most posterior nephrotome of the trunk, a region called *metanephrogenic mesenchyme,* and the definitive kidney is called a **metanephros.** Figure 7.11 shows this succession of kidney tubules and types.

The epithelial tubules forming in the anterior nephrotome connect to the **Wolffian (pronephric) duct.** At about the same time, this duct segregates from the anterior nephrotome. The tip of the duct then appears to grow posteriorly toward the cloaca. Extensive investigations have shown that in some amphibians, the Wolffian duct elongates by a migration mechanism: cell numbers in the wall of the duct increase through mitosis, and the leading tip migrates along a path marked by cues from the overlying ectoderm. But in other species, extension of the duct also involves local recruitment of cells from the nephrotome in the region of the advancing tip. When mesonephric tubules form, they connect to this pronephric duct, which will become the definitive ureter of the adult. All of these developments are more complex in amniotes than in

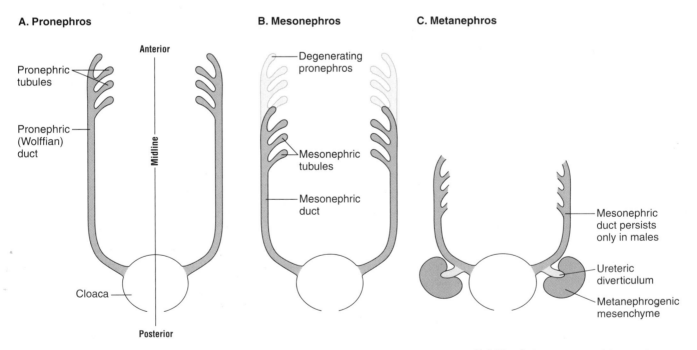

Figure 7.11   The Pronephric. Mesonephric. and Metanephric Excretory Systems Schematic dorsal views showing the arrangement of nephrons and ducts in developing amniotes. The nephrons are depicted as tubules at the anterior end of the nephric duct. The actual complexity of the nephrons is much greater than that shown here. and increases as development progresses from the **(A)** pronephric through **(B)** mesonephric to **(C)** metanephric kidneys. After the pronephros degenerates. the posterior portion of the pronephric (Wolffian) duct persists and is sometimes called the mesonephric duct. Please note that the colors used here are for clarity and do not designate germ layers: all the structures shown are mesodermal.

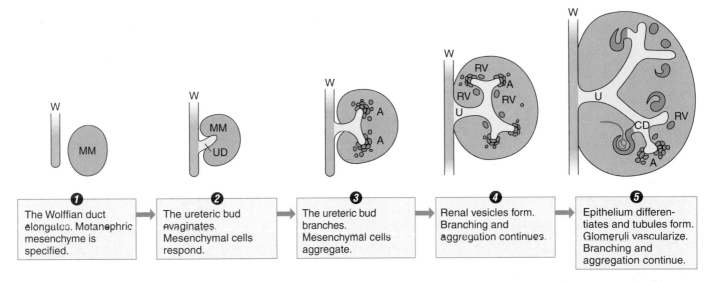

Figure 7.12   The Development of the Metanephric Kidney Schematic diagrams showing the principal activities at the various stages. Abbreviations: A. aggregation of mesenchyme: CD. collecting duct: G. developing glomerulus: MM. metanephrogenic mesenchyme: RV. renal vesicle: UD. ureteric diverticulum: U. ureter: W. Wolffian duct. Please note that the colors used here are for clarity and do not designate germ layers: all the structures shown are mesodermal.

amphibians and fish. A Wolffian duct forms in amniotes, as does a mesonephric kidney, but the mesonephros is transitory. Experiments using the chick embryo have demonstrated that pronephric duct formation requires BMP4, which is secreted by the ectoderm overlying the intermediate mesoderm. Near the region where the Wolffian duct joins the cloaca, a bud diverts from the duct and extends into the metanephrogenic mesenchyme of the posterior trunk. This *ureteric diverticulum* (see Figure 7.11) stimulates the metanephrogenic mesenchyme to "condense." From these two components—the diverticulum and the mesenchyme—the definitive amniote metanephric kidney develops (Figure 7.12): the metanephrogenic mesenchyme will transform into epithelial tubules comprising the filtration and solute reabsorption unit of the kidney, while the ureteric diverticulum will form the collecting ducts of the kidney and the ureter. The interactions between the epithelium of the bud and the mesenchyme were revealed by a classic experiment in embryology; we shall consider them further in Chapter 13.

In amniotes, the pronephros never forms anything but a few poorly developed tubules, and the mesonephros is transitory. The Wolffian duct, even though it no longer serves a function for the kidney, is retained in males to provide some of the plumbing of the reproductive system. In female embryos, the relatively lower testosterone levels cause the Wolffian duct to degenerate.

Another duct, called the **Müllerian duct,** develops in the intermediate mesoderm parallel to the Wolffian duct and is present for a while in the embryos of both sexes. However, the Müllerian duct does not persist in males because an anti-Müllerian duct factor induces apoptosis of its cells. In females, which do not produce this factor, the Müllerian duct persists to form the oviduct, the uterus, and part of the cervix (Figure 7.13).

The Gonads Form in Conjunction with the Nephrotome

The gonads differentiate in mesoderm that is ventromedially adjacent to each nephrotome, in an area called the genital ridge (Figure 7.14). The coelomic epithelium of this ridge proliferates to form "sex cords," which penetrate the mesenchyme. Primordial germ cells (PGCs) invade the area, migrating from elsewhere (as we shall soon discuss). After this point, male and female embryos take different developmental paths. The means by which sex is ultimately determined are complex and often vary in different animals (see Box 7.3).

In the male, the sex cords continue to proliferate, becoming the seminiferous tubules. Sertoli cells differentiate in the tubules and secrete the anti-Müllerian duct factor. Resident PGCs form the spermatozoa, and the mesenchyme of the genital ridge becomes the interstitial tissue of the testis. The seminiferous tubules eventually connect to the Wolffian duct,

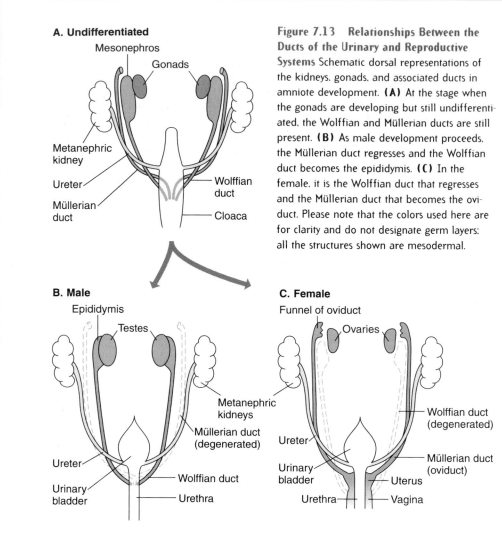

A. Undifferentiated

Mesonephros

Gonads

Metanephric kidney

Ureter

Müllerian duct

Wolffian duct

Cloaca

B. Male

Epididymis

Testes

Metanephric kidneys

Müllerian duct (degenerated)

Ureter

Wolffian duct

Urinary bladder

Urethra

C. Female

Funnel of oviduct

Ovaries

Wolffian duct (degenerated)

Ureter

Urinary bladder

Müllerian duct (oviduct)

Uterus

Urethra

Vagina

Figure 7.13   Relationships Between the Ducts of the Urinary and Reproductive Systems Schematic dorsal representations of the kidneys. gonads. and associated ducts in amniote development. **(A)** At the stage when the gonads are developing but still undifferentiated. the Wolffian and Müllerian ducts are still present. **(B)** As male development proceeds. the Müllerian duct regresses and the Wolffian duct becomes the epididymis. **(C)** In the female. it is the Wolffian duct that regresses and the Müllerian duct that becomes the oviduct. Please note that the colors used here are for clarity and do not designate germ layers; all the structures shown are mesodermal.

which becomes the vas deferens of the male reproductive system.

In the female genital ridge, the first sex cords degenerate and a second set proliferates from the coelomic epithelium. These epithelial growths remain near the surface, forming the prospective follicle cells that surround the oocytes, which develop from the PGCs. In females, the Wolffian duct degenerates, while the Müllerian duct persists to form the oviduct, which collects and transports maturing oocytes.

Curiously, in vertebrates and many other animals, the PGCs do not arise from the developing gonad but migrate there from other locations. In amphibians, the PGCs arise in the endoderm. In mammals, they originate in the yolk sac endoderm and migrate through the developing dorsal mesentery of the gut to the genital ridge. In the chick, the PGCs arise from endoderm anterior to the head, the extraembryonic endoderm; from there they migrate into the developing ridge by means of the developing circulatory system. This long-distance homing of PGCs to the genital ridge is an example of extraordinary selective migration; we shall consider it further in Chapter 13.

Lateral Mesoderm Is Multipotential

Lateral to the nephrotome, the unsegmented mesoderm splits to form the coelom. The signals that impel this separation are not presently known. The mesoderm that is applied to the ectoderm is the **somatopleure,** and will form a mesenchymal layer that encircles the coelomic space. Some body musculature will form here; sclerotome-derived cells in this region will form ribs. Importantly, the outgrowths that form the limbs will arise from the somatopleure. Recall, too, that in amniotes the somatic mesoderm forms part of the amnion and chorion of the extraembryonic membranes, as discussed in Chapter 5.

The mesoderm layer that remains applied to the endoderm is the **splanchnopleure,** or splanchnic mesoderm. The cardiovascular system, heart, blood vessels, and blood cells will arise from this layer. (Some blood vessels of the kidney and body wall will form from the somatopleure.) Also, the smooth muscles of the gut, mesenteries, and part of the lining of the coelom will originate in the splanchnopleure. And, when we consider the endoderm later in this chapter, we shall see that many organs associated with the digestive system receive essential contributions from the splanchnic mesoderm.

DEVELOPMENT OF THE VASCULAR SYSTEM

Hematopoiesis (Blood Formation) Is a Stepwise Process

The earliest blood vessels and erythrocytes of birds and mammals form in the splanchnic mesoderm overlying the yolk sac, outside the body proper of the embryo. Mesenchymal cells form aggregates, called blood islands; these produce a reticulated network in which the peripheral cells of the aggregate form endothelium and the central cells differentiate into

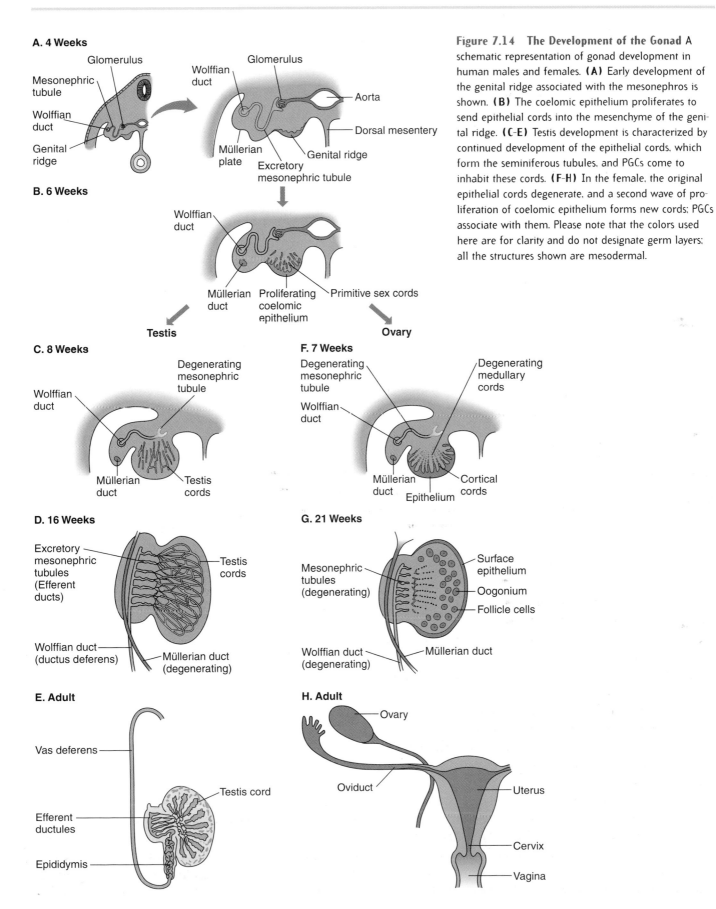

Figure 7.14 The Development of the Gonad A schematic representation of gonad development in human males and females. **(A)** Early development of the genital ridge associated with the mesonephros is shown. **(B)** The coelomic epithelium proliferates to send epithelial cords into the mesenchyme of the genital ridge. **(C-E)** Testis development is characterized by continued development of the epithelial cords, which form the seminiferous tubules, and PGCs come to inhabit these cords. **(F-H)** In the female, the original epithelial cords degenerate, and a second wave of proliferation of coelomic epithelium forms new cords; PGCs associate with them. Please note that the colors used here are for clarity and do not designate germ layers; all the structures shown are mesodermal.

Box 7.3 Sex Determination

During the development of the mammalian reproductive system, both male and female gonads develop from the same genital ridge. Furthermore, at early stages, both males and females possess both Wolffian and Müllerian ducts. What factor determines which pathway of differentiation—male or female—the embryo will follow? And does this pattern of development hold for animals other than mammals?

In mammals, the determining factor is the chromosomal constitution of the embryo. Genes on the Y chromosome, especially the *sry* gene, are essential for development of the testis. Since the female genome is XX and has no Y chromosome, the target genes regulated by the expression of *sry* are not activated in females. This chromosomal sex determination mechanism is called *primary sex determination.* If there are chromosomal abnormalities, or mutations in genes downstream from the primary-sex-determining genes, rare abnormalities such as hermaphroditism can occur.

Normally, the expression of *sry* in males activates pathways resulting in the production of testosterone and the anti-Müllerian duct hormone; lack of *sry* expression in females results in estrogen production. These steroid sex hormones drive *secondary sex determination,* which involves differentiation of the entire reproductive system, including the genitalia and ducts leading from the testis and ovary, as well as other secondary sexual characteristics. We should mention that in adult mammals, steroid hormones with either estrogen-like or testosterone-like activity may be produced in other organs, especially the adrenal cortex. Adult humans of both genders produce both testosterone and estrogen, but in males testosterone predominates and in females estrogen predominates.

The kind of gametes produced in mammals depends in part on the environment of the developing gonad. Hence, it is possible for XX germ cells in an XY testis to begin differentiating into sperm; however, normal sperm do not develop because the process is not completed. Apparently, although gonadal environment is an important determinant in germ cell development, the genetic makeup of the germ cells is also involved.

In other vertebrates, a variety of other factors can affect sex determination, including the ambient environment in which the organism exists. For instance, in many turtles the path of sex differentiation depends on the temperature experienced by the developing early embryo. Eggs developing in the range of 20° to 27°C may be predominantly of one sex, those developing above 30°C predominantly of the other.

Other phyla may use completely different systems of sex determination. In *Drosophila,* for example, primary sex determination is regulated by the ratio of the number of X chromosomes to the number of autosomes (nonsex chromosomes, often abbreviated A). There are six autosomes (two each of chromosomes 1, 2, and 3) and two sex chromosomes (X and Y) in a diploid fruit fly. When the ratio of X chromosomes to autosomes is 1:3 (2 X and 6 A), the sex differentiation is female. When the ratio is 1:6 (1 X, 1 Y, and 6 A), the differentiation is male. (The Y chromosome does not influence the outcome—it is strictly the X/A ratio that counts.) These outcomes imply that there is some kind of counting mechanism that translates into differential gene expression affecting sexual development. Indeed, the genetic and molecular basis of this phenomenon has been worked out. A gene called *sex lethal (sxl)* on the X chromosome is activated when there are two X chromosomes but not when there is only one. When the protein Sxl is made (in females), it leads to a specific splicing event in the transcripts of the *double sex (dsx)* gene that causes cells to follow a female developmental pathway. Individuals without Sxl splice the *dsx* transcript differently, leading to male development. Sex-specific splicing will be discussed in more detail in Chapter 14.

---

erythrocytes (Figure 7.15). A similar process proceeds within the mesoderm in the embryo, forming a network of endothelial tubes arising from mesodermal condensations. The larger tubes—prospective arterioles and venules—will also be surrounded by smooth muscle layers. Most, perhaps all, of the earliest cells in the circulation are erythrocytes that originate in yolk sac blood islands. Very soon, however, the endothelial lining of the dorsal aorta and nearby blood vessels will proliferate to provide additional primitive erythrocytes and stem cells (see Box 6.2 for a discussion of stem cells).

Most internal blood vessels are generated by in situ formation of endothelial tubes, a process called **vasculogenesis.** Some vessels are generated wholly or in part by active immigration of the tip of endothelium into the surrounding tissue, a kind of invasion called **angiogenesis.** This angiogenic sprouting formation of vascular beds is of tremendous importance in adults as well as embryos: it is needed in forming new circulatory beds after surgery; it also occurs in the vascularization of tumors. Vascular endothelial growth factor (VEGF) and angiopoietin are recently discovered growth factors that stimulate angiogenesis. VEGF has been shown to be involved in the vascularization of tumors (by means of angiogenesis), and VEGF is also important for the growth of blood vessels in embryos.

Arteries in the embryo often grow along the same routes taken by nerves, and recent experiments show that the nerves, or the Schwann cells around them, stimulate the growth of blood vessels and their development into arteries. Developing veins do not apparently respond to stimuli from nerves.

Amphibians do not have a yolk sac, so large blood islands form instead in the posteroventral mesoderm; much of amphibian hematopoiesis and vasculogenesis is similar to what occurs in amniotes.

The Principal Sites of Erythropoiesis Change During Development

Three successive changes occur in the cardiovascular system as the embryo matures: the principal sites of hematopoiesis shift from one defined location to another, the types and proportions of cellular constituents of blood change, and the molecular constitution of the hemoglobin in erythrocytes also changes.

Although the liver always plays an important role in *erythropoiesis* (red blood cell formation) in the early vertebrate embryo, the principal sites vary somewhat in different groups. Fish, amphibians, birds, and mammals each have their own definite, and different, succession of erythropoietic sites. In mammals, for example, hematopoiesis typically begins in the yolk sac blood islands; soon thereafter more stem cells are formed in the lining of the dorsal aorta. Stem cells from the aorta (and perhaps the yolk sac) then populate the liver, where erythrocytes for the developing fetus are produced until late in gestation. Finally, stem cells that have populated the developing bone marrow become the definitive source of erythropoiesis for the remainder of the mammal's life.

Mammalian yolk sac erythrocytes are nucleated; they also contain distinctive forms of the constituent polypeptide chains in the hemoglobin molecule, the so-called embryonic forms of α- and β-globins called ζ (zeta) and ε (epsilon). Erythrocytes from the embryonic liver are nonnucleated and contain adult α chains of hemoglobin, but they possess a distinctive, fetal form of β chain denoted γ (gamma). Finally, adult bone marrow is the site of the usual $\alpha\alpha\beta\beta$ tetramer of globin chains constituting the protein of the principal adult hemoglobin in humans. (See Chapter 14 for further discussion of hemoglobins and control of the expression of their genes.) The fetal form of hemoglobin has a higher oxygen affinity than does adult hemoglobin—a useful property for

A.

B.

C.

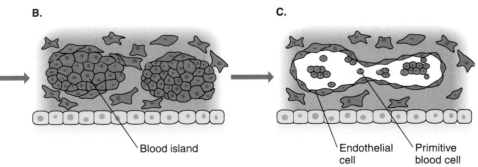

Yolk sac endoderm   Mesenchyme cells

Blood island

Endothelial cell   Primitive blood cell

D.

Figure 7.15   Blood Island Formation The development of blood islands in the yolk sac splanchnopleure of the chick embryo is shown in successive stages: (A–C) cross sections through the yolk sac; (D) a surface view of the embryo surrounded by the area opaca vasculosa. **(A)** Some of the mesenchymal cells begin to form **(B)** tight epithelial clusters. **(C)** The cells in each cluster then form a coherent endothelial covering. Cells in the interior of the cluster become more loosely associated and develop into the first cohort of erythrocytes. **(D)** The developing endothelial tubes anastomose (join together) in netlike fashion.

the embryo, which depends on oxygenated maternal blood for its very existence.

The cells in the earliest circulating blood are all erythrocytes. Only approximately halfway through embryonic development do various classes of lymphocytes begin to appear. As with red blood cell development, there are several sites of lymphocyte production, and in the neonatal animal, formation of the definitive population of lymphocytes is still not complete. Forming the entire gamut of white blood cell types involves the thymus, spleen, bone marrow, and other lymphoid organs. An incredibly complex but immensely interesting and important set of interactions and changes is involved in the development of lymphocytes. This subject—at the heart of the discipline of immunology—is too immense to pursue further in this text.

The Heart Forms in Anterior Splanchnic Mesoderm

The splanchnic layer of the anterior ventral mesoderm forms an endothelial tube, connected to the rest of the developing vascular "tree," that develops a specialized muscular layer called the myocardium. The muscle fibers of the myocardium follow a unique pathway to form cardiac muscle, and the tube located here forms the heart. In amphibians, the leading edge of the ventral mesodermal mantle forms the endocardial rudiment, which hollows out to form the lining of the heart. In amniotes, heart tube formation occurs in splanchnic mesoderm on both sides of the developing gut, and only when the endoderm and its adjacent splanchnic mesoderm fuse in the midline are the two separately forming heart rudiments joined. Figure 7.16 shows this fusion sequence diagrammatically.

Prospective heart cells move through Hensen's node and the anterior streak during gastrulation. These cells may initially occupy an area in the splanchnic mesoderm that is much larger than the area finally contributing to the heart. Do the prospective heart cells migrate within the mesoderm and congregate to define the heart area, or do only the prospective heart cells in the definitive heart area complete their differentiation? Further research is needed for an answer to this question.

The heart first functions as a single tube, with blood entering posteriorly through the sinus venosus, then moving through the atrium and ventricle, and subsequently passing anteriorly out the bulbus cordis (Figure 7.17A). As the heart develops and

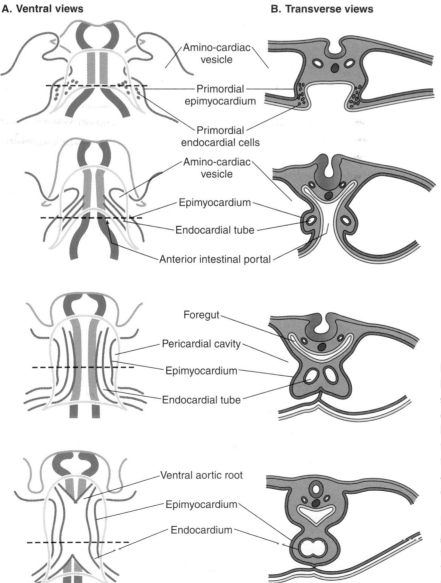

A. Ventral views

B. Transverse views

- Amnio-cardiac vesicle
- Primordial epimyocardium
- Primordial endocardial cells
- Amnio-cardiac vesicle
- Epimyocardium
- Endocardial tube
- Anterior intestinal portal
- Foregut
- Pericardial cavity
- Epimyocardium
- Endocardial tube
- Ventral aortic root
- Epimyocardium
- Endocardium

Figure 7.16   The Bilateral Origin of the Heart Diagrams of chick embryos at successively later stages (*top to bottom*), showing how the heart develops at approximately 25 to 30 hours of incubation: **(A)** the horizontal lines in the ventral views indicate the level of cross section shown in **(B)** the transverse views. The heart forms as two separate amniocardiac vesicles and endothelial tubes (endocardium) on either side of the developing gut (anterior intestinal portal) of the chick embryo. Notice that, as the endoderm invaginates to form a gut at this level, the two heart tubes are brought together, after which they fuse to form a single, unified tube. A muscular epimyocardium surrounds the endocardium.

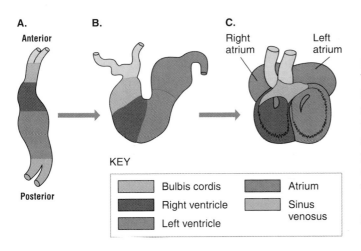

A.

Anterior

Posterior

B.

C.

Right atrium   Left atrium

KEY

Bulbis cordis	Atrium
Right ventricle	Sinus venosus
Left ventricle	

Figure 7.17   The Looping of the Heart During Its Development
This schematic shows how the single tube of the developing mammalian heart loops to adopt its final form. The heart is viewed throughout from the ventral side. **(A)** The bilateral tubes have fused in the midline to form a single tube, with the prospective bulbus cordis and right ventricle situated anteriorly and the sinus venosus and atrium posteriorly. The two posterior vessels feeding into the sinus venosus are the common cardinal and omphalomesenteric veins (see Figure 7.19). For simplicity, these veins and the sinus venosus are not shown in the rest of the figure. **(B,C)** The tube then bends and loops over itself so that the ventricular portion becomes displaced posteriorly.

becomes more complex, it forms a large loop to the right side (Figure 7.17B), then curls over itself so that what was originally the anterior ventricular portion, with its well-developed myocardium, now comes to lie posterior to the atrial portion (Figure 7.17C). Chambers develop within the original simple tube as a result of cell division and migration of endocardial cells into an extracellular matrix that forms between the endocardium (the inner epithelial lining of the chamber) and the muscular myocardium. These endocardial outgrowths form septa that separate the chambers of the heart from one another.

The Anatomical Disposition of the Blood Vessels and Heart Results from Extensive Remodeling

Learning the anatomical details of the developing vascular system used to require a considerable investment of time and energy from embryology students. There were some good

reasons for this. The nomenclature was helpful to aspiring medical students. Furthermore, the anatomical details of the vascular system (together with those of the skeleton) constituted the bedrock of comparative anatomy, which was so important to understanding the evolution of vertebrates. Those reasons still have force, and two more reasons can be added: the remodeling during vascular development illustrates the extraordinarily dynamic morphogenesis that continues after initial organogenesis is complete; it also highlights the crucial role played by growth factor signaling in regulating this morphogenesis. When this remodeling goes awry, the heart or blood vessels may not function properly, endangering the life of the organism. We shall therefore set forth some essentials of this subject, leaving extensive detail aside.

Figure 7.18A shows the initial arrangement of blood vessels in the region anterior to the heart, the basic plumbing of the developing vertebrate. Subsequent development will modify

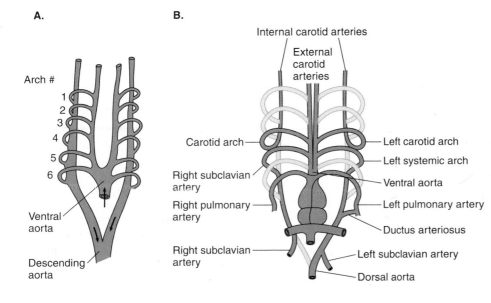

A.

Arch #

1
2
3
4
5
6

Ventral aorta

Descending aorta

B.

Internal carotid arteries

External carotid arteries

Carotid arch

Right subclavian artery

Right pulmonary artery

Right subclavian artery

Left carotid arch

Left systemic arch

Ventral aorta

Left pulmonary artery

Ductus arteriosus

Left subclavian artery

Dorsal aorta

Figure 7.18   The Development of the Aortic Arches Ventral schematic views of **(A)** the six aortic arches found in all vertebrates and **(B)** their further development in mammals. The ventral aorta originating from the heart gives rise to six paired loops, the aortic arches, which course through the mesenchyme of the pharyngeal (or branchial) arches. The third pair of arches give rise to the internal carotid arteries to the head, while the external carotid arteries develop from the remains of the degenerating first pair of arches. The left fourth arch gives rise to the dorsal aorta, which supplies the rest of the body, the yolk sac, and the placenta with arterial blood. The sixth pair of arches develops into the pulmonary arteries.

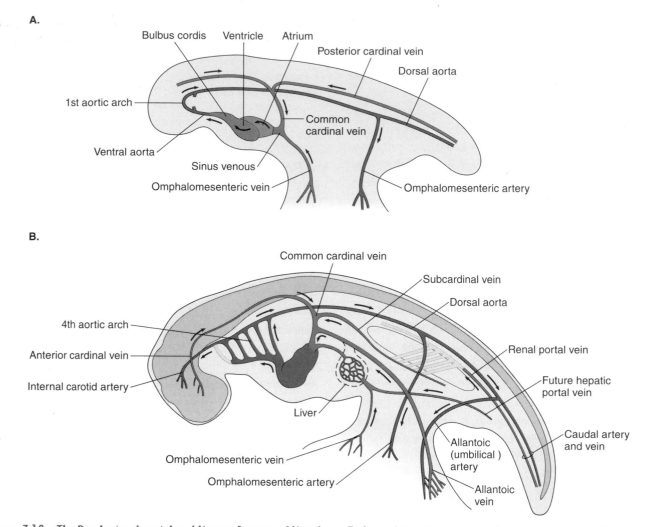

Figure 7.19 **The Developing Arterial and Venous Systems of Vertebrate Embryos** Stylized lateral views of the major arterial and venous vessels of stereotypical amniote embryos. **(A)** An early stage of development. after the heart and early vessels have formed. **(B)** The situation somewhat later. after the branchial arch vasculature has developed and been modified. and vascularization of some of the major organs is under way. Dashed line segments indicate early developing blood vessels that later regress.

this arrangement somewhat, depending on the vertebrate group. In the initial anatomical arrangement, all blood leaves the heart, pumped by the muscular myocardium of the ventricle, through the bulbus cordis (see Figure 7.17A). The bulbus cordis gives rise to the ventral aorta, which runs anteriorly below the developing pharynx. Noticeable ridges, called pharyngeal (or branchial) arches, develop in this prospective head and neck region. The arches contain mesenchyme from the neural crest sandwiched between head endoderm and ectoderm, though arches do form even if cranial neural crest is experimentally eliminated. Between the branchial arches, endoderm is closely apposed to ectoderm that is invaginating to form pharyngeal pouches. In aquatic, nonam-

niote vertebrates, these pouches will perforate, at which point they are called clefts or (gill) slits. In amniotes, the pharyngeal arches never develop into gills.

Developing within the pharyngeal arch mesenchyme are the *aortic arches,* which connect the ventral aorta to the dorsal aorta on each side of the pharynx. In aquatic nonamniote vertebrates, these aortic arches form the capillary bed of the gills. In amniotes, they get extensively remodeled: some of the aortic arches atrophy and disappear, while others develop into major arteries of the head and neck. In mammals in particular—as Figure 7.18B reveals—the left fourth aortic arch persists to form the arch of the dorsal aorta, the major passageway for blood to travel from the heart to the posterior portions of the body. The third pair of arches and anterior portions of the ventral aorta give rise to major arteries of the head: the internal and external carotid arteries. The sixth pair of arches is modified to form the pulmonary arteries to the lungs. The subclavian arteries to the anterior limbs arise from the right fourth aortic arch and the dorsal aorta. The dorsal aorta courses posteriorly from this pharyngeal region to supply blood to the entire posterior portion of the body. Various major branches arise from this dorsal aorta, supplying kidneys, digestive tract, yolk sac (or placenta), and so on. Even after birth, a huge amount of remodeling takes place to produce the very complex vascular tree supplying blood to the mammalian body.

The venous system of amniotes is similarly constructed by modifying a relatively simple structure. An initial framework, shown in Figure 7.19A, collects blood from capillary beds that are forming in all nascent tissues. Venous blood empties into the common cardinal vein and then into the entry chamber of the heart, the sinus venosus. Blood then enters chambers of the atrium, passes through valved portals to the ventricle, and then moves out through the bulbus cordis once again. Venous blood drains the yolk sac or placenta, and all the internal organs including the kidneys and the developing digestive system (Figure 7.19B). Blood from the latter gets diverted through the liver on its way to the common collection point of the common cardinal vein. Extensive remodeling occurs, including dramatic changes at birth, when the lungs start to respire and the yolk sac or placenta no longer function.

LIMB DEVELOPMENT

Limbs Develop from Somatic Mesoderm of the Flank

The anterior and posterior limbs have similar cell types and extensive similarities in overall structure. But they are differently organized—and clearly, in vertebrates they vary in "type" of limb: wings (birds and bats differ), fins, flippers, hooves, and the like. The lateral somatic mesoderm—which is the source of the cartilage, bones, and connective tissue of the limb—proliferates to form limb buds anteriorly and posteriorly (Figure 7.20A). There is now some evidence that the positioning of these limb buds, and hence the instructions to these particular lateral mesoderm cells to follow the path of "limbness," is encoded in a set of genes (hox genes) that direct the overall anteroposterior body pattern. We shall consider these again in Chapter 16.

As the limb buds grow, the **apical ectodermal ridge (AER)**, a ridge of tissue delimiting dorsal from ventral at the tip of the bud, becomes prominent (Figure 7.20B). Elongation of the limb bud occurs; then mesenchymal condensations, which are harbingers of cartilage, appear. Ectoderm and mesoderm differentiate into the skin and feathers (or hairs) of the limb's integument. The limb has an obvious organization in which proximodistal, anteroposterior, and dorsoventral axes are clearly discernible (Figure 7.21). The tissues between the developing skeletal elements of the digits may persist to provide webbing, or may undergo programmed cell

A. Side view of a 3-day chick embryo

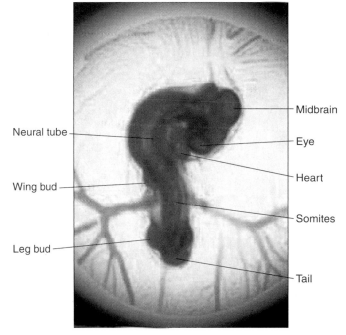

B. Cross section of a chick wing bud

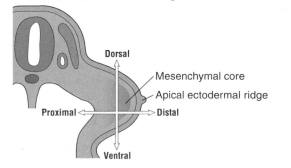

Figure 7.20 The Limb Buds (A) A side view of a 3-day chick embryo: the wing and leg buds are just becoming evident as bulges from the flank. **(B)** A cross section through the embryo at the level of the developing limb bud: note the locations of the developing limb bud and the AER.

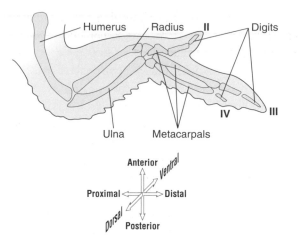

Figure 7.21   The Skeleton of the Chick Wing The major skeletal elements of the right wing of a 9-day chick embryo shown diagrammatically. In the orientation axes, dorsal would lie above and ventral below the page.

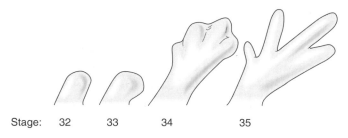

Figure 7.22   Apoptosis of Interdigital Tissue of the Limb Outlines of the leg bud of the chick embryo at successive stages. The regions between the forming digits undergo programmed cell death and sloughing of tissue between stages 32 and 35 (approximately 7.5 to 8.5 days of incubation).

death (apoptosis) during the development of fingers and toes (Figure 7.22).

The Muscles of the Limb Originate in the Myotome

Elegant transplantation experiments using marked cells have shown that not all the mesoderm tissues of the limb arise from lateral somatic mesoderm. Embryos of Japanese quail and of the domestic chicken develop similarly to each other, and transplants can easily be made between them. Yet the cells can be distinguished from each other because the nuclei in quail cells stain more intensely. Figure 7.23 shows the experiment in which somites 15 through 20 have been removed from a chick embryo and quail somites put in their place. In the resulting limb, the muscles are constructed from quail cells, even though the rest of the mesoderm tissues are from the chick. From this experiment, we can conclude that the prospective muscle cells of the limb migrate from the myotome and come to reside in the developing limb bud.

Limb Outgrowth and Organization Result from Tissue Interactions

A complex and interesting set of tissue interactions governs the development of the limb. Some have been known for decades, due to the pioneering work of John Saunders. Molecular details of these interactions are now being provided in an explosion of new information about limb development. The basic tools for defining these interactions have been the customary deletion and transplantation experiments.

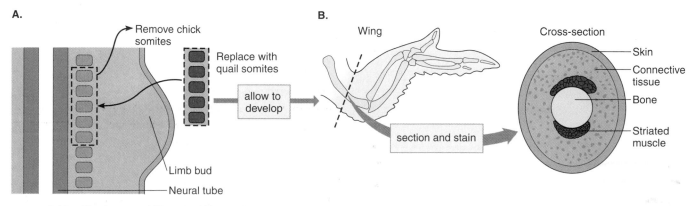

Figure 7.23   The Source of Wing Bud Muscle Cells An experiment designed to determine the origin of cells that form the muscles of the wing. **(A)** Somites on one side in a chick embryo next to the wing bud (somites 15 through 20) are removed and replaced with quail somites of the same number and developmental age. **(B)** The embryo is allowed to develop, and the limbs are examined in cross section for the presence of quail cells, which can be distinguished from chick cells by using appropriate stains. Muscles of the limb clearly derive from quail cells, though most of the rest of the limb derives from chick cells. These results indicate that striated muscle of the limb is comprised of cells that migrated from the somite.

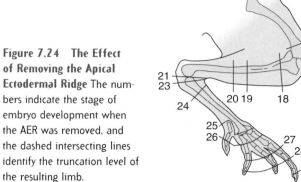

Figure 7.24 The Effect of Removing the Apical Ectodermal Ridge The numbers indicate the stage of embryo development when the AER was removed, and the dashed intersecting lines identify the truncation level of the resulting limb.

Figure 7.24 shows the result of deleting the apical ectodermal ridge, an experiment first carried out by Saunders. Deletion of the AER stops limb outgrowth; the later the ridge is removed, the more distal are the final structures that develop. When the AER is removed early, the stumpy limb that forms has only upper-limb structures. When the AER is removed later, forearm structures develop too. Therefore, the AER is necessary for patterned limb outgrowth. The mesoderm under the AER is an area of mitosis, as successive cell cycles lay down increasingly distal limb structures. The entire proximodistal pattern of the limb is set up in about seven cell cycles (about a 128-fold increase in cell number). Further rounds of mitosis produce a major increase in limb size.

Besides the AER and underlying mesenchyme, two other major sources of interactions help organize the limb. One is the ectodermal covering, which provides signals that organize the dorsoventral organization of the limb tissues. The other is an area of mesoderm known as the zone of polarizing activity (ZPA), which was also discovered in transplantation experiments by Saunders. The ZPA, located in the posterior mesoderm of the limb bud (see Figure 16.19), governs the anteroposterior organization of the limb. We shall return to the limb and its tissue interactions in Chapters 12, 13, and 16 because limb development provides us with important information about the molecular basis of pattern formation.

ENDODERMAL ORGANS

Endoderm Is Gradually Determined Before and During Gastrulation

When and how does endoderm acquire its distinctive identity and regional character? Studies in which portions of the prospective endoderm of the amphibian blastula were excised and placed in culture showed that the explanted cells could differentiate, in vitro, into regionally distinct endoderm tissues. Recent transplantation experiments involving small numbers of marked prospective endoderm cells show that in the mid- and even late blastula, prospective endoderm cells can form mesoderm or even ectoderm if transplanted to those regions. Gradually, however, prospective endoderm cells acquire the ability to retain their original endodermal identity even when challenged by transplantation to an unusual environment. By midgastrula, prospective endoderm cells are determined to form different regional tissues of the gut regardless of their new position. Experiments with various growth factors suggest that members of both the FGF and TGF-β families, as well as a growth factor called VegT, play important roles in the progressive determination of endoderm.

Endoderm Is the Source of the Digestive System and Associated Organs

As a result of gastrulation, the endoderm germ layer lines the archenteron and forms a tube stretching from the anterior to posterior ends of the embryo. The way this tube gets formed in different vertebrate groups varies, of course. Nonetheless, a tube it is, though the anterior and posterior entrances to the tube may form in some species by secondary thinning and apoptosis. For example, the contracted blastopore present in the amphibian gastrula, which will become the anus of the tadpole, is already "open" to the environment (see Figures 4.13 and 4.17), whereas the oral opening for tadpole's mouth forms later, during differentiation of the pharynx (Figure 7.25A), from a perforation at the anterior end of the endodermal tube. In amniotes the midgut remains continuous with the extraembryonic yolk sac; the endodermal tube is blind-ended, both anteriorly and posteriorly, and will only form the cloacal and oral openings many days later.

The endoderm is permanently surrounded along its entire length by mesoderm—head mesoderm anteriorly, splanchnic mesoderm throughout the body. The endoderm will form the epithelial linings of the various portions of the digestive system, sometimes called the alimentary canal. Furthermore, epithelial morphogenesis in concert with the local mesenchymal covering will occur in a number of defined locations along the alimentary canal; these outpocketings will give rise to secretory and endocrine glands of the head and neck, the lungs, the liver and gallbladder, and the pancreas. It should not surprise you that the development of all these organs involves reciprocal tissue interactions between the endodermal epithelium

A.

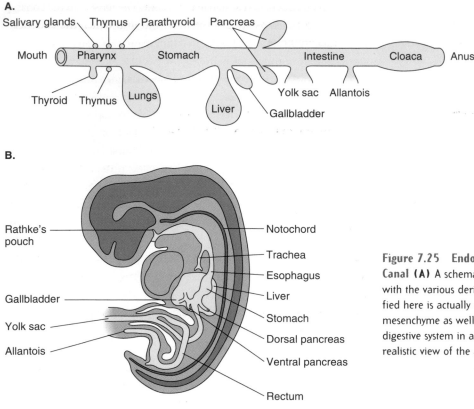

B.

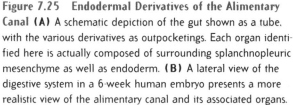

Figure 7.25 Endodermal Derivatives of the Alimentary Canal (A) A schematic depiction of the gut shown as a tube. with the various derivatives as outpocketings. Each organ identified here is actually composed of surrounding splanchnopleuric mesenchyme as well as endoderm. **(B)** A lateral view of the digestive system in a 6-week human embryo presents a more realistic view of the alimentary canal and its associated organs.

and the surrounding mesodermal mesenchyme. Figure 7.25 sketches the various districts of the alimentary canal and associated organs which have endodermal contributions.

The Pharyngeal Region Forms Many Important Organs

As we mentioned earlier in the chapter in discussing the anatomy of the arterial system, all vertebrates develop a series of arches in the wall of the pharynx. The branchial aortic arches develop in the neural crest–derived mesenchyme. The endoderm associated with the arches and the adjoining endoderm of the pharyngeal floor proliferate and invaginate in association with the adjacent mesenchyme to form the thyroid and parathyroid glands, and the thymus, as shown diagrammatically in Figure 7.26. Salivary glands (not shown on the figure) also form from endoderm and mesoderm in this pharyngeal region.

The tip of the archenteron (see Figure 4.13) fuses with ectoderm anteriorly to form the oral plate, which later perforates to form the mouth. Anterior to this oral plate is an inpocketing of ectoderm called Rathke's pouch, which later contacts the diencephalon. Recent evidence suggests that the diencephalon induces the tissue of Rathke's pouch to form the anterior pituitary.

The region immediately posterior to the oral plate is the oral cavity; extending from here to the lung buds is the pharynx. In aquatic nonamniote vertebrates, gills form from the branchial arches and gill slits from the clefts of the pharynx. In terrestrial vertebrates, the whole system is transient and rudimentary—no perforations or gills form—but the pouches between the arches do give rise to adult structures, as shown in Figure 7.26. Pouch I forms the eustachian tube and pouch II the palatine tonsil. The thyroid arises from the floor of the endoderm in the area of these two pouches. Pouches III and IV give rise to portions of the thymus and parathyroid gland. A number of different salivary glands also form from pharyngeal endodermal epithelium evaginating into the surrounding mesenchyme. The salivary epithelium forms tubes that repeatedly branch; eventually, these terminate in blind-ended structures called *salivary acini* (singular *acinus*), which produce the salivary secretions. A complex set of interactions between the salivary epithelium and salivary mesenchyme regulates this branching behavior, as we shall discuss in detail in Chapter 13.

The branchial arches, especially arches I and II, are the loci of formation for the upper (maxillary) and lower (mandibular) jaws. Teeth develop here under the influence of reciprocal interactions between the epithelium and the mesenchyme. A number of growth factors and transcription factors necessary for forming the jaws and the tooth rudiments have been discovered. The alimentary canal continues posteriorly, forming the esophageal portion first.

The Lungs Originate from the Prospective Esophageal Region of the Gut

Just posterior to the pharynx in the floor of the archenteron, a groove develops and deepens, penetrating the surrounding mesenchyme to form a tubular tracheal bud (Figure 7.27). The tracheal bud forms an epithelial tube that branches to form bronchi; these continue to divide, proliferating an epithelial "tree" whose ends terminate in the blind sacs (alveoli)

A.

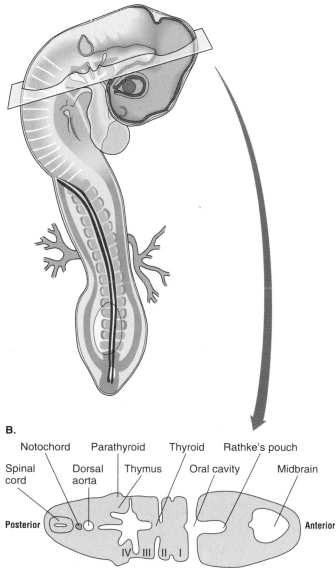

B.

across which gas exchange will occur. The mesenchyme constituting the substance of the lung, and into which the endodermal epithelium grows, plays a dominant role in this branching behavior; some of the genes and growth factors involved are beginning to be identified (see Chapter 13).

The Liver and Pancreas Form from Endoderm of the Stomach and Duodenum

Posterior to the esophagus, the endoderm differentiates to form the specialized cells that constitute the lining of the stomach. Also, endodermal outpocketings in this region form the liver and pancreas. The liver forms as an evagination called the *hepatic diverticulum*, which proliferates into the surrounding mesenchyme; the latter in turn induces this evagination to proliferate, branch, and differentiate as the glandular part of the liver. The base of the liver diverticulum forms the hepatic duct, and a branch of the diverticulum forms the gallbladder.

Slightly posterior to the hepatic diverticulum, in the region of the future duodenum, dorsal and ventral pancreatic diverticula arise, these too projecting into, and interacting with,

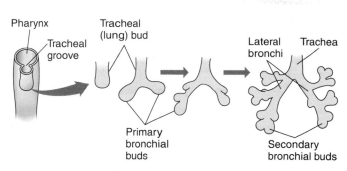

Figure 7.27 The Development of the Lung In terrestrial vertebrates, endoderm in the esophageal region of the gut evaginates to form a trachea. This evagination then successively buds and branches to form the bronchi. Eventually, alveoli form at the termini of the bronchi. All views are ventral and frontal. The branching endodermal epithelium is covered with mesenchyme (not shown).

Figure 7.26 Derivatives of the Branchial Arches (A) A 35-somite chick embryo viewed as a transparent object showing the plane of sectioning. The head is extensively flexed downward (and to the side), so that a single section passes through both the brain and the spinal cord. **(B)** A frontal section that depicts the arrangement of the branchial arches (bulges) and clefts (indentations between them) of the pharynx. Clefts are sometimes called pharyngeal pouches, particularly when no perforation occurs between the arches. Roman numerals designate the branchial arches, and the position of organs associated with these is shown.

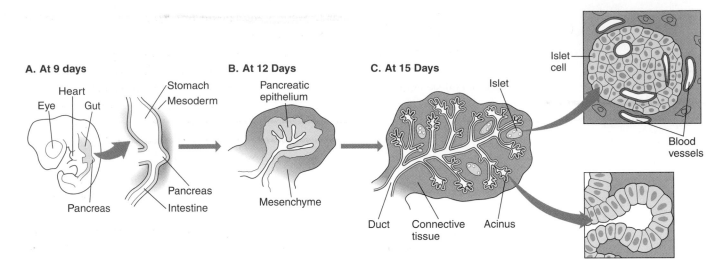

A. At 9 days

Eye
Heart
Gut
Stomach
Mesoderm
Pancreas
Pancreas
Intestine

B. At 12 Days

Pancreatic epithelium
Mesenchyme

C. At 15 Days

Islet
Duct
Connective tissue
Acinus
Islet cell
Blood vessels

Figure 7.28 The Development of the Pancreas in the Mouse Embryo (A) An evagination of the endodermal lining of the gut posterior to the stomach will become the pancreas, which will have both endocrine and exocrine portions. **(B,C)** The endodermal epithelium branches within the surrounding mesenchyme to form the ducts and acini of the exocrine pancreas. Islets of the endocrine pancreas develop between the acini.

the adjacent mesenchyme. Figure 7.28 outlines some of the features of pancreatic development. As the pancreatic endoderm pushes into the mesenchyme, it branches and forms blind pockets called *pancreatic acini*; the acini differentiate into the exocrine cells that will synthesize the zymogens (precursors of peptidases, nucleases, and carbohydrate-hydrolyzing enzymes) involved in digestion. Some epithelial cells of endodermal origin form aggregations of cells with an endocrine function. These *islets of Langerhans* contain at least four cell types, each of which secretes a different pancreatic hormone: glucagon, insulin, somatostatin, or pancreatic peptide.

The development of the pancreas is another classic example of an epithelial-mesenchymal interaction, a subject we shall study in more detail in Chapter 13. Development of the pancreatic rudiment requires signals from the notochord, which are thought to work by suppressing expression of the *shh* gene in the nearby endoderm. Splanchnopleuric mesenchyme is needed for acinus formation, but the epithelium forms islet cells autonomously (that is, without signaling from mesenchyme). Low levels of hormone production by these cells can be detected even before the pancreatic diverticula begin to branch.

The Endodermal Lining of the Alimentary Canal Differentiates

More attention has been devoted in this chapter to discussing the interactions leading to the formation of glands derived from endoderm than to formation of the gut lining itself. Recall that the lining of the gut is highly specialized in a regional fashion. The small intestine possesses several different kinds of specialized epithelia serving different functions, as does the large intestine. The cloacal region is an area of elaborate morphogenesis, which is coordinated with the differentiation of the metanephric kidney; like the metanephric kidney, it has connections to the Wolffian and Müllerian ducts. The urinary bladder also develops from the cloaca. The genitalia differentiate in this area as well. In amniotes, the midgut region is confluent with the endoderm of the extraembryonic yolk sac, which constitutes yet another different kind of specialized digestive epithelium with its associated capillary bed. The hindgut is connected to the outpocketing of the allantois, a crucially important yet transient organ in shelled amniotes (see Chapter 5). In mammals, it is the vascular supply of the allantois that becomes the umbilical artery and vein connecting the placenta to the embryo.

A PREVIEW

This Outline of Organogenesis Provides a Basis for Discussing Mechanisms of Development

This chapter and the preceding one devoted to ectoderm in some sense constitute a short tour of developmental anatomy. We have discussed where things take place, described some of the changes in cellular constitution and overall organ shape, indicated the role of the different germ layers in building the organs, and provided some idea of the overall vertebrate body plan. Some treatment of developmental anatomy, a specialty in its own right and the former bedrock of college embryology courses, is essential for an analysis of the mechanisms of development.

We have avoided a detailed discussion of two very important sets of questions: How do the germ-layer tissues form the actual shapes of the organs and their constituent tissues? And how do the developing organs and their parts get placed where they are? These are the problems of morphogenesis and pattern formation. For instance, how do neural crest cells, ectodermal placodes, branchial arches, pharyngeal endoderm, and many other embryonic structures find their way, adopt their form, and synchronize their differentiation so that the complex of structures we call a head actually comes into

being? Understanding what is going on at that level of synthesis is a tall order.

As we progress through the later chapters, we shall review many of the tissues and organs just discussed in order to get a more complete picture of how all this development is accomplished. At this point, however, it should be obvious that almost all of development after gastrulation is driven by interactions between different cellular "communities." These dialogues between cell groups in turn profoundly influence the behaviors of the cells and the regulation of gene expression within them.

There is one more important stage of development that occurs in many different animal phyla: the dramatic transition from well-developed larva into juvenile. This transformation may occur gradually, or it may involve a rapid transition from one environment to another associated with hatching or birth. Or, in metamorphosis, the entire larva may get remodeled. It is this last process that we turn to in the next chapter.

KEY CONCEPTS

1. The potential of cells in the embryo to form a given structure is best defined through transplantation or isolation experiments, whereas the fate of cells is best determined by observing normal embryos.

2. Vertebrate embryos have a conservative axial organization, proceeding from the midline to periphery: notochord, somite, nephrotome or gonad, and lateral mesoderm (limb buds).

3. Normal tissue and organ formation depends not only on the presence of signaling molecules, but also on the appropriate timing of ligand–receptor interactions. Such communication

can establish feedback loops. The formation of cartilage and bone both illustrate this principle.

4. Some structures and organs that form in the embryo, such as the pronephros, either are transient or serve no ostensible physiological purpose. Others, such as the vascular system, are remodeled extensively by a panoply of developmental processes so that the original form becomes unrecognizable.

5. Some cell types, especially those of the circulation, form in different locations at different times during development.

STUDY QUESTIONS

1. Can you recall some of the interactions, occurring in both directions, between the neural tube and somite?

2. In the last chapter, we learned that Sonic Hedgehog (Shh) helps organize the differentiation of the neural tube. Does Shh influence somite differentiation?

3. The pronephric (Wolffian) duct extends posteriorly. Can you suggest an experiment to help illuminate the mechanism of this posterior elongation?

4. Can you design an experiment to tell whether the hematopoietic cells of the adult chick form from stem cells that arose in the embryonic yolk sac?

SELECTED REFERENCES

Baker, C. V. H., and Bronner-Fraser, M. 2001. Vertebrate cranial placodes I. Embryonic induction. *Dev. Biol.* 232:1–61.

An extensive and up-to-date review of the formation of cranial placodes.

Barranges, I. B., Elia, A. J., Wunsch, K., Angelis, M. K. H., Mak, T. W., Rossant, J., Conlon, R. A., Gossler, A., and Pompa, J. L. 1999. Interaction between Notch signalling and Lunatic fringe during somite boundary formation in the mouse. *Curr. Biol.* 9:470–480.

A report explaining how mouse embryo mutants are used to determine which genes are active in the formation of somite borders.

Chalmers, A. D., and Slack, J. M. W. 2000. The *Xenopus* tadpole gut: Fate maps and morphogenetic movements. *Development* 127: 381–392.

A recent fundamental study of the formation of the gut.

Dahl, E., Koseki, H., and Balling, R. 1997. *Pax* genes and organogenesis. *BioEssays* 19:755–765.

A review of the *pax* gene family, which is involved in organ formation in many different situations, including the eye, kidney, ear, nose, limb, vertebral column, and brain.

Denetclaw, W. F., Christ, B., and Ordahl, C. P. 1997. Location and growth of epaxial myotome precursor cells. *Development* 124:1601–1610.

An embryological study that fate-maps the myotome.

Fishman, M. C., and Chien, K. R. 1997. Fashioning the vertebrate heart: Earliest embryonic decisions. *Development* 124:2099–2117.

A detailed review of the morphological and genetic basis of heart formation.

Garcia-Martinez, V., Darnell, D. K., Lopez-Sanchez, C., Sosic, D., Olson, E. N., and Schoenwolf, G. C. 1997. State of commitment of prospective neural plate and prospective mesoderm in late gastrula/early neurula stages of avian embryos. *Dev. Biol.* 181:102–115.

A research paper using embryological analysis to follow the determination of neural plate cells.

Johnson, R. L., and Tabin, C. J. 1997. Molecular models for vertebrate limb development. *Cell* 90:979–990.

A detailed review of the cellular and molecular basis of limb development.

Kennedy, M. K., Firpo, M., Choi, K., Wall, C., Robertson, S., Kabrun, N., and Keller, G. 1997. A common precursor for primitive erythropoiesis and definitive hematopoiesis. *Nature* 386:488–491.

A research paper using ES cells to determine cell lineages in the blood.

Lechner, M. S., and Dressler, G. R. 1997. The molecular basis of embryonic kidney development. *Mech. Dev.* 62:105–120.

An incisive review of signaling in kidney development.

Nishida, H., and Sawada, K. 2001. *Macho-1* encodes a localized mRNA in ascidian eggs that specifies muscle fate during embryogenesis. *Nature* 409:724–729.

A research paper on the discovery of the muscle determinant in ascidians.

Obara-Ishihara, T., Kuhlman, J., Niswander, L., and Herzlinger, D. 1999. The surface ectoderm is essential for nephric duct formation in intermediate mesoderm. *Development* 126:1103–1108.

A research paper showing that ectoderm secretes BMP4, which is necessary for nephric duct formation.

Rawls, A., and Olson, E. N. 1997. MyoD meets its maker. *Cell* 89:5–8.

A short review on the transcription factors specific and essential for striated muscle development.

Satoh, H. 1994. *Developmental biology of ascidians.* Cambridge University Press, New York.

A comprehensive book on the developmental biology of ascidians.

Snape, A., Wylie, C. C., Smith, J. C., and Heasman, J. 1987. Changes in state of commitment of single animal pole blastomeres of *Xenopus laevis. Dev. Biol.* 119:503–510.

A research paper on the determination of germ-layer cells using classical single-cell transplantation techniques.

Thesleff, I., and Nieminen, P. 1996. Tooth morphogenesis and cell differentiation. *Curr. Opin. Cell Biol.* 8:844–850.

A review of the tissue interactions in tooth development.

Vainio, S., and Muller, U. 1997. Inductive tissue interactions, cell signaling, and the control of kidney organogenesis. *Cell* 90:975–978.

An incisive review of the molecules and cellular interactions involved in kidney development.

Vortkamp, A. 1997. Skeleton morphogenesis: Defining the skeletal elements. *Curr. Biol.* 7:R104–R107.

A paper outlining the role of BMPs in skeleton formation.

Vortkamp, A., Lee, K., Lanske, B., Segre, G. V., Kronenberg, H. M., and Tabin, C. J. 1996. Regulation of rate of cartilage differentiation by Indian hedgehog and PTH-related protein. *Science* 273:613–622.

A research paper on the role of Hedgehog in regulating bone formation.

Yun, K., and Wold, B. 1996. Skeletal muscle determination and differentiation: Story of a core regulatory network and its context. *Curr. Opin. Cell Biol.* 8:877–889.

A review of the genes involved in skeletal muscle formation.

CHAPTER

METAMORPHOSIS

Many animals complete their embryonic development without coming to resemble a young adult. Instead, the embryo in these species forms a **larva,** a transition during which the animal is free-living but sexually immature. Many classes of the different invertebrate phyla include species that employ a larval phase; even some vertebrates, especially amphibians, form larvae. There is a tremendous variety of larval forms, and the transition from larva to juvenile—a young, postmetamorphic adult—is accomplished by a vast number of different developmental programs. The passage from larva to juvenile is often very dramatic, involving a radical reorganization of body form.

The same issues we encountered in embryonic development will crop up in our examination of metamorphosis also: How do cell groups adopt programs of differential gene action? How does coordinated morphogenesis take place? And what higher-order programs regulate the whole undertaking? We shall see from the examples in this chapter that the driving force of metamorphosis is a coordinated program

CHAPTER PREVIEW

1. Many different animals possess a free-living larval stage interposed between the stages of embryo and adult.

2. The change from larva to adult can include a radical reorganization, of both body plan and physiology, called metamorphosis.

3. Insect metamorphosis is controlled by the interplay of the hormones juvenile hormone and ecdysone.

4. Amphibian metamorphosis is primarily controlled by thyroid hormones.

of hormone action. Once again, it is cell–cell communication by means of chemical signals that orchestrates the development. Even though metamorphosis is widespread, extensive research on the mechanisms underlying it is spotty; in many cases, little is known. Consequently, we shall examine metamorphosis in the two best-studied instances: holometabolous insects such as *Drosophila*, and the anuran *Xenopus laevis*.

The existence of larval forms is so pandemic in the animal world that one suspects there must be some biological payoff. Most scientists point to the larva's ability to move over much greater distances than the embryo as an advantage for the species. In the case of amphibians, metamorphosis is part of the transition from the embryo's strictly aquatic existence to a more terrestrial lifestyle.

INSECT METAMORPHOSIS

Molting Is an Essential Part of Insect Development

The embryonic development of *Drosophila*, which we considered in Chapter 3, results in a segmented larva with appendages and muscles to move, a digestive system and feeding appara-

tus to provide nutrients, and a program of further growth and development. The larva is encased in a relatively rigid exoskeleton. The exoskeleton is made of cuticle, a complex noncellular layer that usually contains waxes, proteins, chitin, and calcium carbonate and is secreted by the epithelial surface of the embryo. In order for the insect to grow in size, the restraint of this exoskeleton must be removed. All insect larvae shed their cuticle, usually more than once, in a process called **molting**. The stages of larval development between the molts are often called **instars**. Insect species vary in number of molts.

Furthermore, insects have different strategies for getting from larva to adult. Some insects, like grasshoppers (order Orthoptera) and dragonflies (order Odonata), produce embryos that look approximately like their adult forms, albeit smaller and not fully differentiated. In this kind of larval development, called **hemimetabolous**, molting allows for an increase in size and further differentiation of the different tissues.

In many other kinds of insects, however, including *Drosophila*, the larva appears to be quite different from the adult. After a number of larval molts, in which the larva increases in size, the larva constructs an elaborate external cuticle called a

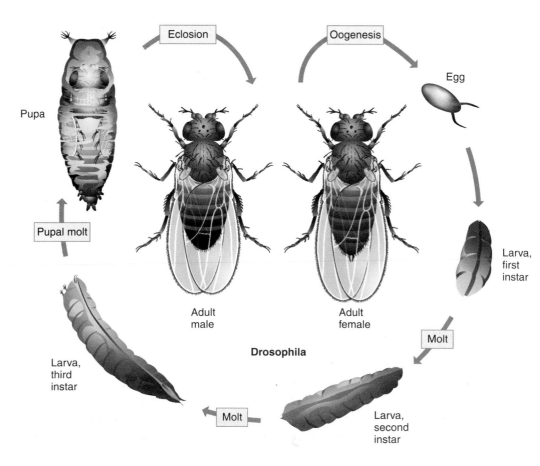

Figure 8.1 The Life Cycle of a Holometabolous Insect In the progression of a typical dipteran (fly), the zygote develops into a larva, then molts through several instars (three in *Drosophila*) before transforming into a pupa. The adult forms within the pupal cuticle and emerges by eclosion. The adult then forms gametes (not shown), which start the cycle again.

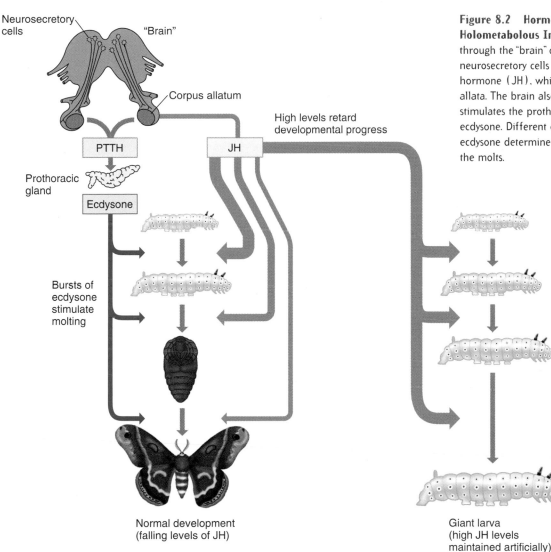

Figure 8.2 Hormonal Circuits in Holometabolous Insects Cross sections through the "brain" of an insect identify the neurosecretory cells that secrete the juvenile hormone (JH), which is stored in the corpora allata. The brain also secretes PTTH, which stimulates the prothoracic gland to produce ecdysone. Different concentrations of JH and ecdysone determine the timing and nature of the molts.

puparium; once in its puparium, the larva is called a **pupa.** The puparium is dark, and the processes occurring within it—**pupation**—are not readily visible. A radical reorganization happens within the puparium, including wholesale apoptosis of many of the pupa's tissues. Nests of cells called *imaginal discs* (introduced in Chapter 3 and discussed at greater length later in this chapter) undergo a complex program of morphogenesis and differentiation that eventually results in **eclosion,** when the adult insect emerges from the puparium. This kind of larval development, called **holometabolous,** is depicted in Figure 8.1 for the life cycle of the moth. Many common insects, such as flies, fleas, moths, and bees, undergo holometabolous metamorphosis.

It is worth noting that in some species the pupa may display a very low metabolic rate, a kind of suspended animation, for a very long time. This condition is called *diapause.* An

environmental signal is needed to release the larva from diapause. An example of this kind of holometabolous metamorphosis is found among the silk moths, where a long cold period followed by warming is required for the silk moth to complete its metamorphosis.

Neurosecretory Circuits Drive Insect Molting

The anterior-most ganglia of the insect nerve cord develop into a complex collectively called the "brain." Figure 8.2 includes a cross-sectional diagram through the brain of a holometabolous insect. Collections of neurosecretory cells, similar to the collections vertebrates have in their hypothalamus, secrete hormones and conduct nerve impulses. One such group of neurosecretory cells secretes a peptide with the tongue-twisting name prothoracotropic hormone, or PTTH for short. This hormone

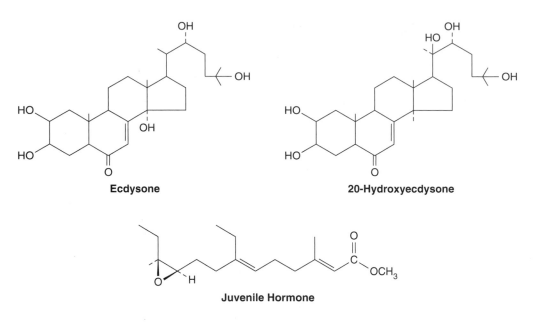

Ecdysone

20-Hydroxyecdysone

Juvenile Hormone

Figure 8.3   The Structure of Hormones Involved in Insect Metamorphosis Chemical structures are shown for two forms of the steroid molting hormone, ecdysone and its 20-hydroxy derivative, as well as for the ecdysone antagonist, juvenile hormone.

stimulates the *prothoracic gland*, situated near the brain, to synthesize and release **ecdysone,** the hormone essential for all molts (Figure 8.3). Ecdysone is more properly called a prohormone, a precursor molecule that must be converted by other tissues, such as the fat body, to another form in order to effect change. The active form, 20-hydroxyecdysone, is then bound by members of a receptor family that are distributed widely in the larval tissues. Figure 8.2 graphically depicts the route of ecdysone action.

The other principal hormone involved in molting, **juvenile hormone (JH),** is synthesized in a pair of endocrine glands called the corpora allata, which are located very near, and attached to, the brain (Figure 8.3). JH is a hydrophobic molecule of the isoprenoid class (unsaturated hydrocarbons assembled from branched, five-carbon subunits). Figure 8.3 shows the structure of JH. By itself, JH does not cause molting, but it does play a crucial role in its regulation. When JH levels are relatively high, ecdysone promotes a larval molt; when JH levels are relatively low, the pupal molt takes place (see Figure 8.2). In some insects, the corpora allata and the prothoracic gland are fused into one compound gland called the ring gland; this is the situation in *Drosophila*.

Insect Molts Are Driven by Ecdysone Production

PTTH secretion, which is probably regulated by both environmental and autonomous signals, produces waves of ecdysone production. As the first wave ends, during the first instar, a small spike in ecdysone concentration occurs, soon followed by a more intense wave of ecdysone release. Stimulated by the

hormone, the epithelial cells of the body surface withdraw from the cuticle and produce a molting fluid containing proenzymes that, after activation, will digest the old cuticle. (A proenzyme is a precursor of an active enzyme that requires some modification—usually the removal of a small peptide by a protease—in order to become fully active.) The epithelium then generates a new cuticle. Because it is distensible, the new cuticle expands as the larva (now in its second instar) grows, until this cuticle, too, becomes hard and inelastic, and the process is repeated again. During the latter portion of the third instar in *Drosophila*, a spike in ecdysone levels again begins the process of molting. But this time, the newly forming cuticle becomes a darkened, hardened cage, the puparium within which the transition from larva to adult will take place (Figure 8.4A).

Many years ago, researchers showed that it is the corpora allata, and the juvenile hormone produced by them, that determine whether the result of a molt will simply be an increase in larval size, or pupation and metamorphosis. Removing the corpora allata surgically during the second instar will cause the next molt to undergo pupation, one full instar early. On the other hand, implanting an actively secreting corpus allatum into a late third-instar larva may in the next molt result in a giant larva rather than a pupa (see Figure 8.2). Puparium formation and pupation are initiated when an ecdysone wave occurs during very low levels of JH, or even in its absence. That was the theory, at least, until very recently. The process may not be that simple. Recent measurements show that some JH is present in late third-instar larvae, and that the precise timing of JH release and the presence or

A. **B.**

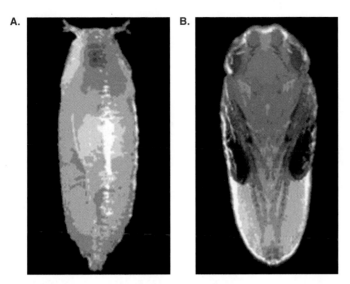

Figure 8.4 Development of the Pupa Dorsal views of **(A)** an early pupa and **(B)** the metamorphosed pupa just before emergence (eclosion) from the pupal cuticle. Note in (B) that the head has emerged and the eyes are visible: bristles on the body are obvious. as are the outlines of the wings.

absence of active JH receptors are also involved in regulating pupation.

What happens inside the puparium? The first wave of ecdysone to occur when JH signals are low causes the pupal cuticle to form and harden; it also leads to further development and subsequent eversion (a turning inside out) of the imaginal discs. A later, second ecdysone wave stimulates eversion of the developing head. This wave also sets in motion a cascade of hormone releases leading to the characteristic events of eclosion—muscular movements and rupture of the puparium—which allow the adult fly to emerge (Figure 8.4B).

Adult Insect Structures Develop from Imaginal Discs

The reorganization that occurs within the puparium, which is driven by high ecdysone levels in the absence of JH activity, is profound. Cells in many of the larval tissues undergo apoptosis and the tissues disintegrate, a process called histolysis. Much of the epithelium and other larval tissues are destroyed. Some larval tissues, particularly in the nervous system and much of the gut, are reorganized. But almost all of the adult's surface structures—such as wings, legs, halteres (balancing organs), and antennae—derive from the imaginal discs, mentioned earlier. These nests of epidermal cells consist of only 30 to 80 cells per disc at the beginning of the first instar; subsequently, these discs will form adult epidermal structures. There are 10 pairs of these discs and a single genital disc. Figure 8.5 identifies seven pairs of the more prominent discs as well as the genital disc. Other nests of cells in the ventral portion of the larva, called *histoblasts*, will develop into internal organs.

The discs and histoblasts divide as the larva develops; a wing disc that originally had 38 cells will possess 60,000 cells

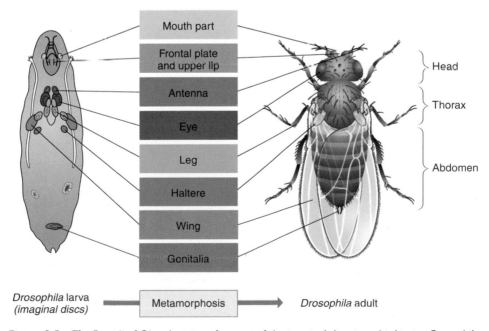

Figure 8.5 The Imaginal Discs Locations for most of the imaginal discs in a third-instar *Drosophila* larva are shown. along with their fates as organs in the adult fruit fly.

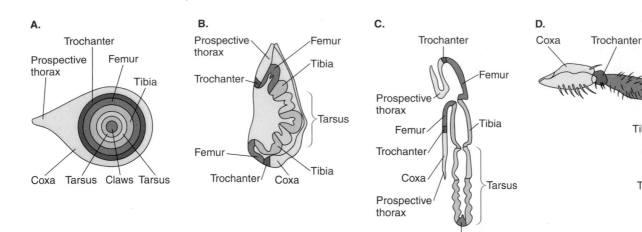

Figure 8.6 Eversion of the Leg Disc During Metamorphosis In all views, the disc and everting leg are oriented such that the point where the leg attaches to the thorax is to the left. **(A)** A diagrammatic view of the surface of a larval leg disc before eversion; note how the distal leg parts are located centrally. **(B)** A longitudinal section through the disc before eversion. **(C)** A longitudinal section of the disc after it has everted and is developing into the definitive leg. **(D)** A surface view of the fully formed leg.

when the puparium forms. By the end of metamorphosis inside the puparium, the differentiation of the discs and nests of histoblasts is complete. And the discs, initially shaped like inverted epidermal sacs, undergo extensive morphogenetic movements to attain their final form. Figure 8.6 shows the eversion of a leg disc to form a leg; notice how the central portion of the disc, when pushed outward, forms the distalmost extremity of the leg, while the more peripheral cells of the disc contribute to more proximal portions of the leg. This remarkable change from disc to leg (or to wing, or to any of the other adult structures) is driven by changes in the shape of the cells in the disc, a subject treated in more detail in Chapter 12.

Ecdysone Works by Directly Influencing Transcription

DNA replication without cell division is common in many tissues of insect larvae, leading to *polyploidy* (multiple sets of chromosomes) or *polyteny* (multiple strands of DNA in a given chromosome). In particular, the salivary gland of the third-instar larva of *Drosophila* possesses highly extended polytene chromosomes aptly named giant chromosomes (Figure 8.7). The salivary gland cells containing them are also large, and portions of the gland may be excised and maintained in nutrient medium. During the 1960s, Ulrich Clever, working in Germany, exposed explanted salivary gland cells containing giant chromosomes to a medium containing small amounts of ecdysone, which had only recently become available in large enough quantities for research. He observed that in certain regions of the chromosomes, the chromatin strands become loosened from one another. This morphological response of giant salivary chromosomes to ecdysone is called "puffing," and the loosened areas are referred to as *puffs*.

Puffs had been observed in living animals, but Clever's results seized the imagination of many biologists; one could actually see the influence of a hormone on chromosome structure and presumably on chromosome function. Decades of research and thousands of experiments later, we know that chromosome puffs are indeed sites of transcription. The puffs represent relatively large regions, each in the neighborhood of 100,000 base pairs or more and which may contain more than one gene.

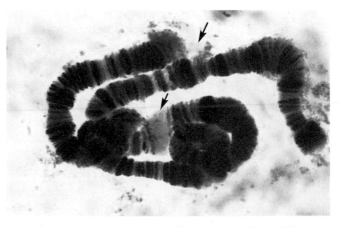

Figure 8.7 Giant Chromosomes of the Larval Salivary Gland A photograph of giant chromosomes dissected from the salivary gland of *Drosophila* during the late third instar. Banding of the chromosomes is clearly visible, and arrows point out some of the puffs.

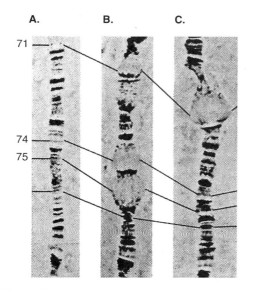

Figure 8.8   Early and Late Puffs The salivary gland of *Drosophila* was cultured in vitro to reveal changes in the puffing patterns of a giant chromosome after ecdysone treatment. Shown here is a region of the right arm of chromosome III. **(A)** No puffs were visible at the outset. **(B)** After ecdysone was added to the medium. the chromosome showed two prominent early puffs at positions 74 and 75. **(C)** Ten hours after the addition of ecdysone. the early puffs had regressed. and a late puff. at position 71. was very evident.

The response of the salivary gland chromosomes to ecdysone is specific. Only certain banded regions of the chromosome show the phenomenon within 30 minutes of exposure to the hormone. These so-called early puffs (there are probably six) are then followed by a series of later puffs (of which there may be about 100) forming at specific chromosomal regions (Figure 8.8). Early and late puffs respond differently to the inhibition of protein synthesis: the early puffs still occur while later puffs do not. The theory, now well established, is that the early puffs encode mRNAs that get translated into proteins necessary for the formation of the later puffs. Thus, ecdysone initiates a specific program of gene action. The program may be modified by the level of JH, which works through mechanisms still not understood. The responding tissues undergo tissue-specific responses.

Some other tissues also have polytene chromosomes, which, while not as large as salivary gland chromosomes, do show puffing. The pattern of puffs in the fat body, for instance, is not the same as in the salivary gland. Even in tissues without polytene chromosomes, such as epidermal cells, we can now follow the synthesis of mRNAs from specific genes, and demonstrate that each tissue responds to ecdysone in a specific way by activating a tissue-specific response of gene transcription.

The Regulation of Insect Metamorphosis Requires an Interplay of Many Factors

This far into the chapter, it should be obvious that metamorphosis in *Drosophila* is a complex program. We know some of the players: PTTH, ecdysone, and JH, plus the receptors for all these hormones. While we have not discussed them, there are also eclosion hormones, which play a part in the emergence of the young adult from the puparium. We are far from understanding how the precise timing and progression of the whole program are regulated. Several of the early-acting genes (*broad complex, e74,* and *e75*) encode transcription factors that then regulate transcription from sites of late puffs, thereby underpinning a sequence of gene expression. The timing of ecdysone and JH release and the levels of their relative titers are certainly crucial. Likewise, these hormones all interact with specific receptors.

A little is known about ecdysone receptors. Ecdysone is a steroid, and like all steroid hormones, it interacts with cytoplasmic receptor proteins after the hormone diffuses across the cell membrane. In *Drosophila*, there are three different known ecdysone receptor proteins. When ecdysone diffuses into a cell, it can interact with any of the different receptor proteins that happen to be there; different cells can have different members of the receptor protein family. In order to be active, the ecdysone–receptor complex requires another protein partner, a product of the gene in *Drosophila* called *ultraspiracle (usp)*. The protein Usp dimerizes with an ecdysone receptor protein to form a heterodimer; the combination of this heterodimer with ecdysone is the active complex in the regulation of transcription (Figure 8.9). It is likely that the amounts of a particular ecdysone receptor protein and of Usp present at any given time are also involved in regulating the timing of effective hormone action. Varying amounts of the three different ecdysone receptor proteins undoubtedly help provide some tissue specificity to the response. Interestingly, *usp* has a counterpart in vertebrates that encodes a receptor

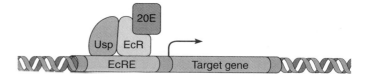

Figure 8.9   Ecdysone Gene Activation Ecdysone (20E) interacts with its receptor (EcR). forming a heterodimer with the co-activator protein Usp. This ternary complex can then activate the ecdysone response element (EcRE) located in the promoter region of a target gene. leading to increased transcription of the target gene.

that interacts with retinoic acid, a vitamin A derivative. We shall consider the action of hormones on transcription in more detail in Chapter 14.

AMPHIBIAN METAMORPHOSIS

The Life Cycle of Many Amphibians Involves Metamorphosis

We have all observed, at least casually, the dramatic change from a tadpole, or polliwog, into a frog. The fishlike larva loses its tail, grows legs, and emerges from the water to become an air breather (Figure 8.10). There is a definite sequence to the morphological changes; limb growth begins early and contin-ues through premetamorphic stages; lungs begin to develop midway through the transition; and tail loss occurs late, dur-ing the so-called metamorphic climax. There are profound changes in the skeleton, especially in the head region, before and during the climax. The long gut of the herbivorous tadpole transforms to the shorter gut of the carnivorous adult. Mem-bers of the class Anura (frogs and toads) show complete meta-morphosis, becoming terrestrial as adults; almost all, however, return to water to lay their eggs. Members of the class Urodela (salamanders and newts) do retain a tail; however, they too form limbs and lungs, lose tail fins and external gills, and like anurans undergo profound changes in their skin structure.

Some urodeles undergo only a partial metamorphosis and retain an aquatic mode of life. For example, the salamander *Ambystoma mexicanum*, which lives in some mountain lakes in

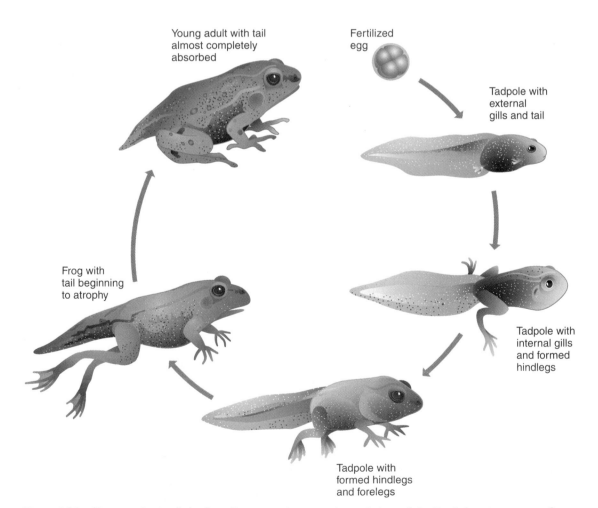

Young adult with tail almost completely absorbed

Fertilized egg

Tadpole with external gills and tail

Frog with tail beginning to atrophy

Tadpole with internal gills and formed hindlegs

Tadpole with formed hindlegs and forelegs

Figure 8.10   Metamorphosis of the Frog Changes in the external morphology of the North American anuran *Rana pipiens*, as it progresses from tadpole to frog. Notice the changes in head shape, limb development, skin pigmentation, and tail resorption.

A.

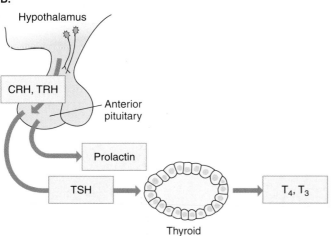

Thyroxine (T$_4$)

Triiodothyronine (T$_3$)

B.

Hypothalamus

CRH, TRH

Anterior
pituitary

Prolactin

TSH

Thyroid

T$_4$, T$_3$

C.

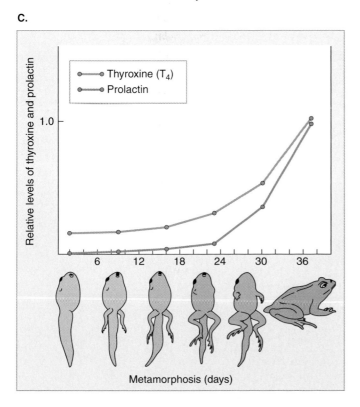

Relative levels of thyroxine and prolactin

Thyroxine (T$_4$)
Prolactin

1.0

6   12   18   24   30   36

Metamorphosis (days)

Mexico, retains into adulthood many larval characteristics, such as external gills and finned tails, and lives its entire life in the lake. Researchers found that this condition in *A. mexicanum* is caused by a failure of the hypothalamus to stimulate the pituitary gland to initiate the hormonal changes driving complete metamorphosis.

When a creature becomes sexually mature in what would otherwise be considered a juvenile or larval state, it is called **neoteny**. Another name used for this phenomenon is *paedomorphosis*. Neoteny is found in many different animal phyla.

Metamorphosis in the Frog Is Driven by Thyroid Hormones

Feeding tadpoles bits of thyroid gland induces premature metamorphosis, an experiment first done in 1912. Surgical removal of the developing thyroid gland of the frog larva was also undertaken early in the 20th century. Such thyroidectomized tadpoles failed to undergo metamorphosis. Later it was discovered that hypophysectomy (removal of the anterior pituitary) produces the same effect. We now know that a combination of environmental factors, especially temperature, and an endogenous developmental program of the brain lead to secretion, from the hypothalamus, of hormones that stimulate the anterior pituitary to synthesize and release **thyroid stimulating hormone** (TSH, also called thyrotropin). TSH controls the thyroid gland and stimulates synthesis and release of **triiodothyronine** (T$_3$, the most potent of the thyroid hormones) and **thyroxin** (nicknamed T$_4$). Most T$_3$ is produced by the deiodination of T$_4$ in various tissues. The structures for these two thyroid hormones and their sequence of actions are shown in Figure 8.11. In the absence of T$_4$ and/or T$_3$, metamorphosis will not occur; the larvae simply grow into large, juvenile aquatic creatures.

Figure 8.11   The Anterior Pituitary and Amphibian Metamorphosis (A) The structures of thyroxin and T$_3$. **(B)** Part of the sequence of hormone stimulating events: The hypothalamus secretes TRH and CRH. TRH causes the anterior pituitary to release TSH, which then stimulates the thyroid gland to synthesize and secrete thyroxin and T$_3$. The pituitary also secretes prolactin, which has multiple effects, some of which are discussed in the text. **(C)** The relationship between morphological changes over time and approximate levels of circulating T$_3$ and prolactin during metamorphosis.

The hypothalamic hormone that stimulates TSH release in mammals (the vertebrates on which most research on endocrine function has been carried out) is **thyrotropin releasing hormone (TRH).** Curiously, TRH is active in stimulating TSH release in frogs only when they are postmetamorphic. In tadpoles, the effective hormone driving TSH release is **corticosterone releasing hormone (CRH),** which also stimulates the release of **adrenocorticotropic hormone (ACTH).** ACTH stimulates the adrenal cortex to secrete corticosteroids. These adrenal steroids have been shown to regulate, at least in part, the production of enzymes that convert T_4 to T_3 in target tissues. Recent evidence indicates that thyroid hormones stimulate the pituitary to produce another hormone, **prolactin,** a peptide that also plays a part in regulating some aspects of metamorphosis, as we shall see shortly.

In summary, during the stages leading up to metamorphic climax, there is an increase in TSH, T_3, T_4, and prolactin (Figure 8.11C); the very high levels of T_3 and T_4 then act on the hypothalamus as part of a negative feedback loop to lower TSH and perhaps CRH production to levels appropriate for juveniles.

Prolactin Antagonizes Some Actions of Thyroid Hormones

Prolactin has many different effects in different vertebrates. In mammals, this hormone aids in the development of the mammary glands and the formation of milk. An experiment in which high levels of mammalian prolactin were injected into tadpoles showed slower metamorphic changes; this result has been interpreted to mean that prolactin in the frog antagonizes the actions of T_3 and T_4. Recent work suggests this action of prolactin may occur because it interferes with the formation of T_4 and T_3 receptors. Furthermore, since we know that prolactin levels increase as metamorphic climax approaches (see Figure 8.11C), it is unlikely that prolactin is a simple antagonist of thyroid hormones.

Though it was once believed that prolactin action is analogous to the role of JH in insects, recent experiments show that this cannot be so. A new technology for producing transgenic tadpoles and frogs employs a variation on the nuclear transplantation procedure discussed in Chapter 1. When the donor nucleus is prepared for transplantation into an enucleated egg, it is bathed in a solution containing the DNA harboring the exogenous gene and various enzymes; the latter help the exogenous DNA enter and become incorporated into the chromosomes of the donor nucleus.

After transplantation of this specially prepared nucleus into multiple host eggs, the transgene is stably incorporated by some of the resulting embryos. Transgenic tadpoles that express prolactin at very high levels have recently been produced by this procedure. Surprisingly, while prolactin interferes with tail resorption and stimulates fibroblast growth in the tail, such transgenic tadpoles still metamorphose at the appropriate time. Juvenile frogs with persistent tails are the result. There is still much to learn about how prolactin and T_3 interact in the living animal and how they orchestrate the changes in timing and the specific hormone actions that are the hallmark of metamorphosis. However, we can conclude that, strictly speaking, prolactin does not act as a juvenile hormone, for it counteracts only some of the many effects of thyroid hormones.

It is hardly surprising, in retrospect, to find two hormones—one acting positively, the other negatively—presiding over a physiological or developmental change. This combination of positive and negative control provides a powerful and finely tuned regulatory circuit. In later chapters, we shall repeatedly encounter this kind of specific interplay between different chemical messengers.

TABLE 8.1 CHANGES DURING ANURAN METAMORPHOSIS

Organ/System	Changes from Larva to Adult
Movement	Tail fins to legs
Respiration	Gills and skin to lungs
	Shift to different hemoglobin types
Nutrition	Diet: herbivore to carnivore
	Gut: lengthy to short
	Skull and mouth: extensive morphological changes
Vision	Retinal pigment: porphyropsin to rhodopsin
Nitrogen excretion	Ammonia to urea
Skin	Epidermis: thin to stratified
	Mucous glands: none to numerous

The Action of Thyroid Hormones Is Tissue Specific

A single hormone, T_3, initiates many changes. For example, muscles in the developing limbs are stimulated to increase in size and to differentiate, while muscles in the tail are caused to wither and disappear. Table 8.1 summarizes the more prominent metamorphic changes induced by T_3; it is probably fair to say that every tissue or organ system is affected by this hormone. Not only are there dramatic morphological changes during metamorphosis; some fundamental metabolic machinery also gets retooled.

For instance, the outer rod segments of the tadpole's retina contain the visual pigment porphyropsin, which is comprised of the protein opsin and the aldehyde of vitamin A_2. Vitamin A_2–based visual pigments are found in the vertebrates that inhabit fresh water. During metamorphosis, the metabolic pathways that provide vitamin A_2 shift to production of vitamin A_1, which is characteristic of terrestrial vertebrates, and the visual pigment in the adult frog's eye is now rhodopsin rather than porphyropsin (Figure 8.12).

The liver of tadpoles possesses the enzymatic machinery for the synthesis of ammonia as the primary excretory product. Ammonia, a toxic chemical, is typically found as the end product of nitrogen metabolism in aquatic animals, an environment in which the ammonia is easily dissolved, diluted, and carried away. During the metamorphic transition, the enzymatic machinery of the tadpole liver changes over to produce urea as the major end product of nitrogen metabolism. This shift entails an increase in the synthesis of the four principal urea-cycle enzymes required for the synthesis of urea from ammonia. Changes of visual pigments and urea cycle enzymes are triggered by an increase in thyroid hormone concentrations.

The response of each tissue, driven by thyroid hormones, is specific to that particular tissue type. Furthermore, the nature and sign of the responses are mostly consistent with a transition from aquatic to terrestrial existence. Through treatment of isolated tissues, researchers have been able to confirm the local and specific nature of the hormone responses. If a pellet that releases thyroxin very slowly is implanted into one eye of a tadpole, the changeover from porphyropsin to rhodopsin occurs much faster in the eye receiving the hormone directly.

An even more dramatic example comes from the ability of the severed and isolated tadpole tail to survive in a simple culture medium for many days. As shown in Figure 8.13, the addition of extremely low levels of T_3 induces macroscopic and cellular changes characteristic of tail resorption during

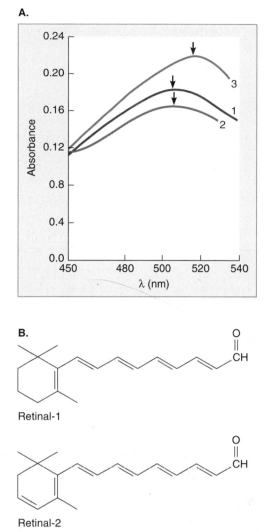

Figure 8.12 The Metamorphosis of Amphibian Visual Pigments
(A) The absorption spectra of the visual pigment extracted from bullfrog tadpoles. Curves 1 and 2 show the spectra for tadpoles treated with thyroxin; the absorption maximum of 505 nm is characteristic of rhodopsin. Curve 3, the spectrum of pigment extracted from untreated tadpoles, has a maximum of about 520 nm, which is characteristic of porphyropsin. **(B)** The structures of retinal-1 and retinal-2, which derive from vitamins A_1 and A_2, respectively. Note retinal-2's additional carbon-carbon double bond (shown in red). Retinal-2 is present in the larval visual pigment, porphyropsin, while retinal-1 is found in the adult pigment, rhodopsin.

A.   B.   C.

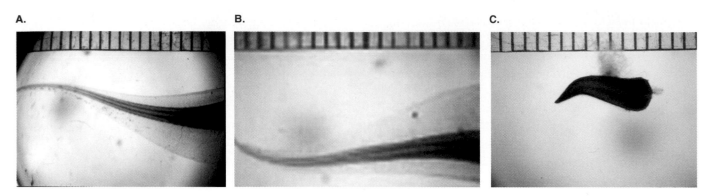

Figure 8.13   The Action of Hormones on Isolated Tadpole Tails Tails of *Xenopus laevis* tadpoles were isolated and placed in a dilute salt solution containing the hormone T_3. The photographs show the condition of the isolated tails after **(A)** one, **(B)** three, and **(C)** seven days in culture. Tails cultured without hormone did not show signs of apoptosis or regression.

metamorphosis: increased keratinization of the epidermis, failure of epidermal stem cells to replace epidermis, increased accumulation of lysosomal proteases (proteolytic enzymes found in the lysosomal vesicles of cells), and apoptosis of muscle cells. If prolactin is added to the medium at the same time as T_3, these changes will fail to occur, demonstrating that both thyroid hormones and prolactin are acting locally and specifically.

The Timing of Metamorphosis in Amphibians Is Partly Regulated by Hormone Levels

What accounts for the sequence of metamorphic changes shown in Table 8.1? A principal factor is the concentration of thyroid hormones, whose action is modified in some target tissues by corticosteroids from the adrenal glands and by prolactin. Thyroid hormone concentrations increase during the progressive changes of metamorphosis, as do levels of corticosteroids and prolactin. Several experiments in which the levels of these hormones have been controlled to some extent clearly show that metamorphic events are induced by different levels of thyroid hormones. Shortening of the intestine and growth of the hindlimbs occur at very low thyroxin levels, while tail regression occurs only at much higher levels.

These kinds of results support the idea that each of the different local responses to the hormones has a threshold concentration. Until the hormone reaches or surpasses its threshold, the local response will not occur. While this model seems reasonable, it does not tell us much about the mechanisms involved.

Hormone Receptors Also Regulate Metamorphosis

We now know that all hormones—including T_3, T_4, and prolactin—act by way of receptor proteins. The threshold concept just mentioned is easier to understand if we restate it in modern terms, by postulating that the concentration and properties of particular receptor types in a given tissue determine, at least in part, the response. The earliest response to thyroid hormones is an increase in transcription of thyroid hormone receptor genes (TR genes), of which there are at least two. Thus, T_3 induces greater levels of its receptor, a positive feedback loop that stimulates increased amounts of its receptor and thus the potential for an increased sensitivity to the hormone.

TR proteins belong to the same class as ecdysone receptors, namely, the steroid hormone receptor superfamily. As with ecdysone, these receptors work as heterodimers, in which each TR protein is joined with a molecule from a class of retinoic acid receptors called RXR (which stands for retinoic acid–like receptor). There is some evidence that prolactin serves to decrease expression of TR genes, which may explain in part why prolactin counteracts some actions of thyroid hormones. While research in this field is ongoing, it seems likely that local tissue-specific responses and the regulation of hormone sensitivity are mainly due to levels of TR proteins, particular TR family members involved, level(s) and type(s) of RXR, and probably other accessory proteins that confer transcriptional specificity (see Chapter 14). We do know that T_3 stimulates transcription of some genes, such as the gene for adult globin, while decreasing transcription of others. We also know that steroid hormone receptor members possess the kind of specificity capable of turning on some genes and turning off others.

METAMORPHOSIS IN OTHER GROUPS

Larval Development Is Widespread

In this chapter, we have concentrated on examples that have been thoroughly analyzed, are dramatic, and are relatively well understood. It is important to remember, however, how widespread metamorphosis is, and how various larval-to-adult transitions are. Most marine invertebrates undergo profound developmental reorganization when they change from planktonic larvae to reproductively competent adults.

We shall have more to say about a few of these cases when we discuss development and evolution in Chapter 17.

Among vertebrates, many groups other than the amphibians experience extremely important developmental reorganizations during their life cycles. Salmon undergo embryonic development in fresh water, then migrate to a marine environment for most of their adult life. This transition from fresh to salt water is accompanied by extensive physiological and morphological changes. The eventual return of the adults to fresh water for spawning requires another set of profound changes, which are known to be hormonally regulated, and sexual reproduction does not occur until these final "finishing touches" are put into place. Many other aquatic vertebrates, notably species of eels, also undergo extensive developmental changes during migration from one habitat to another. The transitions from embryo to reproducing adult are an important aspect of development.

KEY CONCEPTS

1. Development of embryos in many different phyla results in formation of a free-living, sexually immature larva. In order for a sexually mature adult to form, a transition called metamorphosis is required.

2. The release of hormones regulates and drives metamorphosis in holometabolous insects and amphibians, and probably is also crucial in the metamorphosis of other animals.

3. In well-studied cases, a pair of antagonist hormones is involved, one inducing adulthood, the other acting as a brake; this antagonist action gives increased tuning power to the regulatory circuits involved.

4. The responses of larval tissues to metamorphic hormones are specific and unique for each tissue. The responses result from interactions between the hormones and their cognate receptors, which lead in turn to profound changes in gene expression.

5. The nature of local specific responses is determined by the relative concentrations of different forms of a hormone, and the affinity of the receptors for their hormones. Furthermore, the second-messenger machinery (see Chapter 1) and the preexisting state of the chromatin may play roles in tissue specificity.

6. Though the mechanisms involved are not well understood, the regulation of hormone synthesis and release is also partially controlled by environmental cues.

STUDY QUESTIONS

1. Ecdysone and T_3 (triiodothyronine) are chemically very different, and are found in different phyla. Yet they play similar biological roles. Can you comment on how they are similar?

2. How is the character of a molt (larval or pupal) regulated in holometabolous insects?

3. Can you speculate on how the same hormone, T_3, could cause muscle cell apoptosis in the tail and muscle cell growth and differentiation in the limb of a tadpole?

SELECTED REFERENCES

Bayer, C. A., von Kalm, L., and Fristrom, J. W. 1997. Relationship between protein isoforms and genetic functions demonstrate functional redundancy at the Broad-Complex during *Drosophila* metamorphosis. *Dev. Biol.* 187:267–282.

A research paper on the genes involved in puffing and responses of chromatin to ecdysone.

Bender, M., Imam, F. B., Talbot, W. S., Ganetzky, B., and Hogness, D. S. 1997. *Drosophila* ecdysone receptor mutations reveal functional differences among receptor isoforms. *Cell* 91:777–788.

A research paper analyzing ecdysone receptor function.

Berry, D. L., Schwartzman, R. A., and Brown, D. D. 1998. The expression pattern of thyroid hormone response genes in the tadpole tail identifies multiple resorption programs. *Dev. Biol.* 203:12–23.

A paper in which the genes responding to T_3 in tail tissue are identified and their response analyzed.

Buszczak, M., and Segraves, W. A. 1998. *Drosophila* metamorphosis: The only way is USP? *Curr. Biol.* 8:R879–R882.

An analysis of how the ecdysone receptor functions.

Hall, B. L., and Thummel, C. S. 1998. The RXR homolog *ultraspiracle* is an essential component of the *Drosophila* ecdysone receptor. *Development* 125:4709–4717.

A research paper identifying the molecular basis of ecdysone receptor structure and function.

Huang, H., and Brown, D. D. 2000. Prolactin is not a juvenile hormone in *Xenopus laevis* metamorphosis. *Proc. Natl. Acad. Sci. USA* 97:195–199.

An examination of the mechanism of prolactin action through the use of transgenic frogs.

Rose, C. S. 1999. Hormonal control in larval development and evolution—Amphibians. In *The origin and evolution of larval forms*, ed. B. K. Hall and M. H. Wake, pp. 167–216. Academic Press, San Diego.

A scholarly analysis of hormonal regulation of amphibian metamorphosis.

Thummel, C. S. 1995. From embryogenesis to metamorphosis: The regulation and function of *Drosophila* nuclear receptor superfamily members. *Cell* 83:871–877.

A review by a leading researcher of the ecdysone receptors and their relatives.

Thummel, C. S. 1997. Dueling orphans—interacting nuclear receptors coordinate *Drosophila* metamorphosis. *BioEssays* 19:669–672.

A review of the function of genes regulated by ecdysone.

Truman, J. W. 1996. Ecdysis control sheds another layer. *Science* 271:40–41.

A review of cells and hormones specifically involved in the eclosion of the pupa.

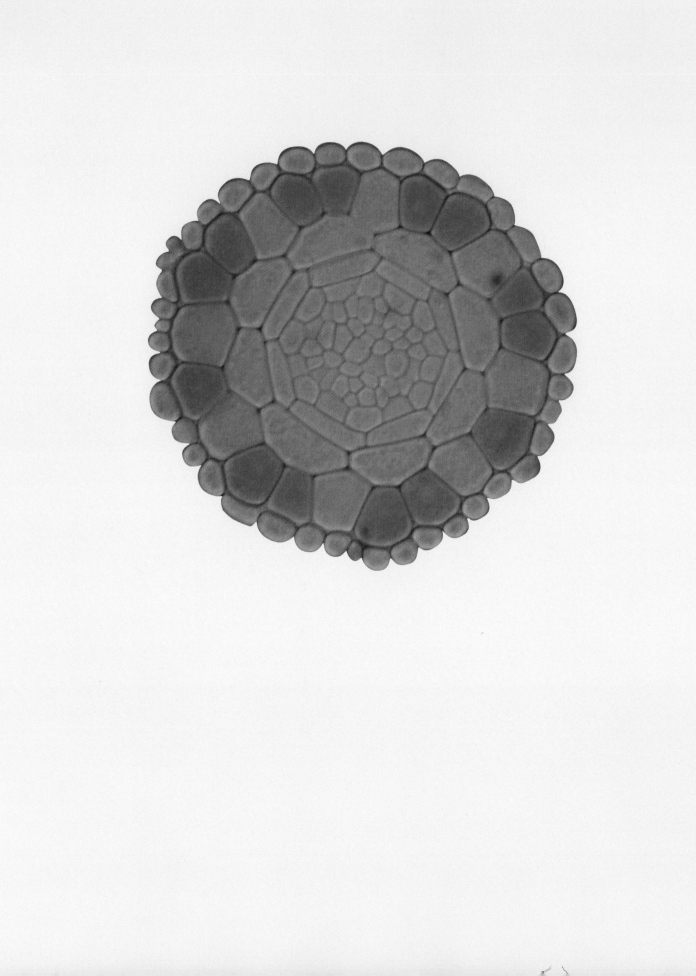

PART FOUR

DEVELOPMENT OF PLANTS

CHAPTER

9

PLANT MERISTEMS

Developmental biology has been a robust and important part of biology throughout the 20th century, but prior to the 1960s the study of development, often called *embryology* by zoologists, did not cross the line between the plant and animal kingdoms. Their grand strategies, only dimly perceived, seemed to be so different. And indeed, there are fundamental differences in emphasis on one tactic or another. On the other hand, the problems of intercellular communication and orchestrated differential gene expression are the same in plants and animals. And amazingly, at the cellular and molecular levels some of the machinery of development clearly antedates the divergence between plants and animals. A deeper inquiry into the principles of development insists on a consideration of plant development as well as animal embryology.

The study of the development of *Drosophila*, mice, and frogs starts with the embryo; this is the place in the life cycle where the body plan is laid out. One looks at an animal embryo and knows the basic organs that will appear. Plants, however, continue to make new organs throughout their life span, and the organs produced later in life may differ dramatically in morphology from those

CHAPTER PREVIEW

1. Plants initiate organs throughout their life span.

2. The center of proliferation and development is the meristem.

3. Meristems are the source of shoots, leaves, roots, and flowers.

4. Leaves initiate in regular patterns from the meristem.

5. Root meristems have their own unique properties.

made earlier. At the center of development, both in the plant embryo and the formed plant, is a developmental center called the *meristem*. For this reason we shall start with a discussion of how plants make organs from meristems in the "adult" plant; an understanding of the meristem will make our subsequent discussion of embryogenesis much richer.

Plant cells are permanently attached to each other through shared **cell walls** (see Chapter 12). These walls form between cells during cell division and, although the walls are superficially similar to complex extracellular matrices that encase some animal cells at the surface, cell walls distinguish plants from animals in many ways. First, since plant cells are encased in their walls, they do not move, and thus all morphogenetic events must be orchestrated in place. Second, cell walls expand, often at some distance in time and space from the cell division that originally laid down the wall. We will see how the constraint placed upon plant cells by their fixed position makes **positional information**—which tells the cell where it is within the greater context of the plant—the guiding force in plant development. It might tell the cell that it is on the surface or in the most internal layer, that it is in a region of rapid division or a region of differentiation, or that it is adjacent to a particular cell type. Positional information contrasts with **lineage-based information,** which is passed on from a mother cell to her daughters. We will see that lineage-based informational cues are rare in plants and that position largely determines a cell's fate.

A typical vascular plant, shown in Figure 9.1, has a shoot that grows above ground and a root that grows below ground. The shoot grows due to the action of a **shoot apical meristem,** which produces leaves, buds, and stem. The root grows due to a **root apical meristem,** which responds to gravity by growing in a direction parallel to that force. Running throughout the plant is a vascular system that transports water and nutrients throughout the tissues. Our discussion in this chapter will move from the shoot to the root. In Chapter 10, we will consider flower formation (flowers are modified shoots) and also reproduction and embryogenesis. Chapter 11 will cover unique aspects of cell differentiation in plants and the signals that direct activities of plant cells.

SHOOT MERISTEMS

We are all familiar with the fact that plants can take a fair amount of abuse as we mow the lawn, prune the rosebushes, and transplant seedlings into the garden. What we may not be aware of is the meristem, the hidden group of cells responsi-

ble for this flexibility (Figure 9.1). Meristems make development of plants fundamentally different than that of animals. Unlike animals, in which the organs for the adult are set out in the embryo, plants continue to generate new organs throughout the life of the plant. Because of this property of meristems, plants can be considered perpetually embryonic. Let us start by considering just what a meristem is and what makes it an indispensable focal point of plant development.

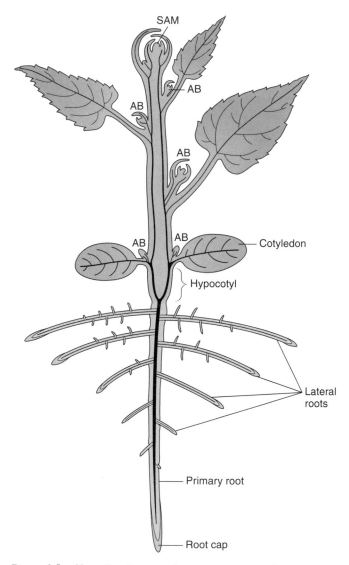

Figure 9.1   Shoot Development A transverse section of a typical flowering plant. Cotyledons (seed leaves) and the hypocotyl (the stem portion between the cotyledons and the root) will be discussed in Chapter 10. Abbreviations: SAM. shoot apical meristem; AB. axillary bud.

Meristems Have a Characteristic Histological Organization

Several histological features are unique to the meristem (Figure 9.2). All meristem cells are capable of dividing, but division rates vary across the meristem. The center of the meristem, or **central zone**, has a low division rate, whereas the surrounding cells that comprise the **peripheral zone** have higher rates. The cells in the central zone tend to be larger and have a smaller vacuole and less cytoplasm than the surrounding, peripheral zone cells. The rate of division in the peripheral zone is roughly equivalent to the division rate of cells in newly initiated organs. Cell divisions in the central zone replenish the meristem, similar to the way stem cells do in animals. Divisions in the peripheral zone contribute more directly to the new organs. Thus, the ultimate fate of a meristem cell depends to a great degree on its location within the meristem.

Meristem cells in flowering plants are organized into **layers** that encompass cells in both the central and peripheral zones. The outer layer or layers are referred to as the **tunica**, and the inner layer is the **corpus.** If there are two tunica layers, they are referred to as L1 and L2; the corpus is then L3 (see Figure 9.2). When the tunica consists of a single layer, it is called L1 and the corpus L2. The layers arise because the tunica cells divide with their walls laid down perpendicular to the surface—in effect, creating a sheet of cells. Cells in the corpus divide with their walls in all directions and thus build a three-dimensional group of cells.

The analysis of chimeras has been useful for understanding the contribution of these layers to the rest of the plant. A *chimera* (see Chapter 3) is a plant that contains genetically distinct cell types. "Variegated" horticultural plants are often chimeras. For example, consider the chimeric plant diagrammed in Figure 9.3, in which the meristem has an albino L2

layer (Figure 9.3A): Each layer contributes to specific tissues of the developing organs (Figure 9.3B): L1 makes the epidermis, L2 makes the bulk of the leaf or flower, and L3 produces a central portion of organs and contributes significantly to the stem. Mixing of cells from different layers rarely happens in the meristem, but it is common during the later elaboration of organs, which explains the jagged boundary between white and green cells in Figure 9.3C. The genetic differences that produce a chimera can arise by mutation or by grafting different genotypes together.

A fascinating and important property of meristems is the ability of their cells to function as a coordinated group. The seemingly well specified patterns can be reorganized if the positional boundaries are altered. For example, in an experiment where a shoot meristem was surgically bisected, two normal meristems formed, each capable of initiating organs (Figure 9.4). This experiment showed that plant cells are able to respond to changes in positional information and do so by organizing a new, intact meristem. A more drastic experiment was carried out by whittling a potato meristem down to 1/20 of its original size. The remaining meristem nub was able to regenerate to its normal size. These experiments document the self-organizing ability of the meristem; when the central zone or peripheral zones are destroyed, new zones with those properties arise.

Meristems Produce Patterns of Organs (Phyllotaxy)

As leaves initiate from a meristem, they are referred to as **primordia**. A primordium is recognizable as a distinct bump on the flank of the meristem, prior to differentiating into a leaf. A numbering system is used to keep track of the stages that leaves go through as they progress in development from primordia to fully differentiated leaves. **Plastochron** refers to the time interval between successive leaf initiations and is used to

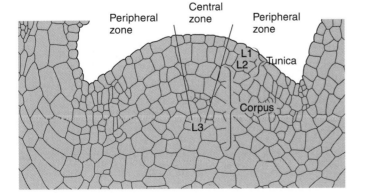

Figure 9.2 Shoot Meristem Organization A section through a shoot meristem shows the discrete layers. The outer layers (L1 and L2) are the tunica. Interior to L1 and L2 is the corpus, or L3. At the center of the meristem is the central zone, which consists of more vacuolate cells that renew the meristem. The peripheral zone surrounds the central zone and is the morphogenetic region. Cells that form organs are derived from peripheral zone cells.

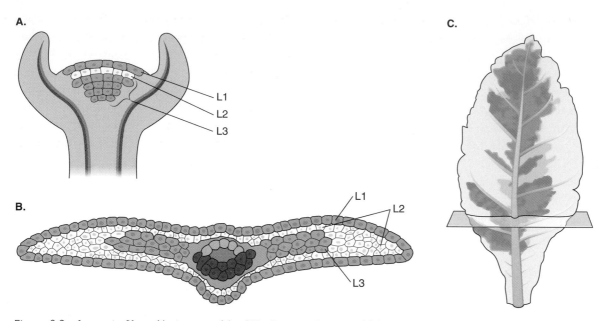

Figure 9.3　Layers in Shoot Meristem and Leaf The layers in the meristem contribute to particular parts of organs such as leaves. **(A)** In this diagram, the cells in the second layer (L2) carry an albino mutation whereas all other cells are green. **(B)** A diagrammatic cross section of a leaf resulting from an L2 chimera. These chimeras suggest that the L2 layer contributes to a significant portion of the tissues. The center of the leaf is green and is thought to arise from the L3 layer. L1 contributes only to the epidermis. **(C)** A typical leaf from an L2 chimera. The lateral extent of the L3 lineage is quite variable.

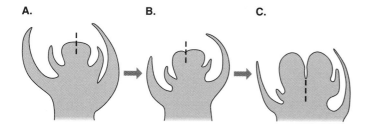

Figure 9.4　The Effect of Bisection on Meristem Development **(A)** A meristem prior to bisection. Note that meristem bisection does not block development. **(B)** An incision was made in the shoot tip of a bean plant. Seven days later, the meristem has reorganized. **(C)** Thirteen days later, two shoots are growing.

describe leaves at different stages. A leaf at plastochron 1 (P_1) is a primordium that has just initiated. That primordium becomes a P_2 leaf when a new primordium forms, and will be referred to as a P_3 leaf when a third primordium initiates. Thus, all leaves move through plastochron stages. The plastochron stage at which a leaf is fully differentiated depends on the species; for example, it is P_{20} for tomato and P_{10} for corn.

Shoot meristems produce primordia in a highly regular fashion. Because the meristem is continually growing as these primordia initiate, the patterning is not as simple as dividing a pie into appropriately sized pieces. As the meristem grows, the initiating primordia are separated from past primordia around the circle as well as in a vertical direction along the shoot axis. The pattern of primordia initiation is termed **phyllotaxy,** and can be either whorled or spiral (Figure 9.5).

Spiral phyllotaxy accounts for some of the more beautiful patterns in nature, such as the head of a sunflower or the arrangement of scales in a pinecone. In spiral phyllotaxy, each primordium initiates a set number of degrees of the circle from the previous primordium. Often this angle is 137.5°, which creates gaps between primordia. Thus, from a top-down view, the fourth primordium forms in the largest gap, which falls between the first and second primordia, and the fifth primordium forms between the second and third primordia (Figure 9.5A).

The other basic phyllotactic pattern is whorled (Figure 9.5B). The common whorled patterns have one, two, or three primordia per whorl. A single leaf in a whorl is referred to as *distichous*, two leaves in a whorl as *decussate*, and three leaves per whorl as *tricussate*. When a single primordium forms in a whorl, the next primordium to form will be initiated from the opposite side of the meristem. When two primordia form in a whorl, the primordia of the next whorl are usually oriented 90° from those in the first whorl.

Most models that attempt to explain phyllotactic patterns focus on concepts such as morphogenetic fields or available space, and thereby consider the impact of the existing organs on formation of subsequent primordia. For example, in one model the previously initiated primordia are thought to pro-duce a diffusible substance that inhibits formation of the next organ, thus allowing a primordium to form only when the distance from previous organs has increased sufficiently. Another view, most strongly championed by Paul Green, posits that organs are not able to form close to existing organs be-cause of biophysical constraints. A useful example is the sur-face of a sunflower. The primordia have a wavelike repetition suggesting that the positions of existing primordia determine the position of the next primordium.

A. Spiral phyllotaxy

B. Whorled phyllotaxy

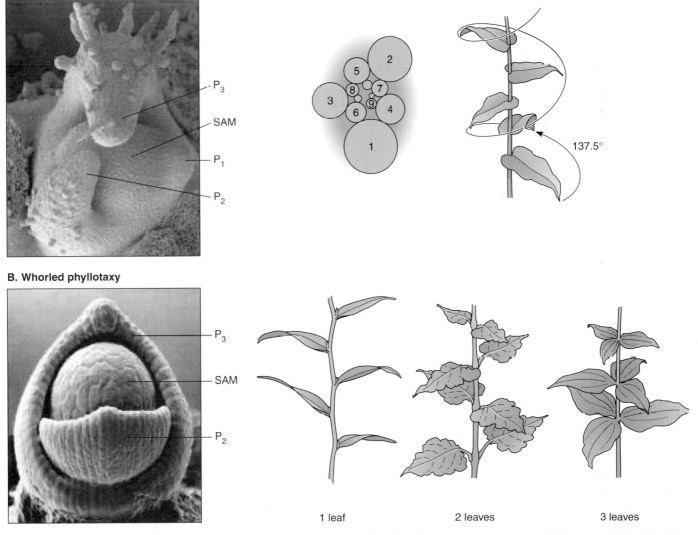

1 leaf   2 leaves   3 leaves

Figure 9.5 Phyllotactic Patterns (A) On the left is a scanning elec-tron micrograph (SEM) of a tomato meristem. The leaf primordia are initiating in a spiral. P_1 identifies plastochron 1, the first leaf from the meristem; P_2 is the second leaf, and P_3 the third. On the far right is a sketch of a plant with spiral phyllotaxy. In the middle is a top-down view of the initiating organs, which are numbered sequentially from 1 as the oldest. Each organ arises in the available space. **(B)** *Left:* An SEM of the shoot apex of corn. The meristem is the ball of cells in the center, with two recently initiated leaves surrounding it. The leaves initiate in a distichous, or opposite, pattern. *Right:* Three different whorled patterns: one leaf (as in corn), two leaves, or three.

A.

Wild type   *abphyl*

B.

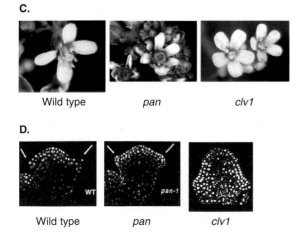

Wild type   *abphyl*

C.

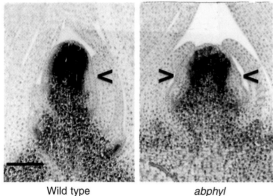

Wild type   *pan*   *clv1*

D.

Wild type   *pan*   *clv1*

Support for the **available space model** (whether such space is physically or biochemically driven) comes from analysis of mutants that make more organs presumably because they have a larger meristem. A mutation in maize called *abphyl* (for "aberrant phyllotaxy") produces a larger meristem and two leaves at a time instead of one (Figure 9.6A,B). In *Arabidopsis* (see Box 9.1), *clavata* mutants make more organs than do wild-type plants and also have larger meristems (Figure 9.6C,D). Unlike *abphyl* meristems, which never make more than two leaves at a time, *clavata* meristems can grow up to 1,000 times larger than wild-type meristems. Size, however, cannot be the only factor determining organ number. Meristem size varies dramatically among different species with the same phyllotactic pattern. In addition, mutants have been described that make more organs but do not have larger meristems; an example is *perianthia* of *Arabidopsis*. (Figure 9.6C,D).

Plants can switch from one phyllotactic pattern to another in response to developmental triggers or surgical manipulation. For example, the shoot apical meristem of maize initiates leaves in a distichous pattern, but switches to a spiral phyllotactic pattern in the production of flowers. Ivy plants produce leaves in a whorled pattern when young but switch to a spiral pattern as the plant matures. Surgical experiments done in the 1930s by M. and R. Snow showed that the phyllotactic pattern could be altered by changing the overall shape of the meristem. They made a transverse cut in decussate apices, which were initiating two leaves at a time. This cut often resulted in two apices that produced leaves with spiral phyllotaxy (Figure 9.7). Occasionally, the surgically imposed spiral phyllotaxy would revert back to decussate. These examples illustrate an inherent flexibility in patterning, but once the pattern is initiated, it is very stable.

Figure 9.6   Mutants with More Organs (A) The *abphyl* mutant of corn makes two leaves per whorl compared with wild-type corn, which makes a single leaf. **(B)** The *abphyl* meristem is larger than the wild-type meristem. Arrowheads point to incipient leaf primordia; note that there are two in *abphyl*. **(C)** A wild-type flower of *Arabidopsis (left)* compared with two different mutants, *perianthia (middle)* and *clavata1 (right)*. Both *perianthia* and *clavata1* mutant flowers make more organs. **(D)** Confocal microscopy of floral meristems shows that *perianthia* meristems are the same size as those of the wild type, while *clavata1* meristems are larger.

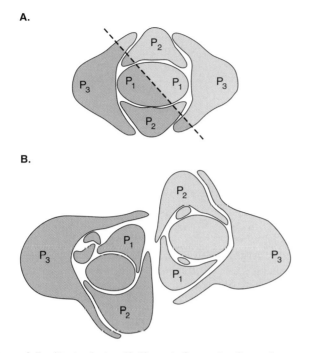

A.

B.

Figure 9.7   Manipulating Phyllotactic Patterning Surgical experiments demonstrate that the phyllotaxy can be altered. **(A)** A meristem with decussate phyllotaxy is cut transversely (*orange line*) to produce asymmetric meristems. The dashed lines indicate the position of the P_1 primordia, which are still attached to the meristem. **(B)** Two shoots develop from this cut meristem. The P_1 leaf at the time of cutting is now P_3. The next couple of leaves to form (P_2 and P_3) arise in a spiral phyllotaxy.

Cell Expansion Is Very Important in Leaf Initiation

So far, our discussion of plant development has neglected the fact that plant cells have walls. The walls have to be strong enough to resist the internal turgor pressure that builds up in the cell due to the water driven into the vacuole by osmotic pressure, but flexible enough to expand so the plant can continue to grow. Plant cells may expand their volume by nearly 20,000-fold. A protein called **expansin** has been identified in plants. Given that animal cells lack cell walls that expand, it is not surprising that this protein is unique to plants. When expansin is added to the cell walls of heat-killed plant cells, it causes the walls to expand in a pH-dependent manner. This expanding effect works on isolated pieces of stem, as well as on paper (which is, of course, made of wood, which is comprised of cell walls), and it works in a catalytic fashion.

Given the consistent phyllotactic patterns in plants, the positions of leaves prior to their initiation can be surmised: A P_0 leaf has not yet formed and a P_{-1} leaf is next in line to develop after the P_0 leaf. When Chris Kuhlemeier placed an expansin-coated bead in the P_{-1} position on a tomato meristem, an ectopic (out of place) organ occasionally formed (Figure 9.8A). Control beads without expansin did not cause any organs to form. The presence of this out-of-place primordium was able to shift the direction of the spiral phyllotaxy from counterclockwise to clockwise. Although the primordium did not develop into a proper leaf (probably because it was not connected into the vasculature of the plant), it is significant

9.1 *ARABIDOPSIS*

Many plant biologists have turned to the weed *Arabidopsis thaliana* for their investigations. *Arabidopsis* is in the mustard family, which includes well-known vegetables such as cauliflower, broccoli, and kale. The small size of its genome (130,000 kilobases, where 1 kb = 1,000 bases), slightly less than that of *Drosophila*, made it possible to sequence the entire genome. The genome has very little noncoding DNA, and a distinct gene occurs about every 5 kb. Its small size allows plant biologists to screen tens of thousands of plants for mutants without using acres of field space. Often the seeds are grown in petri dishes so that particular chemicals or environmental conditions can be provided. *Arabidopsis* automatically self-pollinates and produces copious amounts of seed, thousands per plant.

A.

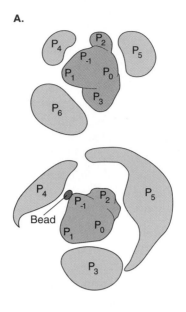

B.

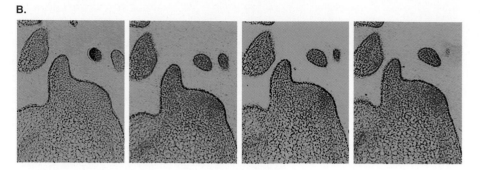

Figure 9.8   Primordium Initiation Through Expansin (A) *Top:* A diagram of a tomato apex that has generated primordia at plastochrons P_6 to P_1. The next leaf to form will be in the P_0 position, followed by a leaf in the P_{-1} position. *Bottom:* The same apex five days after an expansin-loaded bead was placed at the P_{-1} position. A bulge has formed here instead of at the available position (P_0). **(B)** In situ hybridization of expansin gene expression in a tomato shoot apex. Serial (contiguous) sections through the apex reveal expansin expression (*red*) in the P_0 position, which is where the next leaf will form. Serial sectioning is useful for revealing certain structures—in this case, the P_0 leaf—that are visible in only some sections.

that the topical addition of a cell wall protein could affect organ initiation in the meristem. Similar experiments were done with the plant hormone auxin, which we will discuss in Chapter 11. Again, beads loaded with auxin and placed on the meristem were able to affect the phyllotaxy in the shoot. Unlike the expansin experiment, in the case of the auxin application the initiating leaves were normal.

The results of the expansin experiment add to a long-standing debate concerning the importance of the orientation of cell division in leaf initiation. One group of scientists has proposed that changes in the orientation of cell division are essential to the initiation of organogenesis. Other researchers have proposed that changes in the orientation of cell division may accompany organogenesis but are secondary. Support for the latter theory comes from experiments by A. Haber and D. Foard, who showed that a leaf could initiate in the presence of doses of radiation high enough to prohibit any cell divisions whatsoever. Initiation of the leaf was witnessed by a change in growth along the main proximal distal axis. Since expansin first causes cell wall loosening and not cell division, the expansin results support Haber and Foard's interpretation.

If expansin normally plays a role in leaf initiation, then we would expect to find the expansin gene expressed specifically in the cells that will expand to form a leaf. An expansin gene was found in tomatoes that is expressed in the P_0 region of the meristem (Figure 9.8B). Thus, it is possible that one of the first events in making an organ is to express expansin in the P_0 domain, which then loosens the existing walls, making them susceptible to the expansionist forces of

the turgor pressure. Divisions would then follow in the cells whose walls were primed by expansin. This speculation seems reasonable, but rigorous experiments to test this have not yet been done.

We shall soon learn that most of the known genes that act in meristems encode transcription factors or kinases, which we have already learned are crucial for patterning of the animal embryo. By their very function, transcription factors and kinases are the managers or messengers, involved with giving the orders but not actually carrying out the events of differentiation. Eventually, we will want to know exactly how a bump that will form a leaf comes to be at a particular place and time. Finding location-specific gene expression for a protein that acts on wall chemistry, such as expansin, is an exciting first discovery.

Apical Dominance Influences Development of Axillary Meristems

Shoot meristems come in different forms. The shoot apical meristem (SAM) is at the apex of the plant and is responsible for the growth of plants such as maize or sunflowers, which tend to grow primarily upward. **Axillary meristems** are found in the leaf axils and produce axillary buds. A leaf axil is the junction between the leaf and the main stem (see Figure 9.1). A brussel sprout is an example of a tasty axillary bud. Axillary meristems that develop into branches are what give plants their three-dimensional growth habit. (Box 9.2 provides a look into the key role axillary meristems have played in agriculture.) The

9.2 GENES THAT FUNCTION IN THE EVOLUTION OF MAIZE

Maize (corn) is thought to have evolved from a wild grass called teosinte, thanks to the breeding efforts of early agricultural societies in Mexico. To understand how teosinte might have played a role in agricultural selection, we first need to understand the *inflorescence* (flower-bearing) structures in maize. The shoot apical meristem of maize develops into a terminal inflorescence that is referred to as the tassel and bears the male flowers (see accompanying figure). Axillary meristems are mostly suppressed in maize, except for the one or two that become the ear of corn. The female flowers are found on ears and develop into kernels after pollination. Teosinte, on the other hand, is very bushy; its axillary branches are long and terminate in tassels. Only in crowded conditions are the axillary branches of teosinte suppressed. Thus corn has a few, short female branches whereas teosinte has many, long male branches. In corn, the tassel and ear initially have both male and female organs, but in the tassel the female parts abort, and in the ear the male parts abort.

A number of scientists have crossed teosinte and maize, self-pollinated the progeny (the Γ1 generation), and examined the F2 generation for maize-like or teosinte-like traits. From the segregation ratios, John Doebley and others have suggested that just a few genes control the differences between maize and teosinte. In fact, a single gene mutation in maize, *teosinte branched 1* (*tb1*), results in corn plants that look very much like teosinte in their overall plant architecture. In situ hybridization studies (see Box 7.1) have shown that the *tb1* gene, which encodes a putative transcription factor, is expressed in maize axillary meristems throughout their growth. Expression is not detected in the SAM. The gene is present in teosinte but is not expressed normally in axillary branches. Based on the mutant phenotype in maize and the expression pattern, Doebley proposes that *tb1* functions in teosinte to suppress axillary branch growth under crowded conditions, but its function in maize is to make the compact ear that is perfectly designed for harvest and consumption.

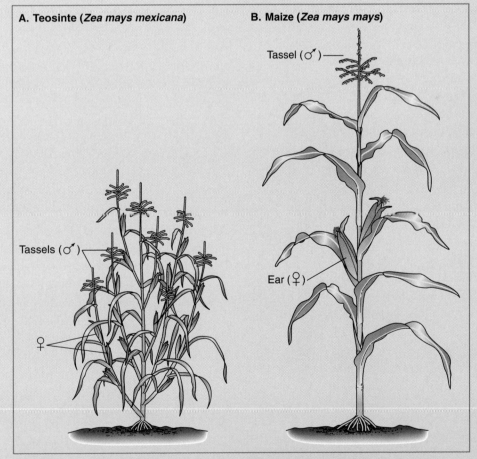

A. Teosinte (*Zea mays mexicana*)

B. Maize (*Zea mays mays*)

Tassel (♂)

Tassels (♂)

♀

Ear (♀)

The Evolution of Maize (A) Teosinte, a wild plant thought to be the ancestor of **(B)** modern maize. Note that the teosinte plant has many long branches, whereas the maize plant has two short branches that end in ears.

SAM often displays **apical dominance,** a phenomenon in which the SAM suppresses growth of axillary meristems. Those who garden are familiar with the experiment of "releasing apical dominance." Plants such as snapdragon or stock (a garden flower in the mustard family) have a single set of flowers, and the plant can get very tall. If the flowering portion is removed to make a bouquet, axillary meristems of the lower leaves grow out and produce more sets of flowers (Figure 9.9).

Adventitious meristems are another type of shoot meristem. **Adventitious meristems** form from surface cells on stems and leaves. Epidermal cells begin to divide, forming an organized meristem. Just as the SAM will suppress expansion of axillary meristems, the SAM often suppresses the initiation of adventitious meristems. One well-studied system is

flax, in which hundreds of adventitious meristems form along the stem if the top of the plant is cut off. The leaves that form on these adventitious meristems have a programmed phyllotactic pattern, in which the first leaf forms on the lower side of the adventitious meristem (that is, the side closest to the root).

In both axillary and adventitious meristems, it is clear that some signals are coming from the shoot tip. Three different hormones, which we will discuss later, play a role in releasing or suppressing axillary growth.

The Meristem Establishes Leaf Dorsoventrality

As new leaf primordia are initiated, previously formed primordia become separated from the meristem and are found at increasingly greater distances from the meristem. As the distance increases, the leaf primordia differentiate and grow in a stereotypical, species-specific manner. Most leaves differentiate downward from the tip, but there are exceptions that differentiate toward the tip. P_1 leaves can appear as a uniform bump, but by the time they are P_2 leaves, a clear dorsoventral pattern is visible (Figure 9.5A). Even cylindrical leaves, such as those found on rushes, begin with a dorsoventral organization. (Plant morphologists and anatomists have historically used the spelling *dorsiventral* instead of *dorsoventral*; however, not all plant biologists use the former, so for consistency we shall continue to use the latter spelling throughout.)

The side of the leaf that faces the meristem is referred to as **adaxial** and can be thought of as the dorsal side. The side facing away from the meristem is **abaxial,** or ventral. The two sides show a number of differences. At a very early stage, such as P_1, the adaxial side has smaller cells with fewer vacuoles than cells on the abaxial side (Figure 9.10A). This dorsoventral pattern is not present in axillary meristems or flower meristems that initiate with radial symmetry. Abaxial-adaxial differences are also reflected in epidermal characteristics, such as the placement of hairs and **stomata** (singular *stoma,* from the Greek meaning "mouth"), which are openings for gas exchange, and in the internal organization of photosynthetic and vascular cells. In some plants, stomata are found only on the abaxial surface. Leaves often have a layer of photosynthetic cells, known as **palisade mesophyll,** that are conveniently placed on the surface facing the sun (Figure 9.10B). Below the palisade layer are mesophyll cells with many air spaces for gas exchange. The vasculature is embedded in this spongy mesophyll, with the water-conducting cells (known as the **xylem**) on the adaxial side and the cells conducting nutrients (the **phloem**) on the abaxial side. Photosynthetic

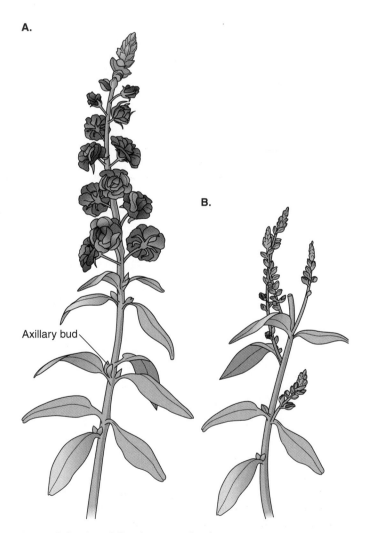

A.

B.

Axillary bud

Figure 9.9 Apical Dominance in Stock (A) The buds on this stock plant are normally suppressed. and flowers are confined to the top of the plant. **(B)** When the stem is cut. axillary buds below the cut grow out to produce flowers at multiple shoot apices.

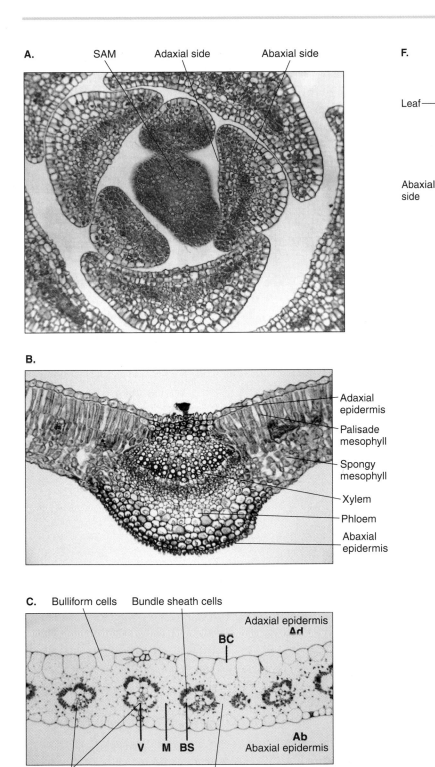

A. SAM   Adaxial side   Abaxial side

B.
- Adaxial epidermis
- Palisade mesophyll
- Spongy mesophyll
- Xylem
- Phloem
- Abaxial epidermis

C. Bulliform cells   Bundle sheath cells
Adaxial epidermis Ad
BC
V M BS
Ab Abaxial epidermis
Veins   Mesophyll

F. P₂   SAM   P₀
Leaf
Abaxial side
P₁   Adaxial side

E.
SC
SC   ← EL
EL →   Leaf flaps
ML   MV   SC   ML

Leaf flaps
Abaxial sides
Adaxial   Adaxial
Abaxial

D.
- Abaxial epidermis
- Vein
- Bundle sheath cells
- Mesophyll
Ab E
V
BS
M

Figure 9.10   Leaf Dorsoventrality (A) A cross section through a vegetative meristem with a recently initiated leaf primordium. The cells on the abaxial side of the leaf are more vacuolated than those on the adaxial side. **(B)** A cross section of a mature leaf reveals palisade cells (specialized for photosynthesis) on the adaxial side. **(C)** A cross section of a maize leaf shows that the adaxial epidermis contains specialized bulliform cells, which swell with water to keep the leaf expanded. The veins are surrounded by bundle sheath cells. **(D)** The maize mutant *leaf-bladeless* occasionally results in radially symmetric leaves. The epidermis has abaxial cell types. **(E)** *Top:* A cross section of a *leafbladeless* leaf shows ectopic leaf flaps, which grow out of the leaf at sporadic intervals. *Bottom:* Leaf flaps support the model that lamina growth results from the juxtaposition of adaxial and abaxial cell types. **(F)** Expression of *filamentous flower* in the abaxial half of the leaf of a wild-type primordium in *Arabidopsis*.

cells in a maize leaf are organized into specialized **bundle sheath cells,** which surround the vein and mesophyll cells between the veins. Within the vein, xylem is adaxial and phloem is abaxial.

The establishment of dorsoventral polarity has been addressed by surgical studies and mutant analysis. Ian Sussex cut potato meristems in such a way that an incipient primordium was isolated from the rest of the meristem. Very often, the isolated primordium developed as a narrow pin with radial symmetry. It lacked the broad flat surface, or **lamina,** typical of leaves. Clearly, the meristem provides some important contribution to the establishment of the dorsoventral polarity as well as to lateral outgrowth.

More recently, mutations have been described in snapdragon, maize, tobacco, and *Arabidopsis* in which the leaves lose their abaxial-adaxial polarity. The mutant plants have abaxialized leaves; that is, features normally on the abaxial surface are also found on the adaxial surface. For example, normal tobacco leaves have stomata on the abaxial surface, but *lam1* mutants have them on both surfaces. Also, *lam1* mutants lack a palisade layer. The adaxial surface of normal maize leaves has specialized hairs and other epidermal cells (Figure 9.10C). These adaxial cell types are missing in *leafbladeless (lbl)* mutants (Figure 9.10D). An opposite phenotype occurs in the dominant *phabulosa* mutation in *Arabidopsis*. The leaves lack abaxial characters; in fact, the leaves have axillary buds forming on both sides of the leaf. The leaves of *phabulosa* mutants are thus said to be adaxialized.

Aside from subserving physiological functions, dorsoventral patterning may be an important developmental cue. Accompanying the lack of abaxial-adaxial characters in these different mutants is an occasional radial symmetry in which the lamina fails to grow out (Figure 9.10D). This finding has led to the suggestion that dorsoventral specification is needed for lamina growth. Support for this idea comes from sporadic patches of abaxial cells on the adaxial side in *lbl* mutant leaves. These mosaic patches are accompanied by the formation of ectopic leaf flaps that grow out of the surface of the leaf (Figure 9.10E). These findings suggest that the juxtaposition of abaxial and adaxial tissues, or the boundary between them, defines the position for lamina outgrowth.

Some of the above-mentioned genes have recently been cloned. Kathy Barton and colleagues discovered that *phabulosa* encodes a protein with a lipid-binding domain and a transcription factor domain. The gene is expressed throughout the leaf at the P_0 and P_1 stages. At about the P_2 stage, expression is confined to the adaxial half of the leaf. In the radial leaves of the dominant mutant, expression is seen throughout the leaf. These results suggest that misexpression of the *phabulosa* gene in the abaxial half of the leaf results in loss of dorsoventral organization. Other genes that also code for transcription factors (such as members of the Yabby family) are normally expressed in the abaxial half of the leaf (Figure 9.10F). The results share some similarities with findings in animal organ development. Some gene products that are required for dorsoventral identity are localized dorsally in the developing limbs of animals and the leaves of plants. Loss of either dorsal or ventral gene expression compromises growth in both the lateral and proximal directions.

ROOT MERISTEMS

Roots do not catch the eye when strolling through a botanical garden; however, they do provide an important avenue for water and minerals to enter the plant. Furthermore, the root offers an excellent opportunity to developmental biologists for studying cell fate. What is responsible for root architecture? In a root's gravity-driven search for water and nutrients, the root grows down and sends out lateral roots. These lateral roots help stabilize the plant as well as increase the absorptive surfaces. Roots can also take on a myriad of other functions, such as food storage (in yams and sweet potatoes), gas exchange (in roots of swamp trees that rise up out of the water), and surface attachment (in aerial prop roots of ivy). The ability to take on these other tasks emphasizes a root's inherent flexibility.

Root Apical Meristems Produce Radial Patterns

Roots have four major zones in the longitudinal direction: a root cap at the tip, a zone of cell division, a zone of cell elongation, and finally a zone of differentiation, where lateral roots and root hairs form (Figure 9.11A). The root cap protects the root tip as it forces its way through soil; the cap cells are constantly sloughed off and rapidly replaced. Root caps are thought to be the **gravitropic** (gravity-sensing) organ. When root caps are chopped off, roots grow in random orientations until the root cap grows back.

In the transverse dimension, roots have approximately six different tissue types arranged in cylindrical progression. Root cap cells surround the tip of the root; then come the epidermis, cortex, endodermis, and finally the pericycle, which surrounds the internal vasculature (Figure 9.11B). Water-conducting xylem cells form a solid core in the center of most roots. The xylem extends outward from the center like the spokes of a wheel. These spokes of immature xylem, referred to as *pro-toxylem poles*, interdigitate with phloem cells. Each of these five layers of the root arises from cell divisions that can be traced back to progenitor cells in the root apical meristem.

The root apical meristem, unlike the SAM, produces cells in two directions, cells that contribute to the root axis and cells that contribute to the root cap. Within the root apical meristem is a group of very slowly dividing cells referred to as the **quiescent center**. In *Arabidopsis*, there are four central cells that are the quiescent center plus four sets of **initial cells**. One type of initial cell produces the epidermis and lateral root cap, a second produces the cortex and endodermis, a third the pericycle and vascular cells, and a fourth the root cap (Figure 9.11B).

These well-defined cell division patterns have led to speculation about the role of cell lineage in defining cell fate. In ablation studies using markers for different cell types, researchers have shown that the final fate of root cells is dependent on their position and not on their lineage. When quiescent central cells were ablated with a laser beam, they were replaced by neighboring vascular initial cells that quickly gained the features of central cells. Similarly, when cortical-endodermal initial cells were ablated, the ablated cell was compressed

and replaced with a pericycle cell. Once again, the pericycle cells acted as cortical initial cells and produced cortical and endodermal cell files (Figure 9.11B). (Files are rows of cells derived from a common initial cell.) Thus, although the orderly files of cells suggest a role for lineage in determining cell fate, experimental evidence has shown that cell fates are dependent on position.

Recently discovered mutants are beginning to help researchers understand how the different cell types are established. The *Arabidopsis* mutants *scarecrow (scr)* and *short root (shr)* lack one of the cell layers between the epidermis and the pericycle. The defects result from a failure in the asymmetric division of the cortical-endodermal initial cell that gives rise to both of these lineages (Figure 9.11B). The resulting layer in *shr* mutants lacks endodermal characteristics, whereas the resulting layer in *scr* mutants has characteristics of both endodermal and cortical cells. The *scr* gene, which encodes a transcription factor, is expressed in the quiescent center cells, the cortical-endodermal initial cells, and subsequently exclusively in the endodermal cell file (Figure 9.12). There is no *scr* gene expression in a *shr* mutant background, suggesting that *scr* requires a wild-type copy of *shr* for expression. Interestingly, RNA transcribed by the *shr* gene is detected in the **stele** (the interior region comprised

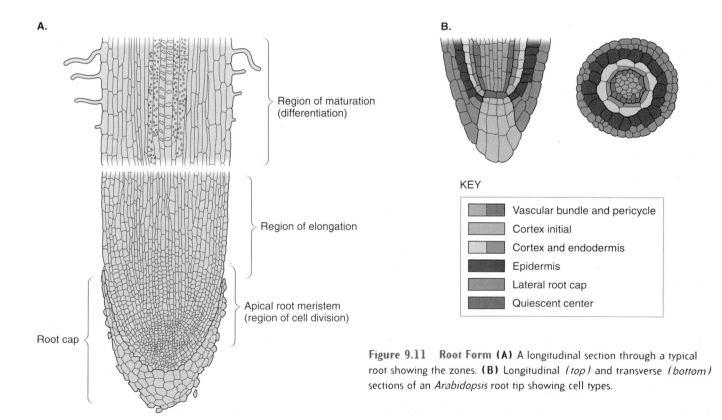

A.

Region of maturation (differentiation)

Region of elongation

Apical root meristem (region of cell division)

Root cap

B.

KEY

	Vascular bundle and pericycle
	Cortex initial
	Cortex and endodermis
	Epidermis
	Lateral root cap
	Quiescent center

Figure 9.11   Root Form (A) A longitudinal section through a typical root showing the zones. **(B)** Longitudinal *(top)* and transverse *(bottom)* sections of an *Arabidopsis* root tip showing cell types.

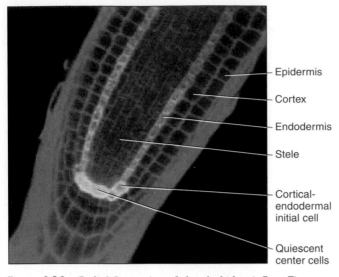

Figure 9.12   **Radial Patterning of the *Arabidopsis* Root** The *scarecrow* gene of *Arabidopsis* is expressed in the quiescent center cells. the cortical-endodermal initial cell(s). and the endodermal cell file.

Lateral Root Meristems Arise from Differentiated Cells

Lateral roots contrast with lateral organs of the shoot (leaves and axillary meristems) in a number of different ways. First, their initiation begins with the redifferentiation of cells located some distance from the root apical meristem. These initial cells giving rise to roots are internal to the epidermis; therefore, a lateral root has to burst its way through the outer primary root layers (Figure 9.13). Second, the temporal pattern of lateral root initiation is not as ordered as the phyllotaxy seen in the shoot; it is more responsive to environmental conditions. Whereas the main root grows toward gravity, lateral roots grow toward nutrients. The density of lateral roots will also increase in an area of higher nutrients.

In *Arabidopsis*, the first sign of lateral root initiation is a cluster of cell divisions within the pericycle cells (Figure 9.13A). The cells undergo a few **anticlinal divisions** (creating new cell walls perpendicular to the surface) before undergoing **periclinal divisions** (creating new walls parallel to the outer surface). The periclinal cell divisions create new layers, and each layer takes on a new fate. For example, cells in the outer layer accumulate starch grains as they differentiate into root cap cells. Experiments in which roots were excised have shown that the lateral root meristem does not become autonomous until three to five cell layers are established. Thus, the

redifferentiated pericycle cells re-create all the layers of the root before the lateral root meristem becomes visible.

Given the fact that there is a ring of pericycle cells, one might ask how a lateral root forms in one place and not another. In most species, lateral roots initiate adjacent to protoxylem poles (Figure 9.13B). These pericycle cells are shorter than other cells in the vertical dimension (Figure 9.13A). If auxin (a hormone to be discussed in greater detail in Chapter 11) is added to the culture medium on which the seeds are germinated, all the pericycle cells in the protoxylem position undergo divisions in an attempt to initiate lateral roots. Thus, it appears that two factors control the spacing of lateral roots: (1) an innate competency that is defined by the position next to the protoxylem pole and is unaffected by additional hormones, and (2) a spacing mechanism that can be partially overcome by the addition of hormones.

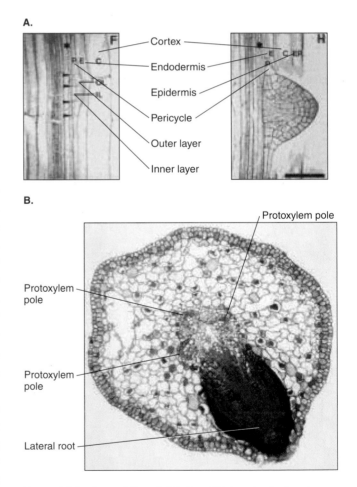

Figure 9.13   **Lateral Root Initiation (A)** Longitudinal sections through an *Arabidopsis* root. *Left:* The first sign of lateral root formation is a series of anticlinal divisions in a few cells of the pericycle layer (*arrowheads*). These cells then divide periclinally to make an outer and an inner layer. *Right:* Additional periclinal cell divisions occur to produce a lateral root. **(B)** This cross section shows a lateral root initiating from pericycle cells at one of the protoxylem poles.

of the pericycle and vasculature), and not in the cells where it is required. We will see in Chapter 11 how the Shr protein acts from a distance and leads to properly differentiated cell fates.

KEY CONCEPTS

1. The fate of plant cells is largely determined by their position.

2. Plant cells can acquire new fates simply by changing their position. If an internal cell divides such that one of its off-spring is in the epidermis, that outer cell takes on the fate of the epidermal cells.

3. Plant organs form from meristems. The shoot meristems are responsible for the patterning of leaves.

4. Root apical meristems differ from shoot apical meristems in that the lateral organs are not produced and organized into strict patterns at the root meristem.

5. Leaves initiate with a dorsoventral polarity that is reflected in histological differences between the adaxial (adjacent to the meristem) side and abaxial side of the leaf. This symmetry may be important for leaf growth.

STUDY QUESTIONS

1. If you bisect a shoot meristem longitudinally without killing the plant, what might you expect?

2. How might you discover what regulates apical dominance?

3. What advantage is there for a leaf to have a dorsoventral organization?

4. Does each cell in the meristem have a specific task based on its lineage? Explain your answer.

5. What defines a leaf?

6. What defines a meristem?

SELECTED REFERENCES

Dolan, L., and Okada, K. 1999. Signalling in cell type specification. *Semin. Cell Dev. Biol.* 10:149–156.

   A discussion of the role of positional information in root development and epidermal cell type specification.

Kaplan, D. R. 2001. Fundamental concepts of leaf morphology and morphogenesis: A contribution to the interpretation of molecular genetic mutants. *Intl. J. Plant Sci.* 162: 465–474.

   An examination of the developmental strategies that diverse plants utilize, with a focus on leaf morphology. This comparative approach provides a useful background for the geneticist trying to understand plant development by examining mutants in a single organism.

Sessions, A., and Yanofsky, M. 1999. Dorso-ventral patterning in plants. *Genes Dev.* 13: 1051–1054.

   A review of some of the genes involved in establishing abaxial-adaxial cell identity in the leaf.

Snow, M., and Snow, R. 1935. Experiments on phyllotaxis. III. Diagonal splits through decussate apices. *Philos. Trans. R. Soc. Lond.* B225:63–94.

   Early experiments that revealed how phyllotaxy can be perturbed.

Steeves, T. A., and Sussex, I. M. 1989. *Patterns in plant development*, 2nd ed. Cambridge University Press, Cambridge.

   Descriptions of many of the classical experiments in plant biology. The book provides a solid foundation for understanding plant biology.

CHAPTER

10

REPRODUCTION IN PLANTS

FLORAL AND INFLORESCENCE MERISTEMS

The shoot apical meristem (SAM) makes leaves continually until it is primed by environmental or developmental cues to make flowers. This switch is under complex genetic control—and no wonder, for a plant needs to make this switch at the time most optimal for its survival as a species. Plants in the Arctic flower as quickly as possible; plants in the desert wait until there is enough rain so that their progeny will germinate; and plants that overwinter wait until spring, their flowering often cued by daylength (to be discussed shortly) or temperature. A meristem that has shifted from making leaves to making flowers is referred to as an **inflorescence meristem (IM).** The IM may still make leaves, but they are usually very reduced in size and referred to as **bracts.**

CHAPTER PREVIEW

1. Daylength regulates the transition from making leaves to making flowers.

2. Although flowers form at the meristem, the signal to form them comes from well-developed leaves.

3. Vascular plants have life cycles with two distinct phases: gamete producing and spore forming.

4. Embryos of flowering plants develop within a portion of the flower.

5. Embryos form the primordial shoot and root meristems.

6. Plants can make embryos from asexual reproduction as well as from fertilization.

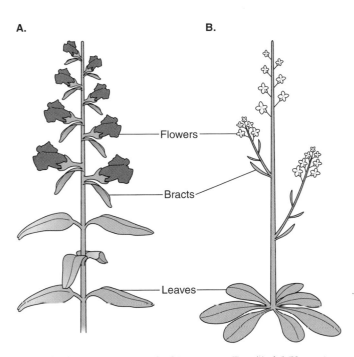

Figure 10.1 Inflorescence Architecture in Two Model Plants A schematic diagram showing the inflorescence in **(A)** *Antirrhinum* (snapdragon) and **(B)** *Arabidopsis* (mouse cress), two commonly used plants for the study of flower development.

Floral meristems are axillary meristems and arise in the axils of the bracts. In some plants, such as *Arabidopsis*, the bracts that subtend the floral meristems are not visible except in certain mutant backgrounds such as *leafy*. Bracts that subtend branches, however, are always visible in wild-type *Arabidopsis* (Figure 10.1). Thus, the switch from making leaves to making flowers is not an either-or decision, but rather a matter of emphasis. During vegetative development, leaves predominate and axillary meristems are often reduced in size. During reproduction, the axillary meristems predominate and leaves take back seat in size.

Flowering Is Regulated by Daylength

Many plants have a **daylength** requirement for flowering, either a shortened day (8 hours) or a longer day (12 to 16 hours), depending on the species. The ability to sense the length of day, as well as the quality of light, is under the control of two families of photoreceptors: phytochromes (which respond to red light) and cryptochromes (one of two known proteins that respond to blue light). We shall consider both kinds of molecules further in Chapter 11. In reality, although we refer to *daylength*, it is often the length of the dark period

that is most important. Experiments done in the 1930s suggested that daylength sensing takes place in leaves. Plants that required short days to flower were exposed to long days except for a single leaf, which was kept in short-day conditions. The exposure of that one leaf to the decreased daylength was enough to shift the plant to flower production (Figure 10.2A). Clearly, a signal went from the leaf to the SAM. The signal was also shown to be graft transmissible; when a plant exposed to short days was grafted onto a plant that had been exposed only to long days, this led to flowering in the long-day-exposed plant (Figure 10.2B).

The importance of the leaf in flowering time is also demonstrated by the *indeterminate* gene in maize. A mutation in *indeterminate (id)* was first identified by Ralph Singleton in a cornfield in 1939. When the other plants were ready for harvest, they found a number of green giants among the shorter plants, which were brown and lifeless (Figure 10.3A). Typical maize plants initiate approximately 18 to 20 leaves before producing a tassel. The *indeterminate (id)* mutants can initiate up to 50 leaves, and the tassel, nicknamed "crazy top," makes little plants in between the flowers (Figure 10.3B). Thus Id, the transcription product encoded by the normal form of this gene is needed not only to control the time to flowering but also to ensure that the meristem makes flowers as opposed to vegetative organs. Joe Colasanti has shown that the *id* gene is expressed in young leaves and not in meristems. Since the switch to making flowers is initiated in the meristem, the finding that *id* is expressed only in leaves and not in meristems supports the idea of a signal moving from leaves to meristem to initiate flowering. Id is probably not the signal molecule itself, but it may regulate the transcription of a movable signal that goes from young leaves through the stem to the meristem.

Another example of how the leaf signals the meristem when to flower comes from maize meristem culturing experiments carried out by Erin Irish. In these experiments, the meristem is surgically removed with different numbers of leaves attached and then cultured. Maize plants normally make 18 leaves. When a meristem that has made 14 leaves is removed from the rest of the plant with two leaves or fewer attached to the meristem, the meristem starts all over and makes another 18 leaves. However, if six leaves are left on the meristem when it is removed, these leaves "tell" the meristem how many leaves it has already made, so the meristem makes only the remaining 12 leaves in culture.

In *Arabidopsis*, more than 20 mutants have been characterized that delay the transition to flowering. These mutations only affect flowering time and do not modify the flower itself.

A.

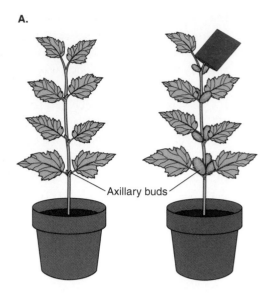

Axillary buds

B.

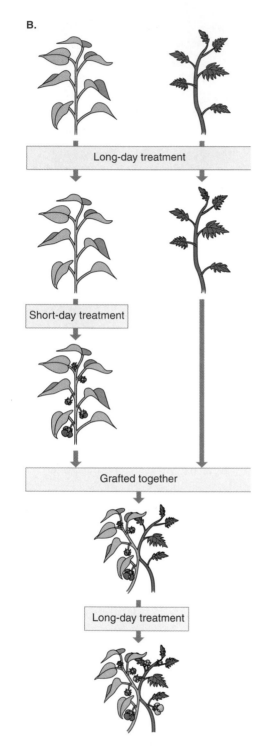

Long-day treatment

Short-day treatment

Grafted together

Long-day treatment

Figure 10.2   Regulation of Flowering by Daylength Two experiments illustrating that the signal to flower moves throughout the plant. **(A)** *Left:* Plants that flower only with short days (that is. with 8 hours of light) will not flower in long days (12 hours of light). *Right:* When a single leaf is covered for 16 hours such that it sees light for only 8 hours. it signals the axillary buds on the rest of the plant to become flowers. **(B)** The flowering signal is graft transmissible. Plants that are exposed to long days do not flower until they receive a short-day induction. When a plant that has not been exposed to short days (*right*) is grafted onto one that has (*left*). the graft induces the long-day-treated plant. while still exposed to long days. to flower (*bottom*).

A.

B.

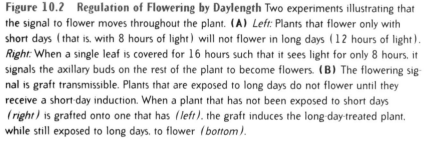

Figure 10.3   Nonflowering Maize Mutant (A) Maize plants with a mutated *indeterminate* (*id*) gene flower long after normal plants do. They make many more leaves that the normal corn plant and are visible as green giants in a field of brown. drying corn plants. **(B)** After initiating many more leaves than normal. *id* mutants sometimes produce a tassel (shown here). The tassel reverts to a vegetative state. with seedlings forming between the flowers; hence the nickname "crazy top."

Genetic experiments show that there are both endogenous and environmental cues, such as daylength or cold treatment, that trigger flowering. We will discuss the environmental cues here. Normally, *Arabidopsis* plants flower sooner when exposed to long days. Mutations in *constans* delay flowering in long days such that flowering time is the same in both long and short days. This finding suggests that the *constans* gene functions in regulating flowering in response to daylength. When extra copies of the wild-type *constans* gene are reintroduced back into the plant by transgenic means, the plant flowers earlier than normal. (See Box 10.1 for how to transform plants with the soil bacterium *Agrobacterium tumefaciens*.) Thus the plant is very sensitive to the dose of this gene product.

Other flowering-time genes were identified due to ecotype variation in flowering time. *Arabidopsis* ecotypes that are found in mountainous ecological niches often require a cold spell before flowering. The requirement for cold is called **vernalization**. Scientists discovered two genes, *frigida* and *flowering locus C (flc)*, that are expressed in these ecotypes and are mutant or not expressed in ecotypes found in warmer climates or used in the laboratory. When these genes are expressed, flowering is prevented. Levels of Flc RNA levels decrease in plants that have been given a cold treatment. Regulation of these vernalization genes is independent of daylength.

Floral Meristems Initiate Floral Organs

Inflorescence meristems initiate bracts and floral meristems, and floral meristems in turn initiate floral organs. At least three different genes have been characterized in *Antirrhinum* (snapdragon) and *Arabidopsis* that are responsible for the switch from an IM to a floral meristem. The *Arabidopsis* genes *leafy (lfy)* and *apetala1 (ap1)* are expressed in discrete zones of the IM that determine where the floral meristem will form. In the absence of either functional gene product, the "flowers" have bractlike leaves instead of sepals, they lack petals, and they exhibit other characteristics of the inflorescence such as spiral phyllotaxy. (We will consider sepals, petals, and other floral organs shortly.)

The *ap1* gene is a member of the large MADS box gene family in plants. Their gene products are transcription factors that regulate many aspects of flowering, from vernalization to initiation of floral meristems and floral organs. In *ap1* mutants, secondary flowers arise in the axils of bractlike leaves (compare Figure 10.4A and B). A striking phenotype arises in double mutants of *cauliflower* (*cal*, a mutation in a related MADS box gene, which, when mutated, has no phenotype of its own), and *ap1*. The *cal/ap1* double mutant (Figure

10.4C) resembles the vegetable cauliflower (Figure 10.4D). Both *cal/ap1 Arabidopsis* mutants and cauliflower contain IMs that produce additional IMs instead of floral meristems, over and over. Marty Yanofsky and colleagues examined the *cal* and *ap1* genes in the wild form of *B. oleracea*, which makes normal flowers with petals (Figure 10.4E), and in the cultivated variety, *Brassica oleracea botrytis* (cauliflower; Figure 10.4D). Interestingly, in cauliflower, *cal* contains a stop codon, whereas the version of *cal* in *B. oleracea* is intact and functional. These results suggest that loss of a functional *cal* gene in *B. oleracea* was selected for in cultivating the cauliflower we eat.

A.

B.

C.

D.

E.

Figure 10.4 Mutations That Change the Food We Eat (A) The wild-type flower of *Arabidopsis*. **(B)** An *apetala1 (ap1)* mutant *of Arabidopsis*. **(C)** An *ap1/cal* double mutant of *Arabidopsis*. As a single mutant, *cal* has no phenotype. **(D)** Cauliflower, a cultivated garden variety of *Brassica oleracea*. **(E)** Wild-type *B. oleracea*.

A.

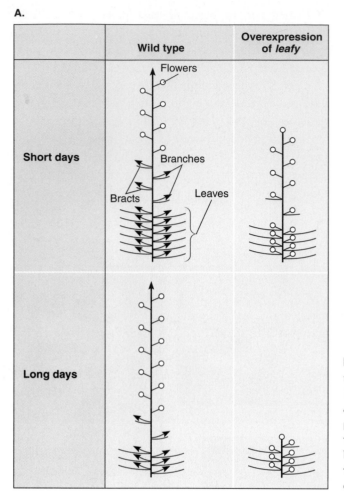

B.

Figure 10.5   The Effect of _leafy_ on Flowering The _leafy_ gene is required in the transition from inflorescence meristem to floral meristem. **(A)** _Left: Arabidopsis_ plants take longer to flower in short days, as shown by a greater number of leaves. _Right:_ When _leafy_ is expressed by a constitutive promoter (CaMV 35S), the plants flower sooner and end in a terminal flower whether grown in long or short days. **(B)** The wild-type plant on the left has not yet flowered, whereas the plant on the right, expressing _leafy_ from a constitutive promoter, has already flowered after the same number of days. Both plants were grown in short days.

A role in regulating floral meristem development is also seen in experiments that express _leafy_ or _apetala1_ from a constitutive promoter (that is, a promoter that causes continual expression in all cells). See Box 10.1 for a brief discussion of promoters. Normally, _ap1_ and _lfy_ are expressed in a group of cells on the flank of the IM that will give rise to the floral meristem. Expression continues as the floral meristem expands. Under control of the constitutive promoter, _lfy_ and _ap1_ are expressed during early vegetative stages and in all cell types. These plants flower almost immediately, thus dramatically shortening the vegetative stage of development (Figure 10.5).

Floral Organs Initiate in Whorls

The vast majority of flowering plants have four distinct whorls of floral organs positioned in concentric circles (Figure 10.6A). The position of each organ in the circle makes sense: on the outside are the **sepals,** which are often photosynthetic and have a protective function. The **petals,** which come next, often serve to attract pollinators. Inside the whorl of petals are the **stamens,** which bear the male gametes inside the _anthers._ And the innermost whorl contains the **carpels,** which house the female gametes or eggs. How does this precise patterning of floral organ position arise, and how do we explain flowers, such as those found on _Camellia_ varieties, which have whorls of beautiful petals but lack sex organs, stamens, and carpels?

Once again, _Arabidopsis_ and _Antirrhinum_ have been useful in identifying genes that function in making floral organs. Homeotic mutations in plants, like those found in animals, result in mutations causing a shift from one organ type to another. (Homeotic mutations and the homeobox domain will be discussed in Chapter 15.) Interestingly, the homeotic mutations in plants are not found in homeobox genes; instead, the majority fall into the MADS box gene family of transcription factors. The mutations that alter identity usually affect adjacent whorls; for example, _apetala3 (ap3)_ and _pistillata_ mutants in _Arabidopsis,_ or _deficiens (def)_ and _globosa_ mutants in _Antirrhinum,_ are missing petals and stamens, and instead have

10.1 PUTTING GENES INTO PLANTS

We are all familiar with galls on trees caused by insect damage. The invading pest somehow tricks the plant into making a home for the insect. Similar galls caused by bacterial infections can be found on numerous plants. During the 1950s, scientists exploring the cause of these galls discovered that the bacterium *Agrobacterium tumefaciens* was producing auxin and cytokinin (two hormones discussed at greater length in Chapter 11). Production of these hormones then triggered unorganized growth in the plant, resulting in galls.

Agrobacterium tumefaciens is extremely adept at getting the plant to do what it wants. The bacterium's DNA encodes a series of proteins, known as Vir proteins, which bring a portion of the *Agrobacterium* DNA, known as the T-DNA, into the plant's nucleus. Both the *vir* genes and the T-DNA are on a plasmid, called the Ti (tumor inducing) plasmid. These steps involve copying the T-DNA section of the Ti plasmid, coating the T-DNA with single-stranded DNA binding proteins that protect it, and targeting it to the plant nucleus (see accompanying figure). The T-DNA includes the genes for the plant hormones as well as genes that encode enzymes to make a food that only this bacterium eats. The food—called *opines*, which are slightly modified amino acids—also signals more bacteria to join the feast.

Having figured out what *Agrobacteria* were up to, plant scientists decided to capitalize on the bacterium's ability to cleanly insert DNA into a plant. Researchers replaced the hormone and amino acid genes in the T-DNA with genes they were interested in studying. Thus, *A. tumefaciens* has become a vector for the transfer of DNA into the genome of plant cells. The fact that almost all plants are responsive to this type of transformation has led to a revolution in agricultural biotechnology.

The opportunity to engineer genes into plants makes it possible to mix and match coding regions and promoter sequences to get the desired experimental effect. The promoter sequences tell the gene when to be expressed, in what tissues, under what conditions, and to what levels. (As we shall see in Chapter 14, it is actually the regulatory regions of the DNA, usually found in or near the basal promoter, that effect this control; for now, we shall simply refer to all this material as "the promoter.") Promoters and coding regions may come from plant or bacterial genes. A commonly used coding region is the β-glucuronidase of *E. coli* that produces a stable blue color when the tissue is incubated in the right substrate. A commonly used promoter in transgenic plant experiments (CaMV 35S) comes from the cauliflower mosaic virus. This virus has cleverly figured out the right sequences to obtain strong expression in most cell types, thus providing constitutive gene expression.

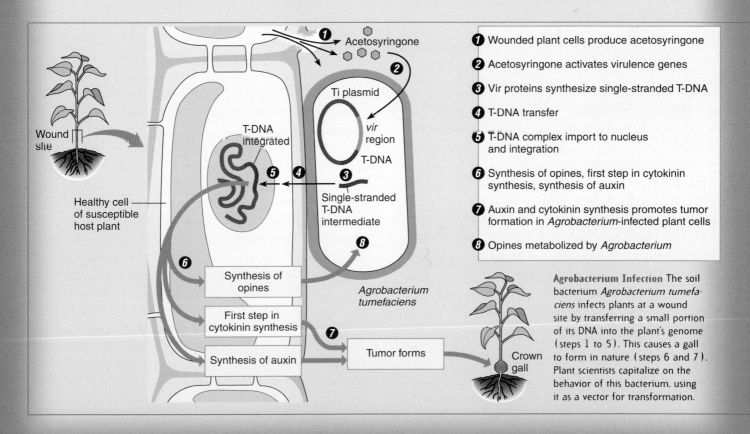

1 Wounded plant cells produce acetosyringone

2 Acetosyringone activates virulence genes

3 Vir proteins synthesize single-stranded T-DNA

4 T-DNA transfer

5 T-DNA complex import to nucleus and integration

6 Synthesis of opines, first step in cytokinin synthesis, synthesis of auxin

7 Auxin and cytokinin synthesis promotes tumor formation in *Agrobacterium*-infected plant cells

8 Opines metabolized by *Agrobacterium*

Agrobacterium Infection The soil bacterium *Agrobacterium tumefaciens* infects plants at a wound site by transferring a small portion of its DNA into the plant's genome (steps 1 to 5). This causes a gall to form in nature (steps 6 and 7). Plant scientists capitalize on the behavior of this bacterium, using it as a vector for transformation.

two whorls of sepals followed by two whorls of carpels. *Arabidopsis agamous* mutants (*ag*) or *Antirrhinum plena* mutants (*ple*) have sepals and petals, but lack stamens and carpels having instead additional inner whorls of sepals and petals. *Arabidopsis apetela2* mutants (and *ovulata* mutants in *Antirrhinum*) have carpel-like sepals, lack petals altogether, but have normal stamens and carpels.

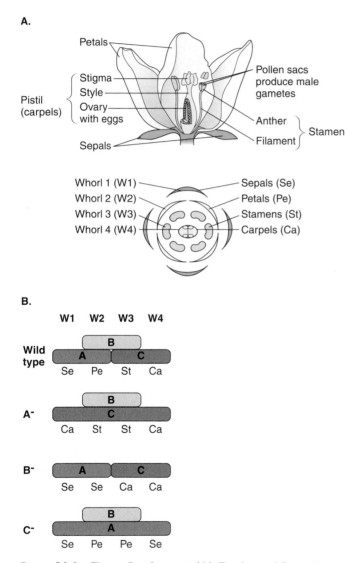

A.

B.

Figure 10.6   Flower Development (**A**) *Top:* A typical flower has sepals on the outside, with successive interior whorls of petals, stamens, and carpels. *Bottom:* A top-down schematic for an *Arabidopsis* flower. Whorl 1 contains four sepals, whorl 2 contains four petals, whorl 3 contains six stamens, and whorl 4 contains two carpels that are fused. (**B**) A schematic of a longitudinal section illustrates the ABC floral organ identity model. Normally, A genes function in the first two whorls, B genes in the second and third whorls, and C genes in the third and fourth whorls. In addition to the wild type, the phenotypes resulting when one of the A, B, or C class genes is nonfunctional are also shown.

The analysis of these mutants and cloning of the genes, which all encode transcription factors, led Elliot Meyerowitz, working with *Arabidopsis*, and Enrico Coen, working with *Antirrhinum*, to devise an elegant model in 1991 that remains robust today (Figure 10.6B). In the model, a series of genes, termed A, B, and C, are required to make adjacent whorls, as follows. A genes are required for sepals, A and B genes for petals, B and C genes for stamens, and C genes alone for carpels. Thus, mutations in a single gene affect two whorls. For example, a mutation in *ap3* (a class B gene) affects both petals and stamens.

An important component of the model is the antagonism between A and C gene products. This becomes clear in the changed phenotypes of mutants in either A or C class genes. In floral meristems of plants with nonfunctional A genes, the C gene product spreads into the domain of the floral meristem where A class genes would ordinarily have been expressed; in mutants without C genes, the opposite occurs (Figure 10.6B). Support for the model has come from double mutants and expression studies. For example, if B and C genes are nonfunctional (for example, an *ap3/ag* double mutant), only A function genes are left. As predicted, the flower lacks petals, stamens, and carpels and is composed of only sepals.

The expression patterns of the A, B, and C genes, for the most part, fall in line with these expectations. The B class genes are expressed in petal and stamen primordia, the C class genes in stamen and carpel primordia. In A class mutants, such as *apetala2* (*ap2*), C class gene expression spreads into the outer two whorls as well, supporting the model of antagonism between A and C class genes. Although researchers have identified additional genes that also function in making floral organs, the model has proven extremely useful in interpreting results in other species.

Expression of the floral identity genes is detectable before the primordia that will become floral organs have developed. Thus we might ask, What regulates the position of expression of each of these A, B, and C class genes? Clonal analysis has demonstrated that lineage does not guide their expression, so they must be responding to some positional cues. Workers have identified additional genes that are expressed in the boundaries between prospective organ whorls (Figure 10.7 and Table 10.1). The *Antirrhinum* mutant *fimbriata* (*fim*) has mosaic sepal–petal organs in the second whorl and sepal–petal–carpel organs in the third whorl. Meristems also pop up between whorls. The wild-type gene products of *fim* and its *Arabidopsis* counterpart, *unusual floral organs* (*ufo*), interact with a class of proteins that participate in cell-cycle progression. These proteins may help establish the boundary between whorls by

TABLE 10.1 FLORAL ORGAN GENES IN *ANTIRRHINUM* AND *ARABIDOPSIS*: SUMMARY OF EXPRESSION DOMAINS AND MUTANT PHENOTYPES

Antirrhinum (Snapdragon)	Arabidopsis	Expression Pattern	Mutant Phenotype
flo	*lfy*	Floral meristems Bracts in *Antirrhinum*	No flowers
fim	*ufo*	Between sepals and petals	Mixed regions
def	*ap3*	Petal and stamen primordia	Petals converted to sepals Stamens converted to carpels
ple	*ag*	Stamen and carpel primordia	No stamens or carpels Extra petals and sepals

affecting cell division rates, for example, which would limit growth between whorls of organs.

Recent experiments have demonstrated that the transcription factor Leafy (Lfy) plays a critical role in directing the gene activity that results in the different whorls. Using an inducible system, Detlef Weigel and colleagues determined that Lfy directly regulates A genes, functions with Ufo to regulate B genes, and is thought to interact with an unknown gene to regulate C genes. Thus, the type of whorl may depend to some extent on which proteins Lfy interacts with.

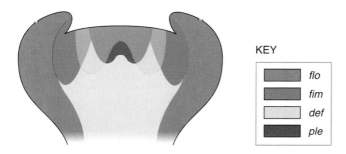

KEY

- flo
- fim
- def
- ple

Figure 10.7 Patterns of Floral Gene Expression in *Antirrhinum* A longitudinal section through a wild-type floral meristem. The *floricala* (*flo*) gene, counterpart to the *leafy* gene in *Arabidopsis*, is expressed in the outer sepal whorl. The *fimbriata* (*fim*) gene is expressed in a ring-shaped domain adjacent to the region of *flo* expression and overlaps (*green*) with the *deficiens* (*def*) expression domain. The domain of *plena* (*ple*) expression overlaps (*orange*) with the domain of *def* expression. Whether cells express only *ple* (a C gene), or *ple* and *def* (B + C genes), determines the type of organ that forms. See also Table 10.1.

THE ALTERNATION OF GENERATIONS: THE HAPLOID-DIPLOID LIFE CYCLE IN PLANTS

In Flowering Plants, the Gametophyte Stage Is Abbreviated

Judging from the fossil record, we can assume that extant plants such as ferns and mosses resemble plants that were present 100 million years ago. Thus we can look at these "primitive" plants for hints about the evolution of reproduction. These hints help explain some aspects unique to plant reproduction, such as the habit of alternating between diploid (2n) and haploid (1n) lifestyles (Figure 10.8A). The diploid stage is called the **sporophyte** because it makes **spores,** the products of meiosis. Depending on the particular plant species, male and female spores may be identical *(homosporous)* or different *(heterosporous)*. These spores, which are haploid, form a distinct entity called the **gametophyte,** which in turn produces female and male **gametes**—the egg and the sperm. In animals and plants, the relationship between spores and gametes differs. In animals, spores become gametes directly without any cell divisions, whereas in plants, the spores undergo cell divisions and develop into the multicellular gametophyte before producing gametes. The gametes fuse during fertilization to produce the sporophyte once again.

A.

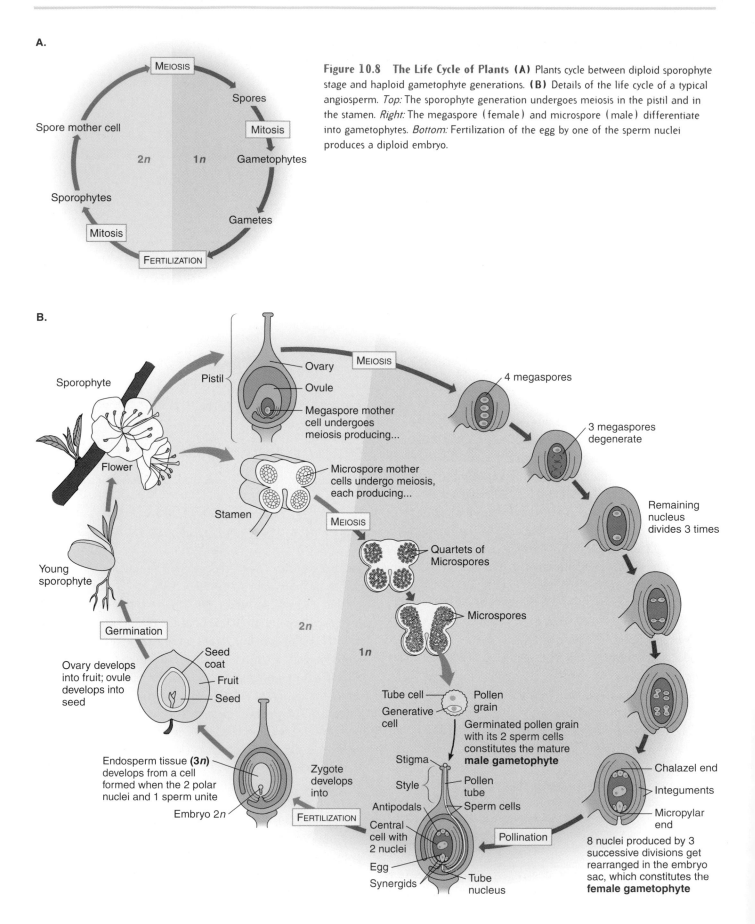

Figure 10.8   The Life Cycle of Plants (A) Plants cycle between diploid sporophyte stage and haploid gametophyte generations. **(B)** Details of the life cycle of a typical angiosperm. *Top:* The sporophyte generation undergoes meiosis in the pistil and in the stamen. *Right:* The megaspore (female) and microspore (male) differentiate into gametophytes. *Bottom:* Fertilization of the egg by one of the sperm nuclei produces a diploid embryo.

B.

When we think of angiosperms, the flowering plants, it is the sporophyte that is the focus of our attention (unless we are undergoing a pollen allergy attack). Unlike the sporophytes of most ferns and fern allies, which may make only one type of spore, angiosperms always make male and female spores.

Male spores, called **microspores,** are made by the stamens. Each microspore develops into a male gametophyte, also known as a pollen grain, a compact and sturdy unit, designed to bring the male gamete to the appropriate female gamete. The microspore begins with a programmed asymmetric division that produces a somatic (or **tube**) cell and a germ-line (or **generative**) cell. Unlike all other cells in plants, the germ-line cell is contained within its sibling tube cell (Figure 10.8B). Pollen grain walls, composed of layers synthesized by both the gametophyte and sporophyte generations, are extremely specialized. The grain's beautiful architecture helps the pollen survive desiccation, float through the air or attach to a pollinator insect, stick to the **stigma** (the receptive surface of the carpels), and hydrate. The architecture of pollen grain walls can be unique enough to identify the particular species.

Female spores, called **megaspores** (Figure 10.8B), are made by the ovules inside the carpels. Although all four male meiotic products (microspores) form male gametophytes, in most flowering plants three of the four megaspores formed by meiosis degenerate, and only the single survivor forms the **female gametophyte.** The surviving megaspore undergoes three rapid mitotic divisions. Cell walls then form, partitioning the eight nuclei into seven cells (Figure 10.8B). Three cells at the **micropylar** end differentiate into the egg cell flanked by two **synergid** cells. The three cells at the **chalazal** (opposite) end of the female gametophyte may continue to divide to form a tissue called **antipodals.** Although their function is not known, they are suspected to be important for nutrient uptake. The remaining cell in the center, the **central cell,** has two nuclei called **polar nuclei.** The entire structure is referred to as the **embryo sac** and comprises the mature female gametophyte. Variations in female gametophyte development can be found; for example, some embryo sacs have many polar nuclei. The female gametophyte of angiosperms is thus buried deep within the flower, unlike the free-living gametophytes of lower plants such as ferns.

A Double Fertilization Is Involved in the Making of a Seed

Fertilization in angiosperms first requires contact between the desiccated pollen grain and the stigma. The stigma,

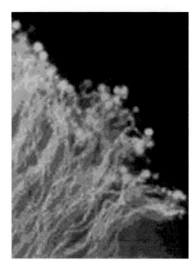

Figure 10.9 Pollen Germination Here pollen grains have landed and germinated on the stigma of a tomato plant.

equipped with a specialized surface to let only the pollen of the same species germinate and grow, sits on top of the female reproductive organ, or **pistil** (Figure 10.8B). A pistil can consist of one to many carpels, which may or may not be fused, and which may produce one to many ovules. The carpels provide protective surroundings for the developing embryo sac or sacs. After the pollen grain lands on the stigma, it absorbs water and extends a growing tip, or **pollen tube,** through the stigma into the **style** of the pistil (see Figures 10.6, 10.8B, and 10.9). In angiosperms, the pollen tube solves the problem of getting the male gamete to the egg, in contrast to ferns, in which the gametes must meet in an aqueous environment for successful fertilization.

The tube cell does all the work involved in making a growing pollen tube, while the generative cell goes along for the ride. The work in making a pollen tube is phenomenal considering the distance a pollen tube must grow to reach the egg (think of a strand of corn silk, for example, which is actually a combination of stigma and style) and the speed at which the pollen tube grows (rates of up to 1 cm/h). The generative cell divides to produce two sperm cells, either during pollen tube growth or before the pollen leaves the *anther* (the portion of the stamen that contains the developing microspores), depending on the species. These two sperm cells will each participate in a fertilization: one will fertilize the egg, and the other will fertilize the binucleate central cell. Embryo sac–specific mutations have demonstrated that a functional embryo sac is required for pollen tube guidance. Since the central cell generally has two nuclei, this fertilization is accompanied by a fusion of the two nuclei with a sperm nucleus, which produces a triploid nucleus. The fertilized egg develops into the

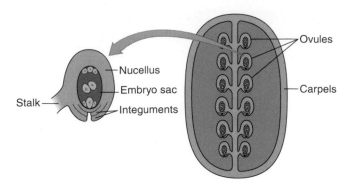

Figure 10.10   Ovule Development Ovules in a typical flower arise from the inner wall of the carpel. The ovule houses the embryo sac. Integuments of the ovule differentiate and grow to surround the female gametophyte and eventually become the seed coat.

embryo ($2n$), and the fertilized central cell becomes the **endosperm** ($3n$), a tissue destined to feed the developing embryo and then die.

To understand what a **seed** is, it is important to think about the position of the female gametophyte within the flower. The megaspore mother cell that gives rise to the female gametophyte is enclosed within the **ovule,** which is contained within the pistil. The ovule is composed of three parts: a stalk for attachment; the **nucellus,** which is nutritive tissue that surrounds the embryo sac; and the **integuments,** which are lateral organs produced by the ovule and grow to enclose the nucellus (Figure 10.10). The tissues are short-lived; the growing embryo sac crushes the nucellus, and the integuments differentiate into the seed coat. Thus the seed—a self-made package containing the products of fertilization—is the last stage of ovule development.

Whether or not a plant produces seeds distinguishes angiosperms and gymnosperms (pines and firs) from plants such as ferns and mosses. We take seeds for granted, but they were a big step in plant evolution, since they proved a successful way to speciate and fill many niches. A seed's built-in nutrition provides an advantage over a spore. In gymnosperms the nutritive source comes primarily from the female gametophyte ($1n$), while in angiosperms the food source is the endosperm ($3n$) and/or the "seed leaves," or **cotyledons** ($2n$). This nutritive supply gives a seed independence during the early stages of germination when the seed absorbs water and begins to grow and develop. Finally, to aid in seed dispersal, seed coats have evolved elaborate shapes, such as burrs and hooks that stick to moving animals, and wings that permit flight on air currents.

Whether the ovule is enclosed or exposed at the time of fertilization is also important, as this distinguishes angiosperms (which have enclosed seeds) from gymnosperms (naked seeds). In angiosperms, ovules form from the inner wall of the ovary (Figure 10.10). In gymnosperms, the ovules form on the upper surfaces of seed scales, such as the spiral ovuliferous scales of a pinecone.

In addition to coordinating endosperm and embryo development, fertilization also induces elaboration of the seed coat and development of the fruit. The **fruit** is a mature ovary that usually surrounds the seed and evolved to entice animals to help disperse the enclosed seeds. Examples from the supermarket include apples, avocados, string beans, and squash. The developing seed produces hormones that promote fruit growth. Once the fruit begins to grow, it can produce its own hormones that maintain the maturation process. Upon fertilization, the surrounding fruit expands to make room for the growing seeds. Without fertilization, the fruit, whether it is a squash or a pea pod, does not expand.

EMBRYO FORMATION

Details of embryo development cover a range of variation, probably as great as that of the mature plants that produce the embryos. In some species, predictable divisions create specific lineages for different tissues; in other species, the division planes appear random, yet organogenesis occurs nonetheless. In all cases, cellular expansion is concurrent with cell division; there is no period in plant embryogenesis equivalent to the period of animal cleavage when successive divisions occur in the absence of cell enlargement.

In most plant species, a period of **dormancy** is imposed during embryogenesis that allows the seed to survive desiccation for long periods of time. There are exceptions, however, such as the red mangrove, in which seeds germinate while the fruit is still on the tree. The number of leaves initiated before dormancy can also vary greatly. Grasses make a number of leaves in their embryos, while orchid embryos show no sign of leaf or root prior to germination. Finally, the first formed leaves, the cotyledons, take on different forms and functions depending on the species. Among angiosperms, dicotyledonous plants (*dicots*) have two cotyledons, whereas monocotyledonous plants (*monocots*) have one. Gymnosperms can have many cotyledons. In all cases, cotyledons often serve as food storage units and absorptive organs.

*In Some Plants, Early Patterning of the Embryo
Is Regulated by Cell Division*

In most plant embryos, the first cell division sets aside the embryo proper from the **suspensor,** which is primarily a supporting tissue (Figure 10.11A). The suspensor is always oriented toward the micropyle (the point of entry for a growing pollen tube) in seed plants, which suggests some maternal instruction from the ovule guides this orientation. The suspensor pushes the embryo up into the ovule and actively absorbs nutrients from the ovary. Eventually, the suspensor is crushed by the growing embryo. Although normally suspensor cells have a transient existence, if the embryo is aborted or arrested, suspensor cells can initiate embryogenesis all over again.

Arabidopsis has stereotypical divisions during embryogenesis. After the first division, which sets aside the embryo

proper from the suspensor (Figure 10.11A), the next two divisions quarter the embryo proper. These divisions are followed by divisions tangential to the previous division planes, in effect establishing the cells that will form the **protoderm,** or epidermal layer. Transverse divisions separate the meristem from cells that will form the **hypocotyl** (the embryonic stem). The orientations of subsequent division planes in this **globular stage** embryo are less predictable. At the **heart stage,** cell divisions and expansion become concentrated in two opposite corners of the apical half to produce the cotyledons (Figure 10.11B). The cotyledons expand as does the longitudinal axis of the embryo. Cells that will form the vascular system of the plant differentiate during the heart stage. Although the root meristem is not distinct until the end of embryogenesis, it can be traced back to early cell divisions in the embryo. The uppermost cell of the suspensor, the **hypophysis,** gives rise to the root meristem in *Arabidopsis.*

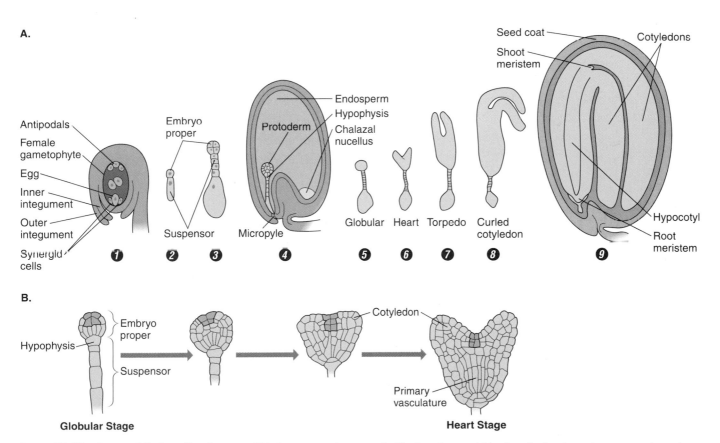

Figure 10.11   Stages of Embryo Development (A) Cross-sectional diagrams of embryogenesis, showing the complete ovule and seed in the left, middle, and right diagrams (1, 4, 9), and just the embryo for the intermediate stages. Embryo development begins with the female gametophyte (1). Following fertilization, the embryo develops within the

ovule. The first division (2) often divides the embryo proper from the suspensor (3). The embryo grows through several stages (5 through 8) to form the seed-born leaves, or cotyledons (9). **(B)** Patterns of cell division in the transition from the globular to heart stages in *Arabidopsis.* The position of the shoot apical meristem is identified in blue.

Plants Can Avoid Sex and Still Make Embryos

A number of plants (around 300 species) produce seeds without undergoing fertilization, a process called **apomixis**. Embryo sacs form from diploid cells, and depending on the species the diploid cell can be either the megaspore mother cell (**diplospory**) or a cell from the nucellus (**apospory**). Figure 10.12 shows a version of diplospory in which the megaspore mother cell skips meiosis altogether, and a type of apospory in which the megaspore mother cell degenerates. In both cases, the diploid cell undergoes the three stereotypical nuclear divisions and migrations found in sexually derived female gametophytes. Synergids and antipodals still differentiate, and the diploid egg cell becomes the embryo. Clearly, a haploid genome is not required to make an embryo sac. The endosperm may develop autonomously or require fertilization. Although the apomictic egg develops into an embryo without fertilization, the endosperm does require fertilization in some species. In such cases, the progeny are exact copies of the mother, because the fertilized endosperm supports seed development only and does not contribute to the germ line.

Clonally exact seeds are of extreme interest to breeders. Most seeds ordered from a catalog are *hybrid seeds*. Made from two genetically distinct parents, hybrid seeds make for more vigorous plants with greater yield. Unfortunately, for the would-be seed saver, the progeny of hybrid seeds are not genetically uniform and can segregate for undesirable recessive traits. If apomixis could be triggered in a hybrid seed, then the progeny would continue that hybrid vigor since there would be no chance for recombination. For this reason, geneticists have sought to identify the genes responsible for apomixis; however, most of the apomictic species are polyploid, complicating genetic approaches.

The embryos that arise from fertilization or from apomixis are found inside the ovary. Plant embryos can also form from somatic tissue that has received the right balance of hormones in a tissue culture experiment, as exemplified by the work of Frederick Steward (see Chapter 1). Somatic embryos also have a suspensor and embryo proper; thus, any instructions from maternal tissues are not an absolute requirement for establishing an apical-basal axis of polarity. These somatic embryos will grow into normal plants but skip the dormancy stage, presumably because they lack a seed coat and endosperm, tissues required for survival as a seed. One of the unique properties of plant cells is their capacity for immortality. Unlike animal cells, which undergo only a limited number of cell divisions, plant cells can be grown in culture indefinitely.

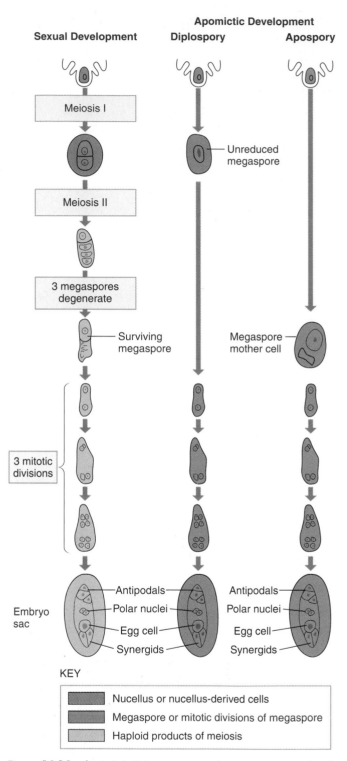

Figure 10.12 Apomixis Two apomictic pathways are compared with the equivalent events in sexual reproduction.

Given that plants have this impressive capacity to regenerate and make embryos, it is likely that numerous genes prevent this from happening out of context. The *pickle (pkl)* gene of *Arabidopsis* keeps roots from making embryos. Roots of *pkl* mutants accumulate lipids normally found in seeds and express embryo-specific genes. If these mutant roots are placed in culture medium without hormones, they will occasionally generate somatic embryos, unlike roots from wild-type plants, which would require hormones. Pkl is a chromatin-remodeling protein. Another example of ectopic embryos comes from *leafy cotyledon1 (lec1)*, also in *Arabidopsis*. The *lec1* mutant was identified by precocious leaf development and lack of dormancy. The transcription factor Lec1 functions to promote embryogenesis, and when it is missing, the embryos go right into seedling development without a seed dormancy stage. When *lec1* is deliberately overexpressed using a constitutive promoter, embryos are induced to form on vegetative tissues such as cotyledons, as shown in Figure 10.13. As one might guess, *pickle* is required to repress *lec1*; thus the mutant *pickle* phenotype may be due to misexpression of *lec1*.

A.

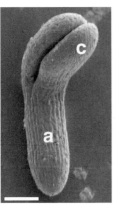

B.

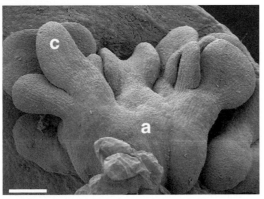

Figure 10.13 Embryos Forming on Leaves Scanning electron micrographs of *Arabidopsis* torpedo-stage embryos. **(A)** The wild type at the torpedo stage. **(B)** Embryos occasionally form ectopically on mature cotyledons when *lec1* is expressed constitutively. *a* is the axis of the embryo. *c* is cotyledon.

The Shoot Meristem Originates in the Embryo

Many embryo mutants in both *Arabidopsis* and maize fail to progress beyond the globular stage. They are blocked in development at earlier stages, prior to pattern formation. These mutations are likely to define genes that are essential for metabolism, such as cell-cycle progression. A more informative class of mutants, for understanding development, survive the globular stage but are blocked at the seedling stage.

A number of genes have been identified in *Arabidopsis* that are required for shoot development. In severe *shootmeristemless (stm)* mutant alleles, only the cotyledons form (Figure 10.14A). In partial loss-of-function *stm* mutants, the shoot recovers from adventitious meristems that arise on the petiole (narrow base) of the cotyledon. These recovery shoots produce fewer branches and the flowers that do form have missing organs, especially the carpels. The gene *stm* is expressed in all shoot meristems but not in root meristems. It is expressed in the late globular embryo in a stripe between the prospective cotyledons (Figure 10.14B). A homologous maize gene, *knotted1 (kn1)*, is also expressed early in the embryo at the presumed site of the apical meristem. Later in embryo development, *kn1* expression is seen in the shoot apical meristem but not in the leaf primordia or the cotyledon (Figure 10.14C). The cotyledon of maize is composed of the coleoptile, a leaflike structure that surrounds and protects the young shoot as it pushes through the soil, and the scutellum, a nutritive tissue. Expression is also seen extending toward the root pole (the position where the root eventually forms). A similar phenotype (cotyledon only) is seen in *kn1* loss-of-function mutants in maize depending on other genetic differences (Figure 10.14A). The genes *kn1* and *stm* encode homeodomain-type transcription factors. Similar to the findings of what happens when *lec1* is overexpressed, overexpression of *kn1* or *stm* results in the presence of shoot meristems on leaves. Once again, we see that differentiation of plant tissues is not permanent; leaf cells have the capacity to redifferentiate to meristem fates.

Other *Arabidopsis* genes that are required for shoot development include *pinhead* and *wuschel*. In *pinhead* mutants, a filamentous structure, or a single leaf, forms between the cotyledons where there should have been a shoot apical meristem (Figure 10.14A). The *pinhead* gene is initially expressed in the globular meristem, and then it becomes restricted to the vasculature and the dorsal side of the leaf primordia. This gene encodes a novel protein that may be involved in translational control. When a double mutant involving *pinhead* and a mutation in a related gene, *argonaut*, is made, the result is an unorganized mass that fails to express meristem-specific

A.

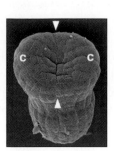
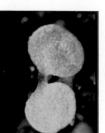

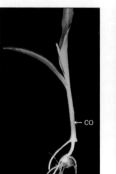
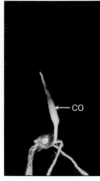

Wild-type
(*Arabidopsis*)

cuc1/cuc2
(*Arabidopsis*)

stm
(*Arabidopsis*)

pinhead
(*Arabidopsis*)

Wild-type
(maize)

kn1
(maize)

B.

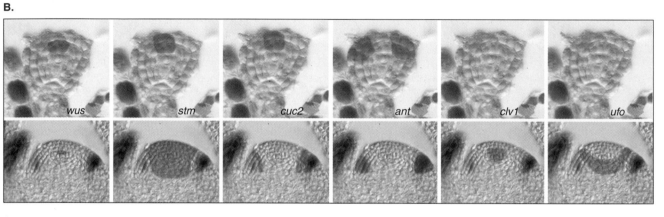

wus   *stm*   *cuc2*   *ant*   *clv1*   *ufo*

C.

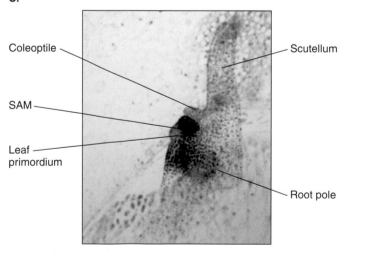

Coleoptile

SAM

Leaf
primordium

Scutellum

Root pole

Figure 10.14   Genes That Function in Plant Embryos (A) Phenotypes of mutations that are blocked at or after the cotyledon stage in *Arabidopsis* and maize, with wild-type seedlings shown for comparison. In *Arabidopsis cuc1/cuc2* double mutants, the cotyledons are fused. The cotyledon of maize plants has a leafy extension, called a coleoptile (co), which surrounds and protects the growing leaves. **(B)** Expression patterns (*blue*) of genes that function in shoot development in *Arabidopsis* during late globular stage embryogenesis (*top row*) and vegetative growth (*bottom row*). **(C)** In a maize embryo, *kn1* is expressed in the shoot apical meristem (SAM) and the shoot axis, leading to the root pole. Kn1 protein is absent from the scutellum (the nutritive part of the cotyledon), the coleoptile, and the leaf primordium.

genes. Related genes have been identified in flies and worms; for example, the *piwi* gene in *Drosophila*, which is required for self-renewal in germ-line stem cells, encodes a similar type of protein.

Similar in phenotype to *stm* mutants are *wuschel* (*wus*) mutants: the *wus* gene encodes a novel homeodomain-type protein and is expressed very early in the apical half of the 16-cell embryo. Expression narrows to just a few cells in the center of the meristem (Figure 10.14B). This inner meristem pattern continues in other shoot meristems as well, such as the inflorescence or floral meristems. The *wus* gene is expressed in an *stm* mutant embryo, and the *stm* gene is expressed in mutant *wus* and *pinhead* embryos at early stages of embryo development. The expression of *stm* and *wus* disappears in the arrested seedlings. Thus these genes seem not to be dependent on each other, at least in the beginning of embryo development.

The expression patterns of these genes and loss-of-function phenotypes suggest that the gene products are required to keep the meristem going. An informative experiment was carried out in which *wus* expression was under control of the *aintegumenta* (*ant*) promoter in transgenic plants. The *ant* gene is expressed in cells of the embryo that will become cotyledons (Figure 10.14B) and in the P_0 region of older shoot meristems. In effect, its expression marks the position of the next lateral organ to form. In plants where the *wus* gene was expressed by the *ant* promoter, cotyledons

failed to form. These results suggest that meristem maintenance and leaf differentiation are opposing fates: a cell can either be a leaf cell or a meristem cell, but not both.

The *pinhead*, *wus*, and *stm* mutants make normal cotyledons, suggesting that these genes are not required for the formation of cotyledons. In contrast, *cup-shaped cotyledon* (*cuc*) mutants in *Arabidopsis* and *no apical meristem* (*nam*) mutants in petunias not only fail in producing organs from the SAM but also have a cotyledon defect. Mutants of *cuc* and *nam* make fused cotyledons (Figure 10.14A), suggesting that one aspect of these genes is to define a domain between the cotyledons. The proteins expressed by these genes comprise a novel class. Initially expression occurs in a region similar to that of *stm*, but later it appears to shift to the peripheral zone of the meristem (Figure 10.14B).

When does the SAM form and when does it develop, even in rudimentary form? It is easy to identify a meristem when it is busy making organs: it is that small group of cells at the center of the action. Meristem identification is trickier without leaf primordia pointing the way, such as in a developing embryo. Given that leaves usually form from a meristem, and that cotyledons are specialized leaves, it is reasonable to assume that cotyledons also form from meristems. The apical half of the globular embryo is the likely candidate for the "embryonic meristem." The fact that many of these "meristem genes" are expressed in the apical half of the embryo suggests that they function in that capacity to initiate cotyledons.

KEY CONCEPTS

1. Changes in daylength trigger flowering in plants. The molecular signal to make flowers moves from leaves to meristem.

2. Cold temperatures are another trigger for flowering.

3. Flowers are organized in four concentric whorls of organs.

4. A number of mutations affect the identity of any two adjacent whorls.

5. Plants alternate between diploid and haploid stages. Both stages involve cell division and differentiation.

6. Fertilization coordinates the development of endosperm and embryo and the growth of the fruit.

7. Embryos form following fertilization and also can form somatically.

8. The globular embryo has specific domains of gene expression, suggesting that developmental compartments form early.

STUDY QUESTIONS

1. What are examples of haploid, diploid, and triploid cell types in flowers?

2. If you have a plant that expresses *agamous* in all four whorls, what will the phenotype be?

3. What extrinsic factors control time of flowering in plants?

4. Why would clonally derived seeds be advantageous to farmers?

SELECTED REFERENCES

Evans, M. M., and Barton, M. K. 1997. Genetics of angiosperm shoot apical meristem development. *Annu. Rev. Plant Physiol. Plant Mol. Biol.* 48:673–701.

A comprehensive discussion of shoot apical meristem development in model organisms such as maize and *Arabidopsis*.

Kaplan, D. R., and Cooke, T. J. 1997. Fundamental concepts in the embryogenesis of dicotyledons: A morphological interpretation of embryo mutants. *Plant Cell* 9 (11):1903–1919.

A look at natural variation in embryo development, in order to understand the underlying processes.

Okamuro, J. K., den Boer, B. G., Jofuku, K. D. 1993. Regulation of *Arabidopsis* flower development. *Plant Cell* 5(10):1183–1193.

A review of how floral organ identity is established.

Samach, A., and Coupland, G. 2000. Time measurement and the control of flowering in plants. *Bioessays* 22:38–47.

A description of the genes identified for control of vernalization and daylength responses. Models are presented for how daylength might regulate flowering through circadian control of gene expression.

CELL SPECIFICATION AND SIGNALING IN PLANTS

CELL SPECIFICATION

The Epidermis Forms Root Hairs, Trichomes, and Stomata

Being on the outside of the plant, the epidermis serves the purpose of interfacing with the environment. This outside position also makes it a great material for the study of plant histogenesis (cell differentiation): Without disturbing the growing plant, one can easily examine cell division and differentiation patterns. Because organs differentiate progressively, from one end to the other, it is possible in a single examination to see all the stages of cellular differentiation.

CHAPTER PREVIEW

1. Cell differentiation in plants occurs as cell division ceases.

2. Plants communicate by fine cytoplasmic connections that penetrate the cell walls.

3. Several hormones have profound effects on plant development.

4. Plants perceive different wavelengths of light, using the information to interpret their surroundings and develop accordingly.

5. Plants have circadian rhythms that regulate their time to flowering, leaf opening and closure, and metabolism.

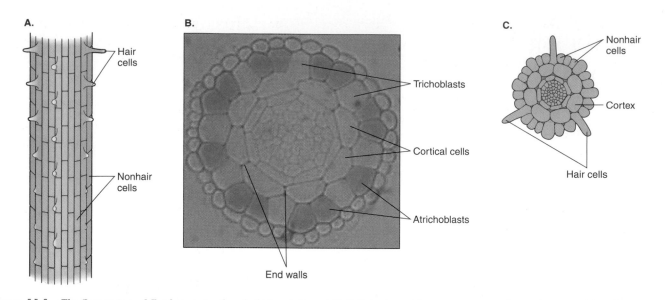

Figure 11.1   The Patterning of Epidermis in the *Arabidopsis* Root **(A)** Cells in the epidermis are organized into alternating files of hair cells and nonhair cells. **(B)** A transverse section through a root stained with Toluidine Blue. The atrichoblasts (cells destined not to become root hairs) stain blue. Trichoblasts, which alternate with atrichoblasts, are adjacent to the end walls between cortical cells. **(C)** A cross-sectional diagram of a root after differentiation, showing the position of hair cells relative to the cortex. The "hair" of a hair cell is not visible in every section.

Root hairs are responsible for most of the absorption of nutrients into the vascular system, as well as providing a means for invading bacteria to enter the root. Root hairs arise from single cells of the epidermis, often in alternating files with epidermal cells that do not form root hairs (Figure 11.1A). Although the root hair does not grow out until the cell is in the zone of differentiation (see Chapter 9, Figure 9.11A), cytological differences can be traced back in the cell lineage to the time the root hair first forms. Hair cell precursors (**trichoblasts**) have smaller vacuoles and take up different stains than **atrichoblasts**, the precursors of nonhair cells (Figure 11.1B). The positioning of the two types of precursor cells—trichoblasts next to two interior cortical cells, and atrichoblasts next to a single interior cortical cell—suggests that signals coming from cortical cells help guide this difference. This hypothesis is supported by experiments in which atrichoblasts were surgically separated from underlying cortical cells, thereby switching their fate.

Arabidopsis mutants exist in which the epidermal cells consist of all hair cells or all nonhair cells. For example, roots of *glabra2* (*gl2*) mutants are completely "hairy"; that is, they lack files of nonhair cells. Normally, *gl2* is expressed in the atrichoblasts. When a constitutive promoter is used to cause expression of *gl2* in all cells, no hair cells form. These results suggest that *gl2* functions to prevent atrichoblasts from developing into hair cells.

Trichomes are the hairs on leaves and stems. They can be single protrusions or multibranched (Figure 11.2A), and in extent they can cover the leaf like a carpet or be restricted to the leaf margin. In nettles, trichomes are responsible for the stinging sensation we get after brushing against the plant; toxins are released when the tips of the trichomes are broken off. In *Arabidopsis*, each trichome is a single cell that becomes highly polyploid. The trichome cell is surrounded by accessory cells (Figure 11.2B), that, despite their close association with the trichome, are not clonally related to it.

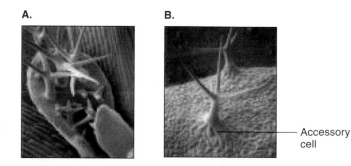

Figure 11.2   Trichomes in *Arabidopsis* Trichomes are hairs on the leaves and stems. **(A)** Trichomes differentiate from the tip of the leaf down. **(B)** In *Arabidopsis*, each trichome differentiates in association with surrounding accessory cells. Trichomes can be branched and are usually polyploid.

At least two genes that regulate the presence or absence of root hairs also regulate trichome differentiation. For example, in *gl2* mutants, the trichomes forming on leaves are diminutive (Figure 11.3). The *transparent testa glabra (ttg)* mutant is similar, having no trichomes on the leaves and lots of root hairs. Ttg protein, which contains a WD40 motif (functioning in protein–protein interactions), positively regulates Gl2. Two other genes operate in these same pathways, but in only one cell type. The *werewolf (wer)* gene is expressed in nonhair cells and required for their differentiation, whereas *glabra1 (gl1)* is expressed only in epidermal cells of the leaf and is required for trichome cell differentiation. Both of these genes encode MYB transcription factors that are 91% similar in the MYB domain. Along with Ttg, Wer and Gl1 positively regulate Gl2. Thus, it is the presence of Wer in the root and Gl1 in the leaf that directs *gl2* expression appropriately.

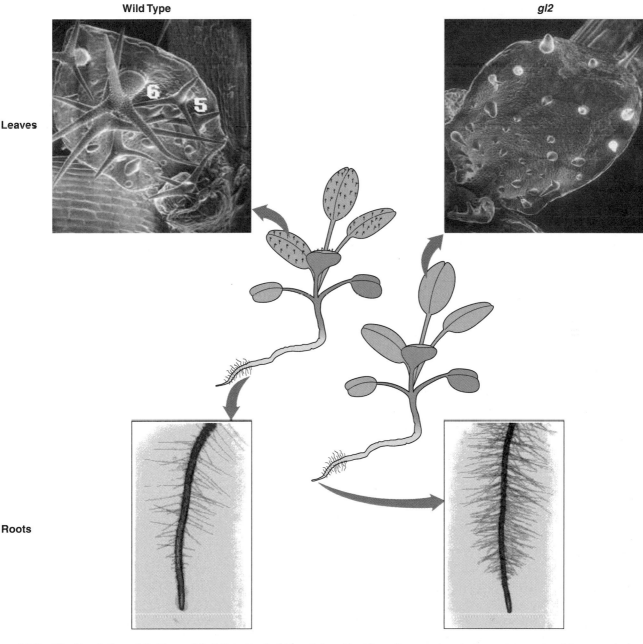

Wild Type

gl2

Leaves

Roots

Figure 11.3   The Regulation of Epidermal Cell Fate in *Arabidopsis* As evident from the scanning electron micrographs of leaves and photographs of roots shown here. mutants in the *glabra2* gene lack developed trichomes on leaves and nonhair cells in the root. Thus a single gene regulates the fate of epidermal cells in both the shoot and the root.

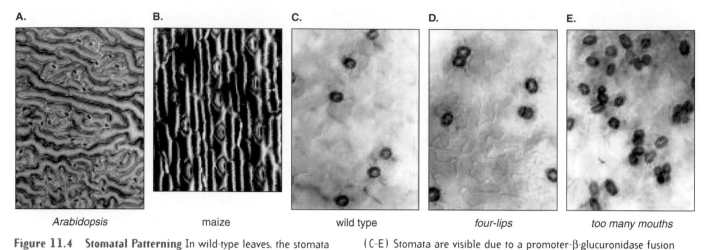

Figure 11.4 **Stomatal Patterning** In wild-type leaves, the stomata are spaced such that there is always at least one cell between each stoma. Scanning electron micrographs of **(A)** an *Arabidopsis* leaf and **(B)** a maize leaf show the arrangements of stomata typical in the leaves of, respectively, dicots (random) and monocots (parallel). (C-E) Stomata are visible due to a promoter-β-glucuronidase fusion that is expressed specifically in guard cells: **(C)** *Arabidopsis* wild type; **(D)** a mutation in *Arabidopsis* that makes adjacent stomata (*four lips*); and **(E)** a mutation in *Arabidopsis* that makes too many guard cells (*too many mouths*).

Another consistent feature on the surface of photosynthetic organs is a specialized opening that allows gas exchange. The opening, or stoma, is formed by two **guard cells** that permit or prohibit the movement of gases and the escape of water vapor (Figure 11.4). Interestingly, scientists have found a correlation between decreasing numbers of stomata and increasing levels of atmospheric CO_2, a trend since the Industrial Revolution. Evolution of the guard cell has been extremely useful for plant survival on land. Plants can survive without trichomes—indeed, mutants such as *glabra* are fertile—but they cannot survive without stomata. Exceptions to this rule are found in the group Bryophyta (mosses, liverworts, and hornworts), nonangiosperms that frequent wet locations. Some members do not have stomata.

In both dicots and monocots, an asymmetric division precedes the differentiation of guard cells. The smaller cell, referred to as the **guard mother cell,** is the direct precursor to the two guard cells. In maize and most other plants, the guard mother cell influences adjacent cells to divide asymmetrically to form **subsidiary cells** (Figure 11.5). These subsidiary cells help the guard cell open and close. Thus, the final stomatal complex consists of a group of clonally related as well as recruited adjacent cells.

An interesting question that arises from studies of epidermal cell types is how the patterns are established. The spacing of stomata is fairly straightforward in a maize leaf given that the cell division patterns that produce stomata automatically separate them from each other. In dicot leaves, the divisions are not as regular, yet the stomata are always separated by at least one intervening epidermal cell (Figure 11.4A, C). Mutants have been isolated in *Arabidopsis* in which the stomata ignore the "no touching rule." Examples are *four lips*, which produces stomata that are touching (Figure 11.4D); *too many mouths*, which produces clusters of stomata in the leaves (Figure 11.4E); and *sdd1 (stomatal density and distribution1)*, which exhibits a two- to fourfold increase in stomatal density. The *sdd1* gene encodes a serine protease that is expressed in guard mother cells. When the gene is expressed at high levels in transgenic plants, fewer stomata form. Thus, Sdd1 is likely to function by inhibiting adjacent cells from developing into stomata.

Spacing is also a consideration with trichomes, which are always separated by intervening cells. The *tryptochon* mutant in *Arabidopsis* produces clusters of touching trichomes (Figure 11.6). Most often, the clusters include the accessory cells that surround a normal trichome.

Differentiation of Vascular Elements Can Occur Without Cell Division

Xylem and phloem comprise the transport system in plants, moving water and ions from the roots to the rest of the plant, and sugars from green, photosynthesizing tissues to storage locations such as tubers and seeds. Water moves through xylem cells, which at maturity lack all cytoplasm and nuclei but have elaborate, specialized cell walls. Sugars are actively transported through phloem cells, which lack nuclei but are metabolically active with the help of companion cells. The

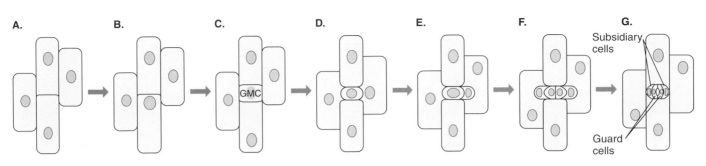

Figure 11.5   Stomatal Complex Formation in Maize (A-C) An asymmetric division in a vertical file generates the guard mother cell (GMC). **(D,E)** The nuclei of cells from both adjacent files migrate toward the GMC and undergo asymmetric divisions. **(F,G)** The resulting cells closest to the GMC are called subsidiary cells. The GMC divides symmetrically to form two guard cells.

vascular pattern of dicot stems is a ring with phloem on the outside and xylem on the inside. Woody plants have a specialized meristem, the **vascular cambium,** which produces these two cell types. When we count the rings on a tree trunk, we are counting the rings of xylem that were laid down each year. When a tree is girdled by having some of its bark removed, the phloem immediately under the bark is destroyed, so the tree dies.

Leaves are connected to the vascular system of the stem via major veins that differentiate during formation of the leaf. Minor veins within the leaf, which differentiate later, connect to the major veins such that all photosynthetic cells are no more than a couple of cells away from a vein. As with the stem, the xylem and phloem have designated positions in a leaf, and these positions are part of the dorsoventral character of a leaf. To determine whether it is position or lineage that determines the differentiation of vascular cells, scientists have examined clonal sectors. (A **clonal sector** is a group of cells in a chimera that are all related because they arise from a single cell; recall from Chapter 3 that a chimera is an organ or tissue composed of cells from genetically distinct sources. In a chimera, cells on either side of the border between adjacent sectors have different genotypes.) Since the clonal sector includes photosynthetic and vascular cells, it is clear that differentiation is determined by position, not cell history.

Vascular differentiation has always been an intriguing subject, given the beautiful patterns of veins in leaves and the fact that differentiated parenchyma cells of the stem further differentiate to form vasculature after a wound. The signaling required for vascular differentiation, however, is difficult to study within the context of the plant. A good system for studying vascular differentiation uses cultured zinnia mesophyll cells (photosynthetic cells full of chloroplasts). These cells are removed from leaves by gentle abrasion and placed into tissue culture. When given the right hormone combination in the culture medium, a percentage of these cells differentiates in vitro into xylem without cell division (Figure 11.7). Experiments in which the cells were exposed to the inducing hormone medium for different lengths of time showed that cells removed from the hormone medium before 48 hours never became xylem; exposure to the inductive medium only after 48 hours also would not induce xylem differentiation. Only when the mesophyll cells were exposed around 48 hours were they capable of developing into xylem.

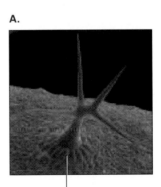

Normal accessory cell

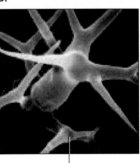

Accessory cell that has developed into a trichome

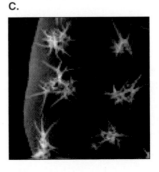

Figure 11.6   Trichome Spacing in *Arabidopsis* (A) A single trichome as in the wild-type plant. **(B,C)** In the *tryptochon* mutant, extra trichomes form from the accessory cells.

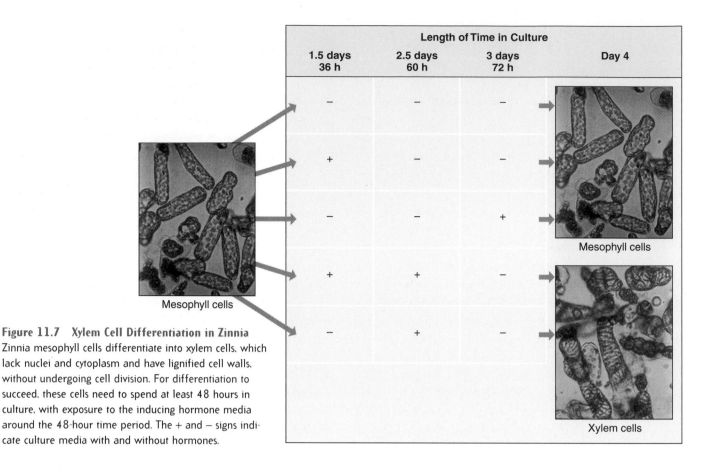

Figure 11.7 Xylem Cell Differentiation in Zinnia
Zinnia mesophyll cells differentiate into xylem cells, which lack nuclei and cytoplasm and have lignified cell walls, without undergoing cell division. For differentiation to succeed, these cells need to spend at least 48 hours in culture, with exposure to the inducing hormone media around the 48-hour time period. The + and − signs indicate culture media with and without hormones.

These experiments provide a handle on understanding the molecules involved in xylem differentiation. The genes that are induced at the inductive time point can be used to examine in vivo differentiation of xylem cells. What remains to be determined is how certain cells get the signal to differentiate into xylem while surrounding cells do not. Likely the hormone auxin is responsible, at least in part, as we shall see later in the chapter.

An interesting aspect of histogenesis is that it marches forward regardless of the condition of the organs. For example, some of the seedling mutants discussed in Chapter 10 are abnormal in morphology. One of these, the *argonaut/pinhead* double mutant, is basically a shapeless mass of cells. Although it does not resemble any plant we have ever seen, it continues to grow and photosynthesize and makes normal stomata and xylem elements. Clearly, signals involved in cell differentiation can be uncoupled from organogenesis.

PLANT CELL COMMUNICATION

Given that positional information plays such an important role in development, there must be mechanisms for plant cells to talk to each other. In this section, we will discuss how adjacent cells communicate with each other, and how long-distance communication is accomplished through hormones.

Plant Cells Communicate Through Cell Walls

The term *cell wall* conjures up an image of an impenetrable fortress, when in effect the wall is more like a tangled hedge. The primary substance in plant cell walls is cellulose, repeating units of glucose that are organized into microfibrils; interwoven with the cellulose are polysaccharides and lignin. During cell division, **plasmodesmata** (sing. *plasmodesma*) form (Figure 11.8). These openings, somewhat reminiscent of gap junctions, are membrane-lined cytoplasmic channels through the cell wall that connect the cytoplasm of adjacent cells. In the middle of each plasmodesma is a strand of endoplasmic reticulum. Plasmodesmata provide **symplastic** continuity between cells, allowing electrical continuity and the flow of small molecules between cells. In contrast, continuity that is **apoplastic** occurs outside the cell's cytoplasm, in the spaces between cell walls.

Plasmodesmata also form between cells that are not related to each other by cell division history. These secondary plasmodesmata often appear in response to environmental or developmental signals. They result from specific cell wall degradation, which occurs from both sides, followed by cytoplasmic fusion.

Experiments were carried out by Pat Zambryski and her colleagues to determine the environmental and developmental control of movement through plasmodesmata. Small fluorescent dyes were loaded into the plant through the phloem of older leaves, and the spread of the dye through plasmodesmata was observed by microscopy. The dye initially moved through the phloem and then passed symplastically through the plasmodesmata. These researchers discovered that the dye moved throughout the shoot apex (the meristem and leaf primordia) of *Arabidopsis* plants when grown under short-day conditions. When these plants were switched from short days to long days—a change that triggers the transition to flowering in *Arabidopsis*—the dye stopped moving throughout the plant. Once the plant had made a number of flowers, the symplastic connection with the apex was regained (Figure 11.9). In a

second experiment, once dye was taken up into developing *siliques* (seed pods), it moved freely through young embryos. At the end of embryogenesis, the symplastic connectivity between the embryo proper and the suspensor was broken.

Plasmodesmatal connections, in addition to changing under environmental influences, also change as cells differentiate. Immature guard cells are symplastically connected to the surrounding cells. As they differentiate, their plasmodesmata close and ultimately degenerate, leaving them isolated. Similarly, all the cells in the undifferentiated zone of the *Arabidopsis* root are coupled by plasmodesmata. The mature root epidermis, however, is isolated from the underlying cells, and the hair cells are isolated from all other cells. Although we do not know the full meaning of these results, they demonstrate that symplastic connections can be dynamic, changing under developmental or environmental stimuli.

Plasmodesmata were once considered a significant barrier to the passage of molecules larger than 1,000 daltons (D). That conclusion changed with the discovery of **movement proteins,** which increase the effective pore size of plasmodesmata, allowing molecules as big as 40,000 D to pass through. Movement proteins are capable of moving between cells and may haul other molecules, such as nucleic acids, along with them. Because they are encoded on viral genomes, movement proteins seem like a clever way for viruses to take over and infect a plant. The movement of macromolecules between cells, however, is not simply a feature of invaders such as viruses. Increasing evidence points to the fact that some plant transcription factors are also capable of *gating* (opening up) plasmodesmata.

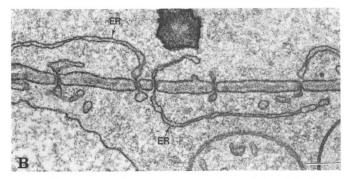

Figure 11.8 Plasmodesmata Between Adjacent Plant Cells Note that plasmodesmata contain strands of endoplasmic reticulum (ER).

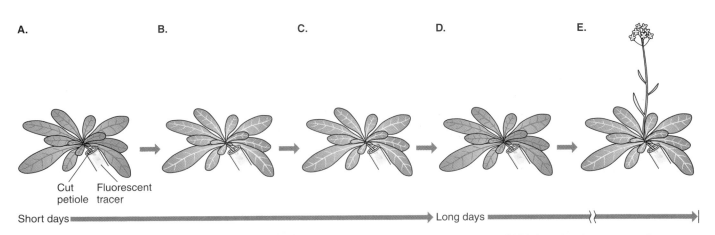

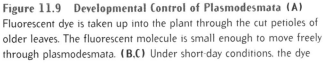

Figure 11.9 Developmental Control of Plasmodesmata (A) Fluorescent dye is taken up into the plant through the cut petioles of older leaves. The fluorescent molecule is small enough to move freely through plasmodesmata. **(B,C)** Under short-day conditions, the dye spreads throughout the plant. **(D)** When the plant is exposed to two long days (which triggers flowering in *Arabidopsis*), the dye stops moving. **(E)** Eventually, when the plant is fully flowering, the dye can be found throughout the plant.

One of the first experiments to test whether plant proteins move through plasmodesmata focused on the meristem protein Knotted 1 (Kn1), mentioned in Chapter 10. Kn1 was a good candidate because it had been shown by clonal analysis to behave nonautonomously; that is, leaf cells with the dominant, mutant version of the gene could affect the phenotype of adjacent, wild-type cells. The movement experiments took advantage of fluorescently labeled Kn1 injected into tobacco leaf cells. The protein moved from cell to cell and increased the pore size of the plasmodesmata, such that other large molecules could sneak through the plasmodesmata in the presence of Kn1.

Viruses have been shown to move nucleic acids somewhat promiscuously; for example, single-stranded RNA viruses may allow the movement of single-stranded RNA of any sequence, but not of double-stranded RNA. In the presence of wild-type Kn1 protein, the sense strand of Kn1 RNA was shown to move from cell to cell but not the "antisense" strand (a strand that does not code for protein). RNA and proteins are not found in all cell types; thus, if these proteins move through the plant, there must be mechanisms that keep them from moving indiscriminately.

The microinjection experiments show that transcription factors can move between differentiated cells but do not prove that movement normally occurs. Microinjection experiments in meristem cells, where Kn1 is normally found, are technically difficult due to the small size of these cells. To address the question of whether movement occurs in vivo, scientists have used chimeras. Using a promoter specific for the L1 layer of the meristem, Allen Sessions, Detlef Weigel, and Marty Yanofsky made transgenic plants in which the *leafy (lfy)* gene (described in Chapter 10) was expressed only in L1 (Figure 11.10A). The DNA construct was placed in a *lfy* mutant background. The scientists were able to ask whether L1-driven *lfy* could make a normal inflorescence. The answer was yes. In addition to normal flower initiation, genes activated by Lfy protein were also activated. Most exciting was the finding that the protein appeared in a gradient from L1 to the inner layers (Figure 11.10B).

The differentiation of endodermal root cells in *Arabidopsis* provides evidence of a role for protein movement in development. The endodermis and cortex arise from an asymmetric division of the cortical-endodermal initial cell (Figure 11.11A; see also Chapter 9). Plants with the *shr* mutation do not carry

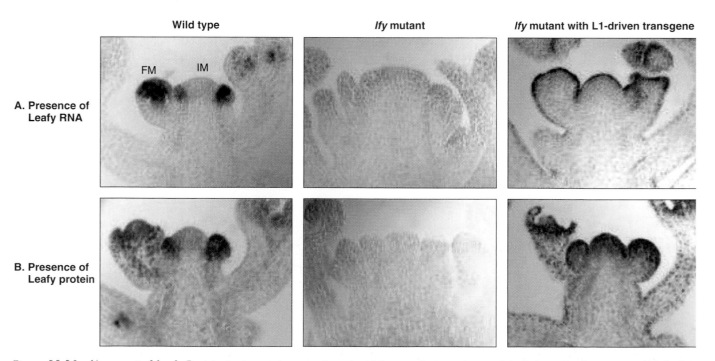

Figure 11.10 Movement of Leafy Protein Leafy protein moves from the L1 layer to the inner layers of *Arabidopsis* floral meristems. **(A)** In the wild type. Leafy RNA is excluded from the inflorescence meristem (IM) and is present in floral meristems (FM). In the *lfy* mutant. Lfy RNA is absent. In a *lfy* mutant plant with a transgene in which *lfy* is expressed from an L1 promoter. Lfy RNA is seen only in the L1. **(B)** In the wild type. Lfy protein is detected in the nuclei of all cells of young floral buds. No Lfy protein is detected in *lfy* mutants. In the same plant as shown in part C. Lfy protein is observed as a gradient. with highest levels in the L1 and L2 layers and decreasing internally: distribution is more even in older flowers.

A.

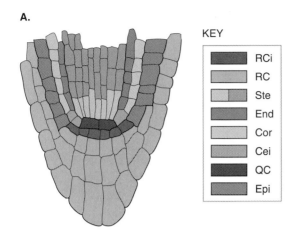

KEY

■	RCi
■	RC
□	Ste
■	End
□	Cor
■	Cei
■	QC
■	Epi

B.

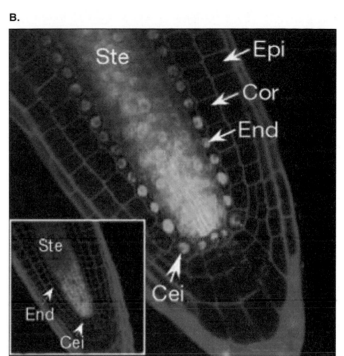

C.

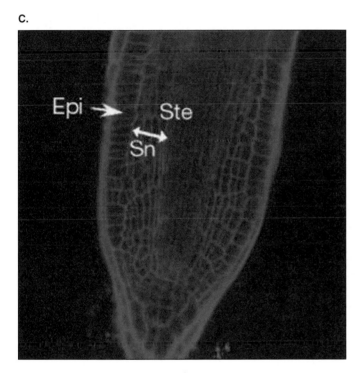

Figure 11.11   Movement of Shr Protein (A) Some of the cell lineages of the *Arabidopsis* root, shown diagrammatically. The cortical-endodermal initial cell undergoes an asymmetric division, leading to separate cell files for the endodermis and the cortex. **(B)** Normally, Shr RNA is expressed in the stele (visualized as green in the inset), while the protein is found in the nuclei of adjacent cell layers in addition to the stele. The protein is here rendered visible by fusion with GFP. **(C)** In transgenic plants, supernumerary layers form in which *shr* is expressed behind an endodermal promoter. Abbreviations: Cei, cortical-endodermal initial cells; Cor, cortex; End, endodermis; Rc, root cap; Rci, root cap initial; Epi, epidermis; Ste, stele; Sn, supernumerary layers.

out this asymmetric division, producing only a single layer with the attributes of cortex. Given that the *shr* gene is required to make the endodermis, Phil Benfey and his students were surprised to find the Shr RNA expressed in the stele, inside the endodermis (see Figure 11.11B inset). Shr protein, however, is localized not only in the stele, but also in nuclei of the endodermis and the cortical-endodermal initial cells (Figure 11.11B). In further experiments, Benfey and colleagues produced transgenic plants with *shr* expressed behind an endodermal promoter. At first, the native protein moved from the stele to the outer adjacent cell layer,

causing the initial cell, as usual, to divide and differentiate into endodermis and cortex. As the endodermis differentiated, however, the *shr* transgene, being under control of the endodermal promoter, was expressed, producing additional Shr protein; this protein, too, moved to the adjacent cell layer, leading to the development of more and more endodermal layers (Figure 11.11C). Thus, it appears that the movement of a protein to adjacent cells leads to an asymmetric cell division and subsequent differentiation. Such a mechanism may exist in other tissues where adjacent layers have different cell fates.

Figure 11.12 Fasciated Ears in Maize The ear on the left is normal; the other two ears are fasciated. A fasciated meristem loses its central growing point and broadens. The mutation is called *fasciated ear 2*.

Plant Cells Communicate Using Kinase Receptors

A phenotype occasionally found in nature is **fasciated** stems. *Fasciated* literally means "to be bound together as in a bundle"; these stems are flattened rather than round, and usually thicker than normal. Fasciated stems result from enlarged shoot meristems that fail to grow from a single tip. Often many more organs form than normal. Fasciated maize ears (Figure 11.12) turn up with some frequency in grocery-store sweet corn. A series of *Arabidopsis* mutants called *clavata* (named for the club-shaped carpels that result) are also fasciated. Many more organs, especially more floral meristems, form in *clavata* mutants than in the wild type. The carpels are often indeterminate. These genes are likely to function in a signaling pathway to limit growth in the meristem.

The *clavata* (*clv1*) gene is expressed in the L3 layer of the central zone in shoot meristems. This gene encodes a membrane-bound receptor kinase with an extracellular domain resembling that of animal hormone receptors. The second mutation in the series, *clv2*, encodes a protein that has a similar extracellular domain as Clv1 but lacks the kinase domain. (The maize mutant shown in Figure 11.12 also encodes a *clv2*-like gene.) And *clv3* encodes a small peptide that may prove to be a ligand that binds to Clv1; *clv3* is expressed in the L1 and L2 meristem cells adjacent to *clv1*-expressing

cells. Genetic analysis suggests that the *clv* genes lie in the same pathway. The expression patterns and protein homologies all support the hypothesis that the gene products interact between cells to organize the meristem and keep it from growing out of control. This hypothesis is supported by the finding that expression of *clv3* at higher levels, using the CaMV 35S promoter, produces a phenotype similar to those of *wuschel* or *shootmeristemless* mutants (see Chapter 10), in which leaves are not produced and the meristem simply quits.

Whether plant cells communicate through plasmodesmata, with receptors and/or ligands, it is clear that the cells coordinate their activities in such a way that an organ functions as a whole, rather than as a group of individual cells acting autonomously. A good illustration of this point is made by the *tangled* (*tan*) mutation in maize. Leaves on *tan* mutants show abnormal cell wall formation (Figure 11.13A,B), yet the overall shape is normal. In fact, cellulose microfibrils that wrap around a cell perpendicular to the direction of expansion of the cell are often aligned with the orientation of leaf growth and not of the individual cell (compare C and D of Figure 11.13). Clearly, the shapes of individual cells are not guiding the overall morphology. Similar to the construction of a house, the positioning of the internal walls does not appear to determine the final form.

HORMONAL CONTROL OF PLANT DEVELOPMENT

It is difficult to think of any aspect of plant development that doesn't involve at least one of the six major plant hormones (Table 11.1). Many of the developmental events discussed previously—such as establishing apical dominance, floral induction, root hair growth, embryo development, root hair initiation, and xylem differentiation—are controlled by hormones. In most cases, more than one hormonal signal is involved and these signals interact. This "cross talk" adds to the complexity. A common way to elucidate such biosynthetic signaling pathways is through the use of genetic screens. Genetic screening involves taking a large number of seeds that have been mutagenized (exposed to agents with a high probability of producing genetic mutations) and observing them for defective growth responses in the presence or absence of hormones. In this way, researchers come to associate a gene with previously established physiological processes.

A.

B.

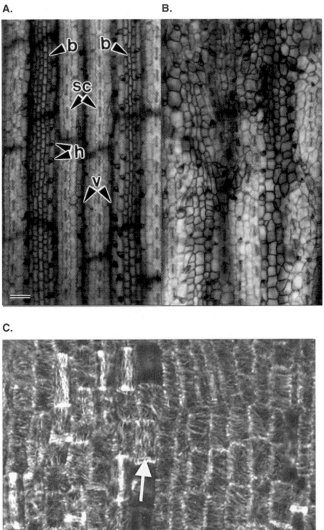

C.

D.

Figure 11.13 The *tangled* Mutation in Maize The *tangled* mutant demonstrates that the orientation of cell division planes is not crucial to organ morphology. A comparison of **(A)** a wild-type and **(B)** a *tangled* (*tan*) leaf surface stained with Toluidine Blue: note that cell files in the *tan* mutant are chaotic. **(C)** Cellulose microtubules (MTs) in wild-type leaves: most MTs are transverse to the cell's long axis. Occasionally, a cell (*arrow*) will have MTs parallel to the long axis of the cell. **(D)** In leaves of the *tan* mutant, MTs are often aligned transversely to the cell's long axis, as in wild-type leaves. In contrast, arrowheads indicate cells whose MTs are better aligned with those of neighboring cells than with each cell's own transverse axis.

The Plant Hormone Auxin Undergoes Polar Transport

Auxin, the first group of plant hormones to be identified by researchers, is responsible for events such as cell elongation; xylem regeneration in a wound; apical dominance; bending toward the light (**phototropism**); adventitious root growth; gravitropism; and the stimulation, by seeds, of fruit development (Table 11.1). Although Darwin was the first to propose that higher plants contained **translocated** chemical messengers (chemicals made in one part of the plant that move to another part), it was Frits Went who carried out the first classical experiments on auxin, in 1926. Went decapitated

TABLE 11.1 THE SIX MAJOR PLANT HORMONES AND THEIR MODE OF ACTION

Hormone or Hormone Family	Structure	Activity
Abscisic acid	S-(+)-abscisic acid	Stomata closure Maintenance of dormancy
Auxin	Indole-3-acetic acid	Apical dominance Cell elongation Phototropism Xylem regeneration Adventitious root formation Fruit development Gravitropism
Brassinosteroids	Brassinolide	Cell elongation Cell division
Cytokinins	Zeatin	Cell division Shoot formation in culture Delay of leaf senescence Release of apical buds
Ethylene	$H_2C = CH_2$	Fruit ripening Root hair growth Abscission Senescence
Gibberellins (GA)	Gibberellin A$_1$	Cell elongation Floral induction Seed germination

shoot tips and placed them each on a block of agar. After a period of time, he placed the blocks of agar on the shoots. He found that a well-centered block caused the shoot to grow straight, whereas an off-centered block caused asymmetric growth or bending. Clearly, something was moving from the shoot tip and was responsible for growth. The substance in the shoot tip was eventually identified as auxin.

The movement of auxin that Went discovered is polar and active. We know this **polar auxin transport** is not due simply to gravity because when isolated stem pieces are turned upside down, auxin is transported toward the original base, that is, away from the pull of gravity. This directionality must be important to development because inhibitors of polar auxin transport stop embryo development at the transition between the globular and heart stages and prevent cotyledon separation. Polar auxin transport inhibitors applied to a plant later in development prevent tropic responses, such as phototropism or gravitropism. Polar auxin transport is also required for flower development and vascular differentiation.

Genetic screens in *Arabidopsis* have led to the isolation of auxin transport proteins such as Aux1 and Atpin1. In this experiment, mutagenized seeds were plated on high concentrations of auxin, which led to decreased root length. The *aux1* mutant was identified by its long roots, which were also agravitropic. Aux1, a protein similar to amino acid permeases, is likely to be an auxin influx carrier, transporting auxin into the cell. The *atpin1* gene was identified by the *pin-formed* mutation, and is required for auxin transport in the shoot. The most striking pin-formed phenotype is a lack of flowers on a naked stem (Figure 11.14A). Sometimes cotyledons are aberrantly positioned or extra cotyledons form.

As seen with the addition of transport inhibitors, vascular development below the leaf is abnormal (Figure 11.14A). The inability of the stem to mobilize the auxin coming from the leaf results in a massive accumulation of xylem cells. The gene *atpin1* encodes a transmembrane transporter that localizes to elongated parenchymatous xylem cells of the stem. Thus, its position is perfect for transporting auxin from the stem toward the root (Figure 11.14B).

Mutations in the auxin efflux protein that functions in roots were discovered from other responses in *Arabidopsis*: insensitivity to the hormone ethylene (*ethylene insensitive roots 1*, or *ein1*), wavy root growth (*wavy6*), and agravitropic growth (*agr1*). All three phenotypes are due to mutations in the protein Atpin2. Similar in sequence to Atpin1, Atpin2 is localized to cortical and epidermal root cells, in particular, to the ends of cells proximal to the shoot (Figure 11.14B). Atpin2 is crucial for gravitropism. Darwin would be pleased to know that his model of polar transport is correct

and that specific molecules have been identified through genetic means.

The Hormone Cytokinin Stimulates Cell Division

The first of the **cytokinins** was initially identified as a growth-stimulating substance present in coconut milk. These discoveries led to its use in tissue culture experiments, where it is often required for cell division to occur. Cytokinins are also associated with shoot formation, loss of apical dominance, and inhibition of senescence (Table 11.1). The biosynthetic pathway of cytokinins has not been elucidated in plants, but a cytokinin-producing enzyme, identified from the soil bacterium *Agrobacterium tumefaciens*, is used experimentally. The recent sequencing of the *Arabidopsis* genome identified a number of genes likely to encode similar enzymes. A link between cytokinins and cell division was demonstrated in *Arabidopsis* through the cell cycle gene *cycD3*. In diploid species, cells cycle from the G1 stage ($2n$), to synthesis (S), to G2 ($4n$), and finally to mitosis (and thus back to $2n$). Transcripts of *cycD3* accumulate in plants prior to synthesis, and may be required for the G1 to S transition. When cytokinins are added to cell cultures or to intact plants, CycD3 accumulates.

What role do cytokinins play in the meristem? To determine this, one would ideally examine the phenotypes of mutants that do not make cytokinins. However, unlike gibberellin and ethylene (two hormones to be discussed shortly), mutants that fail to produce any cytokinins have not been recovered, presumably because cytokinins are essential for cell division. Instead, there are mutants that make extra cytokinins; one of these is the *supershoot* (*sps*) mutant in *Arabidopsis*, which is extremely bushy, with multiple branches from every axil. The gene is expressed in leaf axils and encodes a cytochrome P450. Another *Arabidopsis* mutation, *cre1*, was identified by screening for hypocotyls that were unable to make shoots in tissue culture in the presence of cytokinins. The *cre1* gene encodes a histidine kinase. Tatsuo Kakimoto and his group showed that Cre1 is likely to be the cytokinin receptor itself. Their experiment involved complementing a yeast mutant, as follows. First they found a yeast strain, *sln1*, which lacked a similar histidine kinase; hence the *sln1* mutation is lethal. When they inserted the *cre1* gene into the yeast genome *and* added cytokinins to the media, the *sln1* yeast grew. We still do not know how plants produce cytokinins, but we are closer to understanding how they perceive and respond to it.

Cytokinins, however, are not likely to be acting alone; a number of experiments point to a seesaw effect between

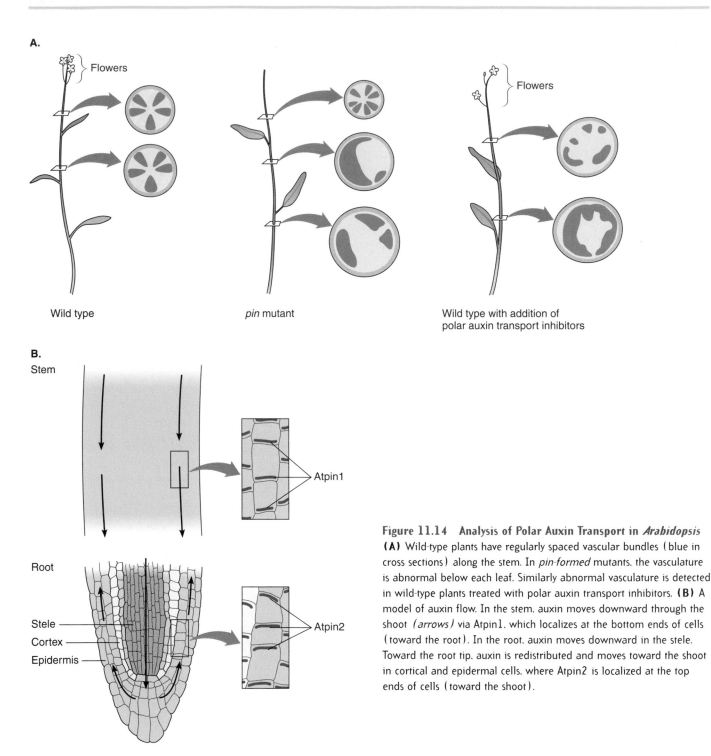

Figure 11.14   Analysis of Polar Auxin Transport in *Arabidopsis*
(A) Wild-type plants have regularly spaced vascular bundles (blue in cross sections) along the stem. In *pin-formed* mutants, the vasculature is abnormal below each leaf. Similarly abnormal vasculature is detected in wild-type plants treated with polar auxin transport inhibitors. **(B)** A model of auxin flow. In the stem, auxin moves downward through the shoot (*arrows*) via Atpin1, which localizes at the bottom ends of cells (toward the root). In the root, auxin moves downward in the stele. Toward the root tip, auxin is redistributed and moves toward the shoot in cortical and epidermal cells, where Atpin2 is localized at the top ends of cells (toward the shoot).

cytokinins and auxin. Whereas cytokinins release the growth of axillary buds, auxin represses it. When tobacco leaves are cut into pieces and placed on a culture medium with a high auxin–cytokinin ratio, roots form along the edges; if the ratio is low, then shoots will form. If levels of both hormones are high, nonorganized growth will occur. In the absence of any hormone, the leaf pieces do not grow (Figure 11.15A).

The Hormone Gibberellin Influences Plant Growth

The principal hormone group responsible for plant height is the **gibberellins** (see Table 11.1). Interestingly, the first gibberellin (often abbreviated GA) was discovered as a compound in fungi that infects rice producing a devastating disease called bakanai (Japanese for "foolish seedling"). These "silly" rice seedlings grow too tall and fall over, drown-

A. Cultures of tobacco leaf discs

No hormone

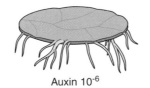

Auxin 10^{-6}

Cytokinins 10^{-6}

Cytokinins 10^{-5}
Auxin 10^{-5}

B. Growth responses regulated by GA

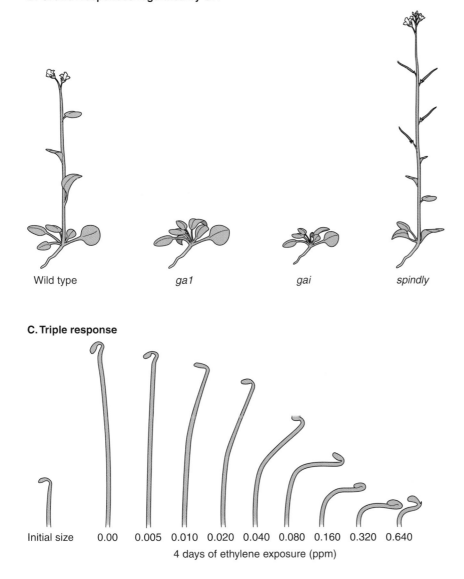

Wild type   ga1   gai   spindly

C. Triple response

Initial size   0.00   0.005   0.010   0.020   0.040   0.080   0.160   0.320   0.640

4 days of ethylene exposure (ppm)

Figure 11.15   Effects of Plant Hormones
(A) Tobacco leaf discs that have been incubated without auxin or cytokinins do not produce organs. In the presence of auxin they make roots, and in the presence of cytokinins they make shoots. In the presence of high concentrations of both auxin and cytokinins, the leaf discs make callus, a nonorganized tissue.

(B) Gibberellin regulates multiple processes including cell elongation and time to flowering, as demonstrated in *Arabidopsis*. *Left to right:* the wild type, a mutant lacking GA *(ga1)*, a mutant that fails to perceive GA *(gai)*, and a mutant that behaves as if it has been treated with too much GA *(spindly)*. **(C)** Pea seedlings exposed to ethylene: Increasing concentrations

of ethylene cause seedlings to become shorter, thicker, and bent over. This phenotype, known as the "triple response," has been used in *Arabidopsis* to screen for mutants that look as if they were exposed to ethylene despite the absence of the hormone, as well as for the converse—mutants that have been exposed to ethylene but respond as if they have not.

ing in the rice field. Eventually, what was thought to be a fungal toxin turned out to be a major growth regulator in plants. Its discovery as a plant compound was facilitated by the analysis of dwarfs in pea and maize, which respond to the compound by growing to normal size. From a combination of recessive dwarf mutations mapping to different loci and various GA intermediates, Bernard Phinney was able to successfully elucidate the biosynthetic pathway of GA biosynthesis in corn. Similar GA-deficient mutants were also identified in *Arabidopsis* and have facilitated isolation of the genes involved.

The phenotype of a dominant dwarf mutation in *Arabidopsis*—*gibberellic acid insensitive (gai)*—is similar to GA-deficient dwarfs (Figure 11.15B); however, there is an important difference: *gai* mutants are defective not in their ability to make GA but in their response to it. In the wild type, high levels of GA serve as a negative feedback mechanism, turning off GA production to make sure the plant does not make too much of the hormone. Since *gai* plants are insensitive to GA, this negative feedback cannot operate; the result is that GA levels are actually elevated, though this is not reflected in the phenotype. The transcription factor encoded by *gai* is similar to Scarecrow protein, which we encountered in relation to root development in Chapter 9.

While dwarf *Arabidopsis* plants are not significant for agriculture, mutations in homologous genes in barley and wheat are used commercially. The dwarfs are less likely to be blown down, and more energy can be put into producing the head of grain rather than the long stem. These GA-insensitive dwarf mutants are also useful because the dwarf character affects the stem and not the fruit. Like dwarf fruit trees that produce normal-sized apples on short trunks, the barley and wheat grains are also normal sized.

Plants need just the right amount of GA; an overdose causes the plants to grow too much, as in the bakanai rice disease. In *Arabidopsis* and barley, *spindly* and *slender* mutants, respectively, look as if they have been exposed to too much GA. Like the *gai* mutants, *spindly* mutants are defective in their response to GA. But whereas the *gai* mutant behaves as if it cannot perceive GA, the *spindly* mutant grows as if the GA response were on at full tilt all the time (Figure 11.15B). This mutant also has very low levels of GA because the negative feedback mechanism that regulates GA biosynthesis is continually turned on.

The Hormone Abscisic Acid Plays a Major Role in Seed Dormancy

Before they germinate, many species require a period of dormancy, which is imposed by various factors. One is a tough seed coat, which prevents water from entering. Another is chemicals in the seed coat or in the seed itself; dormancy often ends after a soaking rain, which is effective in some species in leaching these chemicals away. The major chemical inhibiting germination is the plant hormone **abscisic acid (ABA)** (see Table 11.1).

Just as cytokinins and auxin may counteract each other, GA action is antagonized by ABA. This is well documented in the mobilization of barley endosperm for beer making. When grain is exposed to water, GA levels increase, resulting in the secretion of enzymes from the outer layer of endosperm, which is called the **aleurone.** Unlike the internal layers of endosperm, in a mature seed the aleurone is alive. The secreted enzymes degrade the endosperm, making nutrients available for the embryo. The aleurone cells die soon after. Researchers have isolated protoplasts from the aleurone layer and shown they can survive for weeks in the presence of ABA but die rapidly in the presence of GA. Mutations in both corn and *Arabidopsis* that fail to produce ABA in the seed result in mutants that make *viviparous* seed (seed that germinates precociously, like the seed of red mangrove). Viviparity can have the unhappy consequence of the seed germinating inside its fruit—far from water and thus potentially exposed to desiccation.

Brassinosteroids Are Animal-Like Hormones Found in Plants

Another class of dwarfs has recently demonstrated that plants, like animals, have steroid hormones. The *Arabidopsis* mutants *deetiolated2 (det2)* and *constitutively photomorphogenic dwarf (cpd)* do not respond to GA but instead respond to **brassinolide,** a plant brassinosteroid (see Table 11.1). The gene *det2* encodes a steroid 5α-reductase, which is the first committed step in the synthesis of brassinolide. These steroid reductase enzymes are highly conserved: Det2 can also catalyze the reduction of androgens such as testosterone in humans. Plants that are mutant for the *det2* gene are extremely small. If they are transformed with a normal copy of the gene—either wild-type or human—they synthesize brassinolide and grow to normal height. If the *det2* gene is expressed at high levels, in either wild-type or mutant plants, the plants are larger than normal.

A brassinosteroid receptor was identified by searching for similar-appearing mutants that did not respond to brassinolide. The receptor (encoded by the *det1* gene) contains an extracellular domain with repeated stretches of amino acids, each repeat containing many leucine residues. Similar "leucine-rich repeats" are found in animal hormone receptors, proteins involved in plant disease resistance, and the meristem gene *clv1*. Like Clv1, Det1 also contains a kinase domain. Is it possible that eating certain plants will soon be banned by the Olympic's committee on drug abuse because of the steroids found within?

Ethylene Is a Gaseous Hormone Involved in Fruit Ripening

The plant hormone with the simplest structure is **ethylene**, C_2H_4 (Table 11.1). As a volatile gas, it is the agent responsible for the "rotten apple in the barrel" syndrome; ethylene leads to ripening and triggers more ethylene production, thus acting in an autocatalytic fashion. In addition to ripening, ethylene is produced during wounding, invasions of pathogens, flooding, and senescence. Ethylene is induced by auxin; this explains why the *ein1* mutant, described earlier in the chapter, is caused by defects in auxin transport. Ethylene also affects seed germination, cell elongation, leaf curling, root hair formation, and **abscission** (the natural separation of flowers, fruit, or leaves from the plant). Clearly, ethylene is a hormone for all seasons. Recent work has shown that animals may also use a gas as a hormone, in this case nitrous oxide (NO).

Biochemical analysis has elucidated the *Yang cycle*, the pathway that leads to ethylene production. The ethylene signal transduction pathway, on the other hand, has been elucidated through genetic analysis that included observations of wild-type plants exposed to too much ethylene. The seedlings develop a thick, bent hypocotyl and unopened cotyledons (Figure 11.15C). *Arabidopsis* plants with the *constitutive response (ctr1)* mutation grow as though they have been exposed to ethylene even when they have not; while *ethylene insensitive (ein)* mutants are unaffected by the presence of ethylene.

Putative ethylene receptors have been cloned that encode membrane-spanning histidine kinases similar to some bacterial proteins. In the absence of ethylene, these receptors (a family of four proteins in *Arabidopsis*) block a kinase cascade that leads to an ethylene response. In the presence of ethylene, the receptors no longer block the response. The mutant forms of the receptors, identified through dominant ethylene-insensitive mutations, contain alterations in the ethylene-binding domain. Because the mutant receptors cannot bind to ethylene, they constantly block the ethylene response. A similar dominant mutant phenotype in tomatoes, called "never ripe," has been identified as resulting from a mutation in a related ethylene receptor. From the name, it is not hard to imagine the phenotype of these mutant tomatoes—they are always green and hard.

Other kinases in this signal transduction pathway are Ein2 and Ctr1, the latter an integral membrane protein that is similar to Raf, a human protein involved in second-messenger cascades. Phosphates are passed from the receptors to Ctr1 to Ein2 to transcription factors in the nucleus. Researchers have recovered *ein2* mutants in screens for phenotypes that are resistant to auxin transport inhibitors, cytokinins, and ABA (not features of other ethylene-response mutants). This finding suggests that Ein2 may function in other hormone response pathways.

Given the fact that animals respond to the environment mostly by behavioral changes and plants respond by developmental changes, it is interesting that they often utilize the same type of signaling molecules.

PLANT RESPONSES TO LIGHT

Just as hormones play a pervasive role in all aspects of plant development, so does light. Whether sitting in the shade or sun, waiting to germinate or to flower, plants are always monitoring the quality and quantity of the sun's spectrum (Figure 11.16). The major photoreceptors in plants are the family of pigment molecules called **phytochromes**, which absorb in the red and far-red parts of the spectrum. Daylight is rich in red light, moonlight in far-red light, shade under the leaf canopy is mostly far-red, and the light at the bottom of a lake is mostly red. Phytochrome responses to light include seed germination, greening, stem elongation, flowering, and the induction of gene expression, especially genes involved in photosynthesis.

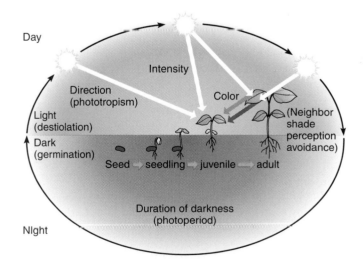

Figure 11.16 Light Perception Plants use light to determine what to do, when to do it, and how. The wavelength and quantity of light that a plant absorbs depends on its immediate surroundings. Information is transferred from the photoreceptors to the regulation of gene activity.

Photomorphogenesis (the development of plant tissues and organs, as controlled by light) has been especially helpful in figuring out how phytochromes work. Seedlings germinated in the dark are pale, with elongated hypocotyls, unexpanded cotyledons, and undeveloped leaves. When these dark-grown seedlings are given a 5-minute pulse of red light, they quickly (within about 4 hours) synthesize large amounts of the proteins involved in photosynthesis. This induction is blocked if the red light pulse is quickly followed by a pulse of far-red light. Induction is also reinitiated if the far-red pulse is followed by another red pulse. Thus the hallmark of a phytochrome response is that it is reversible with red and far-red light. Phytochrome B (PhyB) is likely to play the most significant role in *Arabidopsis* seedling development; *phyB* mutants have long hypocotyls, are pale green, have increased apical dominance, and flower earlier. Similar phenotypes are also seen in *phyB* mutants of cucumber, pea, tomato, and rape.

Chromophores are light-harvesting molecules; they absorb light at particular wavelengths and so appear colored (hence the *chromo-* in their name). Chromophores are covalently attached to the phytochrome proteins. When a phytochrome absorbs red or far-red light, the protein conformation changes. What has eluded scientists for decades is how the light input to phytochrome leads to gene expression. Recent experiments have used transgenic plants to reveal this connection. The phytochrome gene is fused to *green fluorescent protein (gfp)*, a jellyfish gene that encodes a protein that, as we saw in Chapter 6, fluoresces green under UV light. This transgene makes the phytochrome protein visible without killing the plant or interfering with the protein itself. Phytochrome–GFP fusion proteins move into the nucleus in a light-dependent manner. Given a flash of red light, Phytochrome A (PhyA) and PhyB move into the nucleus; however, they move out of the nucleus if that red flash is followed by far-red light. Although the possibility remains that phytochromes act as a cytoplasmically located kinase, Peter Quail and his team have shown that PhyB interacts with a nuclear-localized protein in a light-dependent manner and this protein binds DNA, again in a red-light-dependent manner. The proteins that are bound are also implicated in the greening process.

Also significant are blue-light receptors, which fall into two known groups: **cryptochromes** (named for the fact that they were difficult to find) and **phototropin**. Though the existence of blue-light receptors has been known for a long time—blue-light responses were first described by Darwin—only recently have these molecules been isolated. Phototropin functions in phototropism, particularly the bending of hypocotyls, but also in the blue-light-induced opening of stomata and the migration of chloroplasts within a cell. Phototropin is a serine-threonine kinase protein that is activated by blue light. Two cryptochrome-encoding genes have been studied in *Arabidopsis*: *cry1* and *cry2*. Cry1 primarily regulates hypocotyl elongation while Cry2 primarily regulates the induction of flowering by photoperiod. These two proteins are similar to bacterial photolyases, which function in DNA repair by splitting thymidine dimers, created by UV light, in a blue-light-dependent manner. However, Cry1 and Cry2 are not likely to act through DNA lyase activity. As with phytochromes, chromophores are covalently bonded to the blue-light receptors. Exactly how the Cry proteins transmit the light information is presently not known.

Both phytochromes and blue-light receptors are strongly implicated in setting an internal "clock" that is **circadian** (approximately one day long). Circadian clocks are **entrained** (set) by a light/dark cycle of 12 hours each. Once the clock is set, it tells a plant when to open or close its stomata, when to start increasing photosynthetic gene expression, when to flower, and other important information. The clock keeps these physiological events on schedule, for at least a few cycles, even through a subsequent period of continuous light or continuous dark. Thus, there are two important aspects to a circadian clock: the ability to be entrained, and the ability to persist in the absence of entraining stimuli.

Transcription of a number of genes is under circadian control. For example, the transcript for chlorophyll *a/b* binding protein (Cab), which binds chlorophyll to membranes within the chloroplast, has a circadian rhythm of expression. The *cab* transcript begins to increase in levels even before dawn, in anticipation of the coming day. By this mechanism, *cab* expression is at its peak when the light is strongest. Mutations in *phyB* that result in lack of expression extend the circadian period by a couple of hours, whereas overexpression of *phyB* obtained by using a constitutive promoter shortens the cycle by a couple of hours. Similarly, overexpression of a functional *cry1* gene shortens the period by an hour, and loss of *cry1* results in period lengthening. The effect of *cry* and *phy* mutations on the circadian rhythm is dependent on the wavelength and intensity level of light. Thus, it is likely that phytochrome and cryptochrome genes work together to input a light signal to the clock.

As we end our discussion of cell signaling in plants, it is important to note that, for simplicity, we have painted a picture of hormone and light regulation of development as if these two kinds of processes were separate. However, it is likely that these signals cross paths.

KEY CONCEPTS

1. A number of genes that regulate hair cells in roots also regulate trichome development on leaves.

2. Genetic mechanisms exist to space trichomes and stomata.

3. Transcription factors can move from cell to cell in plants.

4. Plant hormones play critical roles in all aspects of plant development.

5. Phytochromes absorb red and far-red light and translocate to the nucleus.

6. Blue-light receptors regulate phototropism.

STUDY QUESTIONS

1. How would you determine what regulates the spacing pattern in stomata?

2. How would you determine whether a hormone signaling pathway overlaps with a light-signaling pathway?

3. How do plant cells communicate with neighboring cells?

4. A protein that is produced in one cell type is found in an adjacent cell of a different type, where it is not produced. By what mechanism might it move to the adjacent cell?

5. How do plant cells perceive the different qualities of light?

6. What is the signature behavior of a phytochrome response?

7. What are characteristics of GA-deficient mutants?

SELECTED REFERENCES

Bleeker, A. B. 1999. Ethylene perception and signaling: An evolutionary perspective. *Trends Plant Sci.* 4:269–274.

A review of how ethylene signal transduction is an interesting example of the evolution of information-processing systems in plants.

Buchanan, B., Gruissem, W., and Jones, R. L. 2000. *Biochemistry and molecular biology of plants.* American Society of Plant Biology, Rockville, Md.

A comprehensive textbook providing additional information about hormones as well as many aspects of plant biology.

Harmer, S. L., Hogenesch, J. B., Straume, M., Chang, H. S., Han, B., Zhu, T., Wang, X.,

Kreps, J. A., and Kay, S. A. 2000. Orchestrated transcription of key pathways in *Arabidopsis* by the circadian clock. *Science* 290:2110–2113.

A research paper describing the use of gene expression profiling to identify genes expressed in circadian rhythms. Surprisingly, many more pathways are under circadian control than previously thought.

Palme, K., and Galweiler, L. 1999. PIN-pointing the molecular basis of auxin transport. *Curr. Opin. Plant Biol.* 2:375–381.

A review of the *Arabidopsis* gene family whose members encode proteins involved in the polar transport of auxin.

Schiefelbein, J. W. 2000. Constructing a plant cell. The genetic control of root hair development. *Plant Physiol.* 124:1525–1531.

A focus on the genetic analysis of root hair development in *Arabidopsis.*

Zambryski, P., and Crawford, K. 2000. Plasmodesmata: Gatekeepers for cell-to-cell transport of developmental signals in plants. *Annu. Rev. Cell Dev. Biol.* 16:393–421.

A review of different aspects of signaling through plasmodesmata, from the spread of viral movement proteins to plant transcription factors that regulate development.

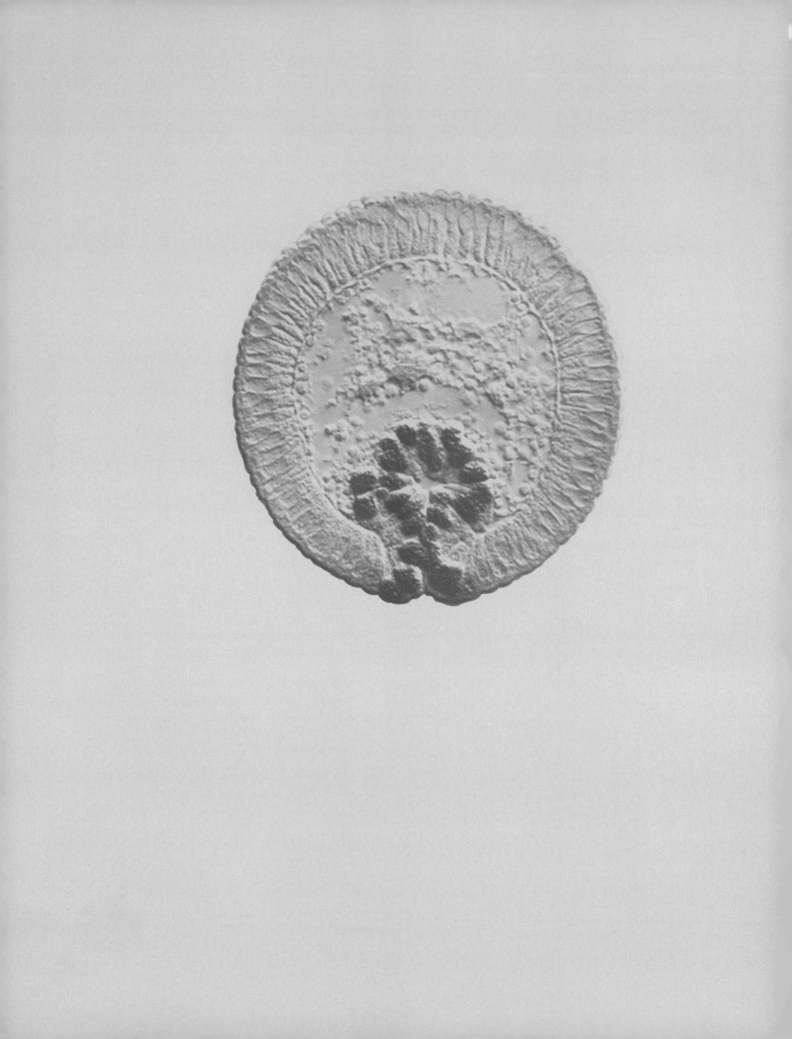

PART FIVE

MORPHOGENESIS

CELLULAR ASSOCIATIONS, ENVIRONMENTS, AND BEHAVIORS

Throughout Part One, we discussed *morphogenesis*, the dynamic changes in cell groups and tissues leading to the changes in shape and form that characterize much of development. Morphogenesis occurs because groups of cells change in one or more of the following ways: cellular shape, cell motility (or more generally, protrusive behavior), rates of proliferation, or adherence to one another (Figure 12.1). These alterations in cellular behavior occur because the interactions between cells and their environment, including other cells, is shifting. The changes may be due either to alterations in the cells themselves, to changes in their environment, or both. Thus, in order to come to grips with morphogenesis,

CHAPTER PREVIEW

1. Mesenchymal cells reside in an extracellular matrix, which is composed principally of collagen, laminin, fibronectin, and proteoglycans.

2. Plant cells are enclosed in a cell wall matrix of cellulose and amylopectins.

3. Epithelial cells are joined to one another by specific junctions.

4. Various molecules are used by epithelial cells for adhesion.

5. Mesenchymal cells employ integrins for attachment to the extracellular matrix.

6. Changes in cellular adhesion, cellular shape, and motile behavior underlie morphogenesis.

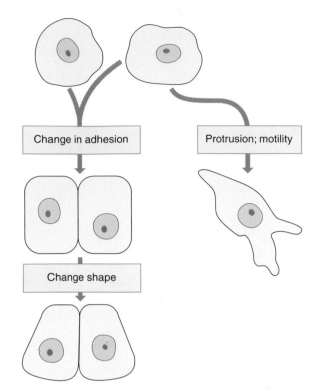

Figure 12.1   Changes in Cellular Behavior Leading to Morphogenesis The diagram illustrates several kinds of behavioral changes that contribute to embryonic morphogenesis. Two cells may adhere to one another, thereby altering their relationships to one another and to the environment. When the changes occur within the adherent cells, the result is changes in tissue shape. On the other hand, one or both of the cells may become motile, thereby moving the cells from one environment to another. Changes in proliferation rates (not shown) can also contribute to morphogenesis.

we have to know more about the biology driving cellular behavior, we have to know something about the environment around the cells, and we have to know the ways cells communicate with this changing environment.

These changing cellular behaviors and cellular interactions sound straightforward, and in a certain sense they are. What makes the scenario complicated is that the underlying mechanisms of adhesion, shape change, mitosis, and motility are themselves complex, and there are still big gaps in our knowledge. There is another reason for caution. Morphogenetic movements do not occur in isolation, nor are changes in gene expression suspended while shape changes take place. In fact, changes in gene expression are often responsible for the changing molecular landscape that drives morphogenetic behaviors. In order to dig more deeply into the principles underlying morphogenesis, we have provided a separate set of chapters concentrating specifically on changing cell

behaviors and environments (Part Five). But it is important to note that this organization is for convenience' sake—the embryo does not make such distinctions. It is not an exaggeration to say that almost every event in the developing embryo influences several other events.

MESENCHYME AND THE EXTRACELLULAR MATRIX

Mesenchymal Cells Inhabit a Complex Extracellular Matrix

The somewhat loosely arranged "filler" tissues of the embryo—the cells inhabiting the spaces between the epithelial linings of the internal and external surfaces—comprise the *mesenchyme*. Though mesenchymal cells may touch, they are not closely adherent to one another. They are also embedded in an extracellular matrix that is highly hydrated and rich in collagen, proteoglycans, glycoproteins, and molecules that are part of the signaling traffic.

Figure 12.2 shows some of the differences between mesenchymal, epithelial, and plant cells; the classification of epithelium or mesenchyme is not applied to plant cells. Recall that plant cells (discussed in Chapters 9 through 11) are surrounded by cell walls, which profoundly influence morphogenesis during development, as we saw in Chapter 10. Later in the chapter, epithelium will be examined in greater detail, while in this section we will concentrate on mesenchyme.

Collagen Is a Major Protein of the Extracellular Matrix

In animals, the most abundant protein of the extracellular matrix is **collagen**, which is actually a family of related molecules, all rich in proline and glycine. There are at least 18 different kinds of vertebrate collagen, and many tissues and organs contain several different forms. They all have regions of tandem three-amino-acid repeats of the form (glycine–X–Y)$_n$, in which Y is very often proline or a proline derivative called hydroxyproline. Table 12.1 gives the composition and tissue distribution of some of the quantitatively predominant forms of collagen.

Figure 12.3 shows some of the features of type I collagen assembly. The mature extracellular molecule is the result of considerable posttranslational modification, including hydroxylation of some of the prolines to form hydroxyproline, and

Mesenchyme

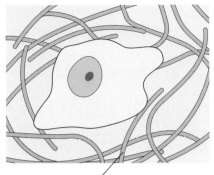

Extracellular matrix

Epithelium

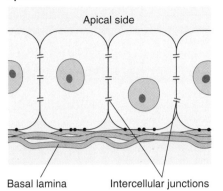

Apical side

Basal lamina   Intercellular junctions

Plant cell

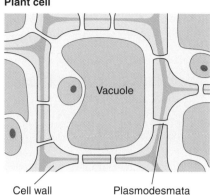

Vacuole

Cell wall   Plasmodesmata

Figure 12.2   Differences Between Mesenchyme, Epithelium, and Plant Cells Three fundamentally different kinds of cells are shown in order to illustrate their main differences. Animals can possess mesenchymal cells, which are surrounded by extracellular matrix.

They also have epithelial cells, which line surfaces. The portion of the cell surface apposed to the lumen is called the apical side, while the opposite surface is termed basal. Epithelial cells are closely adherent to each other, rest on a basal lamina, and possess specialized

junctions, which are discussed in the text. Typical plant cells contain a vacuole, are surrounded by a cell wall containing cellulose fibers, and are connected to each other by plasmodesmata.

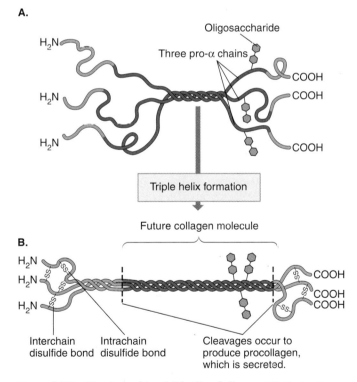

A.

Oligosaccharide

Three pro-α chains

H₂N

COOH
COOH

H₂N

H₂N

COOH

Triple helix formation

Future collagen molecule

B.

H₂N

H₂N

COOH

H₂N

COOH
COOH

Interchain disulfide bond   Intrachain disulfide bond   Cleavages occur to produce procollagen, which is secreted.

Figure 12.3   The Assembly of Fibrillar Collagen (A) By the time the primary polypeptide chains of fibrillar collagen, called pro-α chains, start to intertwine into a triple helix (shown here), they have already been modified by posttranslational attachments of oligosaccharides and hydroxylation of proline. **(B)** Before secretion, the terminal carboxyl and amino portions of procollagen are specifically trimmed. Once modified and trimmed, the procollagen, about 300 nm long, will be secreted and assume its final fibrillar form in the extracellular matrix. It will then be called collagen.

addition of some oligosaccharides. In addition, peptidases remove stretches of amino acids from both the amino and carboxyl ends, intrachain disulfide links form at both the amino and carboxyl ends of the molecule, and interchain disulfide links occur along the whole length; then the precursor chains are assembled into a triple helix. Synthesis, glycosylation, hydroxylation, helix formation, and other modifications occur while still inside the cell, but final processing to form collagen occurs only after secretion of the procollagen molecule.

It may come as no surprise to learn that the extracellular environment can regulate this final assembly of collagen. For example, the mature salivary gland is composed of branching epithelial ducts that end in secretory acini surrounded by mesenchyme. In the development of the gland, the surrounding mesenchymal cells synthesize and secrete the procollagen chains, and the epithelial cells secrete enzymes necessary for final assembly of the collagen. The three chains of the rope-like collagen triple helix are sometimes not all identical, thereby accounting for some of the diversity of fibrillar collagen types. Furthermore, there are many different collagen genes, some of which encode kinds of collagen that assemble into matlike networks instead of fibrils (Table 12.1).

Type IV collagen, for example, consists of two chains of one subtype [α₁(IV)] and one chain of another [α₂(IV)], and both of these differ from the procollagen chains of the type I collagen found in bone (Table 12.1). The basal lamina, which separates epithelial from mesenchymal tissues, is composed of type IV collagen. The basal lamina also has two other important molecules found in the matrix: laminin and fibronectin.

TABLE 12.1 THE MAJOR TYPES OF COLLAGEN

Type	Class	Chain Composition[a]	Kinds of Tissues
I	Fibrillar	$2[\alpha_1(I)] + 1[\alpha_2(I)]$	90% of total: skin, bone, cornea, ligaments
II	Fibrillar	$3[\alpha_1(II)]$	Cartilage
III	Fibrillar	$3[\alpha_1(III)]$	Skin, blood vessels, found with type I
IV	Network	$2[\alpha_1(IV)] + 1[\alpha_2(IV)]$	Basal lamina

[a] Each molecule of collagen is composed of three chains; brackets separate the composition of each chain.

Source: Adapted from T. F. Linsenmayer. 1991. Collagen. In *Cell biology of the extracellular matrix*, ed. E. D. Hay, p. 19. Plenum Press, New York.

Collagen may in some instances act as a signaling ligand as well as a structural component of the matrix. Some receptor tyrosine kinases that are activated by collagen have recently been discovered.

Laminin Is Found in the Basal Lamina

Laminin, as the name implies, is typically found in layers, or laminae, especially in the basal lamina between epithelium and mesenchyme. It is composed of three different protein chains, which are wound around each other in their carboxyl terminal domains. Different regions of the mature laminin molecule possess domains that can bind to other proteins. As shown in Figure 12.4, the B_2 chain has a domain known to interact with type IV collagen. The globular carboxyl terminal region of the A chain is known to bind to a proteoglycan called *heparin.*

While it is one thing to demonstrate that there are potential sites for specific protein–protein interactions in laminin and other molecules of the matrix, it is another to ascertain whether these associations exist and are important in developing or formed tissues. We shall soon see how that issue is being dealt with experimentally.

Fibronectin Is a Common Molecule in the Extracellular Matrix

Fibronectin, a glycoprotein heterodimer found in basal laminae, is another important protein of the matrix. Fibronectin was one of the first matrix molecules to be intensively inves-

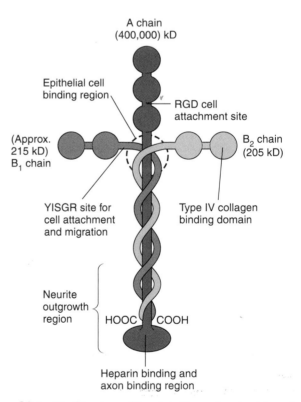

Figure 12.4 The Structure of Laminin The mature laminin molecules with its three chains: A. B_1. and B_2. Potential sites involved in various binding relationships are indicated. RGD (arg-gly-asp) and YISGR (tyr-ile-ser-gly-arg) denote the amino acid sequences (using the single-letter convention for naming amino acids) of two important binding sites.

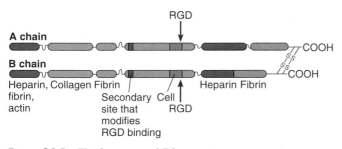

Figure 12.5   The Structure of Fibronectin The mature fibronectin molecule with its two chains, held together by disulfide bonds, each chain approximately 220 kD in molecular mass. Various sites involved in binding fibronectin to cells and matrix molecules are indicated.

tigated; its features are shown in Figure 12.5. Like laminin, fibronectin possesses specific domains capable of interacting with many other molecules of the matrix (such as heparin and collagen). The cell binding domain of fibronectin contains a signature amino acid sequence, the so-called RGD sequence (designating arginine–glycine–aspartic acid). This sequence (or sometimes the related sequence RGDS, which includes a serine at one end) is important for the interaction between fibronectin and membrane-associated molecules of the cell surface called integrins. Fibronectin can serve as a physical bridge between the extracellular matrix and integral membrane proteins, including integrins.

Proteoglycans Are Unusual Protein–Polysaccharide Molecules Inhabiting the Matrix

We have just discussed collagen, laminin, and fibronectin; together they constitute much of the mass of the extracellular matrix. All three of these proteins are **glycoproteins**; that is, they consist of proteins to which individual sugar molecules, or short polymers of sugar molecules called oligosaccharides, are attached.

Another prevalent class of macromolecules found in the extracellular matrix is the **proteoglycans**. These are assemblages of proteins and very large amounts of unusual polysaccharides called **glycosaminoglycans (GAGs)**. The GAGs are composed of repeating disaccharide subunits containing acidic sugars (usually glucuronic acid); the sugars commonly have some hydroxyl groups that are sulfated. These long GAG polymers are very hydrophilic; they bind huge amounts of water and occupy a very large volume relative to their mass. Hence, GAGs can form gels that resist mechanical pressure. The GAGs are usually attached to core proteins (though some GAGs can be present in the matrix without a covalent attachment of protein). The core proteins are attached to glycoproteins that in turn link to a central backbone consisting of an immensely long GAG such as *hyaluronic acid*. This assemblage of hyaluronic acid, linker glycoproteins, core proteins, and GAGs constitutes the proteoglycan, which has a characteristic "bottle brush" morphology. Figure 12.6 shows how the several components link together to form a typical mature proteoglycan.

Proteoglycans are found in large amounts in most connective tissues of the body, and they are highly hydrated. They also occur in the mesenchymal extracellular matrix of the developing embryo. Proteoglycans are an important part of the molecular landscape in which cells of the developing embryo are undergoing morphogenesis. Some of them, for example heparin, bind some growth factors with particularly strong avidity.

It is not easy to isolate and chemically characterize the molecules in the extracellular matrix, and the identification of macromolecules of the matrix that are quantitatively minor is far from complete. However, sparse components may in principle play an important biological role. Our knowledge of the molecular basis of morphogenesis will likely continue to expand as more of these unknown matrix molecules are identified.

A Plant Cell Wall Is an Assemblage of Cellulose and Amylopectin

A fundamental and unique difference between the cells of animals and vascular plants is the existence of a relatively rigid cell wall surrounding the plant cell. Though the precise composition of plant cell walls of different cell types may differ substantially, they are generally composed of long fibrils of cellulose embedded in a matrix of hemicellulose and pectins (Figure 12.7). Cellulose is a polysaccharide composed of a long linear chain of glucose molecules, while *hemicellulose* and *pectins* are a heterogeneous group of branched-chain polysaccharides. These three kinds of molecules are held together by both covalent and noncovalent bonds in a complex assemblage.

The cell wall immobilizes the plant cell encased within it, so that motility is out of the question. Cellular shape changes do occur, however. The cell wall is subject to pressures, including osmotic pressure, which develops because the solute concentration within the cell is greater than in the external environment. The cell wall resists the internal osmotic pressure, called turgor pressure. If the cell wall is elastic—as it is when it is relatively thin and when cellulose molecules, which are not completely cross-linked at this stage, do not completely resist sliding—the cell may expand. This situation is common in newly forming cells of the meristem.

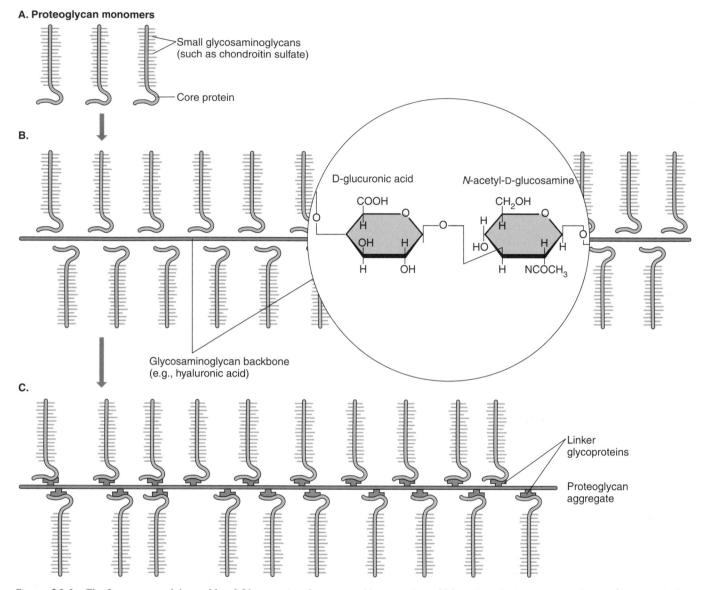

Figure 12.6   The Structure and Assembly of Glycosaminoglycans and Proteoglycans Proteoglycans are complex assemblies of glycosaminoglycans (GAGs) and proteins. **(A)** GAGs, which are polysaccharides composed of repeating disaccharide units, can attach to core proteins to create proteoglycan monomers. **(B)** These units then assemble onto a long GAG, such as the hyaluronic acid chain (composed of alternating D-glucuronic acid and N-acetyl-D-glucosamine) shown here. **(C)** After the monomers are attached with linker glycoproteins, the result is a highly charged, anionic macromolecule with a "bottle brush" structure.

The nature of the expansion depends on the alignments of the cellulose fibrils in the cell wall, and thus the cell may expand more in one direction than another. The fibrils are laid down in a direction dictated by the microtubules present within the cell that synthesized the cellulose. Thus, a newly formed cell, through the orientation of its microtubules, is predisposed to expand more in some directions than in others.

Specific Integral Membrane Molecules Bind to Both Matrix Molecules and Intracellular Proteins

Integrins are integral membrane proteins that serve to link the extracellular matrix with the internal structures and signaling apparatus of the cell. Figure 12.8 diagrams some of the principal features of the integrin family of molecules. They are heterodimers, composed of an α subunit and a β subunit.

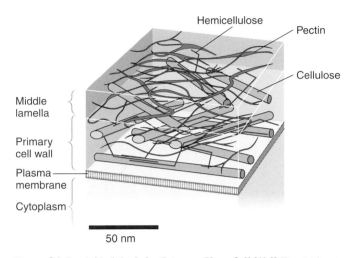

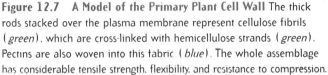

Figure 12.7 A Model of the Primary Plant Cell Wall The thick rods stacked over the plasma membrane represent cellulose fibrils (*green*), which are cross-linked with hemicellulose strands (*green*). Pectins are also woven into this fabric (*blue*). The whole assemblage has considerable tensile strength, flexibility, and resistance to compression.

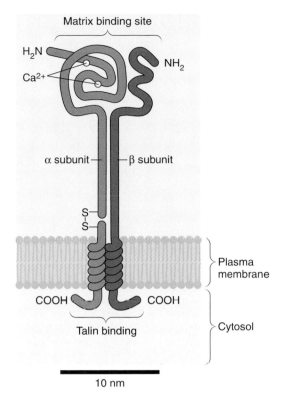

Figure 12.8 The Typical Structure of Integrins Integrin receptors are composed of two subunits. The site for binding to fibronectin or other matrix molecules is indicated. The α subunit is initially synthesized as a single polypeptide; later it is cleaved into a transmembrane portion and an extracellular portion, which are held together by a disulfide bond. The α subunit is about 140 kD; the β subunit, a typical transmembrane protein, is about 100 kD. The carboxyl end of the β subunit interacts with talin and other proteins, which link integrin to the cytoskeleton.

(Portions of integrin molecules are conventionally referred to as subunits, not chains.) Because α and β represent subunit families, each with many known members that can combine in different ways, a huge array of different binding specificities is possible. For example, integrins consisting of β_1 in combination with α_1 or α_2 will bind to collagen or laminin, α_4 or α_5 combined with β_1 will bind to fibronectin, while α_3/β_1 will bind to all three ligands. The formation of functional integrins also depends on the environment; for instance, Ca^{2+} ion is required for proper folding of the α subunit in the dimer.

The various extracellular matrix molecules have specific domains that bind to integrins. Fibronectin, as mentioned previously, possesses the RGD sequence, which is known to bind to its appropriate integrins. Laminin also possesses an RGD sequence for binding to integrin. As shown in Figure 12.8, integrins are integral membrane proteins that pass through the phospholipids of the cell membrane; the intracellular portion is a C-terminal cytoplasmic "tail" that can interact with molecules in the cytoplasm. Several integrin dimers are known to physically link to the cytoskeleton; a current model for this connection is shown in Figure 12.9. Integrins may serve as receptors to stimulate intracellular second-messenger pathways that initiate certain cellular behaviors. Recent evidence shows that the extracellular adhesion domains of integrin are functionally distinct from the cytoplasmic signaling domains. In other words, integrins are multifunctional proteins. Table 12.2 gives examples of the different integrins and their ligands.

Different combinations of integrin α and β subunits result in different specificities of binding that may overlap. Cells may bind to the same matrix molecules using several different integrins, and different kinds of matrix molecules may bind to the same integrin type. Binding may occur between different parts of the extracellular matrix molecule and an integrin. One integrin may mediate adhesion but not stimulate motility; another may do the opposite. Yet another may mediate adhesion changes and protease secretion, while another may only subserve changes in cellular adhesions. Adhesion of cells to a substrate or to the extracellular matrix is necessary for cell motility; integrins function to anchor the membrane to substrate at the leading edge of the cell as crawling takes place. Integrins are also involved in anchoring epithelial cells to the underlying basal lamina.

One powerful way to dissect the functions of the various integrin heterodimers is to selectively delete the gene for a

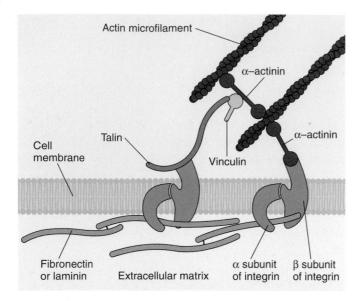

Figure 12.9   The Relationship Between Integrins and the Cytoskeleton A model of how integrins can bind to the extracellular matrix and interact with components of the intracellular cytoskeleton. Not all the cytoskeletal components, which are numerous and complex, are shown. The main intracellular linkage of the integrin dimer is to the actin microfilaments, in combination with the proteins α-actinin, talin, and vinculin.

given α or β subunit. This is currently being done using new and powerful experimental techniques for eliminating a gene or interfering with its expression. You may recall from Chapter 5 that it is possible to "knock out" a given gene in a mouse embryo by homologous recombination. Homozygous mouse embryos lacking various integrin subunits have now been produceed and analyzed. Since integrin subunits may interact with a number of heterodimeric partners and frequently have overlapping functions, analysis of the mutants of a single integrin gene may be very complex. For instance, deletion of the β_1 gene (which can pair with 10 different α subunits) is lethal early in development; the embryos do implant, however. For most types of integrin α subunits, deletion of its gene produces a unique and obvious phenotype, indicating that the different α subunits have functions that are both diverse and essential. However, there are exceptions: deletion of α_1 has little effect; perhaps it is not essential, since the specificity of α_2/β_1 is similar to that of α_1/β_1 for binding collagen and laminin. The results of specific-gene knockout experiments are clearly invaluable for analyzing the functions of adhesion molecules, receptors, and the extracellular matrix.

TABLE 12.2   SOME LIGANDS OF INTEGRIN DIMERS

Major β Subunit	Ligand of Integrin Dimer Subunits	Types of α Subunits
β_1	Collagen	1, 2, 3
	Laminin	1, 2, 3, 6
	Fibronectin	3, 4, 5, V
β_2	I-CAM	2L, 2M
	Fibrinogen	2M
β_3	Fibrinogen	V, 2b
	Fibronectin	V, 2b
β_4	Basal lamina	6

Note: These data illustrate the differing specificities of various combinations of integrin α and β subunits. It is not intended to be a complete list of such combinations and their known ligands. At least 20 different receptors, composed of combinations of eight β subunits and 16 α subunits, are currently known. Fibrinogen is a precursor of fibrin, the predominant protein of blood clots. See Table 12.3 for information on I-CAM.

Source: Data from E. Ruoslahti. 1991. Integrins as receptors for extracellular matrix. In *Cell biology of extracellular matrix,* ed. E. D. Hay, p. 346. Plenum Press, New York.

EPITHELIAL CELLS AND JUNCTIONS

Epithelial Cells Are Joined by Specific Junctions

In contrast to mesenchymal cells, whose immediate environment is the extracellular matrix, epithelial cells are closely adherent to one another. The principal types of epithelial junctions are shown in Figure 12.10. Tight junctions form a circumferential seal between adjacent cells. As a result, the environment on the cell's apical surface (that is, the lumen) is functionally isolated, not only from the basal surface resting on the basal lamina, but also from the cell's lateral surfaces. This provides a barrier to the passage of ions and other small molecules, making transporters (membrane molecules dedicated to transport across the membrane) and channels in the cell membrane necessary for trans-epithelial transport.

Adherens junctions, which are often located just basal to tight junctions, function not only to bind one epithelial cell to another but also to link the binding site to intracellular microfilaments composed of actin. These filaments sometimes form "belts" around the inner circumference of an epithelial cell.

Desmosomes, another kind of specialized adherent structure, are like "spot welds" joining one epithelial cell to another. The integral membrane proteins that constitute the desmosome connect to intermediate filaments of the cytoplasm. Both adherens junctions and desmosomes utilize a class of transmembrane "glue" molecules called cadherins, which we shall consider in more detail later in the chapter. Hemidesmosomes are similar to desmosomes except that they bind intermediate filaments to the basal lamina, not to another epithelial cell.

Gap junctions, on the other hand, are specialized juxtapositions of apposing membranes that allow small molecules (less than about 5 kD) to pass from one cell to another. For instance, it has been shown experimentally that the vital dye Lucifer Yellow (molecular weight, about 500 D) can pass from cell to cell via gap junctions. Gap junctions are formed from very specific structures called connexons, which are assembled from membrane-spanning proteins. When two connexons are aligned, they form an aqueous pore between the two cells.

The term *epithelium* is not used for plant cells. Nonetheless, as we saw in Chapter 11, plant cells can possess extensive cytoplasmic connections, called plasmodesmata, that allow passage of molecules across cell walls (see Figure 12.2). These junctions result from cytoplasmic bridges between sibling cells that are not eliminated when the new cell wall forms between them. Plasmodesmata function much the way gap junctions do in animal cells.

There are also instances in which junctions become unnecessary because whole cells are fused to produce a syncytium. We have encountered several instances of this already, notably in muscle cells.

This armamentarium of epithelial junctions gives the impression that the epithelium is really welded together. In some sense it is, but the welding is very dynamic. The various

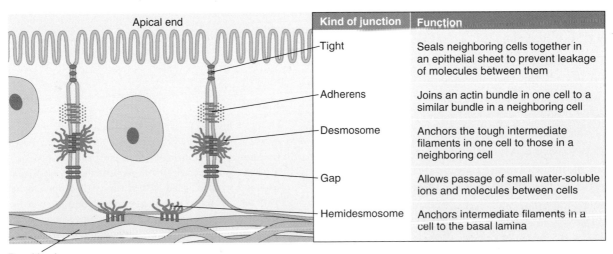

Kind of junction	Function
Tight	Seals neighboring cells together in an epithelial sheet to prevent leakage of molecules between them
Adherens	Joins an actin bundle in one cell to a similar bundle in a neighboring cell
Desmosome	Anchors the tough intermediate filaments in one cell to those in a neighboring cell
Gap	Allows passage of small water-soluble ions and molecules between cells
Hemidesmosome	Anchors intermediate filaments in a cell to the basal lamina

Apical end

Basal lamina

Figure 12.10 Junctions Between Epithelial Cells A schematic diagram of an epithelial cell bordered by two neighboring cells and resting on a basal lamina.

junction types may be broken or disassembled, and then reassembled, with amazing speed. We have already cited many instances of this: During neurulation, neural crest cells depart from the columnar epithelium of the neural plate (that is, they break their epithelial junctions) and adopt a migratory, mesenchymal habit. Subsequently, some neural crest cells, such as those in the adrenal medulla, become a tightly coupled epithelial tissue again. Also, in many creatures, especially marine invertebrates, the early blastomeres produced during embryonic cleavage are loosely held together with extracellular matrix molecules; only during later cleavage stages do typical epithelial junctions develop. These are but two in a long list of such examples.

Receptor Molecules Also Exist as Integral Membrane Proteins in Both Mesenchyme and Epithelium

From the outset, we have tried to emphasize the importance of cell communication as a principal "tool" of development. Cells communicate by means of **ligands**, signaling molecules that interact with specific cell surface proteins called **receptors**. (There are intracellular receptor proteins, too, for ligands such as steroids that can freely permeate the cell membrane.) In Chapter 1 we gave a brief introduction of cell signaling by means of ligands and receptors. We need to elaborate on that discussion in order to understand current experiments dealing with cellular interactions in the embryo.

Figure 12.11 presents the three known kinds of receptors: ion-channel-linked, G protein–linked, and enzyme-linked. All three classes exist in embryos and are essential for cell–cell and cell–matrix interactions. In each of these three classes, the interaction between the signaling ligand and the receptor affects the conformation of the protein in the receptor. This conformation change can have one of several effects. One is that it may directly affect the status of an ion channel, opening or closing a "gate" that allows ions to flow down their electrochemical gradient (Figure 12.11A).

Or the receptor may possess enzymatic domains that become active once the signaling ligand has activated it. There are a variety of ways in which this could occur. One common, important means, shown in Figure 12.11B, is by encouraging the receptor molecules to dimerize, which then allows the intracellular "tails" of the receptor to phosphorylate each other in a specific way (Figure 12.11B). When the phosphorylated amino acid of the dimer is tyrosine, the receptor is designated a *receptor tyrosine kinase (RTK)* (Figure 12.12). In a *receptor serine-threonine kinase*, the phosphorylated amino acid is either a serine or a threonine.

A huge group of RTKs is the recently discovered class of *Eph receptors*. Their ligands, called *ephrins*, are cell-bound, either as transmembrane proteins or tethered to the membrane by a glycolipid anchor. The interaction between ephrin ligands on the membrane of one cell and Eph receptors on the membrane of another can result in a positive (attractive) or negative (repelling) encounter between the cells. The ephrin–Eph system plays an important role in patterning the nervous system. We shall return to them in Chapter 13 when we discuss how efferent neurites migrate within the central nervous system and how neural crest cells migrate.

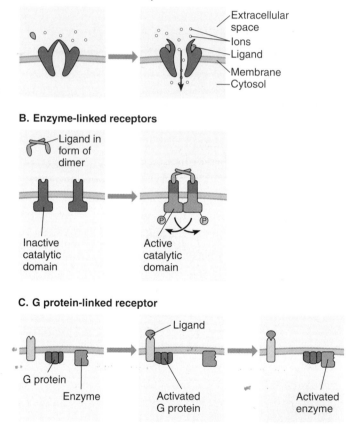

A. Ion-channel-linked receptor

Extracellular space
Ions
Ligand
Membrane
Cytosol

B. Enzyme-linked receptors

Ligand in form of dimer

Inactive catalytic domain

Active catalytic domain

C. G protein-linked receptor

Ligand

G protein

Enzyme

Activated G protein

Activated enzyme

Figure 12.11 The Principle Classes of Membrane Receptors The three main classes of membrane receptors shown schematically. **(A)** Ion-linked channels change their permeability to various ions as a result of interaction with a ligand. **(B)** When a ligand, often in the form of a dimer molecule, interacts with an enzyme-linked receptor, the receptor monomers dimerize, each monomer phosphorylating the cytoplasmic portion of the other. This dimerization in turn activates the catalytic domain of the receptor enzyme. **(C)** When a ligand interacts with a G protein-linked receptor, the G protein is activated (binds GTP), which then allows it to in turn activate another enzyme in the membrane.

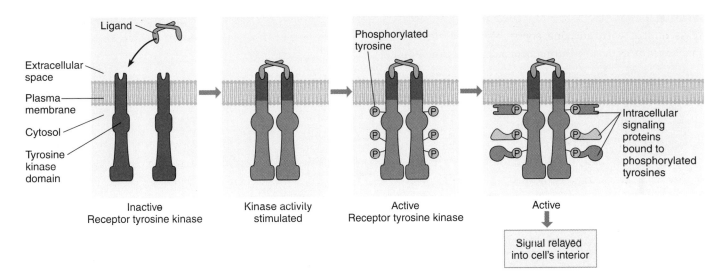

Figure 12.12   Receptor Tyrosine Kinase (RTK) An elaboration of Figure 12.11B, showing in detail the sequence of events that occur when an RTK interacts with a ligand. When the tyrosine of the cytoplasmic portion of the RTK dimer is phosphorylated, its conformation changes. This allows it to interact with several different proteins possessing domains (called SH2 or SH3) that specifically interact with phosphotyrosine.

RTKs and receptor serine-threonine kinases are coupled to intracellular signal transduction systems. As shown in Figure 12.12, a large variety of proteins can potentially participate in various signal transduction cascades. These proteins interact only with the phosphorylated cytoplasmic tails of the receptors, not with the dephosphorylated versions. Hence, phosphorylation provides the link between the extracellular signal (the ligand) and the intracellular world.

G Protein–linked Receptors Are Important in Development

A very large class of receptors activate various enzymes indirectly by means of **G proteins** (Figure 12.11C). G proteins are trimeric membrane-bound proteins that in the active state may bind GTP. When GTP is bound, the trimer can dissociate down to an α subunit with its bound GTP, which then can activate a target membrane-bound enzyme. Review the figure in Box 2.1 (in Chapter 2) for an example of G protein activation of phospholipase C.

In Chapter 2, we discussed how activated phospholipase C effects a hydrolysis of membrane-bound phospholipids, resulting in the formation of diacylglycerol and the phosphorylated hexose called inositol. The triply phosphorylated inositol derivative, inositol 1,4,5-trisphosphate (IP$_3$), is a powerful signaling molecule in itself, causing transient release of ionic calcium from intracellular stores of calcium in the endoplasmic reticulum. Recall from our discussion in Chapter 2 the importance of transient changes in intracellular calcium during the response of egg activation; that Ca^{2+} release is brought about by IP$_3$ release. Many other intracellular events are profoundly affected by calcium ion.

Diacylglycerol, the other molecule produced by phospholipase action, is also a potent activator of a membrane-bound kinase called protein kinase C (PKC). PKC is a serine-threonine kinase that can initiate the phosphorylation of other kinases, which in turn phosphorylate other intermediates. This can lead to profound changes in the regulation of a single gene or of a whole battery of different genes.

Often the target of active G protein is adenylate cyclase, an enzyme that produces cyclic AMP (cAMP) from ATP. As a second messenger in the cell, cAMP regulates a large variety of reaction pathways. Most of these involve cascades of phosphorylation and/or dephosphorylation, which can have several kinds of effects. For example, they may affect enzyme activity, as in the hydrolysis of glycogen to release glucose. Or they may affect gene transcription by altering the activity of transcription factors or their partners.

Receptors, Ligands, and Intracellular Signal Transduction Pathways Are Important in Regulating Development

The molecules and pathways we have just discussed constitute a very brief introduction to a large and important area of cell biology. There are two very important reasons why

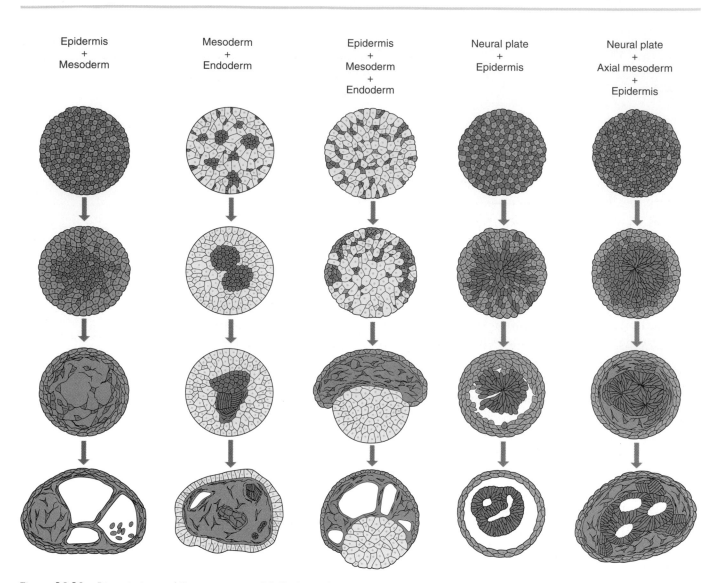

Figure 12.13 Dissociation and Reaggregation of Cells from the Amphibian Gastrula After embryonic tissues were surgically isolated and dissociated to single cells, various types were mixed, and Ca²⁺ was restored to the medium so that the cells could adhere. As described in the text, the individual cells sorted out and reaggregated, adopting locations within the mass that were typical of differentiating tissues.

the student of development needs to understand these molecular mechanisms and know some of the terminology. First, using these mechanisms is how cells and tissues "talk" to each other. This is true not only in mature adult animals responding to their environment, but also in developing embryos, where the conversations between developing tissues dictate which developmental pathways are chosen and how the choices are executed. A great majority of the experiments now being reported in the scientific literature of developmental biology involve identifying different specific members of these communication pathways, and learning how such molecules and pathways regulate particular developmental events.

Second, the more we learn about communication between cells and the intracellular pathways involved, the clearer it becomes that the regulatory pathways within a cell are complex and intersecting rather than linear series of on/off switches. Though our picture is still incomplete, the complexity of how cells communicate and how they regulate intracellular activities is already astonishing. Learning how pathways are functionally interconnected constitutes a real challenge for current research.

CELL ADHESION

Cell–Cell and Cell–Matrix Adhesion Play a Role in Morphogenesis

Now that we have discussed the extracellular matrix, cell junctions, and receptors, we are ready to move on to the important topic of cellular adhesion and its specificity. Even though cell junctions and integrins are vital for adhesion, they are not the whole story. When adhering to one another, cells show specific preferences, a behavioral characteristic that is very important in morphogenesis. The great embryologist Johannes Holtfreter carried out experiments in Germany in the 1930s (and later at the University of Rochester in 1955 with his student, Phillip Townes) clearly showing that cell–cell adhesion is specific for a given cell type.

Holtfreter's approach was essentially to take embryonic tissues apart and let them come back together again. First he dissected out different germ layers of an amphibian gastrula and exposed them to saline solutions lacking Ca^{2+}. Then he subjected the solutions to mild physical shear by gently sucking the tissue up and down through a small-bore pipet. This caused the tissues to dissociate into a suspension of single cells. When the dissociated cells were cultured in a medium containing calcium ion, they reaggregated. The resulting tissues were able to develop some of the structures characteristic of their germ layer of origin.

The surprise came when dissociated cells from two different germ layers were mixed together, again in calcium-containing medium (Figure 12.13). At first the cells associated with one another, but after several hours they apparently "sorted out"— ectoderm cells gradually grouping and adhering preferentially to other ectoderm cells, mesoderm to other mesoderm, and so on. Numerous investigators went on to definitively show that, indeed, the dissociated cell types really do sort out and cells really do have mechanisms for sensing other cells and expressing preferences for adhesion. Furthermore, the reconstituted tissues formed in sorting-out experiments always show specific spatial arrangements with respect to one another. For instance, when prospective epidermis and mesoderm are dissociated and mixed, the two cell types sort out so that the epidermis is always on the surface of the reaggregated mass and the mesoderm internal to it. In addition to germ-layer types, literally hundreds of combinations of different tissue types have been analyzed for their inside/outside preferences during sorting out. Malcolm Steinberg of Princeton and his colleagues

have amassed considerable evidence that the inside/outside order is probably due to differing relative strengths of adhesion. More strongly adherent tissue types tend to occupy more central positions (Figure 12.14).

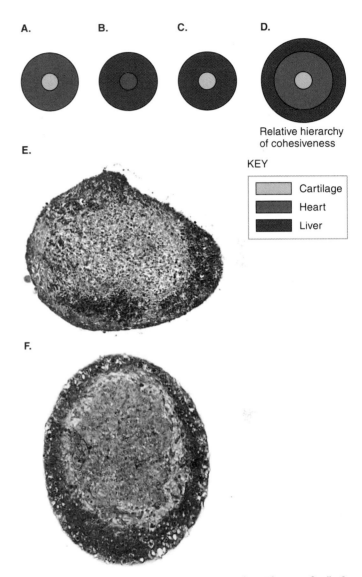

Figure 12.14 The Role of Adhesive Strength in "Sorting Out" of Cells (A-D) Schematic diagrams of sections through tissue masses, showing the arrangement adopted by dissociated cells of cartilage, heart, and liver after sorting out: **(A)** When only cartilage and heart cells are mixed, the cartilage aggregates become central. **(B)** When liver and heart are mixed, heart cells become central. **(C)** And when liver and cartilage are mixed, cartilage cells become central. **(D)** As one would predict from the two-tissue experiments, when all three kinds of tissue are mixed, the cells again sort themselves, with central cartilage surrounded by heart, and both covered by liver cells. This kind of experiment can be used to measure the relative adhesive strength of different cell types, with most strongly adhering at the center, grading to least on the surface. **(E,F)** Photographs of actual mixed-tissue aggregates: **(E)** limb-bud cartilage cells surrounded by pigmented retinal cells and **(F)** central heart cells surrounded by liver cells.

Several Classes of Specific Cell Adhesion Molecules Exist

During the last decade, there has been an explosion of information about the molecules that appear to act as intercellular "glues." These molecules provide for specificity in adhesion in addition to the general, nonspecific epithelial junctions already discussed. Two classes of molecules involved in this selective adhesion are diagrammed in Figure 12.15.

Cell adhesion molecules (CAMs) are a diverse set of proteins, most of which are anchored to the cell membrane, usually by a transmembrane domain or sometimes by a specialized glycolipid anchor (Figure 12.15A). The extracellular portion of the molecule has five looped domains; because each loop bears some similarity to peptide sequences in immunoglobulins, CAMs are considered "immunoglobulin-like" or "Ig-like." Each loop is stabilized by disulfide bonds. Binding between CAM molecules on different cells is usually *homophilic*, that is, occurring between the same kind of molecules. The binding is often independent of the presence of calcium in the medium, but there is great variability. Some kinds of CAMs require

Ca^{2+}, and some CAM interactions are *heterophilic*; that is, the two cells have different kinds of CAMs (see Table 12.3). The extracellular portion of CAMs can be heavily modified by esterification with sialic acid, which is very negatively charged. A current hypothesis proposes that the extent of sialic acid modification alters the strength of homophilic binding between the Ig domains on interacting cells. In this view, CAMs heavily esterified with sialic acid could hinder cell adhesion.

Another large family of intercellular adhesion molecules is the **cadherins**, short for "calcium-dependent cell adhesion molecules" (Figure 12.15B); these molecules always require calcium ions for their function, and their binding is homophilic. Different cadherins are characteristically found in different tissue types. For example, E-cadherins are found in many different epithelial tissues, P-cadherins in the placenta, and N-cadherins in the nervous system; however, the same cadherin type may occur in several different tissues. The cadherins are involved in selective cell adhesions independent of cell junctions. You will remember, however, that cadherins are also an essential part of adherens and desmosome junctions of epithelial cells. On the cytoplasmic side of these junctions,

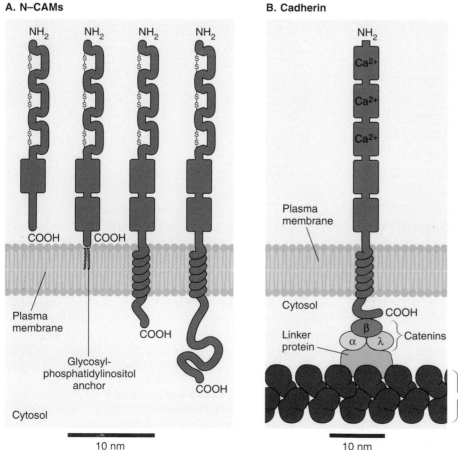

A. N–CAMs

B. Cadherin

Figure 12.15   Cell Adhesion Molecules Schematic diagrams showing the basic structure of two types of cell adhesion molecules. **(A)** All CAMs, not just the four forms of N-CAM shown here, possess five extracellular Ig-like domains held in loops by disulfide bonds. Only three of the five loops are shown. The extracellular portion may be free, linked to the cell membrane by a glycolipid anchor, or pass through the membrane. **(B)** A typical cadherin molecule has a hydrophobic domain that passes through the cell membrane and is anchored on the cytoplasmic side by catenin proteins. The extracellular portion has several Ca^{2+} binding domains. Different types of cadherin proteins (E-cadherin, N-cadherin, and so on) possess somewhat different amino acid sequences.

TABLE 12.3 TYPES OF CELL ADHESION MOLECULES

Class of Molecule (Synonyms)	Binding Mechanism	Ion Dependence	Examples
N-CAM	Homophilic	No	Neural plate
Ng-CAM	Heterophilic	No	Nervous system
I-CAM	Heterophilic	No	Endothelial cells
L-CAM (E-cadherin, uvomorulin)	Homophilic	Ca^{2+}	Blastomeres
A-CAM	Homophilic	Ca^{2+}	Mesoderm, lens, muscle
P-cadherin	Homophilic	Ca^{2+}	Endoderm, placenta
N-cadherin	Homophilic	Ca^{2+}	Central nervous system
EP-cadherin (C-cadherin)	Homophilic	Ca^{2+}	Cleavage stage blastomeres
Integrins	Heterophilic	Varies	Extracellular matrix

Note: This is not an exhaustive list of cell adhesion molecules, but is a summary of some of the principal known types.

Source: Adapted from B. Alberts, D. Bray, J. Lewis, M. Raff, K. Roberts, and J. D. Watson. 1994 *Molecular biology of the cell,* 3rd ed. Garland Publishing, New York, p. 1000. And from L. W. Browder, C. A. Erickson, and W. R. Jeffery. 1991. *Developmental biology,* 3rd ed. Saunders College Publishing, Orlando, Fl., p. 377.

cadherins interact with linker proteins called *catenins,* and thence either to actin or intermediate filaments. So cadherins may function in cell adhesion both as components of junctions and as nonjunctional intercellular "glue."

Table 12.3 lists the cell adhesion molecules we have just discussed. Other types of molecules have also been implicated in adhesion. One diverse class is cell surface–associated polysaccharides that can mediate adhesive functions. Galactose transferases—enzymes involved in the transfer of the sugar galactose—may play a role in adhesion in some systems. There are also ligand–receptor interactions (recall the ephrin–Eph receptor system described earlier in the chapter) that participate in cellular recognition and dynamic adhesions, especially in migratory cells or in the directed outgrowth of neurites. It is probably safe to predict that many more members of each of the different classes of adhesion molecules will be identified in the near future; completely new families of cell adhesion molecules may also remain to be discovered.

The Function of Cell Adhesion Molecules May Be Analyzed in Their Cellular Context

Until recently, it was difficult to definitively implicate cell adhesion molecules and their partners, such as integrins and RTKs, in specific developmental events. All of that has changed with the advent of modern cell and molecular biology. In particular, three powerful experimental tactics have been used: transforming cells in culture with cloned DNA; knocking out gene function in mice; and using reagents, such as antibodies and antisense DNA, that specifically interfere with the function or formation of a given protein. We shall encounter many instances of results obtained with each of these strategies. It is useful at this juncture to illustrate them with a few examples.

Figure 12.16 shows an experiment, carried out by Masatoshi Takeichi and his colleagues in Kyoto, designed to probe the function of cadherins in sorting out. They took L cells, an established tissue culture line of mouse cells that normally do not adhere to one another in tissue culture, and transformed them with cDNA encoding either P-cadherin or E-cadherin. When L cells expressing P-cadherin were cultured, they now aggregated in culture, as did cells transformed with E-cadherin. When E- and P-cadherin-expressing cells were mixed, they adhered preferentially to cells expressing the same type of cadherin, and thus underwent sorting out. In this instance, it is apparent that the homophilic bindings of different cadherin types determined the specificity of adhesion and consequently the sorting out.

The relative amounts of a single cadherin type may also provide a basis for specific adhesions; Steinberg and Takeichi showed that cell lines expressing high amounts of P-cadherin sorted out from a variant of this same cell line that expressed less of the same P-cadherin. Furthermore, as Steinberg originally suggested, the cells expressing more P-cadherin, and therefore presumably adhering more tightly, occupied a position interior to that of their neighbors, which expressed less P-cadherin.

A form of cadherin termed EP-cadherin is expressed by blastomeres of *Xenopus* during cleavage. An antisense deoxynucleotide that hybridizes to the mRNA for EP-cadherin can be injected into the *Xenopus* zygote, which effectively reduces or eliminates synthesis of normal EP-cadherin. The antisense oligonucleotide acts like a loss-of-function mutation. When antisense EP-cadherin DNA is injected into the embryo, the adhesion of the epithelial blastula cells to one another is disrupted. The embryo literally falls apart. EP-cadherin is apparently necessary in *Xenopus* embryos for adhesion of cleavage blastomeres to one other.

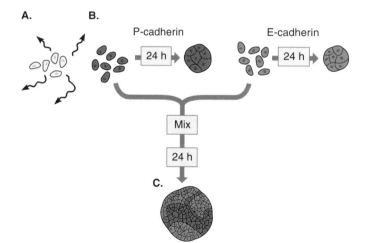

Figure 12.16 Cadherins and Selective Cell Adhesion An experiment carried out by Takeichi and his colleagues, shown in diagrammatic form. **(A)** Mouse L cells are a tissue culture line that does not aggregate and does not possess cadherins. **(B)** When L cells were transformed with plasmid DNA that encodes cadherin of a specific type, such as P-cadherin (*red*) or E-cadherin (*blue*), as shown here, the cells became competent to aggregate with cells of like type. **(C)** When the two types of transformed L cells were mixed with one another and allowed to reaggregate, they sorted out according to cadherin type. The two cell types were distinguished by staining the mass with antibodies against one cadherin type but not the other.

MORPHOGENETIC MANEUVERS

There Are Eight Basic Morphogenetic Movements

Having reviewed the basic characteristics of mesenchyme and epithelium and the various means—general and specific—that cells employ to adhere to one another, we can now grapple with "the real thing": how parts of the embryo change form. For this it is helpful to identify the different kinds of morphogenetic "maneuvers" occurring in the developing organism—but with this caveat: Classifying changes is somewhat misleading because it gives the impression that neat, tidy examples of each kind should be evident during development. That is rarely the case. Nevertheless, having acquired some acquaintance with the complexity of morphogenesis in particular embryos, it is useful to look at morphogenesis abstractly because it helps us to analyze the underlying changes in cellular behavior that drive the morphogenesis.

Figure 12.17 presents the elemental morphogenetic movements diagrammatically: epiboly, intercalation, convergent extension, invagination, involution, migration, ingression, and proliferation. We discussed examples of all of these in earlier chapters, when considering the development of different kinds of embryos and tissues. The purpose of this figure is not only to underline the cell behaviors that drive all of morphogenesis, but also to help you compare and distinguish them.

Cell Motility and Protrusive Activity Involve Attachment to a Substrate

Interaction between the cell surface and surrounding surfaces—whether membranes of other cells, components of the extracellular matrix, or even the plastic of tissue culture dishes—is essential for motility. A cell's motility or protrusiveness in any given situation probably depends as much on contacts with its substrate as it does on the cell's inherent machinery for motility (Figure 12.18A). For instance, when an invaginating surface cell of the lip of the blastopore in the amphibian embryo becomes internalized (a movement that depends on changes in both cellular shape and motility), it now becomes an actively motile cell moving on its new substrate, the internal "ceiling" of the animal cap. Douglas DeSimone and his colleagues at the University of Virginia have shown that this change in migratory behavior requires an interaction between the fibronectin on the roof of the archenteron and the

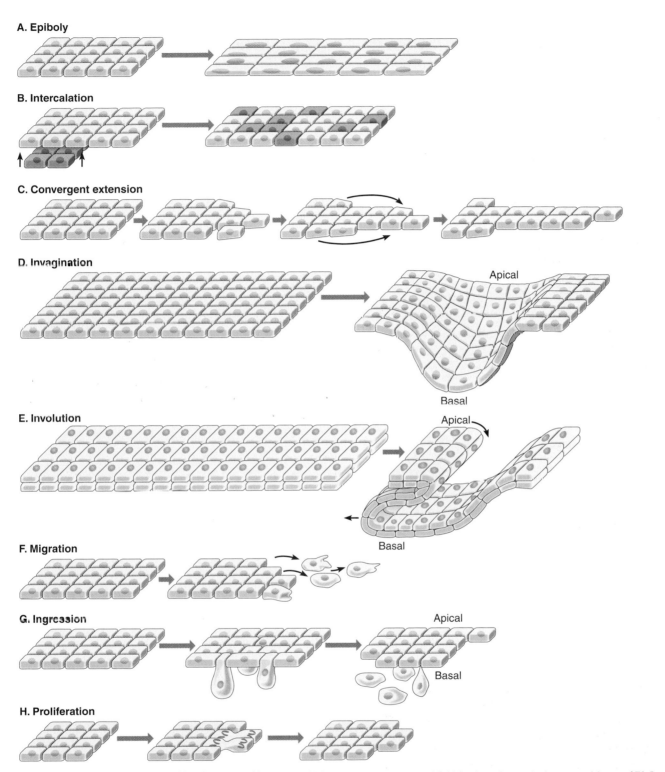

A. Epiboly

B. Intercalation

C. Convergent extension

D. Invagination

Apical

Basal

E. Involution

Apical

Basal

F. Migration

G. Ingression

Apical

Basal

H. Proliferation

Figure 12.17 The Eight Basic Morphogenetic Maneuvers (A) Epiboly occurs when epithelial cells flatten and spread. **(B)** In multilayered tissues. intercalation occurs when nearby cells intermix and spread out. thereby producing a thinner sheet. **(C)** In convergent extension. there is extensive and concerted intercalation of neighboring cells. which can cause elongation of a tissue mass. **(D)** Epithelial cells can invaginate by buckling inward. **(E)** Involution results when epithelial layers invaginate and fold back underneath the original layers. **(F)** In migration. stationary cells begin to move away from the edges of a coherent tissue mass. **(G)** When cells in an epithelial layer detach. move through the adjacent basal lamina. and migrate into the extracellular matrix located basal to the epithelium. the movement is called ingression. **(H)** Proliferation occurs when a small population of cells multiplies at a rate much greater than neighboring cells.

α_4/β_1-integrin on the spreading and migrating leading-edge cells of the involuted marginal zone.

Cell motility is an enormously complex subject, and there is much we do not know. But as each bit of progress in our understanding of motility is revealed, developmental biologists have been quick to apply the new findings in experiments probing how morphogenesis takes place. Thus, in experiments in which the extracellular matrix is altered, or in which integrins and cytoplasmic actins and their associated proteins are altered, there are almost always derangements of morphogenesis. Convergent extension, ingression, involution, and migration are each morphogenetic movements that utilize programmed changes in motility and protrusive activity.

Changes in Cellular Shape Are Crucial in Determining Form

We have already discussed how changes in the cytoskeleton can lead to alterations in cellular shape. During epiboly, cells in a cuboidal epithelium stretch out so that the tissue thins out; such a thinned-out epithelium is sometimes called *squamous* epithelium. When epithelial sheets roll into tubes, as during neurulation, for example, cuboidal or columnar cells become trapezoidal. Experimental interference with the formation of the actin belts encircling the inside apical portion of the cells interrupts their shift to a trapezoidal shape, and consequently stops invagination and involution as well.

A well-studied example of invagination of a sheet to form a tube is the formation of mesoderm in *Drosophila* (Figure 12.18B). Recall from Chapter 3 that a ventral strip of cells in *Drosophila*'s cellular blastoderm undergoes shape changes similar to those of neurulating ectoderm cells in vertebrates, with the result that an internal tube forms. In *Drosophila*, the tube then undergoes a transformation in which epithelial cells become the mesenchymal mesoderm of the embryo. Interestingly, expression of the genes encoding the transcription factors Snail and Twist is necessary and sufficient for the invagination that will form mesoderm in *Drosophila*; how these two proteins act on "downstream" targets to effect the consequent cellular shape changes driving invagination is not yet known, however.

We should recall (from Chapters 9 through 11) that the shape changes plant cells undergo play an important role in their morphogenesis. Since their cell walls prevent them from moving, relative differences in the enlargement of cells along the various axes of a tissue are critical to organogenesis. The rate and plane of cell division may be an important component of morphogenesis.

Rates of Cell Proliferation Influence Tissue Shape

Tissue morphology is also affected by rates of cell proliferation, and by the exact placement of the cleavage furrow in animals cells or the cell division plate in plant cells. Overall cell proliferation is driven not simply by mitosis, though that is obviously important, but also by rates of cell death (apoptosis) and the survival time of newborn cells. This situation is comparable to the problem, faced by demographers, of understanding an unstable population size. Both birth rate and death rate affect population size. In fact, many of the so-called growth factors are really survival factors. Nerve growth factor (NGF), for example, acts primarily to help cells survive (in other words, to postpone apoptosis), thereby increasing the population of NGF-responsive cells by reducing their death rate.

Wherever morphogenesis seems to produce buds or bulges, it is often (though not always) the result of an increase in proliferation relative to mitotic rates in neighboring cells. Thus, the branching epithelial ducts found in salivary glands, lungs, and pancreas all show the highest mitotic rates in the ramifying tips of the ducts. The outgrowth of the limb buds of vertebrates is a clear example of increased mitotic activity. In this instance, the specialized apical ectodermal ridge (AER) secretes growth factors that stimulate proliferation of the underlying mesodermal tissues; the presence of the AER also decreases apoptosis in the underlying mesoderm (Figure 12.18C).

It is important to remember that everywhere in the embryo, both the rate of cell division and the survival time of a given cell are tightly regulated. Sometimes external signals dominate, as when the AER signals the mesodermal core of the developing limb bud to maintain high mitotic rates. In other cases, such as the domains of mitotic activity found in the postgastrula embryo of *Drosophila* (as we shall see in Chapter 13), gene-driven regulation of cell mitotic rates usually dominates.

Cell Division Planes Affect Morphogenesis

The plane of cell division is extraordinarily important in some situations, especially in the early animal and plant embryo. This is obvious when one considers that maternal determinants in an early embryo are localized: the segregation of each determinant to one cell or another is obviously a function of where the cell division plane lies with respect to that localized substance. Some examples illustrate this. The third cell division in *Xenopus* and other amphibians separates yolk-laden vegetal cells from animal cells, and it is the vegetal cells that later influence descendants of animal cells to adopt a mesodermal fate (see Chapter 16). The first division of a plant embryo separating the future suspensor from the embryo proper depends on the proper position of the developing cell division plate. The embryos of many marine invertebrates, whose normal development we have not discussed, show highly stereotyped divisions in early development, and these divisions are responsible for segregation of known or suspected determinants to one cell or another.

A striking example of the importance of a fixed cell division plane occurs during the very first cleavage division in *Caenorhabditis elegans*, the nematode that displays the regular lineage discussed in Chapter 2. Figure 12.18D shows that the first cell division in this creature produces two cells, a posterior one containing unusual granules (called P granules) and an anterior cell lacking these granules. Only cells with P granules can form primordial germ cells, so if the cell division plane is altered, the development of primordial germ cells will be very abnormal.

Two other aspects concerning the role of cell division planes need to be underlined. First, it is the position of the centrosome that determines the division plane, and there is evidence from several different sources that this positioning may be tightly controlled by different genes. For instance, mutations in *Drosophila* affect centrosome positioning after division 14, when the syncytial blastoderm finally becomes cellularized. And in some species of snails, a single gene controls the position of the centrosomes and thus the direction in which spiral cleavage occurs. The direction of the shell's spiral (right- or left-handed) is apparently controlled by the same gene.

Second and contrariwise, sometimes the plane of cell division is of no consequence. For example, dissociated and reaggregated cells of early developing vertebrate organs often go on to produce perfectly respectable and well-organized tissues, even though the positions of their mitotic spindles have been randomized by the dissociation procedure of the experiment.

A.

B.

C.

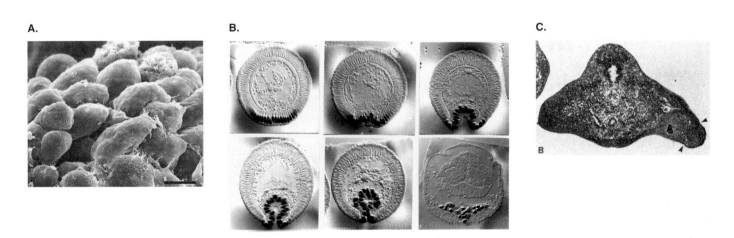

D.

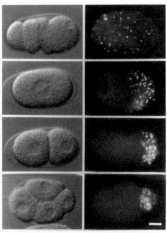

Figure 12.18   Some Examples of Morphogenesis (A) Cell migration: After involution through the blastopore, epithelial cells at the leading edge detach themselves and become actively motile. **(B)** Invagination: The ventral epithelium of this *Drosophila* gastrula shows the inward buckling typical of invagination. The mesodermal cells have been stained with antibodies against Twist, a mesoderm-specific transcription factor. **(C)** Proliferation: In this experiment, the trunk of a mouse embryo possessing small limb buds was excised from the embryo; the surface epithelium and apical ectodermal ridge were removed from the hind limb shown on the right, leaving the limb on the left side undisturbed. The trunk was then placed for 48 hours in tissue culture medium containing the fibroblast growth factor FGF4. Note the greater outgrowth of the right side compared with the left. Apparently, FGF4 penetrated the mesoderm and stimulated proliferation in the right limb bud but could not penetrate the surface epithelium of the left (control) side. FGF4 increased rates of mitosis in the mesenchyme of the denuded limb more than twofold. Other experiments have shown that FGFs can prevent apoptosis in underlying mesoderm, thereby increasing outgrowth even more. **(D)** Segregation of P granules in *C. elegans*. The left set of panels shows *(top to bottom)* the unfertilized, fertilized, two-cell, and four-cell stages. The corresponding right panels show the P granules, stained with an antibody; note that all the P granules segregate to a single P blastomere, the blastomere that will give rise to cells in the posterior portion of the animal (see Figure 2.13).

Cell Adhesion Is Crucial in Morphogenesis

In our discussion about the components of the extracellular matrix, integrins, and cell adhesion molecules, we have already mentioned several examples where compromising the function of any one of them disrupts normal morphogenesis. It certainly follows logically that, for any cellular behavior in which adhesion plays a role (as it does in all eight of the instances shown in Figure 12.17), compromise of any component of the adhesion system, or of any dynamic changes the system or the molecules may undergo during morphogenesis, is going to have an effect. Thus, knocking out integrin functions, perturbing the chemistry of the extracellular matrix, or compromising functions of transduction links between integrin and the cytoskeleton would all be expected to have profound effects. The purpose of knockout experiments is to find out precisely what molecule or molecules are playing what role in a given situation in development.

Table 12.4 presents some of the consequences of knocking out mice genes that encode cell adhesion and extracellular matrix molecules. Since there is some overlap of certain functions, not all knockout experiments are lethal early in development, and they sometimes display late-appearing phenotypes. On the other hand, phenotypes of certain other knockout experiments, such as those of E-cadherin or β-catenin, are lethal early in development.

A striking result using antisense oligonucleotides is shown in Figure 12.19. The function of a maternal store of α-catenin mRNA was abrogated by using antisense technology; this led to a depletion of the α-catenin protein in *Xenopus* embryos. The *Xenopus* blastulae that developed after experiencing this depletion showed very poor intercellular adhesion, due to interference with the development of epithelial junctions. This result is very similar to that obtained after interfering with EP-cadherin function, another component of these junctions.

What Is Cause and What Is Effect in Morphogenesis?

There is a perplexing problem in the analysis of morphogenesis. To illustrate it, let us return to invagination of the neural plate to form a neural tube. Application of the drug cytochalasin to an explanted neural plate in organ culture will disrupt the actin microfilaments and perturb neural tube formation. It is

TABLE 12.4 CELL-CELL ADHESION AND MATRIX GENE KNOCKOUTS IN MICE

Molecule	Viability[a]	Phenotype
N-cadherin	E10–11	Dissociated myocardium
		Somite and neural tube poorly organized
β-catenin	E8–9	No mesoderm
N-CAM	Viable	Defects in spatial learning
Fibronectin	E9–10	No somites
		Vascular defects
β$_2$-laminin	Die 2 weeks postnatal	Defects in neuromuscular junctions
Collagen	E13	Circulatory failure
α$_4$-integrin	E11–14	Placental and heart defects

[a] The "E" notation indicates the day (or day range) of gestation at which the embryos died.

Note: This list is only a small sample of the large number of genes encoding adhesion, extracellular matrix, and integrin molecules that have been examined using gene knockout procedures. A more complete list can be found in Richard O. Hynes. 1996. Targeted mutations in cell adhesion genes: What have we learned from them? *Dev. Biol.* 180:402–412.

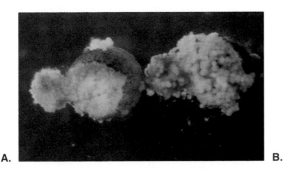

A.　　　　　　　　　　　　　　　　　　B.

Figure 12.19 Depletion of Maternal α-Catenin by Antisense Technology *Xenopus* embryos were surgically opened to estimate the extent of cell adhesion. **(A)** This normal embryo shows adherent cells, whereas **(B)** an embryo that has had its maternal store of α-catenin depleted by antisense technology shows almost no intercellular adhesion.

tempting to conclude that the microfilaments are responsible in this instance for morphogenesis of the neural tube. But disruption of microfilaments is likely to disrupt much of internal cell structure. Do we really know that actin "purse strings" are causing the change in cellular shape? For the purse strings to cause the transformation in shape from columnar to trapezoidal, the lateral walls of the neural plate cells probably must resist excessive plastic deformation and/or be tightly adherent to their neighbors. The basal surface should also be elastic, or a wine-bottle shape might occur. Is the apical microfilament bundle responding to some other underlying change in the cytoskeleton, one that cannot occur when internal cell structure is compromised? Is the shape change observed in neural plate cells what is driving the morphogenesis, or is it a consequence of some other behavioral change? So, even when an experiment indicates that some molecule is involved in morphogenesis, it is difficult to conclude that that particular molecule is really the cause of the observed changes.

To pursue this example even further, even if the microfilaments do drive neurulation, at least in part, do we really understand what is happening? Many related questions arise. Why do the plate cells change shape at that particular time and place? What is regulating the changes in shape, which do not occur uniformly throughout the neural plate? What signals induce the behavioral changes that produce the shape of neural plate cells? Where do these signals come from, and what regulates their timing? Often the results of a simple experiment open up a host of questions that were not anticipated.

Experiments Using Inhibitors of Cell Function Are Useful but Difficult to Interpret

Some of these questions come up in connection with the neural tube example because of the use of an inhibitor in the experiments. It is conventional wisdom, though often forgotten, that no inhibitor is absolutely specific, and that untoward side effects may confound the interpretation of an experiment. In our opinion, any experiment in which inhibitors or drugs are used and the biological event under scrutiny does *not* happen cannot be rigorously interpreted. When something does not happen, you cannot be sure.

However, this does not mean that inhibitor procedures are not worth doing. First of all, such procedures may help point the way, even if they are not definitive. Second, if the inhibitor does, in fact, inhibit what it is supposed to (regardless of what else it might affect) and the biological event in question *does* occur, then you can be pretty sure that the two events are not connected. So, when a drug—colchicine for example—that inhibits microtubule function is applied to a nerve growth cone, and the cone maintains its activity, it is reasonable to suppose that the microtubules are not important in forming the fine filopodia that are characteristic of nerve growth cones as the neurons explore their environment. We shall learn more about neuron outgrowth by means of growth cones in the next chapter.

In this chapter, we have looked briefly at many of the molecules and cell behaviors known to play key roles in morphogenesis. In the next chapter, we shall use this information to revisit some instances of morphogenesis described earlier, in Parts Two and Three. We shall examine in more detail how these molecules are involved in molding shape in the developing embryo.

KEY CONCEPTS

1. Morphogenesis is the suite of cellular behaviors that provide changes in form during development. The cellular behavior itself is ultimately regulated by gene expression and is sensitive to signals in the surrounding milieu.

2. All morphogenesis is driven by changes in cellular adhesion (to another cell and/or to the extracellular matrix); changes in the rate of cell proliferation or in planes of cell division; changes in migratory or protrusive activity; and changes in

cellular shape. Hence, an understanding of the biology of these behaviors is essential to the study of development.

3. Cells in a mesenchyme differ from cells in an epithelium in two fundamental ways: (a) epithelial cells adhere tightly to one another, but mesenchymal cells do not; and (b) mesenchymal cells are surrounded by an extensive extracellular matrix, but epithelial cells are not.

4. The extracellular matrix is not just a passive space filler. The matrix is hydrophilic and may serve as a reservoir for signaling ligands; some molecules of the matrix can themselves act as ligands or cellular attachment points.

5. Integral membrane molecules link the cell to matrix molecules and can transduce signals.

6. Cellular adhesion to the matrix and other cells is subserved by several different classes of molecules and is specific for cell type. Differing strengths of intercellular adhesion help provide specificity.

7. Results of experiments that prevent some developmental event through the use of an inhibitor are hard to interpret.

STUDY QUESTIONS

1. How could you determine whether cell proliferation contributed to invagination of the mesoderm during gastrulation in *Drosophila*?

2. Based on material from this and other chapters, could you devise an experiment to determine whether sorting out occurs when dissociated cells of the developing liver are mixed with dissociated cells of the developing heart?

3. What would be the probable consequences of knocking out the gene in mice for collagen type IV?

SELECTED REFERENCES

Alberts, B., Bray, D., Lewis, J., Raff, M., Roberts, K., and Watson, J. D. 1994. *Molecular biology of the cell.* Garland Publishing, New York.

A popular advanced textbook of cell biology. Chapter 19 covers many of the topics of this chapter in considerable detail.

Gumbiner, B. M. 1996. Cell adhesion: The molecular basis of tissue architecture and morphogenesis. *Cell* 84:345–357.

A review of adhesion machinery from a developmental point of view.

Kintner, C. 1992. Regulation of embryonic cell adhesion by the cadherin cytoplasmic domain. *Cell* 69:225–236.

A research paper that illustrates the power of molecular approaches when applied to analysis of developmental function of an adhesion molecule.

Kofron, M., Spagnuolo, A., Klymkowsky, M., Wylie, C., and Heasman, J. 1997. The roles of maternal α-catenin and plakoglobin in the early *Xenopus* embryo. *Development* 124:1553–1560.

Demonstration of the use of antisense DNA oligonucleotides to study gene function.

Radice, G. L., Rayburn, H. K., Matsunami, H. K., Knudsen, K. A., Takeichi, M., and Hynes, R. O. 1997. Developmental defects in mouse embryos lacking N-cadherin. *Dev. Biol.* 181: 64–78.

A research paper in which generating gene mutants by homologous recombination affords a way of looking at function in development.

Ramos, J. W., Whittaker, C. A., and DeSimone, Douglas W. 1996. Integrin-dependent adhesive activity is spatially controlled by induc-

tive signals at gastrulation. *Development* 122: 2873–2883.

A research paper that explores the molecules necessary for change from epithelial to migratory behavior during gastrulation.

Wolfsberg, T. G., and White, J. M. 1996. ADAMs in fertilization and development. *Dev. Biol.* 180:389–401.

A review of the ADAM class of multidomain proteins, which may function in adhesion, proteolysis, fusion, and signaling.

Yamada, K. M., and Geiger, B. 1998. Molecular interactions in cell adhesion complexes. *Curr. Opin. Cell Biol.* 9:76–85.

A detailed review of how molecular complexes function both in adhesion and signaling.

CHAPTER

13

TISSUE INTERACTIONS AND MORPHOGENESIS

How do the morphogenetic changes brought about by motility, adhesion, cellular shape, or proliferation rate come about? Are the cells changing in accord with some "internal program"? Or is the environment of the cells changing? The answer is both; there is a complex set of signals and responses between a cell engaged in morphogenesis and its environs. What is at issue here is how cells communicate and interact. We discussed some of the machinery for these interactions in Chapter 12. Let us now examine a selected and diverse set of situations in the embryo in which cell interactions are driving and regulating morphogenesis, and thereby see what general principles emerge.

CHAPTER PREVIEW

1. Cells can use receptor tyrosine kinases to guide migration.

2. Neural crest cells use cell-bound and matrix-adherent molecules to help guide their migration.

3. Neurite extension utilizes a growth cone to sense environmental cues.

4. Limb outgrowth involves reciprocal signaling between different specialized regions of ectoderm and mesoderm.

5. Development of lungs, salivary glands, and kidneys requires signaling between tissue layers.

6. Gastrulation is driven by a complex suite of changing cellular behaviors.

7. Convergent extension (repacking of cells within a tissue) can cause substantial changes in the form of the tissue.

We begin by looking at some instances in which the principal morphogenetic tactic is movement of one kind or another. We then discuss some examples where communication between an epithelial and a mesenchymal tissue layer initiates and regulates morphogenesis. Finally, we revisit gastrulation, arguably the most complex instance of morphogenesis, during which a whole suite of morphogenetic movements occurs.

CHANGES IN MOTILE BEHAVIOR

The Homing of Primordial Germ Cells Involves Receptor Tyrosine Kinases

Primordial germ cells (PGCs) may arise from a germ layer both different in kind and far removed from the developing gonad (see Chapter 7). You may recall that in mammals the primordial germ cells arise in the yolk sac endoderm and then migrate to the genital ridge of the mesoderm at the midbody level. In the chick embryo, PGCs have been shown to migrate from extraembryonic endoderm to the genital ridge via the circulation. In mammals it is not clear how much PGC migration occurs by the "overland" route through mesenchyme, and how much (if any) occurs via the circulation. This is an extraordinary example of targeted migration. Not only do the PGCs change from epithelial to motile cells, but they are able to "home" to their correct location over very long distances. A mutation in mice, known for many decades, has gives some clues about how this is done.

The *W* gene in mice has a number of alleles, some of which are fatal in the homozygous condition. These mutant mouse pups show severe anemias, have deficiencies in skin pigmentation, and, because they have no PGCs, are sterile. The *W* gene is now known to encode a membrane RTK (receptor tyrosine kinase) named cKit. This receptor has been shown to be expressed by PGCs. When the ligand of cKit (called Mgf and encoded by the *steel* gene) mutates, it can cause anemia, coat color changes, and sterility. Mgf is present in cells along the migration pathway of PGCs. The complementary expression patterns of cKit and Mgf implicate them in homing. Mutations in the genes for either this receptor or its ligand lead to deficiencies in the directed migration of PGCs, as well as defective migration of melanoblasts arising from the neural crest, and of hematopoiesis. Figure 13.1 depicts this situation graphically.

It is interesting that PGCs adhere in vitro to mesonephric mesoderm and that antibodies against cKit prevent this adhesion. While these results do not prove that cKit and Mgf are responsible for the specific adhesion and migration pathway

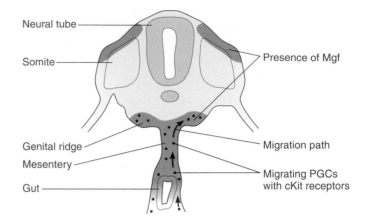

Figure 13.1   The Homing of Primordial Germ Cells A cross-sectional diagram of the mouse embryo (10 to 11 days old) at the level of the genital ridge. Primordial germ cells (PGCs), which originate in the yolk sac, and their pathway of migration to the genital ridge are shown. PGCs express the receptor tyrosine kinase called cKit. Note that the domains where Mgf (the ligand for cKit) is mobilized are congruent with the PGC migration pathway, suggesting that cKit-Mgf interactions are involved in the directional migration. PGCs remain in the genital ridge and do not migrate to the somite. Mgf and cKit are also involved in the migration of melanoblasts and erythropoietic stem cells (not shown here).

of PGCs, they are consistent with it. Recent evidence suggests that a chemo kine (important in immune cell migration) may be involved in PGC migration in some embryos. The general hypothesis is that ligand–RTK interactions may be highly specific, may be involved in migration along specific pathways, and may play a role in some instances of specific cell associations. The evidence from the mouse mutations suggests that the interaction is necessary for migration to occur but does not prove the interaction to be sufficient. And even if Mgf provides a pathway, we do not know how *steel* expression comes to be localized in the first place. Other migrating cell populations, such as the melanoblasts of the neural crest, also show coincident localization of cKit and Mgf, which implicates this ligand–receptor pair in other migrations as well.

The Migration of Neural Crest Cells Is Governed by Several Factors

Little is known about what drives the change from epithelium to mesenchyme, in this instance or elsewhere. We do know that only cells from the edges of the neural plate can form the neural crest. A necessary precondition for forming the neural crest is that the edge of the neural plate be adjacent to prospective epidermal tissue. If a piece of neural plate

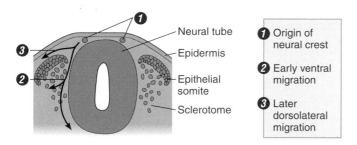

Neural tube
Epidermis
Epithelial somite
Sclerotome

1 Origin of neural crest
2 Early ventral migration
3 Later dorsolateral migration

Figure 13.2 Migratory Pathways of the Neural Crest A diagrammatic cross section of the dorsal portion of a vertebrate embryo. The neural crest cells originate from the dorsalmost portion of the forming neural tube, and then migrate ventrally along the lateral side of the neural tube. Later emigrating neural crest cells will follow a more dorsal pathway to populate the dermis.

not normally giving rise to the neural crest is transplanted so that it abuts some developing epidermis, neural crest development will result. The growth factor BMP4 is the signaling molecule from the epidermis that stimulates crest formation. The origin, migration, and differentiation of the neural crest comprise many classic problems of developmental biology, and have been the subject of a huge amount of research over the years. Several entire books have been written on the subject. Thus, there is an extensive body of knowledge describing the development of this cell population (see Chapter 6).

As the dorsolateral margins of the neural plate begin to fuse in the midline, the crest cells emigrate, then migrate on or near the surface of the newly formed neural tube. Subsets of the crest then follow specific trajectories (Figure 13.2), sometimes over long distances, to their final locations, and a particular pathway of differentiation is realized at that final location. What controls the transition of cells from epithelial to mesenchymal at the time of crest origination? What signals migratory behavior and controls the migratory routes? What regulates the time and place for migration to stop and terminal differentiation to begin? We have already discussed some of the issues about terminal differentiation of neural crest, in Chapter 6. Let us now address issues of the origin and migration of neural crest cells.

It is sometimes presumed that changes in the adhesive properties of cells must take place, certainly a logical inference. Emigrating neural crest cells do show reduced levels of N- and E-cadherins, and later show reduced levels of N-CAM—observations consistent with the hypothesis that reduced adhesivity is important for encouraging crest emigration. It is also interesting that in the reverse case—in which a loose, mesenchymal tissue becomes an adherent

epithelium, it has been demonstrated that activation of cadherin function is crucial for the transition. We discussed experiments in Chapter 12 showing this role of cadherin in adhesion. Other examples of changes in adhesivity can be found in the compaction of the eight-cell stage in the mouse embryo (Chapter 5), and the formation of epithelial myotome in the somite (Chapter 7). It is not certain whether the observed changes in adhesion molecules are causes or effects of the change from epithelium to mesenchyme, or vice versa, or the results of some other, earlier regulatory event. Even though we do not know what might cause changes in the adhesion apparatus, we do know that alterations or perturbations of the apparatus can and do interfere with neural crest emigration.

Cells leaving an epithelium must pass through the basal lamina that underlies epithelial tissues. This has been observed often by electron microscopy, which reveals discontinuities in the basal lamina through which the migrating cells may pass. There is substantial evidence that proteases, especially metalloproteases, may be secreted to help degrade certain components of the basal lamina and other molecules in the extracellular matrix. This activity may be important for allowing the metastatic invasion of cancer cells. Presently, however, there is no substantial evidence on this subject in regard to the neural crest, and we simply do not know how these cells "get loose."

The environment into which the crest cells move as they become actively motile is complex. Early-emigrating trunk crest cells and cranial crest cells move ventrally, while crest cells emigrating a day later may follow a more dorsal route. It has been shown by transplantation studies, however, that the developmental potentials of the early- and late-emigrating crest cells is similar. Hence, their different migratory routes and terminal modes of differentiation are likely due to interactions with the extracellular matrix and other cells. The extracellular matrix that lies along the dorsal route contains chondroitin sulfate proteoglycans that inhibit migration of crest cells through them; this inhibitory activity lessens after the early-migrating crest cells have moved ventrally. Presumably, chemical changes in the dorsolateral extracellular matrix render this matrix a less hostile environment for crest migration during later phases of migration.

Neural Crest Migration Is Sensitive to Ephrins

The ephrin–Eph ligand–receptor system is implicated in crest migration by some striking recent findings. Recall that ephrins (ligands) are bound to cells and that the Eph receptors are RTKs. Neural crest cells originating at the trunk level that migrate ventrally pass through the sclerotomal portion of the adjacent somite, but only through its rostral (anterior) half; these migrating crest cells avoid the caudal (posterior) half of the sclerotome. The presence of ephrin ligands and Eph receptors on migrating crest cells and the sclerotome are shown in Figure 13.3. The receptor Eph B3 is present in the rostral sclerotomal cells and migrating crest cells, and the ligand ephrin B1 is found caudally on sclerotomal cells. There is no spatial overlap between Eph B3 and ephrin B1.

Since ephrin ligands and Eph receptors have broad specificity, it is possible that the Eph B3 receptor on neural crest cells has an inhibitory effect, preventing migration of neural crest cells through this region. To test this idea, migrating crest cells were flooded with soluble ephrin B1 to disrupt normal receptor–ligand interactions. After such treatment, crest cells wandered through both rostral and caudal portions of the sclerotomal somite. In addition, researchers examined the in vitro behavior of crest cells in various situations. They

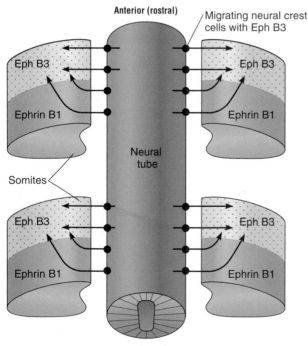

Figure 13.3   The Eph-Ephrin System in Neural Crest Migration A diagrammatic view of the upper surface of a vertebrate embryo with the ectoderm removed. The neural tube is the source of neural crest cells, which migrate ventrally and laterally and traverse the adjacent somites, but only in the rostral portion of each somite. The speckling indicates that Eph B3 is present on the surface of neural crest cells and on cells of the rostral somite. On the other hand, ephrin B1 is present on cells of the caudal somite, and neural crest cells avoid this area.

found that migrating crest cells could wander into an area containing substrate-bound ephrin B1, but then some of their leading protrusions collapsed, preventing further advance into the area. Results similar to these have also been obtained for cranial neural crest cells migrating from the mesencephalic region of the brain. The ephrin system is also involved in regulating whether migrating neural crest cells follow a trajectory into the dermis to form melanoblasts or, alternatively, follow a more ventral trajectory to form neuroblasts. We conclude that one of the components regulating the migratory trajectories of neural crest is the ephrin–Eph system, which appears to work by effectively placing "do not enter" signs in certain tissues. Several components in the extracellular matrix— including fibronectin, laminin, and a proteoglycan dubbed aggrecan—are also known to affect neural crest migration.

Migrating crest cells have well-known destinations (see Chapter 6). When they reach their destination, they lose migratory capability (though we do not yet know why) and then become associated with basal laminae. We know certain locations favor certain types of differentiation, as described in Chapter 6. Once again, the molecules in the environment that cause these changes in cellular behavior and differentiation are not known. Thus, study of the neural crest exemplifies both our present knowledge and our ignorance. The changes in cellular behavior and their specific trajectories are well known, and ascertaining this is no mean feat. Interfering experimentally with the cellular machinery of adhesion and motility results in interfering with normal development of neural crest. We know a lot about the molecules on the cell surface and in the matrix that probably play a role. We are even learning which molecules play specific roles in guiding cell trajectories. We do not, however, understand much about how the changes in cellular behavior that drive morphogenesis are regulated.

Growth Cone Activity Drives Neurite Outgrowth

The outgrowth of nerve fibers from their cell bodies has been a model system for the study of the elements of morphogenesis for a century. In fact, it was the desire to study neurite outgrowth more closely that prompted the great American embryologist Ross Harrison, at Yale University, to devise the technique of culturing tissue. (His French contemporary, Alexis Carrel, independently made similar technical advances.)

Whether in tissue culture or in the embryo, the developing neuron extends protrusions that continually advance over the substrate while the cell body remains anchored to its original location. The advancing tip of the neurite, called the **growth cone,** is an area of considerable motile activity. Extending from the surface of the growth cone are fine, protrusive strands called *filopodia*. Extensive microfilament bundles inhabit the filopodia, giving them structure, though only temporarily: filopodia are very fluid, quickly forming, touching the substrate, and (in most cases) retracting. Those that do not retract continue to adhere, becoming the new extension of the growing neurite. Often platelike extensions of the cytoplasm occupy some of the area between stable filopodia, and this entire

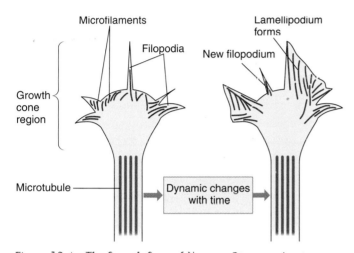

Figure 13.4 The Growth Cone of Neurons Diagrams showing changes in the exploratory growing tip of an extending neuron. Growth cones display narrow filopodia (literally, "threadlike feet"). Sometimes, broad membranous regions called lamellipodia ("platelike feet") appear between filopodia. Filopodia and lamellipodia contain many actin microfilaments. Microtubules are found mostly in the center of the cone and extend into the neurite. The behavior of the microfilaments, microtubules, and membranes is very dynamic.

extended sheet of cytoplasm is called a *lamellipodium*. The dynamic explorations of the filopodia and lamellipodia of the growth cone constitute the pathfinding and extension apparatus of the neuron, and cytoskeletal and other constituents synthesized in the cell body are delivered along this newly laid down "track" (Figure 13.4). Migration of many kinds of cells, not just neurons, involve extensions of filopodia and lamellipodia.

Many different experiments carried out during the past several decades, using various agents that interfere with the cytoskeleton, show that the developing neurite requires the integrity of the microtubule system in order to remain extended. Compromise of the microfilament actin fibers stops filopodial activity and extension. Moreover, modifying the substrate and the neurite's strength of adhesion to it profoundly affects the vigor and direction of neurite outgrowth (Figure 13.5). There is a rich literature on the factors that may be involved in directing neurite outgrowth. For a long time, researchers found it difficult to identify what these molecules are and how they work in directing neurite outgrowth; as we shall soon see, substantial progress has recently been made.

Cell–Cell and Cell–Matrix Interactions May Help Direct Neurite Outgrowth

Experiments using tissue culture, and more recently specific antibodies and inhibitors administered to local regions in the embryo, have led researchers to several ideas about how growth cone extension and direction may be regulated. The main

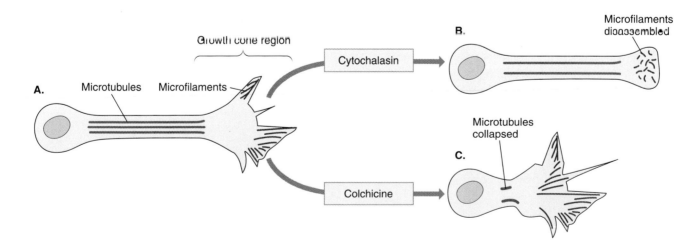

Figure 13.5 The Role of Microfilaments and Microtubules in Growth Cone Extension (A) A neuron in tissue culture is extending its neurite by means of the growth cone. **(B)** If the neuron is treated with cytochalasin, which inhibits microfilament assembly and function, the growth cone ceases its activity but the neurite remains extended. **(C)** On the other hand, if colchicine is applied, the functioning of the microtubules is inhibited, so the extended neurite collapses, retracting into the cell body. However, exploratory behavior of the growth cone persists.

A. Physical: Stereotaxis

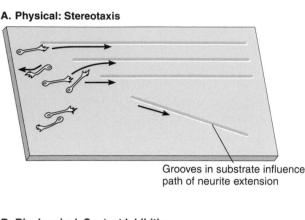

Grooves in substrate influence
path of neurite extension

B. Biochemical: Contact Inhibition

Moving cells make contact → Become paralyzed → Then move in new directions

C. Adhesive: Haptotaxis

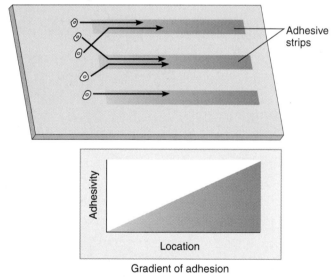

Adhesive strips

Gradient of adhesion

D. Chemical: Chemotaxis

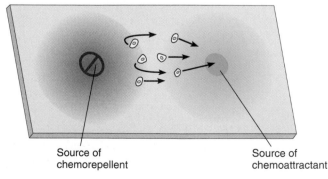

Source of chemorepellent   Source of chemoattractant

hypotheses all center on interactions between the growth cone and the substrate on which it rests and crawls. In brief, these hypotheses are (1) stereotaxis, in which the growth cone senses the physical topography of its surroundings and follows certain physical contours, much as a mountaineer finds the most accessible route to the destination; (2) contact inhibition of motility, in which the growth cone becomes physically paralyzed when it contacts another cell, causing the cone to extend its motile apparatus in a different direction; (3) haptotaxis, in which the growth cone favors the route with the highest adhesive strength; and (4) chemotaxis, in which locally present molecules either attract or repel the activity of the growth cone.

Various of these four hypotheses, which are not mutually exclusive, have been considered for many other instances of morphogenesis that involve cell motility and extension besides the growth of neurons. What evidence is there for these ideas?

Motile cells have been shown to migrate along a groove in a tissue culture plate (Figure 13.6A), so stereotactic behavior is possible in living embryos. However, there is no hard evidence yet to support a role for stereotaxis in the embryo.

In vitro, after two motile cells touch each other, there appears to be a temporary halt in the formation of protrusions at the zone of contact, after which they begin forming from other parts of the two cells that are not in contact. As a result, the two cells move off in other directions (Figure 13.6B). In vivo, contact inhibition as originally described and defined also has not been shown to play a role in embryos. But there is now clear evidence that some growth cones do "collapse" and become temporarily immobile.

The hypothesis of adhesive gradients cannot be easily dismissed, since motile cells, including extending neurites, obviously adhere with some strength to the terrain over which they explore. Experiments in tissue culture dishes show that coating the plastic with a highly adhesive substrate (such as

Figure 13.6 Hypotheses for the Guidance of Migratory Cells The four hypotheses, discussed in the text, concerning how migratory cells are guided. **(A)** In stereotaxis, cells sense the physical topography of their substrate. In this instance, the cells follow grooves. **(B)** In contact inhibition, when motile cells contact each other, they become paralyzed; after a brief period, motile activity starts again, leading to movement in a new direction and separation of the cells that had made contact. **(C)** In haptotaxis, cells follow a gradient of differential adhesion. Here the substrate has been coated with an adhesive chemical so as to create a gradient of increasing adhesivity; the cells are assumed to migrate in the direction of increasing adhesivity. **(D)** In chemotaxis, cells move toward the source of attractant molecules and avoid the source of repellent ones.

positively charged polylysine) may effectively glue cells to the substrate, stopping their emigration or extension (Figure 13.6C). Substrates that are somewhat less adhesive do influence the outgrowth of neurites, whose extension tends to follow paths with somewhat greater adhesivity. Interference with the expression or function of classical adhesion molecules—such as cadherins, N-CAMs, or the integrins with which they interact—can affect neurite outgrowth. This has been clearly demonstrated in *Drosophila*, where mutations in various parts of the adhesion machinery may be introduced. For example, eliminating expression of the gene for Fasciclin II (a relative of vertebrate N-CAM) or of the gene *neuroglian* leads to defects in the guidance and connectivity of certain subclasses of neurons. Such assaults on the adhesion systems do not result in wholesale disruption of the nervous system, however. This suggests that various adhesion molecules share overlapping and cooperative functions.

Other players are likely involved in guidance as well. There is good evidence that many aspects of neurite outgrowth are regulated by both diffusible and membrane-bound recognition and adhesion molecules. Many of these operate as specific **chemoattractants** and **chemorepellents** (Figure 13.6D). In the following sections, we shall explore a few examples of these molecular systems; they may constitute only the tip of the iceberg.

Netrins Can Serve as Chemoattractants

The idea that there are locally acting chemoattractants dates back to the great Spanish neuroanatomist, Santiago Ramon y Cajal. He speculated that in order to orchestrate the connections between the 10^{14} neurons of the human nervous system, there had to be specific chemoattractants to help guide the developing neurons. What researchers needed was a good assay to search for such molecules, and cell culture procedures developed in the late 1980s provided the tools. Using these techniques, J. Davies and A. Lumsden, working in Great Britain, isolated the developing trigeminal ganglion (cranial nerve V) and cultured it in collagen gels. When prospective maxillary (upper jaw) tissue, normally innervated by the trigeminal nerve, was placed in the gel within a few hundred micrometers of the explanted ganglion, efferent neurites grew toward this target tissue but not toward explants of other tissues. Davies and Lumsden supposed these results were due to a specific chemoattractant issuing from the maxillary tissue but not the other kinds of tissue. This conclusion is consistent with the fact that maxillary tissue is normally innervated by the trigeminal nerve.

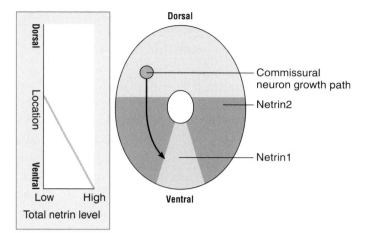

Figure 13.7 The Action of Netrins on Some Dorsal Spinal Cord Neurons A diagrammatic cross section of a developing spinal cord. Netrin1 is a chemoattractant released by the floor plate, while Netrin2, a related attractant, is present throughout the ventral portion of the cord, except for the floor plate. The vertical concentration gradient of total netrins (1 + 2) is shown at the side. Cells possessing the netrin receptor (DCC) respond and grow toward higher concentrations of netrins.

This tissue culture strategy was used by the San Francisco team of Tito Serafini and Marc Tessier-Lavigne, who showed that the floor plate of the developing spinal cord releases a chemoattractant that causes efferent neurites of commissural neurons to grow toward the floor plate. (*Commissural neurons* are so named because their neurites cross the midline of the central nervous system.) The active proteins have been isolated, and the genes encoding them have been cloned. There are two related proteins: Netrin1, located in the floor plate; and Netrin2, found in the ventral portion of the spinal cord excluding the floor plate. Both molecules stimulate the outgrowth of commissural neurons, and thus help explain why their efferent neurites grow ventrally in the spinal cord and come to the midline (Figure 13.7). The receptor for both netrins is encoded by a tumor suppressor gene in humans that belongs to the immunoglobulin family of adhesion molecules. Because the gene is often found mutated in colon cancer cells, it has been named *deleted in colorectal cancer (dcc)*, and the receptor protein is called DCC. (The homolog of DCC in *Drosophila* is Frazzled.)

But there is more to the story. When a floor plate is isolated and the netrin response is compromised experimentally through use of an antibody to DCC, netrin-stimulated outgrowth of commissural neurons stops; however, even though outgrowth ceases, already existent neurites can still bend toward the floor plate. A similar result is obtained using a floor plate from mice bearing a mutation that reduces netrin output. These experiments provide strong circumstantial evidence that there are other chemoattractants in the floor plate in addition to netrins.

If neurites are attracted by netrins to the floor plate, what prompts them to continuing growing across the midline to the other side; that is, how do they become commissural? A recently identified gene called *robo* (for *roundabout*) encodes a receptor, presumably present on the extending tip of the neurite, for guiding efferent neurites, and this receptor is sensitive to a midline cue (a secreted ligand encoded by the gene *slit*). Growth cones have been shown to change their responsiveness to netrins and Slit as they cross the midline. The growth cones of efferent neurites are attracted by netrins to the midline; here they lose their responsiveness to netrins, probably because the receptor of Slit (Robo) binds to the netrin receptor (DCC). As the neurons cross, they become sensitive to Slit and are repelled by Slit. In summary, the growth cone of the commissural neuron has changed from being positively attracted by netrins and insensitive to Slit, to being insensitive to netrins and repelled by Slit. These changes in sensitivity are thought to occur because of interactions between the cognate receptors, DCC and Robo. (Interestingly, *robo* and *slit* are well conserved, being present in both insects and vertebrates.)

In some situations netrins may actually be a chemorepellent. Efferent neurites of the trochlear nucleus (cranial nerve IV, which innervates some ocular muscles) are repelled by netrins in a collagen gel assay. Thus, the very same ligand can act positively in one context, negatively in another. This behavior is actually no different from the action of classical hormones such as thyroxine, which you will recall stimulates limb muscle growth but also causes tail muscle degeneration. It is important to identify which molecular contexts influence a particular ligand to respond positively and which negatively; in the case of netrins, this is unknown. Close relatives of the netrins exist—and not only in other vertebrates: netrin ligands and their receptors function in *Drosophila* and in the nematode worm *Caenorhabditis elegans*.

Semaphorins Are a Large Family of Chemorepellents

Researchers found that a protein isolated from chick brain caused the growth cones of sensory neurons in culture to collapse; that is, both filopodia and lamellipodia retracted. Thus they named the protein Collapsin. When *collapsin* was cloned, sequenced, and compared with other sequences, it proved similar to genes encoding another protein (then called Fasciclin IV, now called Semaphorin I), which acts as a guidance molecule for outgrowing nerve processes in the grasshopper limb.

We now know there is a rather large family of semaphorins; the genes have been cloned, and the amino acid sequences of the proteins have been determined. Semaphorins can exist either as membrane-bound molecules or in secreted form, and are present in many different kinds of animals. They generally act to guide the extension of efferent neurites, whose growth cones are repelled from areas where semaphorins are concentrated. For example, Semaphorin III is found in the ventral spinal cord of vertebrates but not the dorsal cord. When sensory neurons that innervate the skin and extend into the dorsal spinal cord are cultured in vitro, they avoid growing into areas containing Semaphorin III. Sensory efferent neurites from muscle, which normally enter the ventral spinal cord, show no such avoidance of Semaphorin III. Hence, molecules such as Semaphorin III may help control neuron pathfinding by declaring some areas "out of bounds," but only for certain types of neurons. This very large family of ligands interacts with a family of receptors called plexins. Different classes of semaphorins interact with distinct plexin subfamilies, leading to a tangled mess of nomenclature that we shall avoid. Growth cones are probably sensitive to several different attractants and repellents. Growth cone guidance is the outcome of a "tug of war" between different ligands and their receptors.

The Connections Between Retina and Tectum Are Guided in Part by Ephrins

Other adhesion molecules and putative chemoattractants and repellents are being identified and reported in the literature. One can be certain that this area of research will continue to grow.

We cannot leave this subject, however, without discussing the granddaddy of all neuronal connectivity systems. Roger Sperry showed many years ago (and received a Nobel Prize for doing so) that the ganglion cells of the retina—the cells whose efferent neurites form the optic nerve—are precisely connected to the *optic tectum*, a region of the brain found in the roof of the mesencephalon. If the developing eye is rotated or surgically manipulated in other ways, the growing optic nerve fibers nevertheless manage to find their correct targets. It is as if each ganglion cell fiber has an address label on it, and manages somehow to find exactly the right mailbox. The precision of this process has astounded developmental neurobiologists for decades. Figure 13.8 shows diagrammatically what is involved in the eye rotation experiment pioneered by Sperry. Regenerating efferent neurites from the rotated eye connect

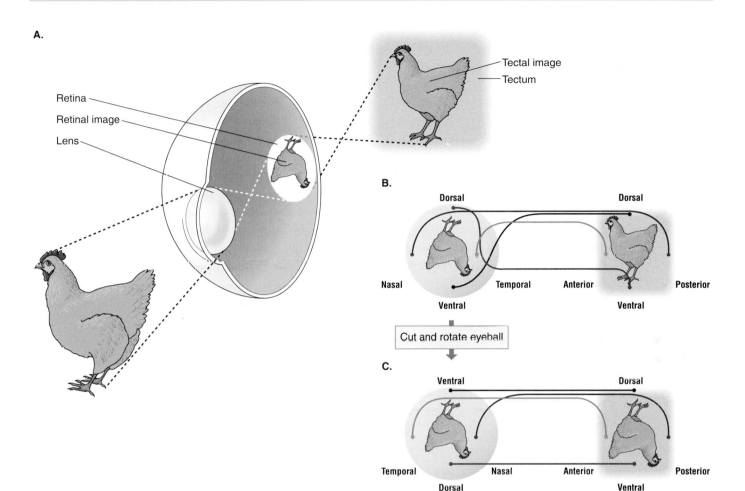

Figure 13.8   The Specificity of Retinal-Tectal Connections The "wiring" between retina and tectum. shown diagrammatically. **(A)** The viewer sees the profile of the chicken. that is. from the side. The image of the chicken is inverted by the lens and projected onto the retina. The pathway of neurons from retina to tectum reinverts the image. **(B)** Point-to-point connections between retina (*left*) and tectum (*right*): dorsal ganglion cells in the retina connect to ventral positions in the tectum. ventral ganglion cells connect to the dorsal tectum. ganglion cells on the nasal side of the retina to the posterior tectum. and ganglion cells on the temporal side of the retina to the anterior tectum. **(C)** In fish and some amphibians. if the optic nerve is surgically cut and the eyeball rotated. the former "correct" connections will be established when the optic nerve regenerates: the former "ventral" ganglion cell neurons still find and connect to the dorsal part of the tectum. and so on. Thus. the tectum now receives an image of the chicken that is upside-down. i.e.. not inverted.

to the same group of tectal cells as in the unrotated eye, even though this produces a perception that is useless to the animal. A frog with such rotated eyes perceives a moth in the upper visual field as if it were in the lower visual field; the hapless frog cannot, therefore, catch any prey.

Ephrins apparently play a role in forging these precise retinal-tectal connections. We have already encountered the ephrin–Eph system, during our discussion of neural crest cells as they migrate through the adjacent rostral somite. So far, 14 Eph RTKs have been identified, and they react with broad specificity to ephrin ligands. As ligands, ephrins divide into two classes. Those of the A class are anchored to the mem-

brane by glycolipid linkages, while the B class ephrins are transmembrane proteins. Studies examining the site of gene expression for Ephs and ephrins have shown fascinating distributions in the developing nervous system, specifically as they appear on the developing optic nerve and optic tectum. Many years ago, Sperry speculated that the affinity of cells for certain specific chemicals could serve as address labels to help guide efferent neurites to the appropriate regions of the tectum; he also suggested that there might be gradients of such address labels, perhaps working as adhesion molecules.

Recent experiments, carried out primarily by Friedrich Bonhoeffer, Uwe Drescher, and their colleagues in Germany, utilized the technique of co-culturing pieces of the optic tectum and retina. During normal development, efferent neurites originating from ganglion cell bodies located in the temporal area of the retina form synapses on cells in the anterior portion of the tectum; these neurites do not grow into the posterior tectum. When cultured in vitro, temporal efferent neurites can grow on "carpets" of cells from either the anterior or the posterior tectum; however, given a choice, these neurites prefer to grow on anterior tectal cells. The Bonhoeffer group showed that this anterior preference was due to temporal neurites being repelled from the posterior tectum, not from their attraction to the anterior. A 25 kD protein was identified in the posterior tectum that was not present in the anterior tectum. The gene for the protein found on posterior tectal cells

was introduced into a line of tissue culture cells (COS cells); temporal efferent neurites also avoided COS cells expressing this protein. It turns out that the 25 kD protein identified by Drescher and Bonhoeffer is an ephrin ligand.

Hwai-Jong Cheng and John Flanagan have independently shown that a gradient of ephrins exists in the tectum, and that a countergradient of Eph, the cognate receptor, exists in the ganglion cells of the retina. The Eph concentration is high in ganglion cells and efferent neurites of the temporal retina, becoming less and less in ganglion cells of increasingly nasal locations. The temporal neurites connect to the anterior tectum, where the cognate ephrin ligand is low. The ephrin concentration becomes higher as one proceeds posteriorly in the tectum, the area avoided by temporal efferent neurites (Figure 13.9). These opposing gradients of ligand and receptor are consistent with Sperry's ideas.

The ephrin system does not completely solve the problem, however. Chemorepellents may help get growth cones into the correct neighborhood, but they do not get the car into the garage of the right house. It seems likely that more adhesion and ligand–receptor "address" systems will be discovered. It is also clear that multiple molecular systems act coordinately to provide enormously detailed information for cell–cell recognition in the nervous system.

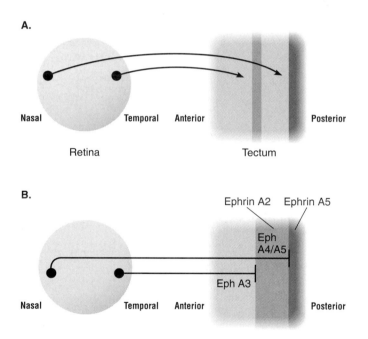

Figure 13.9 The Distribution of Ephs and Ephrins in Retinal Efferent Neurites and the Tectum A model of how Ephs and ephrins might provide address labels for the development of proper connections between the developing retina and optic tectum of the midbrain. **(A)** Efferent neurites from the temporal side of the retina project to the anterior tectum, while efferent neurites from the retina's nasal side project to the posterior tectum. **(B)** Temporal neurites express Eph A3. When they encounter ephrin A2, found in the posterior tectum, they are specifically repelled, and so do not grow into that region. On the other hand, nasal neurites express Eph A4 and A5: they are not repelled by ephrin A2, invading the posterior tectum instead. At the posterior border of the tectum, there is a domain of ephrin A5: the interaction between Eph A4 and A5 on the neurites and ephrin A5 in the tectum prevents the nasal neurites from growing into and beyond this posterior portion of the tectum. Some workers propose that the various ephrins are present in gradients in these expression domains.

EPITHELIAL-MESENCHYMAL INTERACTIONS

Thus far, we have looked at morphogenetic interactions of single or isolated migratory cells or growth cones. Now let us discuss interactions between tissue layers, typically an epithelial layer separated from an adjoining mesenchyme by a basal lamina.

Limb Outgrowth Requires Reciprocal Tissue Interactions

Are the tactics and strategy of morphogenesis different outside the nervous system? Probably not. Once again we see that changes in the extracellular matrix, ligands, and receptors influence adhesivity, cellular shape, protrusive activity, and proliferation to bring about other morphogenetic movements. We return again to the development of limb buds discussed in Chapters 7 and 12, for this is one of the better-understood examples of tissue interaction and pattern regulation.

Limb-bud development also clearly underlines the importance of interactions between epithelium and mesenchyme during morphogenesis. You will recall (from Chapter 7) that the epithelial covering of the developing vertebrate limb has at its tip a ridge of columnar cells (the apical ectodermal ridge, or AER). The initial formation of the AER requires signals emanating from the lateral plate mesoderm, an example of an epithelial-mesenchymal interaction. Recent evidence suggests that the signals from the lateral plate mesoderm may be BMPs. A knockout of the gene encoding a receptor (1A) for BMP signaling in the mouse limb ectoderm disrupted the formation of the AER. (This mouse knockout and other experiments carried out in chick embryos also showed that BMP signaling is important for establishing the dorsoventral axis of the limb.)

The AER, once formed, is important in limb-bud growth, since removal of the ridge stops further outgrowth of the limb. But the precise mechanism by which the ridge stimulates outgrowth is apparently complex. AER removal also stimulates cell death in the underlying mesenchyme. The AER may stimulate mitosis, too, though there is no direct evidence for this. By whatever mechanism, the epithelium of the AER is essential for limb outgrowth.

Several members of the protein family of fibroblast growth factors, among them FGF2 and FGF4, are present in the ridge and can stimulate limb outgrowth. Implanting a bead impregnated with FGF4 can stimulate formation of a secondary AER and consequently a secondary limb outgrowth. However, FGF2 and FGF4 are not present early in limb development when the ridge first forms. A newly identified FGF family member, FGF10, is present very early in the intermediate and lateral plate mesoderm. Shortly before limb buds form, FGF10 becomes restricted to lateral plate mesoderm in the prospective limb area. We know that FGF10 is important in limb-bud growth because implantation of beads soaked in this growth factor can induce a supernumerary limb bud. Both under experimental conditions and in normal development, FGF10 apparently stimulates the appearance of FGF8 in the ridge, just as the ridge is forming. FGF8 can stimulate limb outgrowth by mechanisms that are still not clearly defined, but which probably involve regulation of the proliferation (mitosis minus apoptosis) and even immigration of mesoderm cells from surrounding areas.

If this model is correct, the mesodermal growth factor FGF10 stimulates the overlying epithelium to form the AER, which in turn produces FGF8. FGF8 then leads to limb-bud outgrowth. This dialogue of growth factor interactions between the two tissues is probably the hallmark of developmentally important tissue interactions.

Branching Morphogenesis in the Lung and Salivary Gland Requires Tissue Interactions

A frontal attack on the existence and mechanisms of tissue interactions languished in the late 1930s because of the failure to find the active substances of the Spemann organizer, and because of the disruptions caused by World War II. Soon after the war's end, a new approach started with Clifford Grobstein's work at the National Institutes of Health. Grobstein focused on the development of several internal organs of the mouse, especially the salivary gland, pancreas, and metanephric kidney. (See Chapter 7 for a discussion of the different stages of kidney development.) Other workers extended these studies to virtually every organ possessing epithelial tissues that develop from endoderm, including the thyroid, thymus, intestinal crypts, and lungs. Furthermore, the sweat glands, prostate, and kidney, which develop solely from mesoderm, are similar to endodermal organs in that they develop from a combination of mesenchymal and epithelial layers, and display a pattern of branching. In every instance where epithelium and mesenchyme are separated from one another, neither morphogenesis nor terminal differentiation takes place.

An important question at the time the early experiments were done was whether tissues had to touch in order to interact. The microporous membrane filters that became available in the early 1950s allowed investigators to separate mesenchyme and epithelium and to co-culture them separated by the filter. In some instances, as in the pancreas, barriers preventing cellular contact allowed the interactions to occur. In other cases—for example, the metanephric kidney—the filter seemed to prevent tissue interaction. We can now begin to understand why different results were obtained in different instances. In many examples of ligand–receptor interactions, the ligand is soluble, while in others the ligand is tethered to the cell membrane.

The development of the lung is particularly interesting because the nature of the budding and branching that takes place during morphogenesis depends on the kind of mesenchyme interacting with limb-bud epithelium. Figure 13.10 illustrates the development of the normal branching pattern of lung epithelium and mesenchyme. When the lung mesenchyme covering the tracheal anlagen is experimentally replaced with mesenchyme surrounding the bronchus, a new bud forms in the trachea, which would normally never happen. (Recall from Chapter 5 that anlagen are the earliest identifiable stages of a newly developing organ.) Reciprocally, transplanting tracheal mesenchyme over bronchial epithelium inhibits branching. Transplanting mesenchyme from the lung bronchus to the

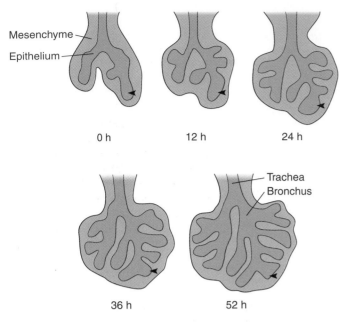

Mesenchyme

Epithelium

0 h   12 h   24 h

Trachea
Bronchus

36 h   52 h

Figure 13.10   The Morphogenesis of the Lung The lung rudiment of a mouse embryo was surgically isolated and placed in organ culture. The growth of epithelium and the relationship between it and the surrounding mesenchyme were examined in culture at various times ranging from 0 to 52 hours. Note that the branching formation is slightly different on right and left sides. Follow the group of cells marked with an arrowhead to get an idea of how a typical branching process occurs.

salivary epithelium will also evoke branching of the salivary gland, but only if fairly large amounts of lung mesenchyme are used.

Scores of experiments of this type have established that there is considerable specificity in the interactions between mesenchyme and epithelium. It is likely that the molecular interactions between ligands and receptors are at the basis of most, if not all, of these effects. Tips of branching epithelial tissues are regions of high mitotic activity; thus, ligands that are mitogenic (that stimulate mitosis), if localized, should stimulate bud growth and perhaps branching. TGF-β (transforming growth factor–beta), EGF (epidermal growth factor) and related molecules, and certain FGF family members have all been shown to affect branching of lung epithelium in culture.

For example, inclusion of TGF-β in an organ culture of lung epithelium decreases bud growth and branching. The receptor for TGF-β, which is found on the epithelium, can, through the use of antibodies, be partially inactivated, resulting in increased branching. The newly discovered FGF family member, FGF10, can stimulate lung endodermal epithelium in organ culture to undergo branching morphogenesis and to differentiate certain biochemical markers of lung epithelium.

FGF10 is localized in the mesenchyme just at the tips of the buds. Furthermore, there is a transgenic mouse strain in which FGF2 receptor function has been compromised by a dominant negative mutation. (See Chapter 5 for a discussion of dominant negative mutations.) These mutants produce apparently normal homozygous mutant embryos, except that their lungs are abnormal, and bronchial branching is severely reduced.

Homologs of the FGF family and its receptors have been identified in *Drosophila* as well. One of the FGF members in *Drosophila* is encoded by a gene called *branchless (brn)*. A null mutation of *brn* results in severe reduction in branching of the developing tracheoles (the airways of the insect respiratory system). In situ hybridization studies show that *brn* is expressed solely in cells surrounding the branching tips of the developing tracheoles. Another gene, *breathless*, encodes an FGF receptor, and when mutated, it too causes defects in branching. The normal version of this gene is expressed in the branching tracheole tips. Interestingly, even though insect tracheoles and vertebrate bronchi are both airways, they are believed not to be homologous structures; yet both seem to rely, at least in part, on specific FGF ligands and receptors for branching morphogenesis (as we shall see when we examine this again in Chapter 17). Other genes are known to help refine and precisely shape the branching patterns (see Chapter 17).

The extracellular matrix molecules and their associated cellular integrins are important participants in many of these ligand–receptor dialogues. Most organ cultures used to study these interactions in vitro require fastidious attention to the kind of matrix used. When lung epithelium and mesenchyme are dissociated, mixed, and reaggregated, the cells of the two tissue types sort out, but the epithelium can undergo morphogenesis and form a polarized epithelium only if the culture medium includes both laminin and heparan sulfate proteoglycan.

Much of what we have learned about the matrix and basal lamina has come from studies of the salivary gland. This, too, is a branching epithelium. The tips of buds are regions of highest mitotic activity; through growth, buds lengthen into tubules of the branching network. Along the sides of already formed tubules there is a heavy deposition of collagen (Figure 13.11). Experiments in which the basal lamina is compromised by enzyme digestion of glycosaminoglycans (GAGs) show that the basal lamina is necessary for maintaining the clefts between buds. The tips of salivary buds show a high turnover of GAGs, while along the sides of tubules the GAGs are more stable, and there is an accumulation of collagen. Antibodies against a characteristic protein found in basal lamina, Nidogen, also interfere with salivary branching.

A.

B.

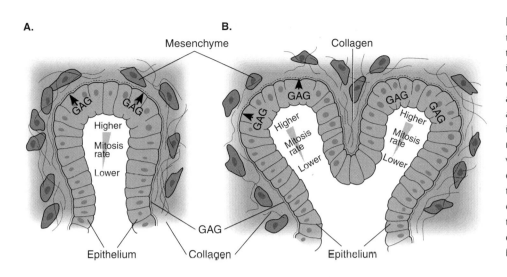

Figure 13.11 The Morphogenesis of the Salivary Gland A model, discussed in the text, for how clefts and branches form in the salivary epithelium. **(A)** A high rate of mitosis occurs at the tip of a bud; GAGs and collagen are deposited near the tip and along the sides. Cleft formation is initiated at the tip, perhaps because of microfilament activity in the tip cells and weakening of the basal lamina there by enzyme digestion. **(B)** As the cleft in the tip deepens, there is extensive deposition of collagen between the two new buds, thereby stabilizing the formation of the cleft. The two "new" tips can initiate branching again.

The kind of branching morphogenesis seen in the lung and salivary gland seems to be regulated, at least in part, by a ligand, such as FGF, being presented in a localized fashion to the mesenchyme. Important signaling ligands such as Wnt and Hedgehog depend on the presence of heparan sulfate proteoglycans in order to function properly. This mode of signaling is transduced by localized and specific receptors that require, or are assisted by, "structural" molecules in the basal lamina or nearby extracellular matrix.

Kidney Morphogenesis Requires Complex Circuits of Interactions Between Interacting Tissues

One of the more intensively studied instances of epithelial-mesenchymal interactions is the metanephric kidney, whose development we outlined in Chapter 7. Current research has identified a substantial number of ligand–receptor interactions occurring during kidney development; as a result, development of the kidney is worth discussing because it illustrates, paradoxically, just how much we do not know. The number and kinds of signaling that apparently take place are complex, and many of the interactions are still not well understood in molecular, or even cellular, terms. The kidney can serve as a useful reminder that just because one molecular player in a cell interaction is provisionally identified, the network of interactions and the means for bringing about morphogenesis can remain murky.

As we discussed in Chapter 7, the Wolffian (pronephric) duct extends posteriorly through the intermediate mesoderm; a branch of the duct, the ureteric diverticulum, forms and extends into the prospective metanephrogenic mesenchyme, initiating metanephric kidney development. Until the gene-

knockout procedure in mice was devised, there was little information on what stimulated ureteric buds to form. The earliest in vitro studies often used tissues from an 11-day mouse embryo, a stage when the anlagen can be cleanly dissected from surrounding tissue, and epithelium and mesenchyme can be separated from each other. By this time, the ureteric bud has already begun to form. We now know that a considerable number of knockout mutations can interfere with ureteric bud initiation, among them genes encoding transcription factors such as WT1, Pax2, and Lim1. WT1 is found in mesenchyme but is needed for outgrowth of the epithelial bud.

While we do not know exactly what WT1 is doing in the mesenchyme or how Pax2 and Lim1 are acting, another knockout experiment has been illuminating. Glial derived neurotrophic factor (GDNF) is a peptide growth factor originally found in the nervous system. But it is also known to be present in metanephrogenic mesenchyme *before* the ureteric bud forms. Furthermore, a knockout of the gene *gdnf* impairs ureteric bud formation. Application of GDNF to ureteric bud epithelium in culture stimulates epithelial growth. The receptor for GDNF is a heterodimeric RTK; a protein called cRet must associate with a partner protein (which may be GDNF-Rα) in order for the receptor to function. A knockout of *cRet* produces an incomplete but severe impairment of ureter formation. It is presumed that the mesenchymal GDNF ligand stimulates ureteric bud proliferation and growth into the mesenchyme; it is not known whether GDNF has any direct guidance function.

The induced bud synthesizes proteoglycans, which are essential for growth and branching of the ureteric bud. There are probably other signaling molecules engaged in these epithelial-mesenchymal dialogues. It is known that proteoglycans can bind signaling molecules such as members of the Wnt family, and are important for them to act. A knockout of a gene in mice encoding an enzyme that catalyzes sulfation of proteoglycans prevents condensation of mesenchyme around the ureteric bud. This results in the absence of kidneys in the newborn mouse. The growing tip of the ureteric bud expresses Wnt family member Wnt11, and there is speculation that Wnt11 is an important accessory signaling molecule in the induction process. Exactly how Wnt11 functions, however, is not clear. Ureteric bud formation and some of the molecular interactions involved are shown diagrammatically in Figure 13.12. For an extended discussion of Wnt signaling, see Chapter 15.

It has been known for over 40 years that the ureteric bud, once formed, can "induce" the surrounding mesenchyme to undergo profound changes. Some mesenchymal cells adjacent to the branching ureteric diverticulum become epithelial, forming the tubular portion of the nephron. This tubule becomes continuous with the collecting duct and ureter proper, the latter forming from the original epithelium of the ureteric diverticulum. We know little about the signaling molecules involved; however, gene knockout experiments implicate FGF2, BMP7, and Wnt11 as potential ligands emanating from ureteric buds. A recent search for molecules secreted by the ureteric bud turned up a cytokine called LIF (for leukemia inhibitory factor); this factor can cause kidney mesenchyme to form epithelial tissue in vitro. The mesenchymal cells destined to form tubules produce specific transcription factors (such as Pax2), cell-bound components of proteoglycans (for example, Syndecan1), and signaling molecules (such as Wnt4). The mesenchyme also secretes soluble proteins that stimulate branching of the ureteric bud, among them a protein called Pleiotrophin. Stromal cells, which comprise the "background" of the mesenchymal tissue surrounding the newly forming nephric tubules, express the transcription factor BF2. Knockouts of either *wnt4* or *bf2* interfere with tubule formation in the mesenchyme, suggesting that the pretubal and prestromal populations of mesenchymal cells interact with one another after they have been influenced by the ureteric bud.

All these protein factors and genes, with their various acronyms, may seem daunting; many of them are summarized in the model of kidney tubule induction shown in Figure 13.13. In general, three points are worth noting, and they apply to the development of any organ: First, there are probably many different signals and receptors. Second, they may act at

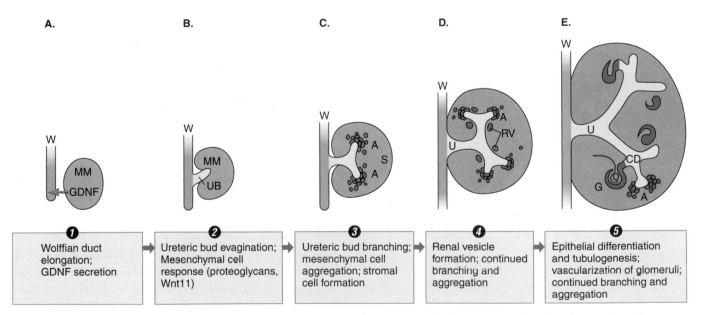

Figure 13.12   Ureteric Bud Formation Growth of the ureteric bud into the surrounding metanephrogenic mesenchyme, shown schematically. **(A)** It is thought that this mesenchyme can stimulate bud formation by secreting GDNF. The nephric duct epithelium receptor, cRet, responds. **(B)** Synthesis of proteoglycans and activation of *wnt11* then follow. **(C)** Mesenchymal cells near the ureteric bud aggregate into epithelial tubules, while the rest of the mesenchyme forms stromal cells. **(D)** The ureteric bud continues to branch, and mesenchyme forms epithelial renal vesicles. **(E)** The ureteric bud now forms the ureter and collecting ducts, which connect to the nephrons forming from the various kinds of epithelial aggregates in the mesenchyme. Abbreviations: A, mesenchymal aggregations forming epithelium; CD, collecting duct; G, glomerulus; MM, metanephrogenic mesenchyme; RV, renal vesicle; S, stromal mesenchyme; U, ureter; W, Wolffian duct.

A. Ureteric Bud Induction

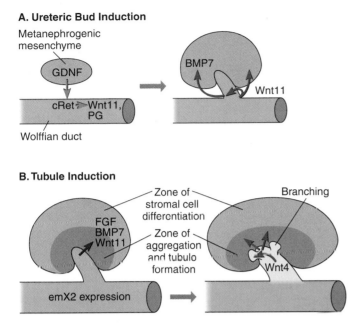

Metanephrogenic
mesenchyme

GDNF

cRet ► Wnt11,
PG

Wolffian duct

BMP7

Wnt11

B. Tubule Induction

Zone of
stromal cell
differentiation

Branching

FGF
BMP7
Wnt11

Zone of
aggregation
and tubulo
formation

Wnt4

emX2 expression

Figure 13.13 The Induction of Kidney Tubules A model for the formation of kidney tubules from mesenchyme. **(A)** The mesenchyme secretes GDNF, which activates the cRet receptor on the Wolffian duct. The duct activates Wnt11 and secretes proteoglycans (PG). This stimulates its own growth and also signals the mesenchyme. BMP7 from the duct also stimulates the mesenchyme. **(B)** The ureteric bud is thought to induce tubules in the surrounding mesenchyme. The gene *emx2* in the ureteric bud may stimulate formation of signals, such as Wnt11, BMP7, or various FGF molecules. These molecules may help pattern the formation of tubules and stroma in the metanephrogenic mesenchyme. Signals from both the stroma and the ureteric bud, including Wnt4, continue to stimulate differentiation of the tubules in the mesenchyme and the ureteric bud.

the same time and place, either synergistically or antagonistically. And third, they are embedded in a complex and ongoing set of interactions between and within the different tissue types. The dynamic network of interactions subsequently influences cellular shape and motility, proliferation, and adhesivity. It will probably seem less complicated once researchers understand the roles of the different molecules more completely.

GASTRULATION REVISITED

Gastrulation in the Sea Urchin Involves Many Changes in Cellular Behavior

We end this chapter by considering gastrulation again. The suite of morphogenetic movements involved in gastrulation is considerably more complex than the examples of morpho-

genesis discussed earlier in the chapter, mainly because so many different complex morphogenetic maneuvers are integrated in order to accomplish gastrulation. There exists a huge amount of descriptive information about gastrulation in the forms that we have already considered—fruit fly, amphibian, and amniote. However, all three embryos are opaque and difficult to observe in real time. Another animal—the sea urchin (mentioned earlier in connection with fertilization)—has the advantage that the optical clarity of the live embryo allows us to actually see what is going on inside. For this reason, it has been a favorite for the study of morphogenesis for a hundred years. Box 13.1 summarizes the overall development of the sea urchin embryo.

Gastrulation in this species is complex and in some sense prototypical, occurring in several separable phases. At the end of cleavage, the hollow transparent blastula is a single cell layer thick, composed of about 500 cells. Thirty-two of these cells, derived from four micromeres generated at the fourth cell division and located in a torus at the original vegetal pole, now ingress from the epithelium (Figure 13.14A). They form a migratory mesenchymal population that will subsequently form the skeleton of the larva. As they transform from epithelium into mesenchyme, these primary mesenchyme cells (PMCs) change their adhesivity, losing their primary affinity for each other and for a gelatinous extracellular matrix surrounding the embryo on the apical side of the epithelium (called the hyaline layer). Cadherins bound to the ingressing mesenchyme undergo endocytosis as the PMCs "poke" their way through the basal lamina, perhaps by digesting it with proteases. The PMCs, with long, actively forming filopodia, apparently explore the basal lamina of the blastula epithelium (Figure 13.14B,C). While there is no evidence for chemotaxis, the PMCs do cease their extensive wandering after a few hours and begin to congregate in special stereotypical locations in the prospective ventrolateral portions of the blastocoel.

This first phase, the emigration of primary mesenchyme, is followed by formation of the archenteron proper. First there is an invagination of a central disc of cells (Figure 13.14B). The ability to follow the changes in cellular shape in real time using recent new cell-marking methods shows that there is some apical constriction of these cells (much like bottle cells in amphibians), which may help drive the invagination. The invagination occurs rapidly, followed by a slower elongation of the archenteron as it pushes across the blastocoel for a distance about two-thirds of the diameter of the blastocoel. The archenteron is transformed from a short, squat structure to a longer, more attenuated one (Figure 13.14C). This results from

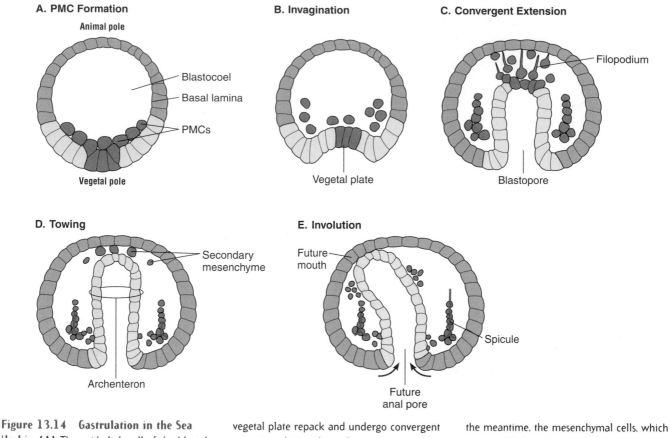

Figure 13.14 Gastrulation in the Sea Urchin **(A)** The epithelial wall of the blastula gives rise to migratory primary mesenchyme cells (PMCs) that emigrate through the basal lamina into the blastocoel. **(B)** The remaining cells at the vegetal pole become more cylindrical and begin to invaginate into the blastocoel. **(C)** Invagination continues as the cells in the vegetal plate repack and undergo convergent extension. driving the archenteron more than halfway across the blastocoel. **(D)** Cells at the tip of the archenteron can form very long filopodia that extend across the blastocoel and seem to explore the blastocoel wall. After contact. the filopodia start to tow the archenteron across the remaining blastocoel space. In the meantime. the mesenchymal cells, which ingressed before the vegetal plate formed. have begun to form skeletal spicules. **(E)** As the towing continues. the archenteron reaches the opposite wall of the blastula. Cells around the blastopore now involute. contributing more cells to the future midgut and hindgut of the archenteron.

the repacking of cells, an example of convergent extension (see Chapter 12). At the same time, many of the cells at the advancing tip of the archenteron become extremely active, forming filopodia. Many of these cells at the tip depart from the epithelium to become secondary mesenchyme cells, which enter the blastocoel and subsequently differentiate in a variety of ways (as pigment and muscle).

Some filopodia from cells resident at the tip actively touch and explore the blastocoel surface. When these filopodia connect to the area of their eventual destination—where the future oral cavity will form—their time of contact with the epithelium is much longer and they eventually fail to disconnect. These filopodia form a kind of cable to help pull the elongating archenteron completely across the blastocoel to the site of oral plate formation. Evidence for this towing role comes from the fact that if these filopodia are severed with a laser beam, the archenteron fails to make the final third of its journey across the blastocoel. When the archenteron tip cells reach the prospective oral epithelium, they become intimately connected to the blastula wall (Figure 13.14D). This area will eventually form a perforation that becomes the mouth of the larva.

But gastrulation is not over when contact between the archenteron tip and the oral epithelium has been made. Recent cell-marking experiments show clearly that cells located outside the archenteron but near its anal side begin to involute, moving around the base of the extended archenteron and then up its sides (Figure 13.14E); involuted cells populate the archenteron, mainly along its posterior (anal) half. There is also some cell division that contributes additional cells to the gut, though archenteron formation will occur even if mitosis is inhibited.

Box 13.1 Sea Urchin Development

Study the accompanying diagram, which outlines the major features of development of the sea urchin. Although the description is brief, it will provide a context for our discussions, not only of sea urchin gastrulation in this chapter but also of transcription regulation in Chapter 14.

Though the egg seems completely homogeneous in appearance and isotropic in its physical properties, a developmental polarity called the animal-vegetal axis is embedded in the architecture of the egg. The polar bodies are produced by meiosis at the animal pole of the egg, and it is this portion of the egg that will produce the ectodermal surface following gastrulation. After fertilization, three equal, orthogonal cleavages occur, to produce eight cells. The fourth cell division results in eight equal-sized *mesomeres* in the animal hemisphere, and four large *macromeres* and four small *micromeres* in the vegetal hemisphere.

Cleavage continues (though the micromeres follow a slower tempo from the larger cells) for another six rounds, for a grand total of 10 cleavage divisions. (There are mitoses after this, but they are not part of cleavage.) The mesomeres give rise to surface ectoderm, which is composed of oral and aboral surfaces (on the same side as, and opposite from, the mouth, respectively). These two ectodermal territories, which have rather different biochemical characteristics, are separated by a ciliary band that helps sweep food particles into the oral cavity. The micromeres generate two different territo-

---

The Major Features in the Development of Sea Urchin Embryos Some of stages of the developing sea urchin embryo: (A-D) surfaces views; (E-G) longitudinal sections. **(A-C)** The cleavage divisions of the two- and four-cell stages occur in planes parallel to the animal-vegetal axis. The third cell division is an equatorial cleavage (orthogonal to the animal-vegetal axis) generating two tiers of cells to produce the eight-cell stage (not shown). **(D)** The fourth division generates three groups of cells: mesomeres in the animal half, and macromeres and four small micromeres in the vegetal half. **(E)** Divisions continue, until 400 to 500 cells constitute a hollow blastula, which has cilia, hatches from its fertilization membrane, and begins to swim. **(F)** The gastrula forms through a series of morphogenetic movements (for more detail, see Figure 13.14). The primary mesenchyme, which arises from the micromeres, begins to secrete a calcareous endoskeleton (spicules). **(G)** The pluteus larva has a complex skeleton, a gut, and an oral and aboral epithelium. Coelomic pouches form on both sides of the foregut; the one on the left of the embryo will contribute much of the material for forming the adult sea urchin after metamorphosis.

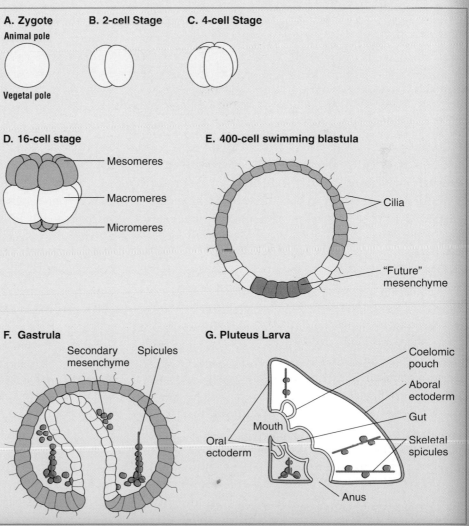

A. Zygote
Animal pole
Vegetal pole

B. 2-cell Stage

C. 4-cell Stage

D. 16-cell stage
— Mesomeres
— Macromeres
— Micromeres

E. 400-cell swimming blastula
Cilia
"Future" mesenchyme

F. Gastrula
Secondary mesenchyme   Spicules

G. Pluteus Larva
Coelomic pouch
Aboral ectoderm
Gut
Skeletal spicules
Mouth
Oral ectoderm
Anus

ries: (1) a central set of eight cells carried inward by gastrulation that come to comprise much of the coelomic pouches; and (2) the primary mesenchyme cells that ingress to form the skeleton.

The macromeres form the subequatorial surface (veg 1) surrounding the vegetal plate (veg 2). Veg 1 and veg 2 are labels used by workers who study these embryos. The vegetal plate forms the gut and sec-ondary mesenchyme; the surrounding veg 1 cells form the late-involuting posterior gut cells and some surface ectoderm around the anal region. The larva that results is thus comprised of the following territories: oral and aboral ectoderm, skeletogenic mesen-chyme, other mesoderm (pigment and muscle cells), coelomic pouches, and gut. Because the larva is thought to resemble a type of easel common during the Victorian era, it is called a pluteus larva (*pluteus* being the ancient Greek word for "easel"). Later, during metamorphosis, the larva becomes part of the plankton for several weeks, and then it can undergo metamor-phosis. The adult sea urchin will arise from a rudiment that forms from some surface ectoderm together with cells of the left coelomic pouch.

Thus, to sum up the activities at the cellular level: gastrula-tion in the sea urchin involves first an epithelial-mesenchymal transformation, then migration of these cells, followed in turn by invagination of the endoderm, convergent extension, filopodial exploration of the oral epithelium, filopodial tow-ing of the newly forming gut, and finally late involution of additional cells for the hindgut.

The challenge for the future is to identify precisely the molecules involved in adhesion, migration, protrusion, and changes in cellular shape, and to learn how they are regulated in this developmental progression. As you might expect, a substantial number of studies have used antibodies and drugs injected into the blastocoel in attempts to interfere with the biological underpinnings of these aspects of cellular behavior. And not unexpectedly, if one interferes with the extracellular matrix assembly, integrin function, or the cyto-skeleton, gastrulation is perturbed. Even in this well-described situation, however, the regulation of concerted changes in cellular behavior is not understood. We do not, for example, know what initiates or sustains the transformation of epithe-lium into mesenchyme; nor what initiates, sustains, and directs the cell shuffling leading to convergent extension; nor what initiates and sustains filopodial exploration of the basal lamina.

Gastrulation in Xenopus Is Also a Multicomponent Process

The morphogenetic movements that bring about gastrula-tion in *Xenopus* and other amphibians are more difficult to observe. But cell-marking experiments have allowed us to trace the kinds and extent of movements, as discussed in Chapter 4. Figure 13.15 summarizes what happens in *Xenopus* during gastrulation, showing bottle cell invagination to form the blastopore lip, involution of surface and adjacent cells to form endoderm and mesoderm, stretching of the surface by epiboly, and motility and repacking of internalized cells.

Modern studies of gastrulation based on analysis of cellular behavior started in 1975 with Ray Keller's reinvestigation of the amphibian fate maps during gastrulation, originally con-structed by Walther Vogt using vital dyes in the 1920s. Keller used very precise dye-marking techniques and arduous analy-sis to produce maps with considerable detail (see Chapter 4). He then began to analyze the tissue movements by making careful morphometric analyses with the use of scanning elec-tron microscopy, and by scrutinizing cellular behavior in various explanted cells in culture. The main insights of this work are twofold: First, gastrulation is a mosaic of differently changing, local cellular behaviors. Second, convergence of coherent cellular layers (that is, narrowing of the blastopore's circumference) and extension (lengthening of the embryo along the animal-vegetal axis) are the main "engines" driving involution and closure of the blastopore (Figure 13.16). Let us consider these processes in some detail.

Invagination is initiated by apical constriction of the bottle cells and subsequent elongation of their cell bodies as they protrude toward the interior (Figure 13.17). The direction of invagination is determined by the physical properties of the cells surrounding the bottle cells. (Evidence for this comes from the fact that bottle cells, when isolated experimentally, do not extend to form a slit but contract uniformly to form only

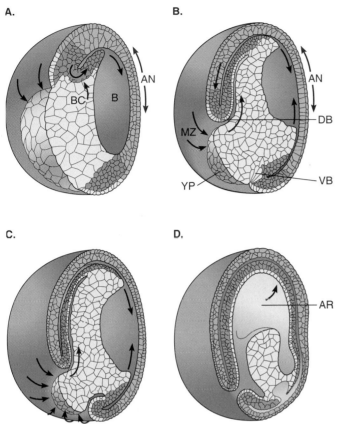

Figure 13.15 An Overview of Gastrulation in *Xenopus* A cutaway view of the major cellular movements that occur during gastrulation in *Xenopus laevis.* Arrows indicate direction of tissue movements. **(A)** Bottle cells begin their movements, resulting in the crescent-shaped blastopore. Cells on the surface undergo epiboly, while involution of the marginal zone cells occurs through the blastopore. **(B)** Convergent extension drives further involution. **(C)** The tip of the involuted marginal zone continues to advance, collapsing the blastocoel. At the same time, convergent extension is closing the blastopore. **(D)** The blastopore closes, the archenteron is completely formed, and prospective endoderm and mesoderm are internalized. Abbreviations: AN, Animal Pole; AR, Archenteron Roof; B, Blastocoel; BC, Bottle Cell; DB, Dorsal Lip of Blastopore; MZ, Marginal Zone; VB, Ventral Lip of Blastopore; YP, Yolk Plug.

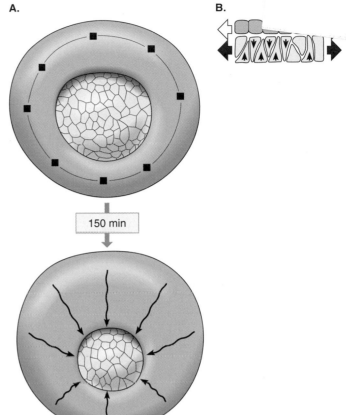

Figure 13.16 Convergent Extension in Amphibian Gastrulation Another view of gastrulation, this time emphasizing the movements of individual cells. **(A)** A view of the blastopore from the vegetal side: squares show individual marked surface cells at the beginning of gastrulation (*top*): 150 min later the marked cells have moved to the positions indicated by the arrows (*bottom*). **(B)** A cross-sectional diagram of the top two layers of noninvoluting surface. Intercalation of the subsurface cells (*thin solid arrows*) causes convergent extension of that layer (*wide solid arrows*). As a result, the surface layer passively undergoes epiboly (*wide open arrows*), expanding and stretching.

a pit.) Next, the apices of bottle cells form a line. By themselves, the contracted apices would be expected to bend the sheet of surface cells in which they are embedded. However, the endodermal core, which is mechanically resistant to pressure exerted by neighboring cells, forces the marginal zone cells outward and toward the vegetal pole. The deep subsurface torus of prospective mesodermal cells is thereby involuted. If bottle cells, once formed, are removed from the embryo, gastrulation will proceed, but the structures normally formed by the anterior archenteron will be missing. Bottle cells disappear by the time gastrulation is well under way. Once in the interior, the former bottle cells respread, deepening the archenteron and eventually forming its anterior wall.

Convergent Extension Drives Involution

If bottle cells are not essential for the early movements of gastrulation by pulling surface and subjacent cells inside, what *does* drive early gastrulation? Careful analyses by Keller and his students, carried out on whole embryos and in vitro, have shown that it is the convergence and extension of the involuting cellular material that causes these early involution movements. (Late in gastrulation, as mentioned previously, convergence and extension take place in the noninvoluting cells around the blastopore, thereby "closing" the blastopore). Convergence occurs by the mediolateral intercalation of cells, a cellular movement discussed earlier (see Figure 4.14B).

A.

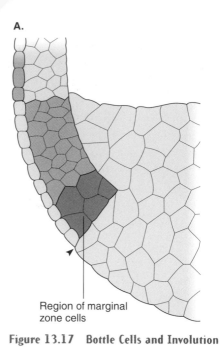

Region of marginal
zone cells

B.

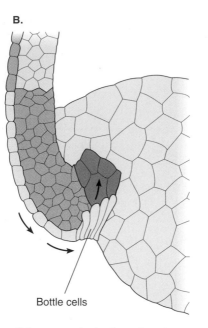

Bottle cells

C.

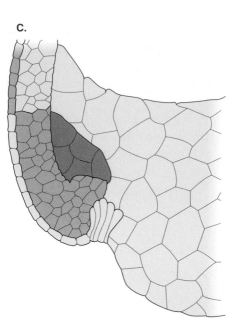

Figure 13.17   Bottle Cells and Involution Changes in shape of bottle cells and the involution of surrounding marginal zone cells. shown diagrammatically. **(A)** Prior to bottle cell formation. the leading edge of the deep involuting mesoderm (*shading*) is adjacent to the site of bottle cell formation (*arrowhead*). **(B)** Bottle cells form, and convergent extension drives the involution of leading marginal zone cells. **(C)** Bottle cells deepen. and involution continues.

Figure 13.18 illustrates mediolateral intercalation in explants of mesoderm from a *Xenopus* gastrula. Large stable protrusions extend medially and laterally; traction is exerted by neighboring cells on one another. Then the cells elongate along the mediolateral axis and intercalate with one another, thus causing the file of cells to become longer and narrower.

There is evidence that a cell surface molecule on the developing mesoderm, closely related to cadherin, is involved in forming the strong intercellular adhesion necessary for convergent extension. Furthermore, a diminution of cadherins may be associated with a lessening of cellular adhesion and the beginnings of cell migration. It is worth noting that in *Drosophila*, there are also dynamic changes of the types and distribution of cadherin molecules during gastrulation. In *Xenopus*, the deeper layers of the surface ectoderm also carry out mediolateral intercalation; however, the cells do not become aligned mediolaterally, and thus, being less attenuated, the cell files extend even more in the anteroposterior direction than does the mesoderm. This ectodermal intercalation depends on prior signals from underlying mesoderm.

The surface ectoderm (of the noninvoluting marginal zone) must somehow spread, but how? Three kinds of movement contribute: the surface cells flatten, they jostle into narrower confines, and the subsurface cells insert themselves into the surface. There is a radial exchange of cells between surface and subsurface, but the net movement produces more surface cells. The surface stretching, or epiboly, thus results from a

Figure 13.18   Mediolateral Intercalation in Mesoderm During Gastrulation These cell outlines were traced from photographs of involuting dorsal mesoderm explanted into a culture. Within the first hour. the cells began to interdigitate. thereby producing convergent extension of the whole tissue. The future anteroposterior axis of the embryo runs from top to bottom of the drawing.

0 min

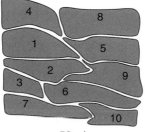

56 min

complex mixture of cellular behaviors. Experiments using a monoclonal antibody to fibronectin show that fibronectin is important for thinning of the blastocoel roof and also for the radial intercalation of involuted marginal zone cells.

Recent experiments have begun to illuminate some aspects of how convergent extension might be regulated. You will recall from Chapter 4 that BMP signaling molecules are important in patterning the amphibian gastrula. BMP concentrations are high ventrally and low dorsally in the gastrula, and play a role in determining what kinds of tissues differentiate along the dorsoventral axis. BMP is an antagonist of the Wnt ligands (probably Wnt11 in this instance). The Wnt receptor Frizzled7 has been shown to be important for correct convergence-extension behavior. Experiments that lead to an overexpression of Frizzled, or to its ectopic expression, can disrupt convergent extension. Thus, the gradient of BMP may have effects on morphogenetic behavior as well as cell specification. Parenthetically, this action of the Wnt–Frizzled–β-catenin pathway probably acts via a distinctive signal transduction pathway that utilizes G proteins and protein kinase C.

Fibronectin Aids Migration of Involuted Cells

The convergent extension machinery is powerful, pulling the surface closer to the blastopore and extending the axis in the anteroposterior direction. The involuted mesoderm that has converged and extended migrates anteriorly, especially dorsally along the blastocoel roof. The active motility of anterior spreading mesoderm puts head, heart, and ventral mesoderm into place. The role of the extracellular matrix, especially fibronectin, in this anterior migration has been suspected for a long time. Fibronectin is not essential for attachment of migrating mesodermal cells to the blastocoel roof. However, it does contribute to this process, as indicated by the looser attachment of mesoderm to roof explants when antifibronectin antibodies are present. The roof cells do secrete fibronectin before mesoderm migration occurs. The central cell-binding domain of fibronectin containing the RGD site is needed for cell spreading and protrusive activity of mesoderm. The growth factor activin, which is secreted by mesoderm cells, stimulates the roof to form fibronectin. Activin also increases the amount of $\alpha_4\beta_1$ integrin that is present; this integrin is needed for the attachment of fibronectin to blastocoel roof cells.

The exquisitely coordinated series of cellular behavior changes just described occurs in *Xenopus*. There are some differences in detail when other amphibian species are examined, but the grand strategy, summarized in Figure 13.19, is similar. Cellular shape changes, driven by apical contraction of the cytoskeleton, have several effects: they lead to progressive bottle cell formation, they mark the site of involution, and they help it begin (Figure 13.19A). The radial intercalation of subsurface cells (first discussed in Chapter 4) occurs; this, combined with cell division, shuttling, and spreading, helps mediate epiboly (Figure 13.19B). Preinvoluting marginal zone cells are driven toward and around the blastopore by a powerful convergence and extension of already involuted

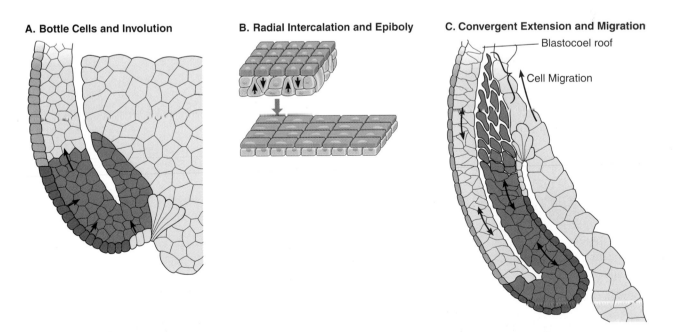

A. Bottle Cells and Involution

B. Radial Intercalation and Epiboly

C. Convergent Extension and Migration

Blastocoel roof

Cell Migration

Figure 13.19 The Strategy of Amphibian Gastrulation A diagrammatic summary of morphogenetic movements in amphibian gastrulation. **(A)** Apical constriction of bottle cells and their shape changes initiate involution. **(B)** Radial intercalation of the subsurface cells causes epiboly of the surface layer. **(C)** Convergent extension of involuted and noninvoluted marginal zone cells contributes to the expansion of involuted marginal zone cells. The leading edge of the involuted marginal zone becomes migratory upon the substratum of the former blastocoel roof.

marginal zone cells; this convergent extension involves medi-olateral intercalation of cells. The leading edge of prospective mesoderm cells migrates actively on an extracellular matrix secreted by the blastocoel roof (Figure 13.19C).

Cells from different areas have specific regional suites of behaviors. For example, noninvoluting marginal zone cells, when transplanted near the lip of the blastopore, do not undergo involution. They run according to a different, autonomously expressed program (Figure 13.20). Conversely, when involuting marginal zone cells are transplanted to a region outside the limit of involution they attempt convergent extension movements anyway, again following their autonomous, cell-specific program.

Some Mutations in Drosophila Disrupt Gastrulation

One way to learn something about the role of a given molecule in the complex morphogenetic movements of gastrulation is to evaluate the effects of mutations in a gene encoding such a molecule. The formation of the ventral furrow in *Drosophila* produces mesoderm, and coordinated cellular shape changes help drive this invagination and subsequent migration (see Chapter 3).

For instance, a putative signaling molecule (encoded by the gene *folded gastrulation*, or *fog*) and a G protein subunit (encoded by *concertina*) are known to be involved in the formation of wedge-shaped cells that probably drive invagination. Mutations of these genes affect ventral furrow formation. Another gene, *drhoGEF2*, encodes a GDP–GTP exchange protein that functions in second-messenger pathways; a mutation of this gene prevents cellular shape changes and thereby prevents invagination. It is hardly a surprise that ligands, receptors, and components of second-messenger pathways are involved in morphogenesis. The specificity of the phenotype of *drhoGEF2* suggests that there are several different representatives of this gene, some of which may be tissue specific. It was previously believed that this GDP–GTP exchange protein was used in many different signaling pathways.

Gastrulation patterns in the sea urchin, *Drosophila*, and *Xenopus* all involve precisely timed, programmed changes in cellular behaviors. What is the "prepattern" and its molecular basis that set up these regional programs of cellular behaviors? We shall explore that question in the next section of the book.

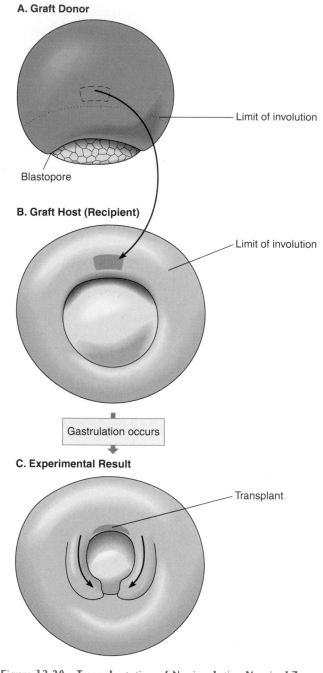

A. Graft Donor

Limit of involution

Blastopore

B. Graft Host (Recipient)

Limit of involution

Gastrulation occurs

C. Experimental Result

Transplant

Figure 13.20 Transplantation of Noninvoluting Marginal Zone Cells The behavior of noninvoluting marginal zone cells after transplantation. **(A)** A patch of the dorsal marginal zone that does not involute (because it comes from a region outside the limit of involution) is transplanted, at the midgastrula stage, to **(B)** a region in the recipient that does normally involute. **(C)** During gastrulation, the host marginal zone cells on either side of the transplant attempt to involute, extending toward the vegetal pole. The transplanted cells, however, do not involute: rather than adopt the behavior of their new location, they retain their former program of morphogenesis.

KEY CONCEPTS

1. Migratory cells can utilize ligand–receptor interactions to guide them along prescribed pathways.

2. Changes in the amount or distributions of adhesion molecules such as cadherins can play a role in epithelial-mesenchymal transitions.

3. The extracellular matrix can provide an environment that is either hostile or permissive for cell migration. The effects of the environment are due to the cooperative or antagonistic action of several different molecules; the same molecular players can play different roles in different contexts.

4. The substrate plays a vital role in cell migration and neurite extension.

5. The movement of cells or cell processes through the embryo can be influenced by chemoattractants and che-

morepellents. The ligands can be diffusible or tethered to the cell.

6. Growth factors can stimulate proliferation, affect the amount of apoptosis that occurs, or both. Localized actions of growth factors can cause localized proliferation and lead to the branching of epithelial tissues.

7. Clefts and bifurcations of branching epithelial tissues can be stabilized in part by the extracellular matrix, especially by collagen.

8. Intercalation of epithelial cells with each other, driven by protrusive activity and adhesive changes (convergent extension), provides the motive force for many large-scale involution movements and extensions.

STUDY QUESTIONS

1. Metalloproteases have been suspected as playing a role in the ingression of epithelial cells through a basal lamina to invade and become part of a mesenchyme. (a) Can you name some instances in amniote development where such an ingression occurs? (b) Devise an experimental strategy to test whether or not metalloproteases affect any of the processes in part a.

2. What is haptotaxis, and how would you evaluate its role in the migration of the mesonephric duct?

3. The pancreatic epithelium of an 11-day mouse embryo is just beginning to bulge into the surrounding mesenchyme. When you isolate a pancreatic anlage and remove its pancre-

atic mesenchyme, the isolated epithelium can be cultured with the following to get different results:

- No mesenchyme: a pancreatic rudiment does not form.
- Lung mesenchyme: a pancreatic rudiment forms.
- Salivary mesenchyme: a pancreatic rudiment forms.

What is your interpretation of this experiment?

4. FGF2, FGF4, FGF8, and FGF10 will all stimulate limb outgrowth. How can you experimentally evaluate which of these, if any, might be involved in the initial outgrowth of the limb bud and establishment of the AER?

SELECTED REFERENCES

Baker, C. V. H., and Bronner-Fraser, M. 1997. The origins of the neural crest. *Mech. Dev.* 69:3–29.

An extensive review of neural crest development.

Chiba, A., and Keshishian, H. K. 1996. Neuronal pathfinding and recognition: Roles of cell adhesion molecules. *Dev. Biol.* 180:424–432.

A review covering many different kinds of molecules involved in neurite outgrowth.

Dodd, J., and Schuchardt, A. 1995. Axon guidance: A compelling case for repelling growth cones. *Cell* 81:471–474.

A review of netrins and semaphorins in growth cone guidance.

Drescher, U. 1997. The Eph family in the patterning of neural development. *Curr. Biol.* 7:R799–R807.

A brief review of the Eph ligands by one of the leading researchers in the field.

Flanagan, J. G., and Van Vactor, D. 1998. Through the looking glass: Axon guidance at the midline choice point. *Cell* 92:429–432.

A discussion of molecules that control whether efferent neurites become commissural or not.

Guthrie, S. 1999. Axon guidance: Starting and stopping with *slit*. *Curr. Biol.* 9:R432–R435.

A review of the ligands and receptors that help control guidance of efferent neurites across the midline of the neural tube.

Holder, N., and Klein, R. 1999. Eph receptors and ephrins: Effectors of morphogenesis. *Development* 126:2033–2044.

A thorough review of this rapidly changing field.

Knust, E., and Muller, H.-A. J. 1998. *Drosophila* morphogenesis: Orchestrating cell rearrangements. *Curr. Biol.* 8:R853–R855.

A discussion of the genes affecting gastrulation in *Drosophila*.

Metzger, R. J., and Krasnow, M. A. 1999. Genetic control of branching morphogenesis. *Science* 284:1635–1639.

A review of the roles of *breathless*, *branchless*, and other genes in tracheole morphogenesis.

Perrimon, N., and Bernfield, M. 2000. Specificities of heparan sulphate proteoglycans in developmental processes. *Nature* 404:725–728.

A general review of the many ways in which proteoglycans may be involved in signaling and morphogenesis.

Perris, R., and Perissinotto, D. 2000. Role of the extracellular matrix during neural crest cell migration. *Mech. Dev.* 95:3–21.

A review of the many factors affecting neural crest migration along different routes.

Redies, C., and Takeichi, M. 1996. Cadherins in the developing central nervous system: An adhesive code for segmental and functional subdivisions. *Dev. Biol.* 180:413–423.

A review, by the discoverers of cadherins, of how cadherins may function to help organize the nervous system.

Schedl, A., and Hastie, N. D. 2000. Cross talk in kidney development. *Curr. Opin. Genet. Dev.* 10:543–549.

A review of the molecules involved in the morphogenesis and differentiation of the metanephric kidney.

Stein, E., and Tessier-Lavigne, M. 2001. Hierarchical organization of guidance receptors: Silencing of Netrin attraction by Slit through a Robo/DCC receptor complex. *Science* 291:1920–1938.

An elegant research paper on the molecular basis of commissural neuron growth cone guidance.

Van Vactor, D., and Lorenz, L. J. 1999. The semantics of axon guidance. *Curr. Biol.* 9:R201–R204.

A review of semaphorins.

PART SIX

REGULATION OF
GENE EXPRESSION

CHAPTER 14

THE LEVELS OF REGULATION OF GENE EXPRESSION IN DEVELOPMENT

The central dogma of molecular biology reduces its most general laws to three simple statements: (1) All the information needed for the maintenance and perpetuation of living cells is encoded in the nucleotide sequences of their DNA. (2) The information in DNA is transcribed into complementary nucleotide sequences in messenger RNAs (a process called **transcription**). (3) The nucleotide sequences in mRNA are decoded and synthesized into the amino acid sequences of proteins (**translation**). While this DNA–RNA– protein trinity does indeed succinctly state how hereditary information is stored and used by the cell, it is insufficient for understanding how an

CHAPTER PREVIEW

1. Transcription is regulated by a myriad of protein–DNA interactions.

2. Control regions of genes act like microprocessors that can integrate many different inputs.

3. Localization, stability, and translation of mRNA all affect gene expression.

4. Posttranslational modification of proteins can serve as a switch to turn developmental pathways on and off.

5. Protein–protein interactions are involved in determining specific tissue functions.

6. Knowing the entire genomic sequence allows scientists to monitor how batteries of genes change expression during development.

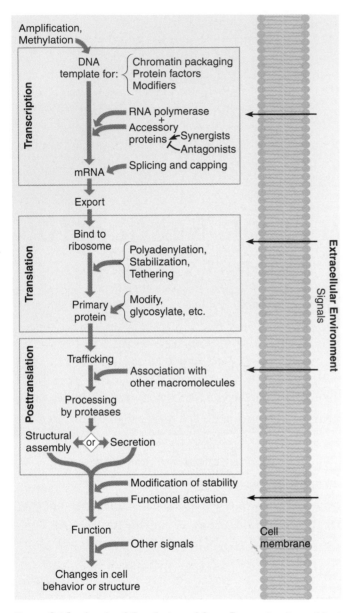

Figure 14.1   Levels of Regulation of Gene Expression A simplified diagram of the way the flow of information—from DNA to RNA to protein to cellular activity—may be regulated. Though shown here as a series of steps, what actually happens is more likely to be a maze of interacting steps than a simple linear chain of reactions.

The appropriate diagram for showing how elements of gene expression are interrelated would probably be a maze, not a simple linear diagram. In other words, the regulation of expression is likely a *network* of different interactions. When different regulatory pathways of reactions share common elements (intersect), such intersections are sometimes termed *nodes of control*. Figure 14.1 gives an overview of gene expression in a cell from a metazoan embryo. The figure has been greatly simplified, not only for didactic reasons, but also because there is so much we have yet to learn. A simple general principle applies here: all phases of gene expression that are capable in principle of being regulated are indeed regulated during development. The sheer variety of subtle tricks that nature employs to regulate gene expression is wonderful and astonishing.

This chapter outlines the fundamentals of how gene expression is regulated, starting at the level of the DNA. A discussion of transcription is followed by a look at protein synthesis encoded by the mRNA, that is, regulation at the level of translation. Finally, we consider the posttranslational level of regulation: how the newly synthesized protein can be modified, transported, and/or assembled, resulting in functional expression of the DNA. Our purpose in this chapter is to lay the groundwork for our analysis, in subsequent chapters, of how the regulation of gene expression governs morphogenesis and differentiation in the real embryo.

TRANSCRIPTIONAL REGULATION

Transcription Uses Chromatin as a Substrate

The study of how genes are transcribed into mRNA made very limited progress until researchers understood that the enzymes and accessory proteins that carry out transcription interact not with naked DNA but with chromatin. Naked DNA probably does not exist in the cell. DNA is complexed with specific structural proteins, the histones, as shown in Figure 14.2; the mass of histone protein in chromatin is roughly equal to the mass of DNA. Most of the chromatin is structured as a string of nucleosomes, each nucleosome consisting of two loops of double-helical DNA wound around eight histones. Many other proteins are associated with this DNA–histone octamer complex; some probably serve a structural role, aid in higher-level folding, and foster specific associations between the chromatin and the laminae of the nuclear envelope. Many other proteins associated with chromatin play important roles in regulating transcription and mRNA processing.

organism undergoes development. Transcription and translation are not the only steps in the expression of the information in DNA. Scores, perhaps hundreds, of steps are required for gene expression to produce functional cellular structures and tissues, and each step is regulated and controlled. Most of these steps, while ultimately encoded in the DNA, are directly or indirectly sensitive to changes in their immediate environment. Multiple factors, some acting positively, others negatively, influence many of the steps.

The DNA–histone–accessory protein complexes are very dynamic. Chromatin (specifically, the nucleosomes within chromatin) is re-formed with each cell cycle, when DNA is replicated, as well as during *turnover*, when histones and other chromatin proteins are degraded, synthesized, or recycled. Large, multicomponent protein "machines" are involved in chromatin assembly and remodeling. Chromatin structure may render entire blocks of DNA—indeed, whole chromosomes— inaccessible to transcription. Making large portions of the chromatin off limits to transcription can involve selective

methylation of the DNA, extensive interaction with histones and other proteins, or both. This inaccessible chromatin is called **heterochromatin,** in contrast to **euchromatin,** the usual, potentially active form of DNA; euchromatin that has been converted into heterochromatin is said to be "heterochromatinized." The two X chromosomes of primate females provide a well-known example: one of the pair is completely heterochromatinized, thereby forming an "inactive X" chromosome, which does not get transcribed; it is the other X chromosome, consisting of euchromatin, that does get transcribed.

The chromatin proteins embracing the DNA may be modified so that the association of protein with DNA can be very dynamic. A host of enzymes in the nucleus add and remove acetyl groups from histones; these enzymes fall into two groups: the histone acetyl transferases and the histone deacetylases. Acetylation of the lysines near the N-terminus (the amino, or NH_2, end) of core histones, especially H3 and H4, modifies the positive charge of the histone and potentially lessens the electrostatic interaction between the nucleosomal histone core and its associated DNA. Histone acetyl transferase activity can be found in several of the transcription factor proteins that are part of the activated RNA polymerase.

Methylation of DNA Can Keep Chromatin Inactive

Let us look more closely at one of the regulatory modifications to DNA mentioned earlier. Certain pyrimidines in chromatin sometimes undergo *methylation*. In particular, a cytosine occurring next to a guanine (a sequence often denoted CpG to distinguish it from the base pair C–G) can be methylated at the 5' position. Indeed, 60% to 90% of the CpGs in adult mammals have methylated cytosines. With a gene-specific probe, it is fairly easy to diagnose the methylation status of that gene. The restriction enzyme MspI cuts the sequence

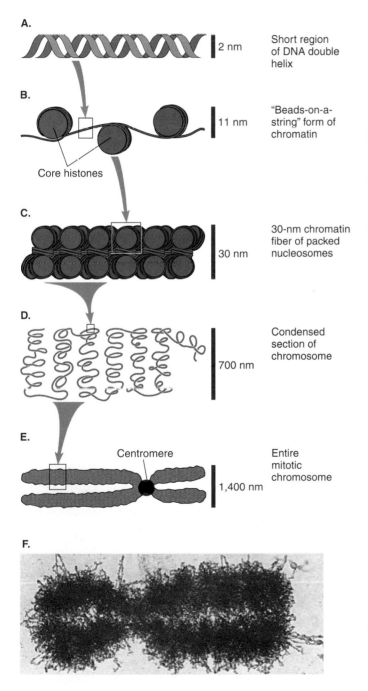

A.

2 nm — Short region of DNA double helix

B.

Core histones

11 nm — "Beads-on-a-string" form of chromatin

C.

30 nm — 30-nm chromatin fiber of packed nucleosomes

D.

700 nm — Condensed section of chromosome

E.

Centromere

1,400 nm — Entire mitotic chromosome

F.

Figure 14.2 Protein-DNA Interactions and Chromatin (A) The double-stranded DNA helix is wound around an octamer composed of four pairs of histone molecules to form **(B)** precisely spaced nucleosomes. **(C,D)** The basic nucleosomal DNA is complexed with other proteins that modify its conformation so that higher orders of looping are facilitated. Extensively looped chromatin may interact by means of other proteins with special domains of the nuclear envelope (not shown). The chromatin structure is very dynamic, changing during the cell cycle; chromatin also gets modified during the activation and suppression of transcription of various genes. During mitosis the chromatin is even more tightly compacted, forming **(E)** the familiar chromosome. As a result of all the looping and compaction of chromatin, the chromosome is now about 50,000 times shorter than the extended length of the DNA double helix. **(F)** An electron micrograph of a human chromosome is shown for comparison.

CCGG whether it is methylated or not, while Hpal, another restriction enzyme, cuts this same sequence only when it is unmethylated. Thus, by comparing the patterns of digestion using these two enzymes, one can evaluate whether a particular cytosine is methylated or not.

DNA prepared from chromatin derived from the mouse embryo at the blastula stage has a very low level of methylation. On the other hand, DNA from chromatin derived from later-stage embryos shows that a number of specific, previously unmethylated CCGG sites have become methylated. If the enzyme that carries out cytosine methylation is deleted in a gene knockout experiment, the mouse embryos will die before extensive organogenesis can take place. Genes that are active only in certain tissues, such as the gene for muscle actin, must be demethylated before the gene can be expressed during terminal differentiation.

It is not clear just how methylation causes or stabilizes the repression of transcription; two possible means are by altering chromatin's configuration or by binding general repressor proteins. Nor is it clear how widespread methylation is as a mechanism for finely tuning transcriptional control. However, its role in generally maintaining the repression of transcription is well established for amniote embryos. Curiously, DNA methylation is not present in *Drosophila*, so precise control of tissue-specific transcription obviously does not universally require methylation.

Methylation Imprints Genes in Mammals

Successful mammalian development requires both a male and a female pronucleus to contribute a haploid genome. Both are needed because the male and female contributions to the zygote are not functionally equivalent. As we pointed out in Chapter 5, this nonequivalence is due to the methylation differences in a subset of genes in the two sexes; 21 instances of this kind of methylation nonequivalence are currently known.

Some genes are expressed during development only if they reside on chromosomes contributed by the male, a situation known as imprinting (see Chapter 5) For example, *igf2* (the gene for insulin-like growth factor 2) is actively transcribed from chromatin originating from the male haploid chromosome set contributed at fertilization but not from female-derived chromatin. This is because *igf2* is more heavily methylated during oogenesis than during spermatogenesis. Differential methylation can work the other way around, too, with a gene being active in female-derived chromatin but not in the corresponding male chromatin.

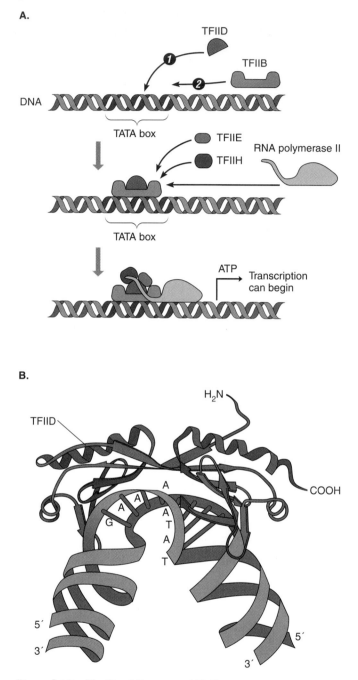

Figure 14.3 The Basal Promoter (A) The stepwise interaction between various transcription factor proteins (TFIID, -B, -E, and -H) and the RNA polymerase enzyme, as they assemble on the DNA of the basal promoter. The actual number of components and the complexity of the interactions have been simplified in this diagram. The interaction of TFIID with the TATA box actually induces a distortion in the DNA not shown here. **(B)** A ribbon diagram of DNA, to which the TATA binding protein (TFIID) is bound, thereby inducing a kink in the DNA. This may be important for organizing the assembly of other transcription factors on the promoter.

Methylation is not a permanent genetic change. The previous methylation patterns are erased during early stages of gametogenesis in both males and females, to be reinstated once again during later stages of gametogenesis.

RNA Polymerase Must Associate with General Transcription Factors to Become Functional

There are three different types of the enzyme RNA polymerase in eukaryotic cells; each type is composed of several subunits and transcribes different classes of genes. The RNA polymerase comprised of its different subunits is often called a holoenzyme. Type I polymerase specializes in transcription of the ribosomal RNA genes that produce three of the four structural RNAs (the 5.8S, 18S, and 28S molecules) that eventually comprise the completed ribosome. (See Box 4.1 for a review of how ribosomes are constructed.) Type II polymerase transcribes almost all other genes, thereby producing mRNAs (and some small RNAs of the splicing machinery that removes introns). Type III is specialized for the transcription of transfer RNA, the 5S ribosomal RNA, and some other small structural RNAs.

Although necessary, RNA polymerases are not sufficient for regulating transcription. In contrast to bacteria, transcription in eukaryotic cells also requires many other proteins, often called *general transcription factors*, which help position the polymerase complex correctly; they probably also participate in several other functions: regulating histone–DNA interactions in the vicinity of transcription, "loosening" the double helix, and allowing the polymerase transcription assemblage to proceed from the basal promoter. The **basal promoter** (or simply **promoter**) is the specific DNA sequence upon which the RNA polymerase and general transcription factors assemble; the promoter adjoins, or overlaps with, the actual transcription initiation site.

The comparison of transcription to the flow of a stream of water has led to the use of the terminology "upstream" and "downstream." Most genes in eukaryotes have a short sequence, about 25 nucleotides upstream from the transcription initiation site, called the TATA site or TATA box, since it is composed of adenines (A) and thymines (T). When the general transcription factor TFIID interacts with DNA at the TATA box, the DNA becomes kinked, providing a "signature" that may help other proteins and RNA polymerase to bind appropriately to the promoter (Figure 14.3). There are also TATA-less promoters; these are thought to have other, less easily identified, signature sequences for docking the poly-

merase holoenzyme and its associated general transcription factors on the promoter.

Activator and Suppressor Proteins Regulate the Initiation of Transcription

Activator (sometimes called enhancer) proteins and suppressor proteins are known from studies of genes in bacteria and bacteriophage. In these organisms, suppressors and activators usually interact with the genome at or very near the promoter, preventing or enhancing access of the polymerase to the promoter. In eukaryotic cells many, perhaps most, genes are regulated by protein–DNA interactions; many of these interactions take place hundreds, sometimes thousands, of base pairs away, either upstream or downstream, from the promoter. Just how these remote interactions between regulatory proteins and regulatory DNA affect activities at the promoter is not understood in detail; there is good evidence from studies on several different genes that the activator or suppressor protein comes into physical contact with, and thereby influences activities at the polymerase promoter site; a diagrammatic model of this is depicted in Figure 14.4.

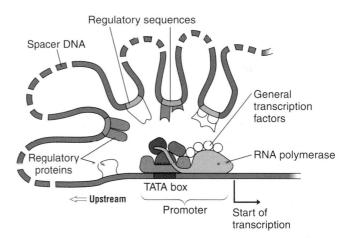

Figure 14.4 Gene Regulatory Proteins and the Basal Promoter A model for how transcription factors, interacting with binding sites for regulatory proteins located at a distance from the promoter, might influence transcription. In this case, activating transcription factors *(open symbols)* and suppressing transcription factors *(solid symbols)* interact with upstream DNA. The complexes of regulatory proteins bound to their target sequences may be able to loop back so that they interact with general transcription factor proteins at the promoter. The distance between a regulatory sequence and the promoter may be hundreds or thousands of bases in length. Furthermore, there may be many different regulatory sites with bound transcription factors (as shown here and also in more detail in Figure 14.6), all of which loop to make contact with the promoter.

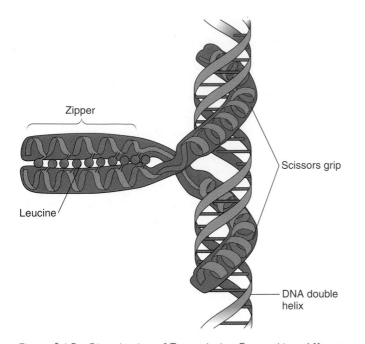

Figure 14.5   Dimerization of Transcription Factors Many different types of transcription factors must form homo- or heterodimers in order to be active. Shown here is an interaction for one such class of transcription factors, the so-called leucine zipper (or bZIP) type, which are homodimers. Each monomer possesses a domain (in this instance at one end) with a sequence in which a leucine occurs every seven amino acids; this spacing creates a hydrophobic face because all the leucine residues are aligned in the same direction on the α helix of the protein. The two hydrophobic faces can bond to one another, establishing not only the basis for dimer formation, but also providing the dimer with a structure in which the polar, positively charged domain of each monomer can interact with the DNA double helix.

There may be many different activators or suppressors acting on a given promoter; having different combinations of factors provides a way to regulate the specific activity of a single gene or small group of genes. Thus, the regulatory DNA sequences and their interactions with activators and suppressors function like a microprocessor to integrate influences and regulate transcription. As with a microprocessor, many different inputs may be integrated and then relayed as an apparent simple "yes" or "no" command. The activator and suppressor proteins may themselves be very complex. For example, they may be active when phosphorylated, inactive when dephosphorylated; or alternatively, active when dimerized with the identical (or similar) molecular partner, inactive when monomeric. A given activator protein can serve as an input element affecting a single gene or several. And events outside the cell that influence signal transduction pathways can, and do, influence transcription of a given gene by their action (for example, by phosphorylation) on a particular activator or suppressor.

In the remaining chapters, we shall encounter many instances of activators and repressors regulating transcription, but let us consider a concrete example for the moment. In the case of many hydrophobic ligands—such as steroid hormones, thyroid hormone, and retinoic acid (derived from vitamin A)—their receptors can act as transcriptional activators when stimulated by the appropriate ligand. There is a whole family of related **transcriptional activating factors (TAFs)**, which respond to stimulation by such ligands. The ligand activates the TAF by inducing a conformational change in it, which in turn allows it to enter the nucleus. The TAF then can bind specifically to an enhancer regulatory element on the DNA, and, when dimerized with a partner, can stimulate transcription. The receptors for all-*trans*-retinoic acid, thyroid hormone, and vitamin D_3 require interaction with a receptor (called RXR) in order to bind and activate the genes containing the response elements for retinoic acid, thyroid, or vitamin D_3, respectively. Retinoic acid (see Chapter 16) and thyroid hormones (see Chapter 8) both play powerful roles in normal development. It is obvious that for such hormones and ligands to play a role, both the appropriate receptor and its dimerizing nuclear partner must be available in active form in the cell.

Other families of transcription factors also employ dimerization in order to be active. Members of one such family possess an amino acid motif called a *leucine zipper*; the leucines in the protein helix of the monomers interact to produce an active dimer (Figure 14.5).

The endo16 *Gene of Sea Urchin Embryos Illustrates How Regulatory Sequences Act Like a Microprocessor*

The arrangement of regulatory DNA elements that participate in governing the transcription of a given gene activated during development can be complex. Figure 14.6 provides a schematic of the regulatory elements of a gene found in the sea urchin embryo. (For an outline of this animal's development, review Box 13.1.) Dubbed *endo16* because it is expressed in endoderm cells, it encodes a protein thought to play a role in cell adhesion in the developing gut of the embryo. The developmental expression of *endo16* is complex. Endo16 first appears in all cells of the developing archenteron, but as the gut forms its different regional parts, *endo16* expression becomes restricted to the midgut, which eventually forms the stomach. The upstream regulatory region of the *endo16* promoter extends over 2,300 base pairs. At least 30 different sites in the DNA interact with 13 different transcription factors. Twenty of these sites interact with a single transcription factor called SpGCF1.

Eric Davidson and his colleagues, who pioneered this type of analysis (which is based on the use of *reporters;* see Box 14.1), have shown that the DNA–protein interactions are organized as modules (Figure 14.6). These modules have been defined operationally by creating mutations or deletions in different DNA regions and analyzing their effects on the expression of the gene, as monitored by a reporter gene. For example, module A, next to the basal promoter, has a positive function and must be occupied with its cognate TAFs for transcription to occur. At the other end of the regulatory region, module G, which also has a positive function, boosts expression of the gene when modules A and B are active. Modules C, D, E, and F serve various functions that help regulate the pattern of spatial expression of *endo16*. When modules C through F are active, they help to keep the gene turned off in territories where *endo16* should not be expressed. The integration of a large number of different inputs from other modules into module A is the reason why this kind of control is often compared to a microprocessor.

Each module may have several different sites that interact with (bind to) proteins. Some of these sites, such as the one binding the protein SpGCF1, are present in many different modules. Other sites are unique to that particular module, and each module has one or two of these. The widely distributed SpGCF1 sites have been shown to promote the looping of chromatin, and thus might be involved in bringing distal sites (module G, for instance) closer to modules A, B, and the basal promoter, where transcription initiation is occurring.

This sort of complex modular organization of the regulatory elements is probably very common. The Davidson group has carried out detailed analyses of other sea urchin genes that possess similar features. Well-analyzed genes in *Drosophila* (some of which we will explore further in Chapter 15) also have long, complex regulatory regions.

The type of control of transcription we have been discussing here is often termed *combinatorial.* The different modules, or control elements, may individually have both positive and negative effects on transcription (see module B in Figure 14.6). In addition, some elements may act synergistically, others antagonistically. Furthermore, not all elements of a particular regulatory region of a given gene may play a role simultaneously; some may be involved only at certain times. This variability allows the control to be very precise and extraordinarily responsive to changing conditions. For example, growth factors can engage receptors and activate second messengers, which in turn participate in a cascade of phosphorylation events that regulate the activity of different transcription factors.

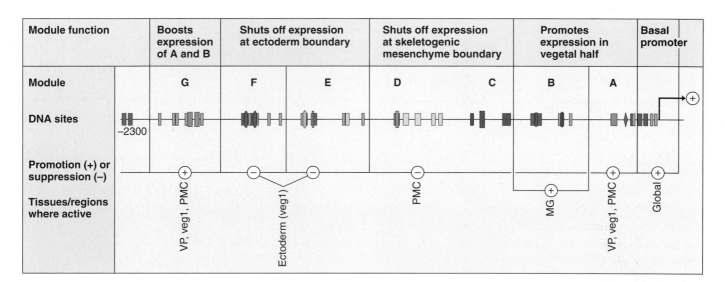

Figure 14.6   Regulatory Modules of the *endo16* Gene in the Sea Urchin A schematic of the regulatory region of *endo16*. which begins 2.300 base pairs upstream *(left)* and extends to the transcription start site *(right, with elbowed arrow and plus sign)*. Sites of specific DNA-protein interactions are indicated by various colored symbols along the line: SpGCF1 sites. which probably allow chromatin looping. are shown by light green rectangles. Some interaction sites involve unique transcription factors (indicated by distinctive symbols). Above the line are noted the various functional modules (A-G) into which the upstream DNA has been dissected. The bottom two rows indicate where in the embryo the various modules are active. and whether the activity is activating (+) or suppressing (−). Abbreviations: MG, midgut: PMCs, primary or skeletogenic mesenchyme cells: veg 1. the layer of ectoderm cells overlying VP: VP, vegetal plate.

Box 14.1 Using Reporters to Study the Regulation of Transcription

Many of the experiments that probe the regulation of transcription in different genes depend on use of a powerful experimental tool called **reporter genes**, or sometimes **reporter constructs**. To serve as a reporter, a gene must encode a protein that is not found in the cells being studied, and is easy to assay. To analyze a gene of interest, its promoter plus any possible regulatory regions (but not the encoding region itself) are fused to the encoding region (but not the promoter) of the reporter gene; expression of this construct is then observed after its introduction into living cells. The cells being used for the studies considered in this book are usually those in an embryo.

A favorite reporter is the gene for chloramphenicol acetyl transferase (CAT). In bacteria, this enzyme acetylates, thereby inactivating, the antibiotic chloramphenicol. Eukaryotic cells do not possess this gene. Hence, to create a reporter gene for examining the regulatory structure of a eukaryotic gene, its regulatory regions (including the promoter) are fused to the CAT gene's encoding region. If this construct can be introduced into a eukaryotic cell, and the cell shows CAT activity, one can be certain that it was the promoter from the gene under consideration that caused transcription of the CAT gene. CAT is easily assayed quantitatively; its activity can also be visualized by the use of in situ hybridization (see Box 7.1).

In addition to CAT, some common reporters are exons that encode the fluorescence-

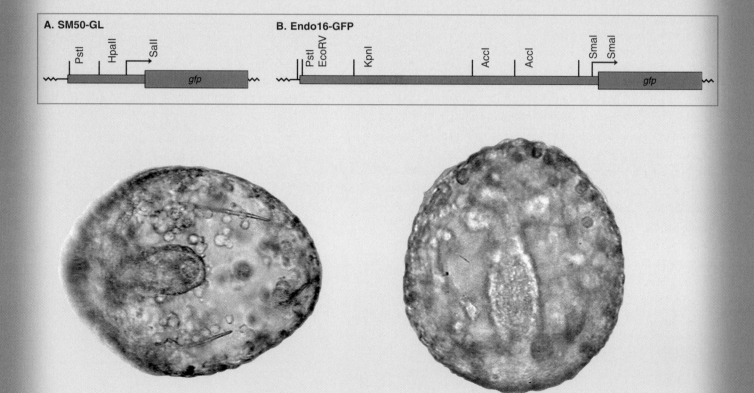

Examples of Reporters for Analyzing Transcription Regulation *Top:* Schematics of the DNA constructs created by fusing the upstream regulatory regions from two different genes, expressed during sea urchin development, with the protein-coding region for green fluorescent protein (GFP). Various restriction enzyme sites are indicated, and an elbowed arrow indicates the start of transcription. *Bottom:* Photos of postgastrula embryos into which these promoter–reporter constructs were injected at the fertilized egg stage.

By chance, only some of the embryonic cells received the construct. **(A)** The sea urchin gene *sm50* is expressed only in cells making skeleton in the embryo. Because GFP diffuses through syncytial filopodia connecting all the cells, *all* the skeletogenic mesenchyme cells have been labeled (as indicated by the green fluorescence). **(B)** The *endo16* gene is expressed only in cells of the developing gut. Because there is no syncytium in this case, only cells that received the construct show the fluorescence.

producing enzyme luciferase, the green fluo-rescent protein (GFP), and β-glucuronidase and β-galactosidase. All of these proteins are easily assayed and may be visualized in single cells by their ability to emit light (luciferase, GFP) or be located by histochem-istry (β-glucuronidase). The accompanying figure shows two examples of using the fluo-rescence of GFP as a reporter. Suppose we fuse the regulatory region from a gene

expressed only in spicule-producing cells of sea urchin embryos to a GFP reporter, and then inject this construct into a fertilized sea urchin egg. When the embryo develops its skeletal spicules, GFP fluorescence will be strictly localized to the primary mesenchyme cells that form the skeleton (see figure, part A). Similarly, if we take the regulatory region from a gene that is expressed only in the gut of sea urchin embryos, the injected

construct will "drive" reporter expression only in gut cells (part B).

By removing or altering various sequences in the regulatory DNA driving the reporter, one can examine the functional anatomy of the regulatory DNA. This was the approach taken to collect much of the data used to construct the functional representation of the regulatory modules of *endo16* shown in Figure 14.6.

Transcription of the ß-Globin Gene Family Is Regulated by a Complex Remote Control Element

We shall encounter many situations where genes are known to have complex regulatory regions and where control may be combinatorial. We discuss only one other example at this point, this time involving a whole family of genes.

Hemoglobin is a protein composed of two copies of two different kinds of polypeptide chains, termed α and β. Both α and β chains are encoded by different gene families. Using the example of humans, we shall examine changes in the β gene family, though there are changes in expression of α family genes as well. The β-globin chain of hemoglobin changes during development (see Chapter 7). The liver is the predominant source of red blood cells during early develop-ment of the fetus. The hemoglobin present in the erythro-cytes produced in the liver is composed of two α-type and two β-type chains. The β-type chains result from the expression of the ε-globin genes, which are very similar to the β-globin genes and are members of the β-globin gene family. Later in development, the spleen and bone marrow become important sites of erythropoiesis. Erythrocytes pro-duced here contain fetal hemoglobin, which has a distinctive β-type chain called γ-globin, encoded by two different γ-globin genes, $γ_1$ and $γ_2$. Near the time of birth and there-after, expression of the γ-globin genes diminishes, while expression of the genes encoding the definitive adult β-globin chains occurs. (In addition to β-globin, there is a minor adult β-type chain called δ-globin.) Thus, the temporal order of expression of genes during development is ε, $γ_1/γ_2$, and β/δ.

Members of the β-globin gene family are all located on human chromosome 11, and are arranged in the same linear sequence as the order in which they are expressed (Figure 14.7). Each member of this gene family has a regulatory region controlling its own individual promoter. Another control ele-ment for the β-globin gene family, the *locus control region (LCR)*,

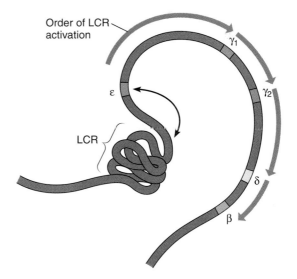

Figure 14.7 A Model for β-Globin Gene Transcription The DNA of the β-globin family gene region is represented as a curved line, and its locus control region (LCR) as a multilooped chromatin upstream from the five different β-globin genes. The LCR region can interact sequen-tially with promoters of each of the five different genes, thereby regu-lating the order of their expression. The double-headed arrow between the LCR and ε indicates the first interaction, which can be followed later in development by interactions with $γ_1$ and $γ_2$, and then with β and δ.

is located a considerable distance away, about 50 kilobases upstream, from the nearest member, the ε-globin gene.

The chromatin structure of the β-globin LCR is different in erythroid (red blood cell) precursor cells than in other types of cells. Some researchers believe that this distinctive structure is the basis for its tissue specific function, though the exact mechanism is not known. The distinctive structure of the LCR can be shown by exposing chromatin from erythroid precursors to very dilute solutions of deoxyribonuclease (DNase). When DNA is packaged into nucleosomes, it is somewhat protected from DNase digestion. However, the DNA in the LCR region possesses specific sites that are extraordinarily susceptible to digestion by DNase, even at extremely low concentrations of the enzyme. Such sites, known as DNase-hypersensitive sites, are often characteristic of regions of chromatin where nucleosomes are not present.

When the active LCR interacts with the regulatory sites for one of the genes—say, for ε-globin—that gene can be transcribed. Since no two members of the β-globin gene family interact with the LCR at the same time, the other four members cannot be transcribed while the ε-globin gene is being expressed. The same is true when the LCR interacts with any of the other family members. The order in which the different β-globin family genes are expressed during development is apparently due to the progression of LCR interactions with the regulatory regions of those β-globin gene family members. Note from Figure 14.7 that this interaction shifts from the gene closest to the LCR to genes increasingly remote from it. We cannot yet explain this pattern. Nor do we understand why the LCR region is activated in erythroid precursor cells and not in others. Recent experiments suggest that LCR activity is increased when some histone H3 molecules in the LCR domain become highly acetylated.

It seems likely that other genes outside the β-globin family domain must somehow regulate the activation of the LCR and its accessibility to different regulatory regions. Indeed, regulatory regions similar to the β-globin LCR have been discovered for domains of other genes, such as human growth hormone.

While study of the globin genes has been important, researchers have wanted to explore the role(s) of many different genes. In recent years, our knowledge of human genes has made spectacular advances because of the Human Genome Project. Indeed, this project is providing a flood of new information about many different organisms. To keep track of all these data and facilitate their use in research, new disciplines and technologies have been invented, some of which are described in Box 14.2.

TRANSLATIONAL REGULATION

Posttranscriptional Steps Must Occur Before Translation Can Happen

It is important to remember that transcription in eukaryotes is followed by a series of regulated events, all of which must take place before protein synthesis is possible. The nascent mRNA in the nucleus is modified at its 5′ end so that it has a methylated guanosine "cap," which is essential for subsequent binding to the ribosome. Then the 3′ end of the primary transcript is polyadenylated as transcription termination occurs, and thereafter the modified primary transcript associates with other proteins and small nuclear RNAs. The latter engage in the removal of transcribed introns that are present in the primary transcript. These introns must be spliced out to produce a mature mRNA, with its encoding sequence of exons.

Some genes are regulated by *differential splicing* to produce variant mRNA transcripts from the same gene. Differences in splicing choices can have profound effects. For example, the gene cascade that controls sex determination in *Drosophila* is regulated by a series of differential splicings. Figure 14.8 depicts the exons of the *double sex* gene (*dsx*), which is the last gene of the sex-determination cascade of *Drosophila*. The *dsx* gene can encode a male-specific Dsx protein (exons 1, 2, 3, 5, and 6) or a female-specific version (exons 1 through 4). Male-specific Dsx blocks female-type differentiation, while female Dsx blocks male-type differentiation. Furthermore, the splicing of *dsx* transcripts is controlled by other genes that act earlier in the sex determination pathway and which also engage in different splicing patterns in developing prospective males and females. Regulation of differential splicing is now known for several other genes besides those involved in this sex-determination cascade.

While still in the nucleus, the 3′ end of the primary transcripts of almost all genes (histone mRNA being an exception) undergo a posttranscriptional addition of a polyadenylated (poly A) tail. The poly A + mRNA, as it is termed, must also move from the nucleus to the cytoplasm. How it does so is currently a subject of intense investigation. It is becoming apparent that export of poly A + mRNA does not occur by diffusion; rather, an elaborate export machinery, possessing considerable specificity and subject to extensive regulation, is involved.

Regulation of mRNA Translation During Development Is Common and Involves Various Mechanisms

Translation requires not only available mRNA synthesized by transcription, but active machinery—the ribosomes and

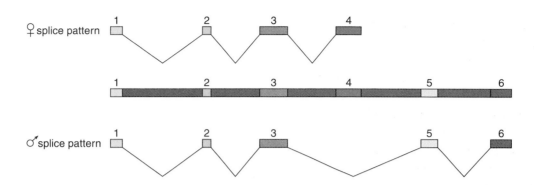

Figure 14.8 Differential Gene Splicing in *Drosophila* Splicing choices of a single gene can have enormous consequences. as shown here for the development of gender in *Drosophila*. The "map" in the center shows the arrangement of exons (numbered) and introns (unnumbered) in the *double sex (dsx)* gene. In females, splicing of *dsx* transcripts joins exons 1 through 4. while in males, exon 4 is skipped altogether and exons 1 through 3 are linked with 5 and 6. Each of the two different versions of the resulting Dsx protein in turn suppresses expression of the genes producing the sexual characteristics of the opposite sex. The differential splicing of *dsx* is more complex than shown here: it is part of a pathway that includes differential splicing of three other genes, earlier in the pathway, whose gene products eventually dictate the splicing of *dsx*. The whole pathway of sex determination is a cascade of regulated differential splicing.

enzymes for polypeptide initiation, elongation, and release. In principle, changes in the rate of translation during development could be regulated at any of these steps. Subtle changes have been detected in initiation and elongation rates during different developmental situations. The presently known key aspects of translation that are regulated during development seem to be mRNA availability and stability. The details of this regulation vary widely in different situations. In the text that follows, we examine the principal known mechanisms of this regulation, which include (1) masking of mRNAs, so that they cannot be translated; (2) varying the length of the poly A tail; and (3) localizing mRNAs within the cell.

Translation of mRNAs Made During Oogenesis Is Regulated

The first discovery of translational regulation in development was the change in rates of protein synthesis in the eggs of some marine invertebrates following fertilization. Figure 14.9 shows some data from sea urchin eggs. Within minutes following egg activation, the rate of protein synthesis increases at least 20-fold. This increase is due not to new mRNA synthesis, but rather to the formation of new polyribosomes from preexistent mRNAs, preexistent ribosomes, and the necessary enzymes. (Polyribosomes, sometimes called polysomes, are strings of ribosomes on mRNA.) The egg has everything it needs, but it is somehow unable to utilize its stored mRNAs and translation machinery until jump-started by fertilization. Messenger RNA can be extracted from both unfertilized and fertilized eggs, and both kinds of mRNA can be translated in vitro by a protein-synthesizing system derived from reticulocytes (immature red blood cells from mammals).

Eggs from some species do not undergo such a dramatic increase in overall rates of protein synthesis (the rate increase is modest in mouse eggs, for example), but the phenomenon

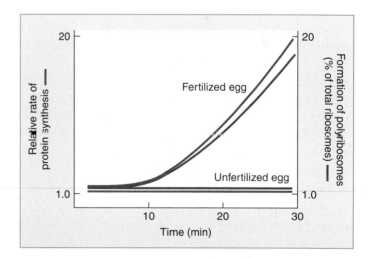

Figure 14.9 Increased Protein Synthesis in Fertilized Sea Urchin Eggs A graph showing the rate of protein synthesis in the egg. as measured by the number of amino acids incorporated into the growing polypeptide as a function of time following fertilization. Within a few minutes of activation, the rate of synthesis increased such that by 30 minutes following fertilization there was a 20-fold increase in protein synthesis. In a parallel experiment, the formation of polyribosomes was also followed as a function of time. Note that the rate of increase conforms to approximately the same kinetics as the increase in protein synthesis.

is relatively common. Control of translation of premade mRNAs has been closely studied in *Xenopus*, for example. One can also observe similar increases in translation rates in other instances where metabolic dormancy plays a role, such as the germination of certain spores, or the resumption of gastrulation in dehydrated brine shrimp eggs.

Messenger RNAs Can Be "Masked" and Polyadenylated

In *Xenopus* oocytes, some mRNAs are associated with specific proteins, which, when bound to specific sequences in the 3′ untranslated region (3′ UTR), act as powerful inhibitors of translation. These inhibited mRNAs are sometimes called *masked messengers*. Unmasking of them requires removal of these inhibitory proteins, which happens as a consequence of the metabolic and structural revolution in the egg following activation.

In many species, including sea urchins, *Xenopus*, and the mouse, a dramatic change in the length of the poly A tail accompanies the changing translation rate. As masked messengers are unmasked, the poly A tail can elongate, sometimes doubling, or more, in length. This modification of mRNA is a cytoplasmic reaction, and it is dynamic: phosphorylated adenosines (AMP) are both added to and removed from the tail, but a net lengthening prevails. The function of a poly A tail is now known to be complex. Substantial evidence indicates that longer poly A tails may work in combination with other factors to increase the rate at which translation is initiated. Furthermore, longer poly A tails act in concert with specific mRNA-binding proteins to confer additional stability on the mRNA, thereby increasing the concentration of mRNA and of the protein products translated from it. In the *Drosophila* egg, many (but not all) of the stored mRNAs, such as Bicoid mRNA, undergo a lengthening of their poly A tails before translation occurs in the syncytial blastoderm.

Cytoplasmic polyadenylation of maternal mRNAs (mRNAs synthesized in the oocyte before fertilization), although part of the suite of changes that result in an upsurge in mRNA translation in zygotes, is probably not itself the trigger that increases translation rate. In sea urchins, poly A lengthening occurs slightly *after* the upsurge in translation rate. It has been shown quite convincingly that the enzymatic machinery for initiating translation in sea urchin embryos works more efficiently within a minute or two after egg activation. This increase in translational efficiency and the unmasking of

mRNAs probably combine to increase protein synthesis after fertilization of sea urchin eggs. Similarly, in mouse eggs, the mRNA for FGF1 (fibroblast growth factor 1) receptors is relieved from translational suppression during final maturation of the egg; this mRNA also gets polyadenylated, but this by itself does not enhance its translation.

Messenger RNAs May Be Localized to Specific Parts of the Cell

Many examples are now known in which specific mRNAs are restricted to one part of the cell or another. This localization is crucial in development, of course, since it provides a mechanism by which an unequal distribution of information can be transmitted, one daughter cell receiving more copies of a given mRNA (and hence of the protein encoded by it) than the other; or even receiving all the copies.

Localization is important during the early cell divisions of the zygote, but it occurs later in development as well. One of the first reported instances of intracellular localization of mRNA showed that the stored mRNA for histone H1 is located in the maternal pronucleus of sea urchin eggs. This stored H1 mRNA can be translated only after the nuclear envelope is disassembled at mitosis and the stored mRNA is released into the cytoplasm. As a second example, recall from Chapter 3 that the important morphogen for determining anterior structures in *Drosophila*, Bicoid mRNA, is localized to the dorsoanterior portion of the unfertilized egg, where it remains tethered during the syncytial blastoderm and cellular blastoderm stages. This localization plays an essential role in establishing a gradient of Bicoid protein during early development.

Nanos mRNA Is an Example of Localized mRNA

You will recall from Chapter 3 that during *Drosophila* development a gene called *nanos* is transcribed in nurse cells. Nanos mRNA is transported during oogenesis into the oocyte, where it is localized near the posterior pole. Nanos mRNA is translated into Nanos protein, which prevents the translation of Hunchback mRNA. Nanos has a partner in this translational repression called Pumilio. When Pumilio protein is bound to a specific sequence in the 3′ UTR of Hunchback maternal mRNAs, Nanos protein can bind to this, to form a ternary complex; this leads to the deadenylation of Hunchback mRNA, followed by its subsequent degradation.

Recent technological breakthroughs now enable researchers to estimate relative levels of gene expression for vast numbers of genes simultaneously. The approach was fueled by the multinational Human Genome Project, which has involved not only sequencing of the entire human genome but also spin-off efforts on *Drosophila*, *C. elegans*, and the mouse. Genomes of other creatures are also having their genomes sequenced. Since the sequences of many different mRNAs and their cognate proteins are now known and recorded in databases, researchers can identify many DNA sequences as particular genes. All this activity has spawned a new scientific field dubbed *genomics*, an umbrella term that encompasses not only sequencing of genomes but also searching through the sequences (using computers) for meaning: which sequences encode what protein, or what regulatory region, and what are the characteristics of such regions? Computer-aided analysis of DNA sequences is sometimes called gene discovery.

The successful use of automated robots to carry out much of the tedious work involved in sequencing DNA has emboldened researchers to use robotic methods for creating microarrays of sequenced clones of genomic DNA or of cDNAs (which are equivalent to mRNAs). Arrays of extraordinarily small (50 to 200 μm in diameter) dots of DNA are placed on glass microscope slides or wafers of silicon. These microarrays (sometimes called chips) are then used as targets for the hybridization of fluorescently labeled complex probes.

Use of this powerful new technology is growing rapidly. The most common application is probably for monitoring relative levels of gene expression. In a typical experiment, RNA is obtained from the two sources to be compared. Let us compare, for example, the relative levels of mRNA present in the developing vertebrate lens at the epithelial placode stage and in the completely invaginated lens vesicle. RNA is extracted from these tissues and converted into cDNA by reverse transcription. Then cDNA from each source is labeled with

a different fluorescent dye, by attaching the dye to nucleotides in the DNA. The two types of cDNA are mixed, then hybridized to the DNA microarray containing the entire genome. A laser-scanning microscope records the ratios of the two different fluorescent peaks for each dot on the microarray, and the data are stored in a computer. Thus, the researcher can ascertain the amount of mRNA present that is encoded by a particular genomic sequence. One can look for patterns of expression of specific genes, or changes in relative expression levels of large numbers of genes.

The technique can also be used for many other purposes, such as clinical evaluations of different kinds. For instance, the DNA of a subject can be hybridized to a microarray of known mutant genes to determine whether a particular mutation is present. The challenge of analyzing large amounts of data has spawned the new field of bioinformatics. Two among the large number of websites with information on this new approach are: *www.nhgri.SONM/SIR/LCG/15K/HTML* and *www.genome.stanford.edu*.

Since Nanos is present in high concentrations posteriorly, Hunchback mRNA is not translated there. Hunchback, which is a transcription factor, participates in transcription of target genes that result in anterior structures; therefore, proper posterior development requires suppressing formation of Hunchback protein. It is regulation at the level of translation by Nanos protein that results in a sharp gradient of Hunchback protein, which underlies this crucial feature of anteroposterior patterning. Figure 14.10A,B summarizes this interaction between Nanos and Hunchback.

Not only does Nanos protein work to regulate the translation of some mRNAs; Nanos mRNA itself is translationally regulated in an important way. A platoon of other genes (*oskar*, *staufen*, *valois*, *vasa*, and *tudor*) is involved in localizing Nanos mRNA to the cytoskeleton at the posterior pole. The proteins of these five genes accomplish this localization by each interacting with a specific sequence in the 3′ UTR of Nanos mRNA. However, a significant amount of Nanos mRNA remains present in the anterior and middle regions of the syncytial cytoplasm of the zygote; in other words, Nanos

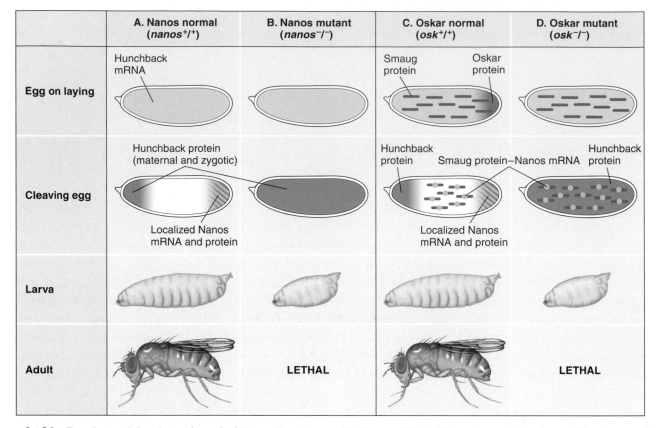

	A. Nanos normal (*nanos⁺/⁺*)	B. Nanos mutant (*nanos⁻/⁻*)	C. Oskar normal (*osk⁺/⁺*)	D. Oskar mutant (*osk⁻/⁻*)
Egg on laying	Hunchback mRNA		Smaug protein / Oskar protein	
Cleaving egg	Hunchback protein (maternal and zygotic) / Localized Nanos mRNA and protein		Hunchback protein / Smaug protein–Nanos mRNA / Localized Nanos mRNA and protein	Hunchback protein
Larva				
Adult		LETHAL		LETHAL

Figure 14.10   Translational Regulation by and of Nanos The interactions between Nanos and several other proteins are shown diagrammatically in a series of experiments utilizing different mutations. The several gene products involved are distinguished by color, with mRNA shown as a lighter shade of the same color used to denote its protein; in addition, individual mRNA transcripts are indicated as dots, protein molecules as rods. **(A)** In the normal situation, Nanos protein represses the translation of Hunchback mRNA posteriorly, allowing posterior development. **(B)** A mutation of *nanos* allows Hunchback mRNA to be translated posteriorly, so Hunchback protein blocks posterior development.

(C) Nanos mRNA itself is under translational control: In the normal situation (*oskar⁺/⁺*), Nanos mRNA is present throughout the cytoplasm, and a repressor protein (Smaug) binds to its 3' UTR, preventing translation of Nanos mRNA. However, Nanos mRNA, which is localized by the protein Oskar to the posterior end of the egg, is relieved of this repression by Smaug, so Nanos can be translated posteriorly. **(D)** If, however, *oskar* is inactivated by mutation (*oskar⁻/⁻*), Nanos mRNA is not localized. This would result in abnormal development because of the failure of Nanos to repress translation of Hunchback mRNA posteriorly.

mRNA is not localized exclusively to the posterior pole. But only the Nanos mRNA found there is translated. Outside this privileged domain, translation of Nanos mRNA is repressed by Smaug. This protein binds to a specific sequence in the 3' UTR of Nanos mRNA, thus hastening degradation of the latter. Hence, unlocalized Nanos mRNA is not translated, ensuring a sharp gradient of Nanos protein (Figure 14.10C,D).

A decade ago, many biologists believed that regulation of transcription could and would suffice to explain how gene expression regulates development. We now know that this opinion to be naïve; a variety of modes of regulation of translation is often essential.

We should note here that Nanos protein not only functions to repress Hunchback mRNA, and thereby ensure proper posterior development; Nanos function is also required for the proper determination and migration of primordial germ cells (PGCs) to the gonad. PGCs are often said to be "transcriptionally quiescent," meaning only low levels of transcription occur within them. Nanos protein represses translation of mRNAs of other genes; if Nanos function is missing, transcription of some genes apparently starts too soon and development of PGCs is compromised. This function of Nanos for orderly development of PGCs is also displayed by the equivalent of Nanos in *Caenorhabditis elegans*. Furthermore,

Pumilio is necessary for Nanos function in PGC development in *C. elegans*, just as it is in *Drosophila*.

POSTTRANSLATIONAL REGULATION

Protein Modifications Can Be Nodes of Developmental Control

In their primary structure (their amino acid sequence), proteins contain instructions for intracellular trafficking and modifications. An N-terminal sequence signals whether a protein can dock at and/or pass through a membrane system. In addition, certain amino acid signatures—some terminal, others nonterminal—dictate the localization of a protein to various organelles or metabolic pathways. For example, there are sequences that localize proteins to the nucleus. Other sequences are essential for N- or O-glycosylation. Enzymes can methylate or acetylate ε-amino groups of lysine; phosphorylate (and subsequently dephosphorylate) OH groups of serine, threonine, or tyrosine residues; or attach fatty acids at the C-terminus (the carboxyl, or COOH, end of a protein's primary sequence)—to list a few of the more prominent modifications.

Furthermore, we now know that many proteins, perhaps most, are part of structured macromolecular complexes. This fact is obvious for the so-called structural proteins, such as myosin and actin in myofibrils, and for the collagen molecules arranged in orthogonal bundles in the cornea. It is not so obvious for proteins that are not part of morphologically well defined organelles. Most enzymes and soluble proteins participate in protein–protein interactions. These intermolecular interactions are important for the function of the aforementioned enzymes and soluble proteins. Posttranslational modification of either or both of the protein partners in these interactions affects the structure and function of both partners and potentially could play an important role in development. Here we wish to emphasize the importance of posttranslational modification; in a subsequent section, we shall consider further protein–protein interactions.

The Hedgehog Ligand Is Posttranslationally Modified

The Hedgehog family of growth factors has a number of members, some of which we have already met. In Chapter 6, we discussed Sonic Hedgehog (Shh), important in forming the neural tube and limbs. In Chapter 7, we encountered Indian Hedgehog (Ihh), which plays a role in bone formation. Shh and Ihh are vertebrate homologs of the first discovered and founding member of the family—plain old Hedgehog (Hh)—which is found in *Drosophila* but not vertebrates. Hh is an important signaling ligand in the formation of boundaries for body segments and other patterned elements during development. (We shall discuss this signaling circuit in Chapter 15.)

Hh possesses a signal peptide that causes it to be inserted into the lumen of the endoplasmic reticulum and processed as it moves through the Golgi apparatus. After leaving the Golgi apparatus, Hh is further processed in an unusual reaction in which it cuts itself in two (autocatalytic cleavage): The C-terminal portion of the unprocessed protein brings about an internal cleavage between residues 257 (glycine) and 258 (cysteine), thereby producing the C-terminal fragment (which is 25 kD) and an N-terminal fragment (19 kD). During the cleavage, an esterification reaction adds cholesterol to the carboxyl end of the N-terminal portion, converting this fragment into the active ligand. These reactions are shown in Figure 14.11. Our understanding of this protein's structure is supported by some experimental evidence. It has been possible to produce truncated N-terminal Hh peptides by purely biochemical means, and to show that they possess Hh signaling capability when applied to tissue culture cells.

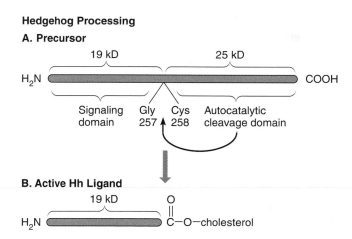

Figure 14.11   The Processing of Hedgehog Protein Schematics of Hedgehog (Hh) between its amino and carboxyl ends. **(A)** The 25 kD carboxyl portion of the precursor polypeptide, in an autocatalytic reaction, cleaves itself off between the glycine and cysteine at positions 257 and 258. **(B)** This produces a 19 kD amino terminal portion, which is the active form of the protein, and a 25 kD C-terminal protein, which has no further function and is not shown. After cleavage, cholesterol is esterified to the newly created carboxyl group of the N-terminal portion. The N-terminus also has palmitic acid (a 16-carbon saturated fatty acid) attached to it (not shown).

The C-terminal peptide shows no such signaling activity. These results confirm that it is the 19 kD N-terminal domain that possesses the signaling capability.

Esterification of Hedgehog Can Possibly Restrict Its Diffusion

This kind of autocatalytic protein cleavage was once thought to be exotic and unusual, but many other examples are now being discovered. Is there some specific utility that this autocatalytic cleavage provides? It is difficult to generalize since the proteins subject to autocatalytic cleavage are diverse in structure and function. In the case of Hedgehog, the cleavage is followed by an esterification, and indeed, cleavage is required for the esterification to occur. So an important question, at least for Hedgehog, concerns the role of esterification. One current hypothesis is that esterification with cholesterol helps to control the diffusion of Hh. This in turn builds up high local concentrations of the ligand, thus limiting its developmental effect to specific regions of the embryo.

There is some evidence to support these ideas. Let's examine the various parts of the hypothesis. The esterification of the carboxyl group of the N-terminal ligand with cholesterol serves to tether the ligand to the cell because cholesterol has a tendency to insert itself into the lipid bilayer of the cell membrane. An examination of *Drosophila* epithelial cells expressing *hh* shows that the ligand is concentrated in small granules near the basolateral cell surface.

The importance of controlling how far Hh ligand diffuses is buttressed by its known interactions with other genes and proteins in the embryo. Hh signaling in *Drosophila* is important for maintaining the expression of *wingless* (*wg*) in responding cells; moreover, *wg* is expressed only in cells of the anterior wing compartment immediately adjacent to *hh*-expressing cells in the posterior compartment of the wing disc. (Null mutants of *wingless* have no wings, but the gene is also important earlier in development for proper segmentation in *Drosophila*, as we shall discuss in more detail in Chapter 15.)

It is also known, however, that even though Hh acts upon the immediately neighboring cells of the wing disc, the ligand can diffuse some six to eight cell diameters from the cells of the posterior compartment that release it. What could account for this variation in diffusibility? Another recently discovered gene—*dispatched* (*disp*)—encodes a protein whose function is apparently to liberate the cholesterol-modified Hh from the lipid bilayer of the cell membrane, though it is not currently known just how Disp does this. After release from the cell membrane, Hh (still modified with cholesterol) can interact with sterol-sensing sequences that are part of the receptor for Hh (called Patched), which is located in the receiving cell. Studies of another gene, *tout velu*, a gene which encodes an enzyme needed for proteoglycan synthesis, shed light on Hh activity at this stage. Mutations in *tout velu* can affect the diffusion of Hh to which cholesterol is attached; in the absence of Tout Velu, Hh does not diffuse at all. Thus, Hh diffusion can apparently be modified by molecules, such as proteoglycans, that are present in the extracellular environment. This research area is moving rapidly.

Hh signaling is powerful and important, and it operates in several different developmental situations; we shall consider it further in Chapter 15. The local concentration of Hh is important, so factors that control the extent of its diffusion from a localized source can be expected to have important developmental consequences, which is why autocatalytic cleavage followed by esterification with cholesterol is of such interest.

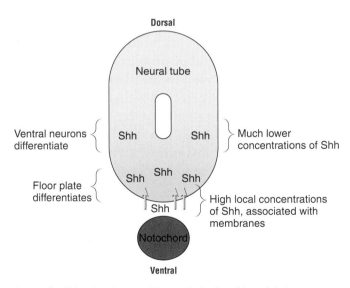

Figure 14.12   Gradients of Sonic Hedgehog Ligand A diagrammatic cross section through the developing vertebrate notochord and spinal cord. The notochord is the source of high local concentrations of Sonic Hedgehog (Shh), which remains mostly associated with cell membranes due to cholesterol esterification and possibly other posttranslational modifications. Floor plate-type neuronal cells differentiate under the influence of high Shh levels. Somewhat lower levels of Shh, found a little more dorsally, are sufficient to activate differentiation of ventrolateral-type neurons.

Diffusion of Sonic Hedgehog Is Important for Vertebrate Development

In the case of the vertebrate neural tube (where Sonic Hedgehog is the ligand), experimental results have proved interesting: A synthetically produced, truncated N-terminal peptide of Shh was applied in vitro to the developing neural tube. High concentrations of the engineered Shh resulted in the formation of floor plate neurons, whereas lower concentrations produced motor neurons. This experiment supports the situation thought to obtain in the embryo: During development, the notochord is normally the source of Shh signal. We would expect to find high concentrations of active Shh near the notochord because Shh is kept close to the cells of origin by the same means as shown for Hh in *Drosophila*—the tendency of cholesterol to favor the hydrophobic neighborhood of the phospholipids of the cell membrane. Indeed, the floor plate, which shows a high concentration of Shh, lies immediately adjacent to the notochord. The ventrolateral portion of the neural tube, where motor neurons form, is a little farther away from the source of Shh, so that is where we would expect to—and do— find much lower concentrations of Shh. Figure 14.12 shows this distribution.

Genes related to *disp* and *tout velu* will likely be found to be important in establishing concentration gradients of Shh in vertebrate tissues. Though more details will surely be revealed, it is already clear that posttranslational processing of the Hh family ligands is crucial to their diffusion and effective action during development.

Vg1, a Localized Ligand Implicated in Early *Xenopus* Development, Has to Be Processed to Be Active

In Chapter 16, we shall explore the molecules and mechanisms involved in patterning the body of the *Xenopus* embryo. Let us anticipate this discussion by illustrating the importance of the posttranslational processing of one of the molecular players thought to be involved in this patterning. The molecule is Vg1, which may play a role in forming mesoderm.

First some background. Both fibroblast growth factor (FGF) and activin (a member of the TGF-β family) are present in the early embryo. FGF is moderately active, and activin strongly active, in eliciting the formation of mesoderm in animal cap explants. But neither FGF nor activin is localized in the region of the Nieuwkoop center, which is known to be the source of mesoderm-inducing signals during cleavage. On the other hand, a suitable candidate for a mesoderm inducer from the Nieuwkoop center is the signaling molecule Vg1, also a member of the TGF-β family. Vg1 mRNA is transcribed in the oocyte and localized in the prospective endoderm that forms the Nieuwkoop center. In order to be active, Vg1, like the other members of the TGF-β family, must first be proteolytically cleaved from a precursor molecule and must then dimerize with a partner. However, for a while all experimental attempts to find evidence of biological activity by Vg1 in the early embryo failed to bear fruit.

To understand this conundrum, some clever experiments involving proteolytic cleavage were devised. Figure 14.13 shows the molecules involved. Douglas Melton and his colleagues at Harvard University reasoned that perhaps the proteases necessary for converting pro-Vg1 (the precursor of Vg1) to the active form are themselves either localized or kept in some precursor form; the researchers assumed it was the lack of processing of pro-Vg1 that was confounding their experiments. To circumvent this, they constructed a synthetic hybrid mRNA. The protein encoded by this mRNA consisted of the carboxyl terminal portion of Vg1 coupled with the amino terminus from a TGF-β family member called BMP2. (Chapter 16 will explore the role BMPs play in controlling gene expression in greater detail.) When the hybrid mRNA was injected into the embryo, it was translated to form a *fusion protein*; this precursor was then proteolytically cleaved by cells of the *Xenopus* embryo to produce a form of mature Vg1 almost identical in sequence to the natural molecule. This activated Vg1 proved very potent for inducing dorsal mesoderm in animal cap explants. It could even rescue a *Xenopus* embryo in which the development of dorsal mesoderm had earlier been completely inhibited by UV irradiation.

We do not yet know whether Vg1 is an important part of the Nieuwkoop center. What we do know is that posttranslational modification of precursor proteins is crucial. As pointed out earlier, there are many different kinds of modification. We should not be surprised to find that many developmental processes are regulated to some extent by control of the modification process itself.

The Assembly of Proteins into Macromolecular Complexes Constitutes Another Level of Regulation of Gene Expression

Proteins, which are encoded in the genes, assemble into multiprotein complexes, and these complexes may assemble with other macromolecular entities to construct the machinery of

A.

B.

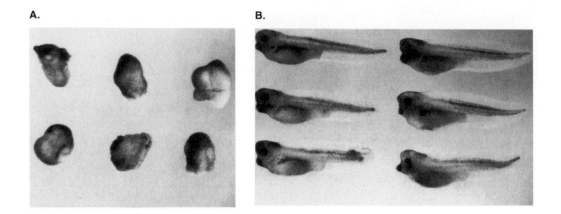

C.

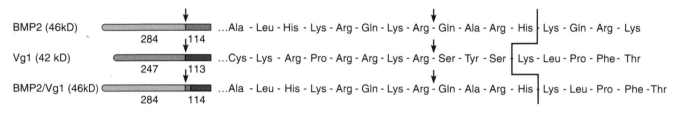

Figure 14.13   The Action of Vg1 Before and After Processing in *Xenopus* *Xenopus* embryos were irradiated with UV light to the vegetal pole at the one-cell stage. This treatment stops cortical rotation and development of dorsal axial structures. When the embryos reached the eight-cell stage, their vegetal blastomeres were injected with mRNA encoding either **(A)** Vg1 or **(B)** a fusion protein of both BMP and Vg1. The embryos receiving Vg1 mRNA continued to look like uninjected UV-irradiated embryos (not shown); Vg1 had no effect on their development. In contrast, the embryos receiving the hybrid mRNA showed almost normal development; presumably, this is because the BMP2/Vg1 fusion protein was easily processed by specific proteolytic digestion that made mature Vg1 ligand available. **(C)** Schematic diagrams of Vg1, BMP2, and the fusion construct before processing. Arrows separate nonligand domains on the left from the ligand domain for the mature proteins on the right; numbers underneath indicate their length (number of amino acids). To the right of each whole protein schematic is the amino acid sequence of the region of BMP-Vg1 fusion. Arrows indicate proteolytic cleavage sites. The jagged line indicates the position of suture between carboxyl BMP2 and the amino Vg1.

the cell. A dominant idea in this field has been the concept of *self-assembly,* which originated out of studies on how bacteriophage are constructed. The proteins of the phage head, once produced, assemble without further ado into complex, structures capable of encapsulating the phage genomic DNA; after this, the phage are ready to infect bacteria. The term *self-assembly* is misleading, however, for phage head construction occurs only under specific conditions of ionic strength, pH, and divalent cation content—all of which, of course, are stringently regulated by living cells.

Recent studies of eukaryotic cells have revealed a complex set of proteins (called *chaperones*) whose function is to guide the correct folding of proteins. The assembly of organelles—such as the cytoskeleton, mitochondria, the mitotic appara-

tus, and so on—is not completely understood; what we do know is that it is a complicated, multistep process, subject to regulation. So, while the concept of self-assembly is useful, it should not cause us to neglect the conditions (such as pH or the presence of a particular chaperone) that are necessary in order to complete terminal differentiation. For instance, to understand how the lens of the eye is formed, we need to know not only about synthesis and posttranslational modification of the major lens proteins (crystallins), but also about the activity of accessory proteins, chaperones for proper folding, and regulation of the ionic milieu. Self-assembly by itself is inadequate for understanding the formation of cellular structures. We know that chaperone synthesis and activity, and ion flow into and out of the cell, are strictly regulated. Thus, the eventual function of any protein is dependent in some way on the activities of other proteins, and in some sense is regulated by those activities.

Differentiation of Skeletal Muscle Illustrates the Importance of Protein–Protein Interactions

The construction of a skeletal muscle fiber involves many steps over an extended time period. The founding cells have to be determined for entering the mesodermal pathway and adopt "muscleness" as a fate. This determination is brought about by cell–cell signaling, which results in differential gene expression (the subject of this and subsequent chapters). Thereafter, a series of important decisions, some of which we discussed in Chapter 7, must be carried out: the myoblasts have to be retired from the cell division cycle, they need to develop the capability of fusing at their bipolar ends to create a syncytium, a muscle-specific cytoskeleton then has to develop, and the muscle cells must acquire a sarcoplasmic reticulum (endoplasmic reticulum in muscle cells adapted for calcium storage). The proteins of the contractile machinery that comprise the myofibrils have to be synthesized and assembled. The thick filament composed of myosin is precisely arranged in the middle of each sarcomere unit; surrounding the thick filament are thin filaments formed from actin molecules. The actin is anchored into a Z disc at the end of each sarcomere. The actin filaments are decorated with tropomyosin, which interacts with troponin. It is this tropomyosin–troponin interaction that confers sensitivity to calcium ion on the contractile apparatus. One cannot simply mix all these proteins together to get a muscle fiber. Before the muscle can contract, there must be precisely timed synthesis, posttranslational modification, and localization of the proteins, followed by regulated construction of the organelles.

Though these terminal steps of differentiation are more visible than the early ones, we still understand very little about how they take place, not only for muscle but also for most differentiated cell types. This puzzle constitutes a challenge for future research.

Molecular and Cellular Turnover Contributes to Posttranslational Control of Gene Expression

With the exception of DNA, all the macromolecular constituents of a cell are subject to degradation: molecules *turn over.*

Messenger RNA molecules may have a lifetime ranging from less than a minute (as with the mRNA encoding δ-amino levulinic acid synthetase, which participates in the synthesis of heme in the liver) to several days (such as hemoglobin mRNA in a human reticulocyte). Thus, mRNA "lifetimes" can greatly affect how much protein is synthesized, and thereby affect how much protein is present. For example, regulating the lifetime of histone mRNA molecules is very important for the embryo's ability to carry out cleavage divisions during early development.

Protein molecules are also subject to degradation, and are constantly being replaced. Indeed, there is an elaborate cellular machinery dedicated to the removal of misfolded and damaged proteins. But even correctly folded and functioning proteins are constantly being destroyed and renewed. Biochemists and physiologists have known for a long time that some protein molecules, such as hemoglobin, may persist many, many weeks after synthesis. On the other hand, heart muscle cells replace most of their proteins in a week or two, and nerve cells are continuously renewing proteins present in the neurites. There are many situations during development that require removal of active proteins. Not much is known yet about the exact mechanisms of protein destruction during development; because of its significance, however, it is likely to become an important area of research.

Of course, the cell as a whole is also subject to senescence and death. Programmed cell death, or apoptosis, is important in early development, as we saw in Chapter 6. Two notable regions where apoptosis is important in development are the spinal cord and limbs. Many growth factors, such as NGF, are important either for stimulating mitosis or for maintaining the cell in a postmitotic state. Molecular turnover and apoptosis have a significant influence on the development of an embryo.

In this chapter, we have discussed the principal levels of regulation of gene expression, ranging from transcription at the level of DNA, to assembly of multimolecular complexes that form organelles and tissues. In the next two chapters, we shall turn to situations in the developing embryo—some familiar, some new—where various tactics are combined to bring about patterns of differentiation in the embryo. And again we shall consider the question posed at the beginning of this text: how is differential gene expression at specific times and places orchestrated so as to produce a correctly patterned organism?

KEY CONCEPTS

1. In eukaryotic cells, there are many steps that comprise gene expression—ranging from DNA transcription, to protein assembly into cellular structures—and all are subject to regulation in all cells. Hence, each step is a potential control point for effecting or influencing differential gene expression during development.

2. Three control points, or nodes, are especially important for differential regulation: transcription, translation, and posttranslational modification.

3. Transcription is often governed by regulatory proteins interacting with specific DNA sites at some distance from the basal promoter. These interactions may activate or repress efficient initiation of transcription.

4. Translation of mRNA can be regulated in two important ways: It can be prevented outright by specific proteins binding to the 3′ UTR of the mRNA. Translation can also be limited to particular locales in the embryo by having the mRNA be spatially tethered to certain intracellular regions.

5. Posttranslational modification of proteins, including assembly into macromolecular complexes, is often essential for their proper activity.

6. The cellular machinery that modifies proteins may itself be regulated or may selectively influence differential gene expression.

STUDY QUESTIONS

1. An analog of cytidine is 5-azacytidine (5-azaC), which is taken up by cells and incorporated into their DNA. The presence of the azo group (as this -N=N- group is called when standing alone) at the 5′ position makes it impossible to methylate this analog. There is a line of tissue culture cells (10T3-1/2) which, when exposed to 5-azaC, will differentiate in vitro into muscle cells. What explains this result?

2. What consequences for the development of *Drosophila* would arise from producing mutations in the gene for RNA polymerase III?

3. Suppose you could create a reporter gene to examine the function of various portions of the regulatory regions upstream of the *endo16* gene in sea urchin embryos. Predict the outcome of the activity of a reporter in the embryo that is driven by *endo16*'s basal promoter and includes modules A, B, E, and F. (Refer to Figure 14.6, which diagrams the regulation of this gene.)

4. The *oskar* gene is one of several necessary for localization of Nanos mRNA to the posterior pole of the *Drosophila* oocyte. What would be the effect of an *oskar*$^{-/-}$ mutation on development of the anteroposterior axis of *Drosophila*?

SELECTED REFERENCES

Curtis, D., Lehmann, R., and Zamore, P. D. 1995. Translational regulation in development. *Cell* 81:171–178.

An excellent review of the roles of localization, polyadenylation, and other mechanisms on translation and development.

Davidson, E. H., and 24 other authors. 2002. A Provisional Regulatory Gene Network for Specification of Endomesoderm in the Sea Urchin Embryo. *Develop. Biol.* 246:162–190.

An example of a model of complex networks of gene action

Ingham, P. W. 2000. Hedgehog signaling: How cholesterol modulates the signal. *Curr. Biol.* 10:R180–R183.

A summary of new developments in understanding how the diffusion of Hedgehog protein is controlled.

Johnson, R. L., and Tabin, C. 1995. The long and short of Hedgehog signaling. *Cell* 81: 313–316.

A review of the role Hedgehog protein gradients play in intercellular signaling in the embryo.

Kadonaga, J. T. 1998. Eukaryotic transcription: An interlaced network of transcription factors and chromatin-modifying machines. *Cell* 92:307–313.

A review of how the structure of chromatin affects the process of transcription.

Lemon, B., and Tjian, R. 2000. Orchestrated response. A symphony of transcription factors for gene control. *Genes Dev.* 14:2551–2569.

A detailed review of how transcription factors and co-activators are organized in the nucleus.

McMahon, A. P. 2000. More surprises in the Hedgehog signaling pathway. *Cell* 100:185–188.

An analysis of recent developments in ascertaining how Hedgehog protein diffuses from its origin to target cells.

Parisi, M., and Lin, H. 2000. Translational repression: A duet of Nanos and Pumilio. *Curr. Biol.* 10:R81–R83.

A review about how Nanos protein represses translation of Hedgehog mRNA.

Perler, F. B. 1998. Protein splicing of inteins and Hedgehog autoproteolysis: Structure, function and evolution. *Cell* 92:1–4.

A scholarly review on the subject of protein autocatalytic domain excision, of which Hedgehog processing is an example.

Porter, J. A., and 10 other authors. 1996. Hedgehog patterning activity: Role of a lipophilic modification mediated by the carboxy-terminal autoprocessing domain. *Cell* 86:21–34.

A research paper describing the processing and cholesterol esterification of Hedgehog protein.

Rosenthal, E. T., and Wilt, F. H. 1987. Selective mRNA translation in marine invertebrate oocytes, eggs and zygotes. In *Translational regulation of gene expression*, ed. J. Ilan, pp. 87–110. Plenum Press, New York.

A review of translational-level regulation at fertilization in a variety of marine eggs.

Siegfried, Z., and Cedar, H. 1997. DNA methylation: A molecular lock. *Curr. Biol.* 7: R305–R307.

A short review and key references to the subject of DNA methylation.

Struhl, K. 1998. Histone acetylation and transcriptional regulatory mechanisms. *Genes Dev.* 12:599–606.

A review of how the addition and removal of acetyl groups influence the transcription machinery.

Thomsen, G. H., and Melton, D. A. 1993. Processed Vg1 protein is an axial mesoderm inducer in *Xenopus*. *Cell* 74:433–441.

Experiments using BMP2/Vg1 fusion proteins to test Vg1 function.

Wijgerde, M., Grosveld, F., and Fraser, P. 1995. Transcription complex stability and chromatin dynamics in vivo. *Nature* 377:209–213.

Experiments that bear on how the locus control region controls the dynamics of β-globin gene expression.

Yuh, C.-H., Bolouri, H., and Davidson, E. H. 1998. Genomic *cis*-regulatory logic: Experimental and computational analysis of a sea urchin gene. *Science* 279:1896–1902.

A research paper analyzing how the modules of the *endo16* gene work.

CHAPTER

15

DEVELOPMENTAL REGULATORY NETWORKS I: *DROSOPHILA* AND OTHER INVERTEBRATES

In preceding sections of the book, we undertook to describe how several different kinds of organisms develop. Then we examined the cellular behaviors that underpin morphogenesis during development. And in the preceding chapter, we explored aspects of how gene expression is regulated. Now it is time to put this information together in order to discern the overall strategy and real-life tactics used in living embryos. This is the task before the contemporary researcher and student alike—to apply our newly acquired knowledge of the way cells work to understanding the miraculous undertaking of how real embryos construct an organism from an egg. In this and the next chapter, we shall make a stab at this enterprise.

CHAPTER PREVIEW

1. Asymmetric cell divisions involve preexisting localization of proteins in the mother cell.

2. Some intercellular signaling is inhibitory.

3. The body segments of *Drosophila* are established by a cascade of complex interactions between proteins and the control regions of certain genes.

4. Homeotic selector genes govern the differentiation of different segments.

5. Development of the insect wing shows how intercellular signaling can affect gene expression.

GENERATING NONEQUIVALENT CELLS

Developmental Networks Are Complex

At the heart of our inquiry is how cells in the embryo become different from, and communicate with, one another. This is where we started in Chapter 1, and ultimately where we shall finish. Figure 15.1 summarizes this central set of issues. Two nonequivalent sibling cells arise from a precursor cell that is originally nonpolarized but becomes polarized before mitosis. Several kinds of polarities—in receptors, in cytoskeletal architecture, in transcription factors, and/or in signal production capabilities—may exist in the precursor cell (Figure 15.1A). These possibilities are neither mutually exclusive nor exhaustive. It is also possible that sibling cells in the early embryo are identical, and that their nonequivalence arises when they come to reside in different microenvironments.

The nonequivalent division can result in cells communicating with one another by utilizing their differences as a basis for communication. The messages sent may be simple (one ligand) or complex (a cocktail of ligands), stimulatory or repressive, quantitatively strong or weak, or sent at various times. On the other hand, the receiving cell or cell population may have different receptors and co-receptors; timing of the presence of receptors is regulated too. Signals from other places may bombard the receiving cell, and they may be synergists or antagonists of local signals. Some of these variables are shown in Figure 15.1B. The receiving cell must interpret the received message or messages using complex second-messenger pathways. These decoded, modified messages can result in output on the part of the receiving cell; common examples of output include changes in cytoskeletal behavior, and activation of transcription of a gene or set of genes (Figure 15.1C).

Because Figure 15.1 is generalized and abstract, it under-represents the true complexity of most cellular interactions and responses in the embryo. But in essence, development depends on generating nonequivalent cells and then regulating the sending and receiving of messages between them.

We turn next to examining some instances where asymmetric cell division produces two nonequivalent daughter cells with profound developmental consequences. These examples will illustrate the importance of the cytoskeleton in establishing asymmetry and the importance of regulated signaling between the nonequivalent cells. In the remainder of the chapter, we revisit the *Drosophila* embryo to explore how the early asymmetric placement of transcription factors and receptors, discussed in Chapter 3, serves as a foundation for

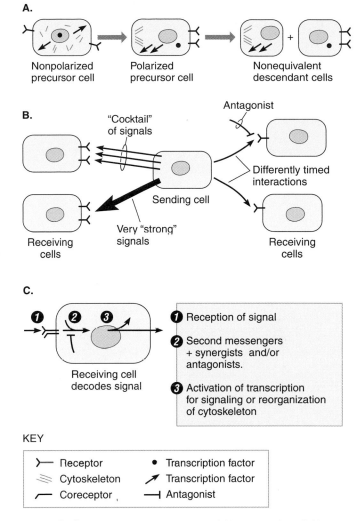

Figure 15.1 Asymmetric Divisions and Nonequivalent Cells A diagrammatic representation of several different ways in which asymmetric cell division can initiate signaling. **(A)** A nonpolarized cell can become polarized and then give rise to progeny that contain different amounts of a given transcription factor, receptor, ligand, or cytoskeletal structure. **(B)** A cell sending a signal by means of either a diffusible or a tethered ligand can have different effects on two receiving cells. The receiving cells may receive the signals at different times, or outside signals (antagonists and/or synergists) may be involved in signal reception. Furthermore, different second-messenger transduction pathways (*not shown*) could be present in different receiving cells responding to the same ligand. **(C)** The receiving cell decodes and transduces the signal by (1) receiving it; (2) engaging second-messenger pathways, which may be affected by synergists, antagonists, or both; and then (3) altering transcription rates or reorganizing the cytoskeleton.

establishing details of the body plan of the embryo. We finish the chapter by concentrating on one region of the body, the wing, to illustrate how signaling between nonequivalent cells operates during organ formation and differentiation.

Asymmetric Cell Divisions in Yeast Give Us Clues

Yeast cells are among the simplest eukaryotic cells, yet they display the phenomenon of nonequivalent cell division. Along with some prokaryotes, *Caenorhabditis elegans*, and *Drosophila*, yeasts have provided us with extraordinarily detailed genetic knowledge. When a yeast cell buds off a new daughter cell, the daughter retains the same mating type as its mother. However, after undergoing this mitosis, the mother cell routinely changes to the opposite mating type. In yeast, which propagates asexually as haploids, mating (and consequent formation of diploid cells) can occur only between cells of opposite mating type. Mating-type conversion thus has the happy outcome of putting cells of opposite type close to one another, thereby facilitating mating (Figure 15.2A,B).

Due to the power of genetic analysis in yeast, a considerable amount is known about the molecular basis of mating-type conversion. At its heart is the replacement of one mating-type gene (say α) with the gene encoding the opposite type (call it *a*), which until the conversion exists in a cryptic, untranscribed state elsewhere in the genome. This gene conversion (*a* replacing α) is governed by the presence of an endonuclease called the Ho protein. The *ho* gene is actively expressed only in mother cells; hence only they can switch mating types. A large number of mutations in other genes affect the specific expression of *ho* in the mother cell.

The key regulator of mating type conversion seems to be a protein called Ash1p, which represses Ho mRNA synthesis. (Ash1p stands for asymmetric synthesis of Ho.) Ash1p mRNA is localized to the daughter cell, where it is translated (Figure 15.2C). Ash1p prevents Ho expression there, probably by binding to the *ho* promoter in the daughter cell. A large number of genes affecting the cytoskeleton can interfere with the localization of Ash1p mRNA in the daughter, including a gene encoding a form of myosin. Ultimately, it is a polarization of the cytoskeleton that controls mating-type conversion.

The site of bud formation, and hence of the daughter cell, is also complex but well studied. In haploid yeast, the site is marked by a spatial cue in the cytoskeleton, and is located adjacent to the previous bud site. Thus, in yeast, the asymmetry that results in cellular polarization is brought about by an inherent polarity embedded in the cytoskeleton. It is this

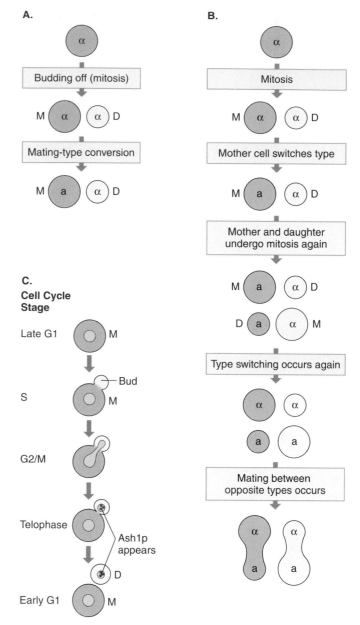

Figure 15.2 Mating-Type Conversion in Yeast (A) The mother cell (M) undergoes switching after mitosis, but the daughter (D) does not. **(B)** Mating-type conversion facilitates the presence of competent mating partners in the same vicinity. **(C)** The Ash1p protein is localized to daughter cells during mitosis. Ash1p accumulates during telophase and remains in the daughter cell nucleus through early G1 of the next cell cycle, after which it disappears. The daughter cell can then become a mother cell.

polarity that is responsible for generating the difference between mother and daughter cells.

Asymmetric Cell Division in the Early C. elegans Embryo May Result from Cell Signaling

Now we examine a very different situation, in the embryo of a simple invertebrate, where the emphasis is on how the sending and receiving of a signal are regulated. One part of a cell receives a signal while another part does not; when mitosis separates this cell into two parts, the two daughters are quite different from one another as a result of the asymmetry of the signaling. The nematode worm *C. elegans* undergoes stereotyped, essentially invariant cell divisions. (See Chapter 2 for an earlier discussion of this embryo.) The 558 somatic cells present in the hatched embryo all descend from five founder cells. The AB, C, and MS cells give rise to a mixture of 518 somatic cells of different types: neurons, epidermis, muscle, and so on. The D cell results in 20 body-wall muscle cells, and the E cell gives rise to the 20 cells of the gut. A sixth cell, P_4, produces the germ line. Figure 2.13 diagrammed the early divisions giving rise to these cells; for your convenience, we give the information again (Figure 15.3).

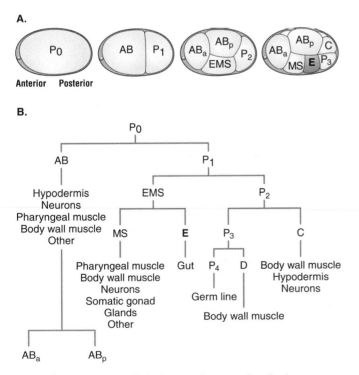

A.

Anterior   Posterior

B.

Figure 15.3   Reprise of the Lineage Diagram for *C. elegans* **(A)** The one-, two-, four-, and eight-cell stages of a *C. elegans* zygote, viewed from the side. **(B)** The lineage diagram, identifying cell names and fates.

Because the cell divisions in *C. elegans* are so regular and invariant, it was once thought that each cell must be autonomously programmed from the outset. Now that experimental manipulations of these early embryos have been carried out, we know this is far from the case. Several important asymmetric cell divisions occur during early cleavage, and their cellular and genetic basis, including cell signaling, is beginning to be understood. The first division of the egg (P_0) is nonequivalent; the P_1 cell contains P granules and other components, and only P_1 can give rise (through several cell divisions) to P_4, and thence to germ cells. The site of sperm entrance dictates polarization of the fertilized egg, and at least six genes (called *par* genes, for "partitioning defective") help organize the cytoskeleton and polarize the cell. It is the cytoskeleton, then, that sets up P granule distribution. Some of the *par* genes are known to involve the cytoskeleton. As with mating-type switching in yeast, cytoskeletal polarity plays an important role in generating nonequivalent cells in *C. elegans*.

Let us focus on the second cell divisions. The AB cell gives rise to an anterior daughter (ABa) and a posterior daughter (ABp), and P_1 gives rise to the P_2 and EMS cells (Figure 15.3B). In the early 1990s, while a graduate student at the University of Texas, Bob Goldstein was able to show that a brief, transitory interaction between the EMS and P_2 cells results in polarization of EMS, so that it subsequently divides (the third cell division) into two cells, E and MS, with very different developmental capabilities. Removing P_2 early after the second division eliminated formation of the E cell from EMS. Restoring P_2 so that it touched any part of the EMS cell allowed the E cell to form, resulting in subsequent development of the gut.

Some of the molecular basis for this interaction is now known, and several genes that are important for the influence of P_2 have been identified. For example, *mom2* is active in P_2 and encodes a signaling glycoprotein similar to members of the well-known Wnt signaling family. Wnt members, you may recall, play important roles in the development of many other animals, including fruit flies, frogs, and mice. Another gene, *mom5*, must be expressed in the receiving cell, EMS. Mom5 is homologous to the *Drosophila* Wnt receptor Frizzled. A consequence of the signal Mom2 being received by Mom5 is that the accumulation of a transcription factor called Pop1 is downregulated in the posterior end of the receiving cell. This unequal presence of Pop1 carries over to the two daughter cells: Pop1 is present in higher concentrations in the anterior daughter than in the posterior daughter in a number of different asymmetric cell divisions in *C. elegans*. In fact, a high level of Pop1 may specify an anterior fate. (Note from Figure 15.3 that MS is anterior to

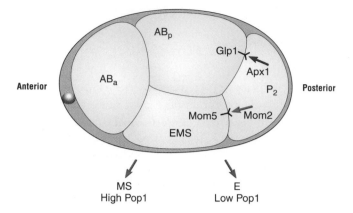

Figure 15.4   Blastomere Interactions at the Four-Cell Stage in
C. elegans The P$_2$ cell signals the adjacent EMS cell (ligand Mom2
interacts with receptor Mom5). The EMS cell will divide into an anterior MS cell and a posterior E cell. Only the E cell will have been influenced by the Mom2–Mom5 interaction, leading to a reduction in the
transcription factor Pop1. P$_2$ also signals the adjacent ABp cell to adopt
its correct fate (by sending ligand Apx1 to interact with receptor
Glp1).

E in our present example.) When P$_2$ is eliminated, the signal to
downregulate Pop1 in a portion of the adjacent EMS cell is not
received. In that situation, no E cell can form, so both daughters will be MS cells. Figure 15.4 summarizes the molecular
interactions between P$_2$ and EMS.

We shall meet homologs of Pop1 in *Drosophila* (Pangolin)
and vertebrates (TCF and Lef), where they act as effectors of
Wnt signaling. The asymmetric cell division of EMS is
brought about by an asymmetric presentation of a ligand.
The part of EMS that interacts with the Wnt ligand becomes
E; the part of EMS that does not interact with it becomes MS.
The asymmetric first division, which involves cytoskeletal
polarity, thus results in a patterned ligand presentation that
produces another nonequivalent cell division.

P$_2$ not only influences EMS development. It also influences
the ABp cell to follow its "correct" developmental path. A ligand
encoded by the *apxl* gene is presented by P$_2$ to ABp, which
possesses the receptor encoded by the *glpl* gene (Figure 15.4).
This event helps to explain the known embryological facts:
ABa and ABp can both give rise to the same kinds of tissues.
Only after one of them is "touched" by the adjacent P$_2$ cell
does it adopt the ABp fate, giving rise to the distinct array of
tissues known to develop normally from ABp.

Thus, the same cell, P$_2$, presents a different ligand to two
different neighbors possessing different cognate receptors,
thereby signaling each receiving cell to enter its appropriate
pathway. What makes this situation extraordinarily interesting is that several years earlier, the homologs of these genes

from *C. elegans* were identified as crucial developmental regulators in *Drosophila*, where the ligand homologous to Apx is
encoded by the gene *delta* and the receptor homologous to Glp1
is encoded by the gene *notch*. As with Wnt, the Notch signaling
pathway is found in many embryos in different phyla.

There is another interesting twist to this story. The mRNA
for Glp1 is present in all the cells of the embryo, but Glp1, a
membrane receptor, is present only in the ABp cell. The Glp1
mRNA is translated in blastomeres located anterior to P$_2$,
but not the posteriorly situated P$_2$. We conclude that the
ability to suppress translation of Glp1 mRNA must also be
asymmetrically distributed to posterior cells!

Asymmetric Divisions of Neuroblast and Sensory Organ Precursors Utilize Cytoskeletal Cues

A final example of how asymmetric divisions are established
comes from studying the formation of neurons in *Drosophila*.
Once again, the details that we know (and we need to know
more) have been gleaned through genetic analysis coupled
with traditional embryology. To establish the various kinds
of cells found in the peripheral and central nervous systems,
a precursor cell gives rise to distinct progeny by means of
clearly asymmetric mitoses (Figure 15.5).

Let us consider some of the details in these two lines of
neural development. You will remember from Chapter 3 that
the cells in the ventrolateral ectoderm of the embryo delami-
nate into the interior during gastrulation; these delaminated
cells are *neuroblasts*, cells that form the neurons and glia of the
ventral nerve cord. The neuroblasts produce a protein called
Numb, which becomes localized to the basal side of the cell.
When the neuroblast divides, separating the mother neuroblast into an apical daughter neuroblast and a basal *ganglion
mother cell (GMC)*, the store of Numb is partitioned preferentially to the GMC. In order for the cell division to produce
apical and basal progeny, the cell division plane must rotate
90° (Figure 15.5A). The GMC then divides to form two neurons. (When Numb is not localized, the plane of cell division
in the mother neuroblast fails to rotate 90° and neurons do
not form.) The daughter neuroblast accumulates Numb again,
and once more the protein is asymmetrically partitioned to
the GMC. Numb is known to antagonize Notch activity, a
subject we shall soon discuss.

The body of an adult fruit fly is covered with sensory bristles, each composed of a peripheral neuron, a hair cell, a
sheath cell, and a socket cell. These four cells are derived
from a *sensory organ precursor (SOP)*, a cell located in the surface
epithelium of the pupa. During pupal development, the SOP
undergoes two asymmetric cell divisions; with each division,

A. CNS Neuroblasts (NB)

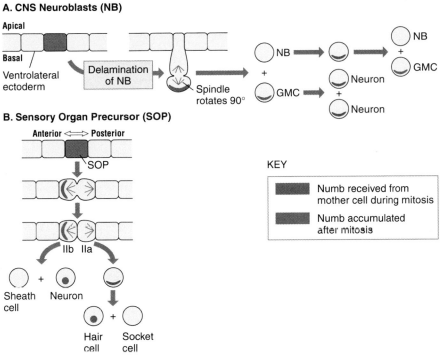

B. Sensory Organ Precursor (SOP)

KEY

▬ Numb received from mother cell during mitosis

▬ Numb accumulated after mitosis

Figure 15.5 Asymmetric Cell Divisions in the Embryonic Ectoderm of *Drosophila* (A) An idealized section through the surface ectoderm of the embryo. The neuroblast cell with Numb delaminates. the mitotic spindle rotates. and an apical neuroblast (NB) stem cell and a basal ganglion mother cell (GMC) are formed. Numb is partitioned to the GMC. which then divides to form two neurons. The neuroblast accumulates Numb again, then divides to form a GMC (containing Numb) and another neuroblast (initially not containing Numb). The neuroblasts give rise to the neurons of the ventral nerve cord and supporting glial cells. **(B)** An idealized section through the surface ectoderm of the pupa. The SOP cell accumulates Numb. which is localized as a crescent on the anterior side of the cell. The anterior (IIb) cell that receives Numb gives rise to one cell with Numb (a neuron) and one cell without Numb (a sheath cell). The posterior (IIa) cell. which did not receive any Numb from the SOP during formation. nevertheless starts to produce Numb. This Numb. too. is localized. so that one daughter (a hair cell) receives it while the other daughter (a socket cell) does not.

its store of Numb is apportioned to only one of the two daughter cells (Figure 15.5B). When the SOP divides, the anterior cell (called IIb) receives the Numb while the posterior cell (called IIa) does not. Note that in this instance, the asymmetry is aligned within the plane of the surface epithelium, and does not require rotation of the mitotic spindle. IIb then divides to form a neuron, to which Numb is partitioned, and a sheath cell, which receives no Numb from IIb. IIa, initially without any Numb, accumulates Numb on its own, and then divides to form a hair cell (which receives Numb from IIa) and a socket cell (which does not).

How is Numb localized? Let us examine what is currently known for the case of apical-basal localization in the neuroblast. Since the neuroblast is embedded in an epithelium before it delaminates, it already has an intrinsic apical-basal polarity; this polarity presumably provides some cue associated with membrane organization, the cytoskeleton, or both. Recent evidence indicates that the cue is a complex of several proteins (one with the colorful name Bazooka) that are localized near the apical membrane. One member of this complex is Inscuteable, a multidomain, multifunctional protein localized to the apical membrane during the prophase and metaphase of the cell division of the delaminated neuroblast. Inscuteable is needed for correct spindle rotation and for localizing Numb. It is also required for localizing *prospero* gene products (mRNA and protein) to the basal portion of the cell, that is, to the

cytosol of the prospective GMC. (Prospero is a homeodomain-containing protein that regulates transcription of genes involved in neuronal development.) Inscuteable function is mediated by certain adaptor proteins: Pon (short for Partner of Numb), Miranda, and Staufen. These relationships are diagrammed in Figure 15.6. Recent experiments have shown that *inscuteable* expression can be regulated by *snail*, which is a target of the maternal transcription factor Dorsal. Hence, it appears that researchers are close to delineating the pathway linking asymmetries in the egg to asymmetric cell divisions in the ventrolateral ectoderm giving rise to neurons.

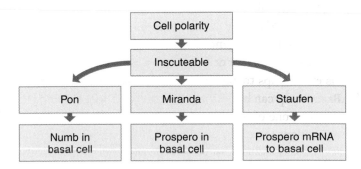

Figure 15.6 The Inscuteable-Numb Pathway An outline of the various steps in localizing Numb and Prospero to the prospective daughter basal cell in an asymmetric cell division. Pon is a recently discovered protein that associates with Numb and is necessary. at least in some instances. for its appropriate localization.

It is probably not surprising that the machinery for localizing Numb in the SOP is somewhat different from that utilized in neuroblasts. In this instance, the asymmetry is expressed not by an apical-basal difference, but along the anteroposterior axis of the embryo within the plane of the epithelium. Null mutations of *inscuteable* do not affect Numb localization in the SOP. The Numb always forms a crescent in the SOP so that the anterior daughter receives most of the Numb. However, mutations of *frizzled* (which encodes the receptor for Wnt ligands) have been shown to disrupt Numb localization, implicating Wnt in planar polarity. Attempts to discover the details of this "Wnt planar-polarity" pathway and the molecular players involved are a very active area of research; several proteins are known, among them a form of cadherin and the cytoplasmic protein Disheveled (Dsh). This pathway is distinct from the canonical Wnt pathway shown later in the chapter (Figure 15.28) and does not utilize β-catenin.

Thus, an elaborate set of molecular interactions is necessary to localize Numb. These interactions constitute a kind of transduction apparatus that translates a membrane or cytoskeletal polarity into a protein localization. The localization can then lead to an asymmetric cell division, *the* crucial step in embryonic development.

Inhibitory Signaling Between Cells Is a Common Mechanism in Development

Genetic studies in *Drosophila* support the idea that one function of Numb is to inhibit reception of signals via the Notch pathway. We shall meet the Notch system in several different contexts; this is a good place to become acquainted with it. The gene *notch* was named for a mutation first found in *Drosophila*, so called because some of its alleles cause notches to appear in the wings of heterozygotes. Notch turned out to be at the heart of a widespread and important signaling system that uses the principle of **lateral inhibition** (Figure 15.7A). Lateral inhibition acts as an insurance policy to maintain the separate identities established by asymmetric divisions. When *notch* function is eliminated in *Drosophila*, for example, the result is an embryonic lethal. The ectoderm that normally would give rise to the central nervous system and ventrolateral ectoderm (see Figure 3.12) forms no epidermal surface in the null *notch* mutant, only neuroblasts. This mutation became a key to understanding how some cells form neurons and others epidermis.

Many instances of lateral inhibition using the Notch pathway are known, such as development of the genitalia in *C. elegans*,

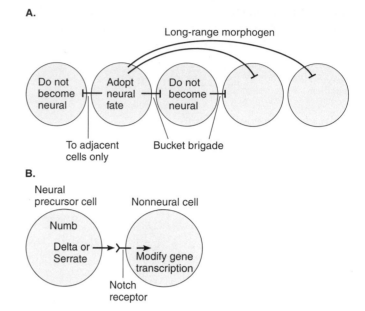

Figure 15.7   Lateral Inhibition (A) As with other kinds of signaling, lateral inhibition can involve one or more kinds of interactions: the signal may be passed to adjacent cells only; it may be passed directly from cell to cell in "bucket brigade" fashion; or it may diffuse over longer distances, bypassing intermediate cells. **(B)** Shown here is an example of lateral inhibition utilizing the Notch pathway in *Drosophila*. The cell to the left has adopted a neural fate and possesses Numb. This neural precursor cell utilizes the ligand Delta or Serrate to signal the neighboring cells via its receptor, Notch. The activation of Notch then results in the repression of proneural genes; as a result, this cell cannot adopt a neural fate but must remain an epidermal precursor. The cell on the left might also possess Notch receptors, but activation of Notch in this cell is suppressed by Numb. Because the cell on the right does not have Numb, it receives an effective signal to Notch. The intracellular events subsequent to Notch activation, which are complex, are discussed later in the chapter in connection with wing development.

assignment of cell fates in the sea urchin embryo, and proper differentiation of T (thymus-derived) cells in the immune system of mammals. Furthermore, there are many examples in the classical embryological literature in which differentiation of one cell type precludes a neighbor from undergoing the same differentiation. Presumably, lateral inhibition is going on in these instances, too, but the molecular basis is unknown.

Let's take a closer look at how lateral inhibition operates in the Notch pathway for differentiation of neuroblasts from ventrolateral ectoderm (Figure 15.7B). When an ectodermal cell with Notch function directly contacts a cell containing either Delta or Serrate (both are ligands for Notch), its destiny is restricted—it must remain an epidermal cell. A similar situation exists for the case, discussed earlier, of the IIa and

IIb cells arising from the SOP cell (see Figure 15.5B). IIa and IIb both possess the Delta ligand and Notch receptor, but Notch receptor function is inhibited in IIb by Numb. Thus, the Notch signaling system does not determine a cell's fate; rather, it acts as a receptor of inhibitory signals, so that once a cell has adopted a neural or proneural fate, the neighboring cell expressing Notch is restrained from adopting the same fate.

Both Notch and Delta are membrane-bound proteins; the interacting cells must touch for the communication to occur (Figure 15.7B). The Delta/Serrate ligand may be proteolytically cleaved, and Notch itself is cleaved in a number of steps to generate the active intracellular form. The active form of Notch interacts with other proteins that suppress transcription of proneural genes. This guarantees that the Notch-expressing cell, when activated by a ligand, will not form a neuroblast. Later in the chapter, in our discussion on wing development, where it will be shown that Notch can also act as a transcriptional activator, we shall discuss in further detail how Notch activity is transduced.

Notch signaling is both important and complex; many other proteins interact with Notch and modify its signaling, and much of our understanding about its function is still murky. However, we can say that one consequence of Numb localization is to establish Delta–Notch signaling; this signaling can then maintain an orderly distinction between nonidentical sibling cells.

The Machinery Used to Establish Asymmetry Is Complex, Yet Its Use Is Widespread

The apparatus used to initiate and maintain asymmetry resulting in nonequivalent cells is complex. This is probably because localizing materials to only one daughter cell is not simple. However, the result is essential. As we pointed out in Chapter 1, the fact that cells generated by mitosis have the same genetic information logically requires that mechanisms exist for establishing nonidentity between sibling cells. The examples we have discussed in this chapter—mating-type conversion in yeast, gut development in *C. elegans*, and neuroblast formation in *Drosophila*—all possess ways to ensure that some crucial gene product (protein, mRNA, or both) goes to one daughter cell and not to the other. This asymmetric distribution can lead to regulated signaling between cells, as in the case of Numb suppressing the Notch signal pathway. Or regulated signaling may itself play a crucial role in the polarization of a precursor cell that generates an asymmetric mitosis, as in the case of the P_2 cell influencing the EMS cell in *C. elegans*.

The details of signaling regulation also vary in different organisms. Some of the molecular players we have discussed are unique to a given organism or a particular function. For instance, Ash1p has been found only in yeast and seems to operate only in the assignation of mating types. On the other hand, Staufen is involved in protein localization in several different contexts, such as the sequestering of Bicoid in the *Drosophila* egg and of Prospero during development of the fruit fly's nervous system. Some mechanisms, such as the Wnt signaling pathway utilized by the P_2 cells of *C. elegans*, seem well nigh universal, used again and again in different contexts, both within a given organism and in many different organisms.

ESTABLISHING THE SEGMENTS IN *DROSOPHILA*

Morphogens Initiate Differential Gene Expression

When we discussed development of *Drosophila* in Chapter 3, we considered the gradients of maternal transcription factors and signaling pathways that establish the body axes of the embryo. The gradients of Bicoid, Nanos, and Dorsal and the localization of the receptor Torsolike form the basis for the anteroposterior and dorsoventral organization. A summary of these findings—a recap of this discussion from Chapter 3—is given in Box 15.1. It should be clear that asymmetric cell divisions (for example, into germ and nurse cells) and regulated signaling (as with Gurken or Torsolike) are the foundation of the patterned organization of the zygote.

But *Drosophila* is a segmented creature (Figure 15.8). How do the initial gradients of Bicoid, Nanos, and Dorsal come to specify 14 different segments, each possessing dorsoventral polarity, an essential step in making a fly? The general answer is known, though some details remain to be elucidated. And that story of how segments are formed is one of the triumphs of modern biology: it has transformed our view of developmental biology and energized the whole field. It is the impetus for writing a new generation of textbooks. At its heart are sequences of interactions between transcription factors and intercellular signaling. We have chosen to call these regulatory pathways "networks," since they are not always linear. Most of this and the following chapter will be devoted to them. In fact, these so-called networks are the molecular basis of what embryologists have called epigenesis.

The story originates from mutants that turned up in the genetic screens carried out by Wieschaus and Nusslein-Volhard.

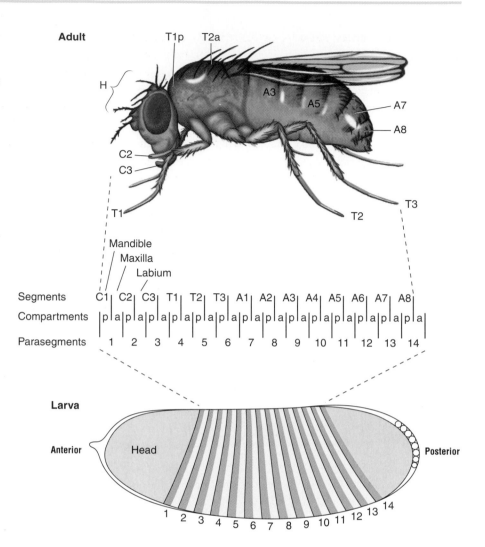

Figure 15.8 Segmental Organization of the *Drosophila* Larva A reprise of Figure 3.13: The segments of an adult fly are labeled above and the corresponding parasegments of the embryo below, with their p (posterior) and a (anterior) compartments identified. (Naming the portions of the parasegment p and a—a somewhat confusing convention—indicates their eventual orientation in the *adult* segment.) There are 14 main parasegments in the embryo plus a head (H) and tail region.

The maternally deposited gradients of transcription factors (see Box 15.1) activate a set of **gap genes,** so named because a null mutation results in a large gap in segmentation in the larva. Gap genes are expressed during cleavage cycles 9 through 14, before cellular boundaries are established. Gap genes all encode transcription factors, and the mRNAs and proteins encoded by gap genes are relatively unstable. As a result of their action, broad body regions from anterior to posterior are specified. Then a two-step process takes place. First, certain **pair-rule genes**—these, too, encode rather unstable transcription factors—lay down seven broad cellular stripes around the circumference of the embryo. Then expression of a suite of **segment polarity genes** divides each of the broad stripes in two, producing 14 narrower stripes, each with an identifiable anteroposterior polarity. Some segment polarity genes are transcription factors; others encode ligands and receptors. The 14 identical segments now acquire specific identities through the action of **homeotic selector genes,** often referred to simply as homeotic genes. These are all of the *homeobox class,* and in different combinations the transcription factors they encode specify the identity of different segments.

An overall flow diagram of this cascade of gene action is shown in Figure 15.9. It will serve as the basis for our more detailed analysis of how this all works, which follows.

Gap Genes Establish Seven Broad Stripes

Figure 15.10 outlines the regional distribution of the products of the zygotically active gap genes: *tailless, huckebein, giant, hunchback, krüppel,* and *knirps.* This group of genes has two essential functions. First, they interact with each other and with products of the terminal (posterior and anterior) axial systems to establish seven broad stripes in the embryo. Second, these striped domains set up conditions for regulating the pair-rule genes, each of which is active in patterning the seven repeating stripes. During this process, broad regions and gradients of gene expression come to be expressed in sharply defined patterns, resulting in a set of stripes. We shall see that the regulation works by both activating and suppressing the transcription of pair-rule genes.

How is gap gene expression regulated? Let us examine what happens anteriorly. Bicoid mRNA is sharply localized; thus there is a gradient of Bicoid protein that is high anteriorly and fades as one moves posteriorly. As mentioned previously, Bicoid stimulates *hunchback* transcription, which thereby creates a gradient of zygotically synthesized Hunchback protein. (Bicoid also stimulates another gap gene, *giant*, to which we shall return presently.) The posterior extent of Hunchback is regulated by how far the Bicoid gradient extends.

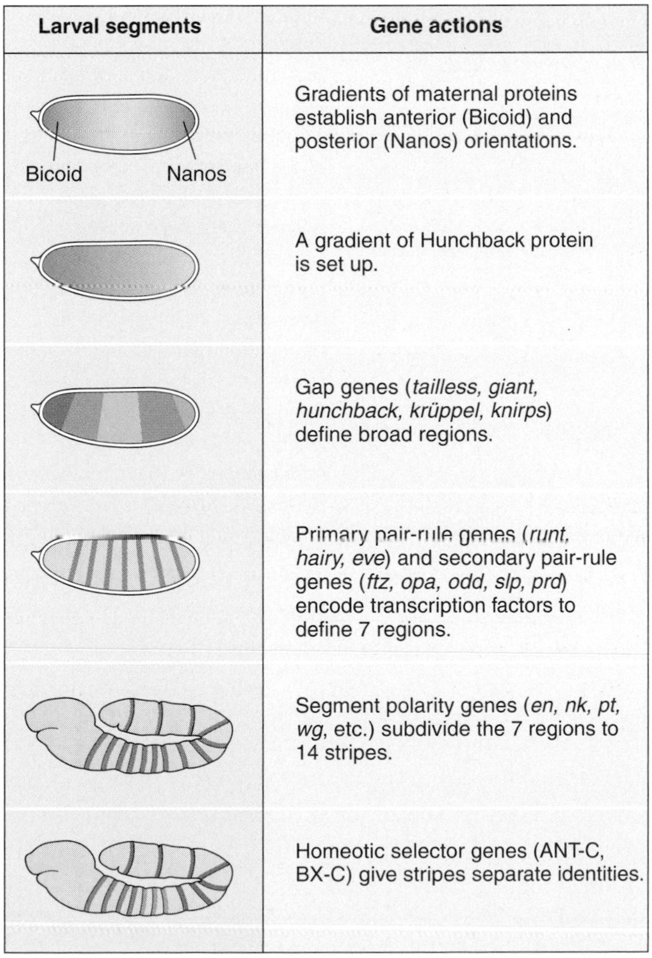

When the number of maternal copies of the *bicoid* (*bcd*) gene is increased artificially, more Bicoid mRNA is transcribed, and thus more Bicoid protein is made. The greater amount of Bicoid diffuses farther posteriorly, so the limit of Hunchback moves posteriorly also (Figure 15.11).

There are also Hunchback binding sites in the control regions for several of the other gap genes; moreover, Hunchback may have an activating or suppressing effect on transcription, depending on the precise DNA binding site and other protein–DNA interactions occurring in the area. As one researcher has put it, Hunchback acts like a kind of molecular vernier: because of its precise affinity for different DNA binding sites, Hunchback creates thresholds for different effects of protein–DNA interactions.

A reasonable model for Hunchback action is based on knowledge of the expression domains of the gap genes, as shown in Figure 15.10. At high levels of Hunchback, *krüppel* (*Kr*) is repressed, so the anterior border of *Kr* expression is set by high levels of Hunchback. Surprisingly, somewhat lower levels of Hunchback are needed for *Kr* to be expressed, so where the concentration of Hunchback decreases, *Kr* is turned on. Then, as Hunchback levels become even lower more posteriorly, *Kr* is turned off because of insufficient Hb. Thus, Hunchback does act like a vernier—first repressing, then activating, and finally dropping below levels of sufficiency for *Kr* expression. There are also sites on the genes *knirps* (*kni*) and *giant* (*gt*) having different affinities for Hunchback. Above certain levels of Hunchback concentration, *kni* and (in the posterior domain) *gt* are repressed; below those levels, the two genes are active. You can see by looking at Figure 15.10, then, that the gradient of Hunchback is crucial in establishing expression domains for *Kr*, *kni*, and posterior *gt*. Expression of *huckebein* (*hkb*) and *tailless* (*tll*) is controlled by the terminal genes *torso* and *torsolike*, which activate a cascade resulting in inhibition of a protein (called Groucho) that suppresses *hkb* and *tll*. Hence, Hkb and Tll are able to specify terminal structures by a kind of double negative—inhibition of a suppressor. Table 15.1 summarizes these interactions, as well as other gap gene relationships to be discussed shortly.

But there is an anterior domain of *gt* expression as well, and this is a region where Hunchback levels are very high. You may remember that Bicoid is a positive regulator of *gt*, so even though Hunchback represses *gt* anteriorly, Bicoid stimulates it, which may be why there is an anterior expression domain for *gt*. Expression of *gt* is repressed by *tailless*, a gene activated by the terminal system, so *gt* expression is absent from the terminal regions of the embryo.

Larval segments	Gene actions
Bicoid Nanos	Gradients of maternal proteins establish anterior (Bicoid) and posterior (Nanos) orientations.
	A gradient of Hunchback protein is set up.
	Gap genes (*tailless, giant, hunchback, krüppel, knirps*) define broad regions.
	Primary pair-rule genes (*runt, hairy, eve*) and secondary pair-rule genes (*ftz, opa, odd, slp, prd*) encode transcription factors to define 7 regions.
	Segment polarity genes (*en, nk, pt, wg*, etc.) subdivide the 7 regions to 14 stripes.
	Homeotic selector genes (ANT-C, BX-C) give stripes separate identities.

Figure 15.9 The Regulatory Cascade for Segment Formation in *Drosophila* This is a broad summary of the sequence of gene actions that give rise to the different segments of the larva and eventually of the adult.

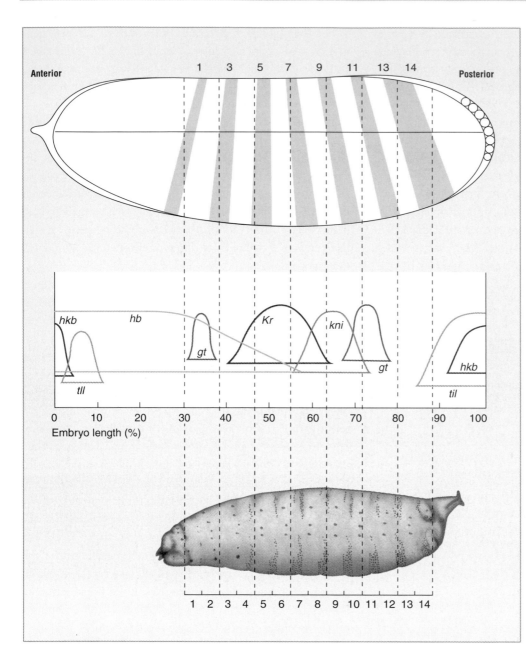

Figure 15.10   The Expression Domains of Gap Genes *Top:* A side view of the blastoderm just as it is beginning to form cells. Odd-numbered stripes indicate the approximate anterior borders of the odd-numbered segments (the result of *eve* expression), while stripe 14 indicates the approximate posterior border of segment 14. *Bottom:* A larva, also seen from the side, with a scale to identify the 14 parasegments. *Middle:* Graphs of the approximate expression domains of various gap genes. The relative levels of the various gap proteins are indicated by the height of the curves (*colored lines*). The dashed lines running vertically through the figure show how different parts of the figure (blastoderm, expression domains, embryo length, and larva) correspond. Abbreviations: *gt,* giant: *hb,* hunchback: *hkb,* huckebein: *kni,* knirps: *Kr,* krüppel: *tll,* tailless.

An extremely important strategy is at work here. Expression of one gene may affect the expression of another. This works because the protein made by one gene has an affinity for the regulatory DNA of the target gene, and so interacts with it. The specific affinity may set a threshold for the primary interaction; secondary interactions—other proteins either interacting with DNA nearby, competing for the same site, and/or interacting with proteins that modify the affinity of the primary protein for DNA—can influence the effectiveness of the primary interaction and determine whether it is stimulatory or repressive. Thus, borders of gene expression are set.

This strategy is used for interactions of maternal factors, such as Bicoid and Dorsal. In the case of Dorsal, the strength of its affinity for various regulatory sequences in target genes, as well as whether the action of Dorsal is activating or suppressing, is important for organizing the pattern of differentiation along the dorsoventral axis. Dorsal binds directly to the regulatory DNA of *twist,* thereby activating it. Twist and Dorsal both stimulate *snail* in such a way that a very sharp border of *snail* expression determines the limit of prospective mesoderm cells. Dorsal also binds to *zen* and *dpp* and represses them, thereby restricting expression of these two genes to prospective dorsal structures. The strategy is also used for

BOX 15.1 ESTABLISHMENT OF AXIAL POLARITY IN *DROSOPHILA:* A SUMMARY

A. The **terminal system** specifies the acron and telson. When this system is nonfunctional, the termini of larvae are missing, and head and abdomen are expanded.

Genes and Gene Products: (1) Torsolike, a ligand placed by follicle cells in the vitelline membrane and perivitelline space, activates (2) the receptor Torso, which in turn activates (3) *tailless,* a transcription factor gene. Tailless represses (4) the transcription factor gene *giant,* keeping it turned off at the termini.

B. The **anterior system** specifies the head and thorax. Mutants are headless and have an expanded abdomen: the terminus becomes a telson instead of an acron.

Genes and Gene Products: Nurse cells deposit (1) Bicoid mRNA in the anterior portion of the egg. Bicoid protein then activates the expression of (2) *hunchback* and (3) *giant,* both of which encode transcription factors. While *giant* is activated by Bicoid (and by itself), it is repressed by Hunchback and Tailless, resulting in a broad stripe of *giant* expression.

C. The **posterior system** specifies the abdomen. Null mutants have no abdomen and possess an expanded thorax.

Genes and Gene Products: (1) Nanos, a maternal mRNA made by nurse cells, encodes a protein that represses translation of (2) Hunchback mRNA, thereby producing a gradient of Hunchback and eliminating it from the posterior. The genes (3) *krüppel,* (4) *knirps,* and (5) *giant* encode transcription factors that interact with each other, and with Hunchback and Tailless, to produce different regions with high concentrations of the different gap gene transcription factors.

D. The **dorsoventral system** controls the dorsal or ventral character of the body. Most mutations (except *cactus$^{-/-}$*) eliminate or perturb ventral development.

Genes and Gene Products: Expression of (1) *gurken,* a gene of the oocyte, results from asymmetry in where the oocyte nucleus is placed, and signals (2) the receptor Torpedo in the follicle to specify a dorsal identity. (3) Pipe, Windbeutel, and Nudel can form complexes only on the ventral side, where they interact with several proteins, including (4) Easter, which is a protease that activates (5) Spätzle, which in turn activates (6) its receptor, Toll, only on the ventral side. Toll signals (7) the kinase Pelle, which phosphorylates (8) Cactus. Cactus is an inhibitory binding protein that interacts with (9) Dorsal, a transcription factor kept in the cytoplasm by Cactus. When phosphorylated, Cactus releases its hold on Dorsal, which can then move into the nucleus; there it activates the transcription factor genes *snail* and *twist,* and represses the transcription factor gene *zen.* Also, when present in the nucleus at low concentrations, Dorsal allows (10) the gene *dpp* (short for *decapentaplegic*) to be active. The gradient of Dorsal is what dictates dorsoventral polarity.

TABLE 15.1 SUMMARY OF SOME IMPORTANT GAP GENE INTERACTIONS

Protein		Genes Whose Expression Is	
Name	Levels	Activated	Repressed
Bicoid	High	*hunchback, giant*	
Hunchback	High		*krüppel, giant, knirps*
	Intermediate	*krüppel*	
	Low		*krüppel*
Dorsal	High in nucleus	*twist, snail*	*dpp, zen*

A.

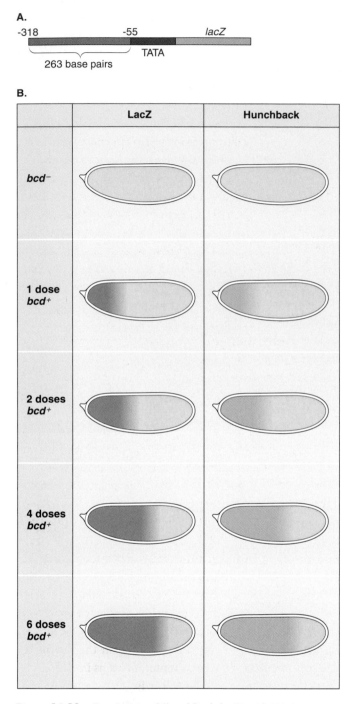

B.

	LacZ	Hunchback
bcd⁻		
1 dose *bcd⁺*		
2 doses *bcd⁺*		
4 doses *bcd⁺*		
6 doses *bcd⁺*		

Figure 15.11   Regulation of Hunchback by Bicoid (A) A reporter construct was created by attaching a β-galactosidase reporter *(lacZ)* to 263 base pairs of DNA from the *hunchback* enhancer sequence. This construct was then stably introduced into fruit flies. Genetic manipulations were carried out to change the number of copies of the *bicoid* gene *(bcd)* in female flies carrying the construct. **(B)** *Left:* As the number of copies of *bcd* increased, so did the extent of expression of the reporter. *Right:* Embryos were also stained with an antibody to detect Hunchback, encoded by the endogenous gene *hb*, which responded similarly.

interactions among the gap genes and between gap gene products and their downstream targets, the pair-rule genes. Furthermore, this strategy of regulation by utilizing differential protein–DNA affinities is not just a peculiarity of *Drosophila* (though that is the instance in which it was first worked out and from which we have gleaned the most information). The strategy is used again and again, in almost every developmental situation, in almost every well-studied organism.

Pair-Rule Genes, Activated by Gap Genes, Create Seven Repetitive Stripes

The transcription factors encoded by gap genes and Bicoid help regulate the promoters of genes that establish the anteroposterior sequence of seven circumferential stripes. This is an important step on the road to forming the 14 repeating segments (along with a posterior terminus containing the genitalia and anal plate) that comprise the basic body plan of *Drosophila*. All eight pair-rule genes encode transcription factors. Three of them, called the primary pair-rule genes, are expressed somewhat early on; these are *runt, hairy,* and *eve* (for *even-skipped*). Because *eve* has been intensively studied, we will follow the course of its behavior as an example of primary pair-rule gene expression. The gene *eve* is first expressed in a wide region covering most of the embryo. Then a single, broad, fuzzy band appears, which narrows and sharpens to become the first of seven "stripes." The remaining six stripes then appear, pretty much at the same time. Gradually, they become more defined, attaining sharper and sharper borders. Figure 15.12 shows this progression in the case of *eve; runt* and *hairy* show similar but not identical progressions of expression.

Eventually, both the primary and secondary pair-rule genes are expressed in seven stripes of activity. Figure 15.13A shows the signal that reveals the expression domains for several of them: *runt, hairy, ftz* (for *fushi tarazu*), and *eve*. Even though each of the eight genes is expressed in seven stripes, the borders of their expression domains do not exactly align. Instead the domains overlap, thereby creating complex repeating patterns of expression, as shown in Figure 15.13B. For example, *ftz* is expressed in the posterior half of each of the seven stripes, *runt* is expressed in the middle, and *eve* in the anterior half. Later in the chapter, we shall see how within each *eve* stripe, Eve activates *engrailed* (*en*), which then leads to *wg* expression in neighboring cells. Expression of *en* marks the anterior boundary of each parasegment, while *wg* expression marks the posterior boundary of each parasegment.

A.

B.

C.

D.

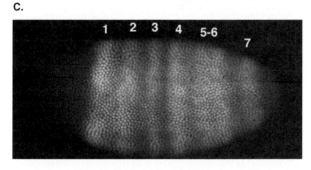

Figure 15.12   The Progression of *eve* Stripes An antibody against the protein Eve was used to stain embryos at different stages: anterior is to the left in all images. **(A)** In a cleavage-stage embryo, all the nuclei are stained. **(B)** At early stage 14, the nuclei in the posterior two-thirds of the embryo show strong staining. Recall that nuclei have moved to the surface and cells are forming: hence, staining of the embryo will appear more intense at the surface than in out-of-focus cells in the interior. **(C)** At midstage 14, the pattern of stripes is becoming visible. Not all stripes are equally stained. **(D)** As stage 14 progresses, cellularization of the blastoderm is completed and the seven stripes become clearly delimited.

Pair-Rule Genes Have Complex Regulatory Regions

How is this pattern of striped expression attained? The basic mechanism is similar to how gap gene expression domains are set up. The transcription products of the regulatory genes upstream in the developmental sequence—that is, the maternal genes (*bicoid* anteriorly and *caudal* posteriorly) and gap genes—interact with enhancer and suppressor regions on the target pair-rule genes. In other words, each primary pair-rule gene has a large, complex control region containing dozens of different sites at which various gap and maternal gene transcription factors are bound. Whether the binding of a given factor at a particular site activates or represses transcription depends on many factors: the particular transcription factor, the exact binding sites occupied, and which transcription factors are bound to other sites of the control region. It is the exact *pattern* of transcription factors binding to the control region that sets the conditions for expression.

The primary pair-rule genes possess clusters of sites that, when occupied, activate expression. There can be seven such clusters for a given pair-rule gene, each cluster responding to a different combination of gap gene transcription factors, so that the pair-rule gene is turned on at different places in the embryo. Suppressive sites are sprinkled throughout the long, complex regulatory region, providing opportunities for turning off gene expression in cells between the stripes.

Once pair-rule gene expression gets going, interactions help "sharpen up" the stripes. For example, *eve* enhancer modules (regulatory elements in the DNA that turn on *eve* expression) for stripes 2, 3, and 7 are located in the first 5 kilobases (kb) upstream from the transcription start site (Figure 15.14A); each enhancer responds to a different combination of gap gene products. Eve is produced in this manner in the early-appearing stripes. Together with Hairy and Runt, Eve then acts on another upstream region of *eve*, located at about −6 kb; this "late enhancer" region then turns on expression of *eve* in sharply delimited domains in each of the seven stripes (Figure 15.14B). What is going on here is that the anteroposterior pattern of expression of the several gap genes is employed to create a repeating series of *eve* expression domains. Similar stories can be told regarding activation of the other seven pair-rule genes, though details differ.

The Gene *eve* Is Regulated by Both Activating and Inhibitory Interactions

Let us look in a little more detail at just one of these primary pair-rule genes to discern how the striped arrangement

A.

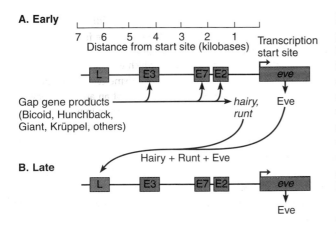

runt

hairy

fushi tarazu (ftz)

even-skipped (eve)

B.

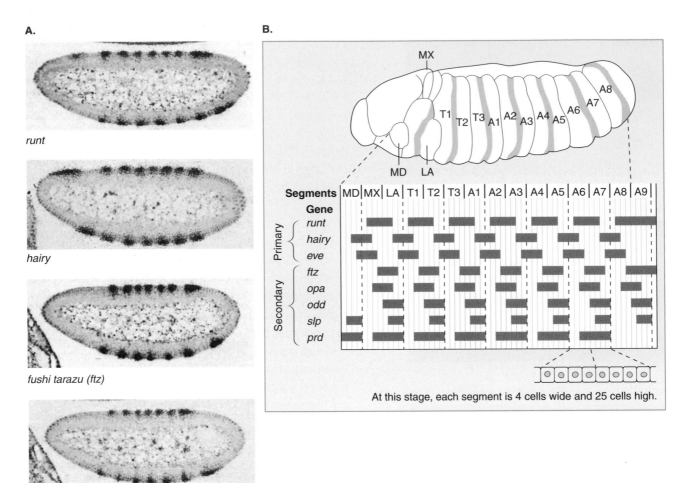

At this stage, each segment is 4 cells wide and 25 cells high.

Figure 15.13   The Expression of Pair-Rule Genes (A) The striped patterns of expression are shown for four of the eight pair-rule genes at the cellular blastoderm stage (stage 14): *runt, hairy, ftz,* and *eve.* **(B)** Comparison of expression domains is sometimes better appreciated when viewed diagrammatically. Here bars are used to designate the zones of repetitive expression for all eight pair-rule genes. The thoracic (T1 through T3) and abdominal (A1 through A8) segments are labeled on the diagram of the larva, which is shown to indicate the fate of the cells engaged in expression of pair-rule genes at stage 14. At the cellular blastoderm stage, each future adult segment is only four cells wide anteroposteriorly and 25 cells high ventrodorsally. Other abbreviations: LA, labial; MD, mandibulum; MX, maxillary.

A. Early

7 6 5 4 3 2 1   Transcription
Distance from start site (kilobases)   start site

Gap gene products
(Bicoid, Hunchback,
Giant, Krüppel, others)

hairy,
runt

Eve

B. Late

Hairy + Run + Eve

Eve

Figure 15.14   Early and Late Promoters of Some *eve* Stripes The *eve* promoter has modular binding sites (shown here as blocks E2, E3, and E7) that control activation of *eve* expression in stripes 2, 3, and 7. (The modules for regulating stripe 1 and stripes 4 through 6 are located some distance downstream from the *eve* coding region.) **(A)** Modules 2, 3, and 7 first interact with combinations of proteins encoded by maternal genes (*bcd* and *hb,* which activate) and proteins encoded by gap genes (such as *Kr* and *gt,* which repress). Activating and suppressing interactions compete with one another to determine whether *eve* will be turned on or not. In the regions in the embryo corresponding to prospective stripes 2, 3, and 7, the concentrations of transcription factors allow activation, so *eve* transcription starts. **(B)** Together with Hairy and Runt, the Eve protein generated by these earlier activated stripes (2, 3, 7) activates a "late enhancer" sequence (denoted L). This enhancer (also called an autoregulatory region) helps activate transcription of *eve* in the other four stripes as well as sustaining transcription in the early stripes.

A.

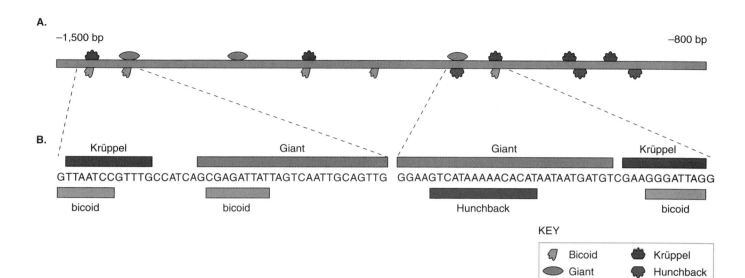

Figure 15.15   The Structure of the Regulatory Module of *eve* Stripe 2 The binding sites for various transcription factors in this regulatory module are shown diagrammatically, with expression-inhibiting sites above the line and activating sites below the line. **(A)** The relative binding positions for the four proteins Krüppel, Giant, Bicoid, and Hunchback.

(B) The exact binding sites for regions where two different factors can bind to a particular gene. These sites were determined by footprinting. It is obvious that positively acting factors (Bicoid and Hunchback) and negatively acting factors (Krüppel and Giant) can compete for certain sites.

develops. Figure 15.15 shows the binding sites on *eve* for several proteins; the sites are in a region 800 to 1,500 base pairs upstream from the transcription starting point. Using the technology of reporter gene expression, researchers found this region to drive expression in stripe 2. When this stripe 2 region was examined by means of footprinting (see Box 15.2), it was shown to possess large numbers of activating binding sites for Bicoid and Hunchback, as well as inhibitory sites for Giant and Krüppel (Figure 15.15). In general, activating and inhibitory sites are often adjacent or overlapping, and there are often multiple copies of the same site close to one another. These complex regulatory elements allow extensive protein–protein interactions and competitions between transcription factors and co-activators. Researchers can take advantage of this behavior to make and test specific predictions.

Specific elements and combinations of elements can be isolated, linked to a reporter gene, and placed in *Drosophila* tissue culture cells to see whether the artificial regulatory element works or not (see Box 15.2). Or reporters can be introduced into embryos in which the genes in the host embryo have been modified by appropriate genetic manipulations. It is then possible to see, against a "neutral" background, how the reporter behaves. Figure 15.16 shows just such an experiment, in which the regulation of the *eve* control

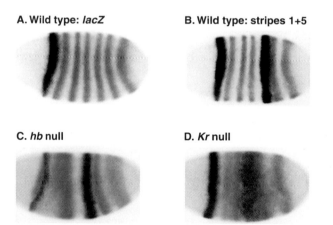

Figure 15.16   The Function of Regulatory Modules for Several *eve* Stripes Reporter constructs are used to study the function of regulatory modules. The *eve* promoter for stripes 1, 5, or 1 + 5 together was fused to the coding region for *lacZ*, which encodes β-galactosidase. Thus, wherever this promoter is active, the enzyme can be detected (*blue*). The presence of Eve is detected by use of an antibody against that protein (*orange*). Wild-type embryos are shown in which the stripe 1 enhancer drives **(A)** the reporter, and **(B)** the enhancers for stripes 1 and 5. **(C)** When the reporter construct for stripes 1 and 5 is expressed in a null mutant of *hunchback*, *eve* expression is compromised in stripe 1 but remains intact in stripe 5. **(D)** On the other hand, when this same construct is introduced into a null mutant of *krüppel*, stripe 5 is compromised (as are other stripes), but expression in stripe 1 remains intact.

regions for stripes 1 and 5 has been rendered visible by use of a reporter. The results of many such experiments support the general model we have been exploring.

One outcome of the complexity of these regulatory systems is that they can be extraordinarily sensitive to slight changes in concentrations of partner gap gene transcription factors. For example, a twofold increase or decrease in Hunchback concentration may change the status of an enhancer from "on" to "off." Gentle gradients of transcription factors can thereby define sharp edges of domain expression.

The enhancer that drives *eve* in stripe 3 is located some distance from the basal promoter, approximately 4,000 base pairs upstream (Figure 15.14). The arrangement of regulatory elements for stripe 3 is different from that of stripe 2. It can respond to lower levels of Hunchback, Bicoid, and Giant in the presence of higher levels of Krüppel because of the number and arrangement of control elements in the DNA for these particular transcription factors.

In addition to their function in establishing segments, some pair-rule genes are also expressed in other places at later times during development. Some maternal genes, such as *bicoid*, are employed during early development only. In contrast, the *ftz* gene—a secondary pair-rule gene expressed in cells where *eve* is not active—is also expressed in certain neurons during development of the nervous system. This is not a unique example; most transcription factors can serve several different organized gene-regulatory modules, which then can exert their effects in completely different developmental events.

Segment Polarity Genes Subdivide the Seven Stripes

The seven broad stripes created by patterns of pair-rule gene expression regulate the segment polarity genes. Some of the segment polarity genes encode transcription factors. Others encode ligands and receptors for intercellular signaling. Segment polarity genes are expressed in a repeating pattern that subdivides the expression domain of each of the seven primary stripes into two subdomains. These 14 subdomains then become the 14 parasegments of the first-instar larva (as discussed in Chapter 3), which forms the basis of the body plan. Since the pair-rule genes are expressed repetitively, a repeating subdivision can depend on a mechanism that works reiteratively from anterior to posterior end.

Figure 15.17 diagrams the expression domains of four key segment polarity genes. (For the record, the other genes of this class are *cubitus interruptus, hedgehog, fused, armadillo, gooseberry,* and *pangolin.*) Note that each broad pair-rule gene stripe has an anterior domain expressing *eve* and a posterior domain expressing *ftz.* The segment polarity gene *engrailed (en)* is expressed in a file of cells congruent with high levels of Eve or Ftz; thus each broad stripe now possesses two thin En stripes. Between the two *en*-expressing stripes are groups of cells expressing *naked (nkd), patched (ptc),* and *wingless (wg).* Cells expressing *wg* and *en* are always next to one another. Each segment polarity gene is expressed in a narrow portion (at first only one cell "thick") of each parasegment. Current ideas propose that alternating domains of *eve* and *ftz* play an important

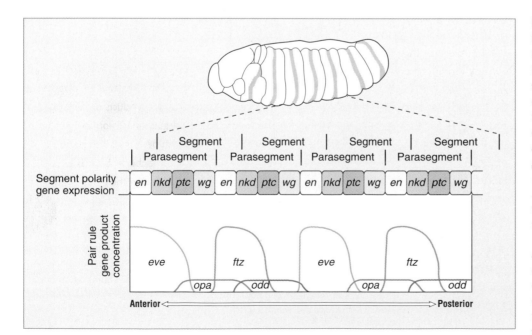

Figure 15.17 The Expression Domains of Some Segment Polarity Genes Each of the seven broad stripes of pair-rule gene expression sets up conditions for repetitive expression of certain segment polarity genes. Diagrammed here are two of the seven broad pair-rule stripes, and expression of several pair-rule genes within these domains: *eve, ftz, opa (odd-paired),* and *odd (odd-skipped).* These work in combination to activate repetitive expression of the segment polarity genes *engrailed (en), naked (nkd), patched (ptc),* and *wingless (wg).* Each of these is expressed in a single vertical file of cells; thus, the parasegment is four cells wide at this stage. Note that the even and odd parasegments are specified by different combinations.

BOX 15.2 METHODS FOR STUDYING GENE INTERACTIONS

The interactions among gap genes that we have just described were analyzed using genetic manipulation in *Drosophila*, an important way to probe regulatory circuits among genes. For example, a null mutation can be created in one gene to see what happens to the expression of another.

What about studying the actual interaction of these proteins with the DNA of control regions? Several methods are available. One is the *electrophoretic migration shift assay*. Extracts of embryos, or extracts of nuclei prepared from them, are mixed with radioactive DNA derived from the control regions of different genes, and the result is then run through a gel by electrophoresis. Whenever a protein in the extract has interacted with the DNA, the mobility of that piece of DNA will be slowed down. Note that, although this assay shows whether a protein and a piece of DNA under investigation do interact, it does not tell us whether the interaction is functionally important—or even whether it occurs in the living cell. The test does provide a clue, however.

Another, similar test involves mixing the protein extract with

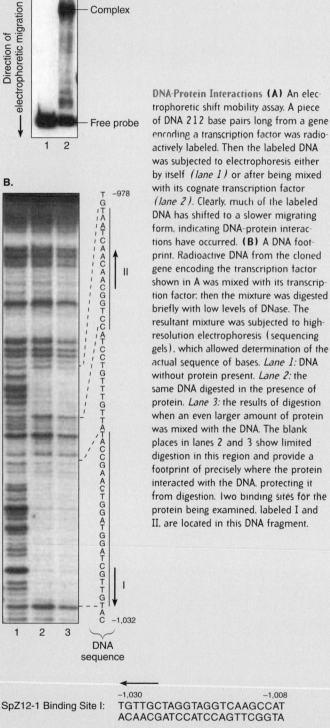

DNA-Protein Interactions (A) An electrophoretic shift mobility assay. A piece of DNA 212 base pairs long from a gene encoding a transcription factor was radioactively labeled. Then the labeled DNA was subjected to electrophoresis either by itself *(lane 1)* or after being mixed with its cognate transcription factor *(lane 2)*. Clearly, much of the labeled DNA has shifted to a slower migrating form, indicating DNA-protein interactions have occurred. **(B)** A DNA footprint. Radioactive DNA from the cloned gene encoding the transcription factor shown in A was mixed with its transcription factor; then the mixture was digested briefly with low levels of DNase. The resultant mixture was subjected to high-resolution electrophoresis (sequencing gels), which allowed determination of the actual sequence of bases. *Lane 1:* DNA without protein present. *Lane 2:* the same DNA digested in the presence of protein. *Lane 3:* the results of digestion when an even larger amount of protein was mixed with the DNA. The blank places in lanes 2 and 3 show limited digestion in this region and provide a footprint of precisely where the protein interacted with the DNA, protecting it from digestion. Two binding sites for the protein being examined, labeled I and II, are located in this DNA fragment.

radioactive DNA and then partially digesting the DNA with a DNA nuclease. Protein that is interacting with a specific region of DNA will often partially protect that area by interfering with its digestion by the nuclease. This protection can be seen by carrying out a sequence analysis of the partially digested DNA. The signature of a specific region protected by a protein interaction is called a *footprint*. The accompanying figure shows examples of shift assays and a footprint.

Finally, surrogate cells can be used to test for possible protein-DNA interactions. First, lines of tissue culture cells are transformed with DNA; then the action of the transgene (the transformed DNA) is studied in the tissue culture cell. For example, when the promoter of a gap gene, such as *krüppel*, is linked to a β-galactosidase reporter, and the reporter and the gap gene *hunchback* (*hb*) are introduced into tissue culture cells, the effects of the products of *hb* on the regulatory region of *krüppel* can be ascertained by noting where the reporter is expressed in the tissue culture cells.

A.   B.   C.

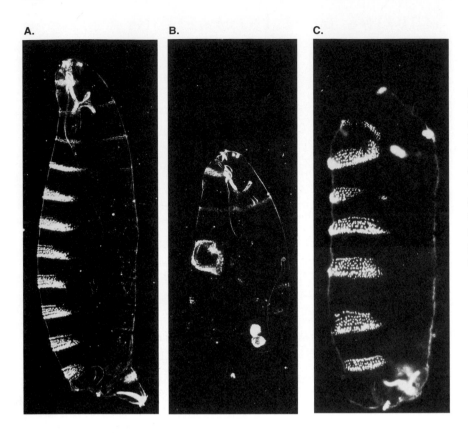

Figure 15.18   Ventral Denticle Belts These first-instar larvae of *Drosophila* have been fixed, rendered transparent, and photographed from the lateral side to show the arrangement of denticle belts in the anterior portion of each segment. The anterior end is at the top. **(A)** A wild-type larva. **(B)** A mutant of the posterior-acting maternal gene *nanos*, in which all abdominal segments are missing. **(C)** A mutant of the anterior-acting maternal gene *bicoid*, which has no head: the most posterior structure of the embryo, the telson, has developed in its place.

role in allocating cells to an anterior (odd-numbered) or posterior (even-numbered) parasegment carved from each pair-rule gene domain, though other pair-rule genes are also involved.

The anterior and posterior portions of a parasegment have a morphological counterpart in the larva. The anteroventral side of each parasegment possesses small hairlike projections called denticles; the posteroventral portion of the parasegment has no denticles and is naked (Figure 15.18). In *wingless* mutants, denticles are present all over each parasegment, the denticles in the posterior portion having a polarity opposite to that of the denticles in the anterior portion (that is, the two regions are mirror images of each other). When *wg* is overexpressed in all cells (using an inducible promoter for the transgene), the entire ventral surface is naked. Strongly expressing mutant alleles of segment polarity genes are eventually lethal.

Expression of the Segment Polarity Genes Provides Permanent Markers for Parasegmental Boundaries

Before considering how the identical parasegments come to differentiate along separate pathways, we should mention that many of the genes discussed so far are only transiently expressed. However, expression of *engrailed* and the other segment polarity genes provides more or less permanent markers of the segmental boundaries. How is continual expression of these genes maintained? The regulatory regions of *engrailed* (to take only one of these genes) is huge and complex, involving both upstream and downstream DNA, and the analysis of its organization is still incomplete. Nonetheless, we do know that *engrailed* has an autoregulatory region that participates in a positive feedback loop: Engrailed protein acts back on *engrailed* by binding to a feedback enhancer that keeps the expression of the gene turned on.

Intercellular Communication Involving Engrailed Keeps Its Expression "On"

There is a second important, albeit complex, way that continual expression of *engrailed* is maintained. It comes about by a ligand–receptor interaction with the neighboring cell that expresses *wg* (Figure 15.19). The *en*-expressing cell not only stimulates its own *en* genes; it also stimulates *hedgehog* to synthesize the ligand Hedgehog, which acts on the neighboring *wg*-expressing cell. In this cell, a pathway involving the genes *ptc* and *smoothened (smo)* suppresses *wg*. Hedgehog inhibits this *ptc–smo* pathway, allowing the ligand Wingless to be secreted; Wingless is subsequently received by the *en*-expressing cell, where Wingless is a positive signal for *en* expression. Thus, the loop between *en*-expressing and *wg*-expressing cells is maintained, each reinforcing its neighbor to do its own separate thing.

This *en–wg* signaling pathway turns out to be involved in many developmental pathways, not only in *Drosophila* but also in vertebrates. It is a very active area of research, and we shall have much more to say about it later in the chapter. But let us move on now to another important issue in the cascade of gene actions that govern early *Drosophila* development: how the 14 identical parasegments differentiate from one another in order to develop into the different body parts.

HOMEOTIC SELECTOR GENES AND PARASEGMENT IDENTITY

The Bithorax Complex Dictates Thoracic and Abdominal Segment Identity

A fascinating collection of mutations in *Drosophila* result in "monsters" in which a particular segment changes its character and becomes like another segment. Such mutations affecting the thorax and abdomen have been explored for several decades, especially by Ed Lewis and his associates working at the California Institute of Technology. Some of these mutants had extra pairs of wings, or no wings at all, or halteres converted to wings, or abdominal segments converted to thoracic tissues. His work led to the insight that genes of this type are at the top of another hierarchy of genes that controls what a segment will become. Lewis called these

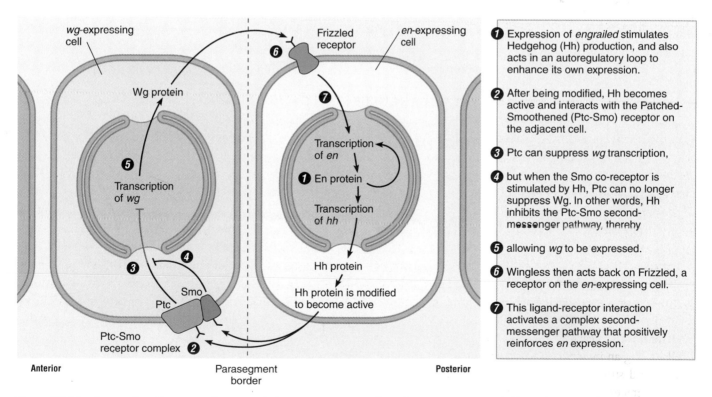

1 Expression of *engrailed* stimulates Hedgehog (Hh) production, and also acts in an autoregulatory loop to enhance its own expression.

2 After being modified, Hh becomes active and interacts with the Patched-Smoothened (Ptc-Smo) receptor on the adjacent cell.

3 Ptc can suppress *wg* transcription,

4 but when the Smo co-receptor is stimulated by Hh, Ptc can no longer suppress Wg. In other words, Hh inhibits the Ptc-Smo second-messenger pathway, thereby

5 allowing *wg* to be expressed.

6 Wingless then acts back on Frizzled, a receptor on the *en*-expressing cell.

7 This ligand-receptor interaction activates a complex second-messenger pathway that positively reinforces *en* expression.

Anterior   Parasegment border   Posterior

Figure 15.19 Interactions Between *wingless* and *engrailed* Shown here are two cells at a parasegmental border: the *en*-expressing cell at the anterior end of one parasegment and the adjacent *wg*-expressing cell at the posterior end of the next parasegment. Expression of *engrailed* stimulates Hedgehog (Hh) production, and also acts in an autoregulatory loop to enhance its own expression. After being modified, Hh becomes active and interacts with the Patched-Smoothened (Ptc-Smo) receptor on the adjacent cell. Ptc can suppress *wg* transcription, but when the Smo co-receptor is stimulated by Hh, Ptc can no longer suppress Wg. In other words, Hh inhibits the Ptc-Smo second-messenger pathway, thereby allowing *wg* to be expressed. Wingless then acts back on Frizzled, a receptor on the *en*-expressing cell. This ligand–receptor interaction activates a complex second-messenger pathway (to be discussed later in the chapter) that positively reinforces *en* expression.

genes homeotic (from the Greek, *homeos*, meaning "similar"). For this work, Lewis shared the Nobel prize with Eric Wieschaus and Christiane Nusslein-Volhard.

An analysis of the various mutations that Lewis studied showed that when the function of one homeotic gene is eliminated, the segment in which it is ordinarily expressed loses its more posterior identity and becomes identical to a more anterior segment. The **bithorax complex (BX-C)** studied by Lewis illustrates this point. The BX-C complex is comprised of three genes—*ultrabithorax (ubx)*, *abdominal A (abdA)*, and *abdominal B (abdB)*—all of which encode proteins. When the entire region is deleted (thereby removing the function of all three genes), all the abdominal segments and the third thoracic segment are altered so that they now possess the character of the second thoracic segment (T2) of the adult. This creature dies, of course. When only *ubx* is deleted, the third

thoracic and first abdominal segments (T3 and A1) vanish, both to be replaced by T2 segments. Deletion of other genes in the complex results in comparable morphological transformations in which posterior regions adopt the fate of more anterior regions. These kinds of morphological transformations are sometimes called *homeotic transformations*. Strictly speaking, the homeotic genes are influencing the development of the embryonic parasegments, which, as we have seen, overlap with but are not identical to the morphological adult segments.

Figure 15.20 shows the registration of parasegments and segments and the sites of gene action for the BX-C complex, along with those of another homeotic gene complex, ANT-C (which we shall discuss later in the chapter). The BX-C and ANT-C gene clusters taken together are often termed **HOM** genes.

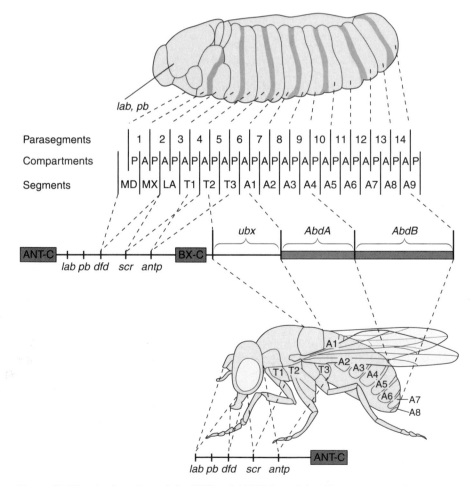

Figure 15.20   An Overview of the BX-C and ANT-C Complexes The segments and parasegments of a larva and an adult fly, shown in linear order from anterior to posterior end, with corresponding transcription modules and expression domains of the homeotic gene complexes ANT-C and BX-C.

The BX-C complex is huge, occupying over 300 kb on the third chromosome. Each of the three genes in the complex has several alternative promoters and multiple exons, and different alternate splicing patterns are utilized. Thus, each gene can encode several transcripts, so there are different versions of the Ubx, AbdA, and AbdB proteins. These proteins are transcription factors, all of which have a similar region called a **homeodomain,** which is 60 amino acids long and binds to DNA. The portion of the gene encoding the homeodomain is called the **homeobox,** and because its sequence varies slightly from one homeotic gene to another, it is often referred to as "the homeobox motif." There are other genes that also contain the homeobox motif; such "homeobox genes" do not confer segmental identity and thus are not *homeotic* genes.

The control regions of the three genes of the BX-C complex are extraordinarily complex, perhaps the most complicated of any gene complex so far known. Of the 300 kb section in which the exons of *ubx*, *abdA*, and *abdB* are embedded, over 80% consists of cis-regulatory regions (control DNA located in the same chromosome as the gene it regulates) and introns. Given the number of regulatory sites, it is no wonder all the details of the regulation of the BX-C genes are not yet known. But we do know of many mutations involving these regulatory regions that produce distinct homeotic effects; in these mutations, the primary sequences of the encoded proteins are not affected, only the regulation of transcription of the exons encoding the proteins.

The structure of the BX-C complex is shown in Figure 15.21. One important feature of this arrangement of genes, still poorly understood, is that the linear order of the three genes in the DNA is the same as the linear order of their spatial expression in the embryo; thus, the expression domain for *ubx* is "anterior" to that of *abdA*, which in turn is anterior to the expression domain of *abdB*. It is not clear whether there is some deep functional meaning to this linear arrangement, but since this congruence of gene order and expression domain is seen in many different animals, it is presumed to be important. For example, when the DNA is manipulated so that the entire *ubx* gene region (including its cis-regulatory control elements) is moved to another location, segment identity is nevertheless specified in a normal way. On the other hand, scrambling the positions of different cis-regulatory regions for a given gene within the BX-C complex can produce monsters. Clearly, the relative positions of these finely tuned control regions is critical for normal development.

The homeotic genes were termed "selector genes" by the Spanish geneticist Antonio Garcia-Bellido. He proposed that

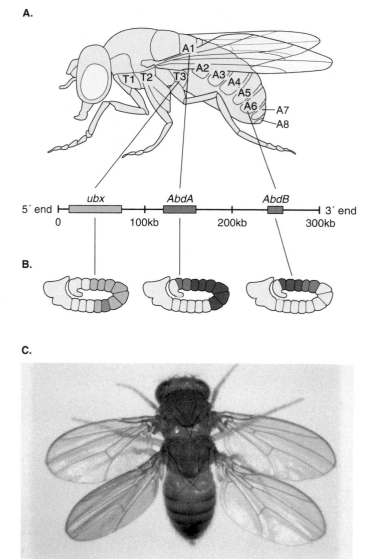

Figure 15.21   The Organization of the BX-C Complex (A) The line below the adult fly represents the DNA transcription units for the three genes of the bithorax complex. The individual genes are transcribed from right to left. The exact borders of *abdB* are not yet known. Each of the three genes has introns. Regulatory DNA governing the transcription of these three genes is present in the DNA beyond both the 3' and 5' ends of each gene. The lines from the DNA to the fly identify the anterior-most segment affected by mutations in this particular section of DNA. **(B)** The expression domains of *ubx*, *abdA*, and *abdB* have been projected onto embryos of the extended germ-band stage. **(C)** The *ubx* gene in this fly was rendered null by imposing several mutations in the regulatory regions of *ubx*. The third segment, normally possessing halteres, now has well-developed wings.

A.

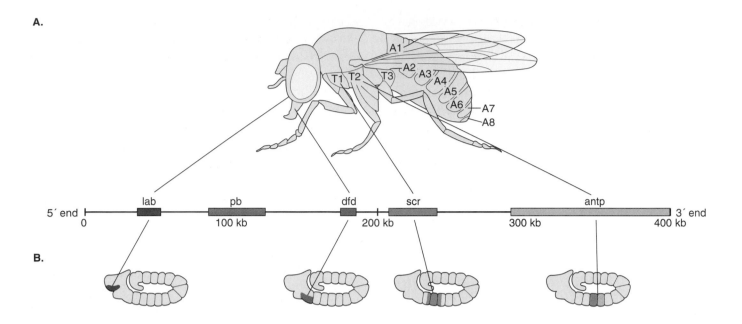

B.

C.

Figure 15.22   The Organization of the ANT-C Complex (A)
The line represents some 350 kb of DNA showing the relative
positions of genes in the antennapedia complex. **(B)** The expression domains of four of the genes in the ANT-C complex—*labial*
(lab), *deformed (dfd)*, *sex combs reduced (scr)*, and *antenna-*
pedia (antp)—are shown projected onto an embryo of the
extended germ-band stage. **(C)** A scanning electron micrograph
of a fly possessing a null allele of the *antp* gene. This head-on
view shows two legs emerging from the head where antennae
would normally be found. The faceted eyes are posterior to the
ectopic legs.

the homeotic genes act by selecting a set of target genes, with
different sets responding to different combinations of selector
gene products such that each set specifies the morphogenesis
and differentiation of a particular segment (or body region in
some other animals that do not possess clear segments). This
idea has proved to be largely correct.

The Antennapedia Complex Controls Anterior Segment Identity

Genes of the BX-C complex specify identity in the abdomen
and T3. Another complex of homeobox genes controls segment identity in more anterior regions, the labial and maxil-

lary regions of the head and thoracic segments T2 and T1. Just as with the BX-C complex, the **antennapedia complex (ANT-C)** is comprised of several large, complex genes arranged on the chromosome such that the linear order reflects the antero-posterior order of their expression in the body plan (Figure 15.20). When all five genes of the ANT-C complex are deleted entirely, the condition is lethal, and the labial and maxillary portions of the head, as well as segments T2 and T3, adopt a T1 fate. When *antennapedia (antp)* alone is deleted, only T2 and T3 adopt a T1 fate, resulting in a leg appearing where an antenna would have been. The genes of the ANT-C complex, along with the null *antp* mutation just described, are shown in Figure 15.22.

Other Homeotic Genes Specify Head and Posterior Parts

When all the ANT-C genes are deleted, the anterior portions of the head (that is, structures anterior to the labial and max-illary portions of the head) are still correctly specified. The function of other homeotic genes is necessary for the development of these parts. Segmentation is virtually impossible to see in the anterior head area. Yet there are three transient stripes of *engrailed* expression in this region (in addition to the 14 body stripes already discussed); presumably, these reflect a primordial segmental organization of this region. Two homeobox-containing genes (*ems*, for *empty spiracle*; *otd*, for *orthodenticle*) and a zinc finger–containing gene (*buttonhead*) are known to be essential for segment specification in this region. There is evidence that another homeobox gene, *distalless*, is also involved in anterior segment specification.

As for the posterior end of the embryo (the region after segment A8, and including the anal plate and genital discs), the gene *caudal*, among others, is essential for specification.

How Do the Homeotic Genes Function?

How do the homeotic genes attain region-specific expression, how is this regional expression maintained, and what do the proteins encoded by homeotic genes do in their expression domains? The idea here is that earlier-acting genes, such as gap genes, not only activate pair-rule genes but also set up conditions for regional expression (through interactions with cis-regulatory regions) of homeotic genes. For example, expression of *krüppel* is necessary for *antp* to be expressed (Figure 15.23). If the expression domain of *Kr* is altered, then the early expression domain of *antp* will be

changed accordingly. As development proceeds, the boundaries of expression domains for homeotic selector genes become sharper and sharper, probably as a result of interaction between the encoded proteins and the regulatory regions of other homeotic genes.

The function of homeotic selector genes is to define the anteroposterior identities of body regions. These regions have been called *compartments* by scientists working on *Drosophila*. Once a compartment has been defined, a cell within that compartment is dedicated to forming some structure normally originating in that compartment; it can no longer form a structure arising from another compartment. Furthermore, each cell in a given compartment has an adhesive affinity for all other members of that group; it will not cross over the border between one compartment and another. Hence, a compartment is a unit of specific cellular affinities, as well as an address for the kinds of differentiation that may be undertaken. Figure 15.20 shows how compartments relate to segmental and parasegmental boundaries.

Segment or compartment identity seems to be controlled by a dominance relationship between the control regions of the different homeotic selector genes. Genes that are expressed in compartments located posteriorly in the animal, such as *abdB*, dominate and control the expression of genes expressed in compartments situated more anteriorly. This is probably

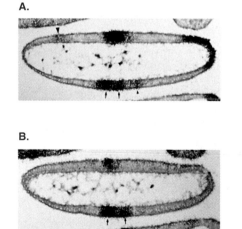

Figure 15.23   The Expression of *antp* Dictated by Krüppel (A) A sagittal section through the blastoderm showing the position of *Kr* expression as indicated by hybridization to a radioactive probe for Kr. **(B)** The expression of *antp*, detected in the same way, in a wild-type fly. The expression of *Kr* can be experimentally altered by transforming the fly with a *Kr* gene driven by a different promoter: if the expression domain of *antp* is thereby altered, the location of *Kr* expression is correspondingly altered.

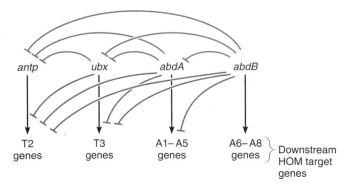

Figure 15.24   Posterior Dominance of Homeotic Selector Genes
Shown along the top are four HOM genes: *antp* and the three BX-C genes. Genes that are regulated by (in other words. are "targets" of) the different HOM transcription factor proteins and are expressed in various segments are shown along the bottom. Arrows indicate the anterior limit of expression for each HOM gene. Lines with blocked ends indicate the repressive effects of posterior HOM genes on anterior ones. as discussed in the text. In particular. posterior genes act on anterior segments. both by inhibiting transcription (*blue lines*) and by blocking protein activity (*red lines*). For example. *abdA* can repress transcription of *ubx* and *antp*. as well as interfere with the activity of Antp and Ubx proteins. But *abdA* cannot influence the more posterior expression of *abdB*.

accomplished in two ways: first, the transcription factors encoded by posterior genes, such as *abdB*, act directly on the control regions of more anterior genes, such as *abdA*, to stop their transcription. Second, the transcription factors encoded by posterior genes also compete with and antagonize the effects of more anterior gene products on their normal target genes. These relationships are diagrammed in Figure 15.24.

This model derives from experiments that alter the expression domains of particular homeotic selector genes. For example, suppose we microinject into an early embryo an engineered piece of DNA containing the *ubx* exons and an inducible promoter. One convenient kind of inducible promoter is called the "heat shock" promoter because it turns on the genes to which it is linked when ambient temperatures rise above a certain level. So if the heat shock is administered to a *ubx* transgene driven by this promoter, then *ubx* will be expressed everywhere, not just in the T3 segment. And when *ubx* is turned on everywhere, all homeotic selector genes that are normally expressed anterior to T3 are now inhibited, giving segments anterior to T2 a T2 identity. Segments posterior to the original T2 segment are normal, even though *ubx* is expressed posteriorly; AbdA and AbdB apparently prevent Ubx from wreaking havoc posteriorly.

Thus, one thing these homeotic selector genes do is to interact with each other to define addresses. The other action must be to *select* genes for expression. One possibility is that homeotic genes govern a hierarchy of other genes that eventually accomplish the specific, local programs of differentiation in the different segments. Researchers have found the search for these target genes, which is still under way, rather difficult. As of this writing, 19 different target genes have been identified, and more are being added to this list. Five of these genes generate molecules that participate in basic cell processes; these include specific forms of tubulin for microtubule assembly, a centrosome component, and a molecule involved in intercellular adhesion. Eight of the known targets produce transcription factors that are no doubt involved in regulating genes even further downstream in the developmental process. For example, it was recently discovered that *antp* antagonizes two other homeobox genes, *extradenticle* and *homothorax*, which are important for antenna differentiation as well as the differentiation of other organs. And five target genes, among them *dpp* and *wg*, produce components of signaling pathways that were discussed earlier in the book.

Another possibility (not mutually exclusive with the previous scenario) is that homeotic genes could act directly on several different steps during differentiation of a particular region of the body. In this scenario, the homeotic gene would act repeatedly at particular times and places, rather than govern differentiation from the top of a hierarchy. But this second possibility has seemed unlikely because it would require that the activity of the homeotic protein be regulated by intrinsic differences in the target cells. Experiments carried out by M. Rozowski and Michael Akam at Cambridge University support the idea that the homeobox protein Ubx works to regulate bristle development in the leg by this repetitive, context-dependent mechanism. You will recall that the sensory organ precursor (SOP) discussed earlier in this chapter divides asymmetrically two times to give rise to a sensory bristle, a neuron, and supporting cells. Rozowski and Akam were able to cause *ubx* to be expressed at times and places different than normal. Ubx is present in the legs of segment T3. These legs do not possess bristles like those found on the legs of segment T2. Rozowski and Akam discovered, by controlling the time and place of *ubx* expression, that Ubx was acting repeatedly in T3. Ubx could inhibit formation of the SOP, and could inhibit directly each of the asymmetric divisions giving rise to the distinctive T2-type leg bristles. These responses to Ubx were restricted to small numbers of cells and to particular times. These researchers concluded that Ubx-mediated repression must involve partner molecules present in the target cells that are not present in other cells. In other words, the action of Ubx depends on the context in which it appears.

We still do not have a complete picture of how any homeotic gene or combination of homeotic genes brings about specific morphogenesis and differentiation of a given segment or compartment. It is obviously complex, and it is important to stress again that so many multicomponent interactions are involved that a linear chain of reactions is probably not a suitable model. There are other important interactions we have not explored in this chapter. For example, many transcription factors encoded by homeobox genes require partner molecules encoded by other genes. These partners may not even contact DNA, but are nevertheless essential for the transcription factors to act, at least in certain contexts. The gene *extradenticle*, mentioned above, interacts with other homeodomain proteins so as to allow them to bind to their target sites. Another complex but vital system of genes is the polycomb group; their function is to keep the expression of homeotic selector genes turned off where it is not appropriate in a particular domain. The gene *polycomb* encodes chromatin proteins, which presumably alter chromatin configurations and restrict access of transcription factors to certain potential target sites in the genome.

PATTERNING THE WING

Wing Development Is Governed by Cellular Interactions

Thus far, we have concentrated on how regulated signaling between nonequivalent cells determines overall body-plan assignments. Similar tactics are used to regulate the differentiation of particular compartments and organs, about which a huge reservoir of new information is accumulating.

A particularly well understood example is the development of the *Drosophila* wing. As you will recall, the wing develops from an imaginal disc in the second thoracic segment, an address specified by genes of the bithorax homeotic complex (along with other interacting genes). The 20 to 40 original wing disc cells number about 200 by the middle of the second-instar phase, a time when crucial developmental decisions are being made that organize the future wing blade. By the early pupal stage, the wing disc has about 50,000 cells; distinctive bristles and vein patterns are forming, patterns essential to researchers for diagnosing the effects of experimental interventions that result in aberrant wings.

Anteroposterior (A/P) Patterning Depends on Intercellular Signaling

The wing has anterior and posterior compartments, which derive from parasegmental progenitors. As pointed out earlier, cell clones in one compartment never cross the boundary between compartments in normal embryos. Recent genetic experiments investigating a cell membrane protein called Capricious implicate it, along with others, in establishing different intercellular affinities for cells in anterior and posterior compartments. Posterior compartment cells express *engrailed*, which may serve as both a pair-rule and a homeotic selector gene; the border of the *engrailed* expression domain marks the A/P compartmental boundary. Cells expressing *engrailed* also express the signaling molecule Hedgehog; this may remind you of the relationship between *engrailed* and *hedgehog* expression during the establishment of parasegmental borders. The Hedgehog signal is received in adjacent anterior compartment cells, where it interacts with a membrane receptor (Patched) and a partner (Smoothened). Usually Patched inhibits *dpp*, but this interaction with Hedgehog removes the inhibition, allowing *dpp* to be expressed in a narrow band of cells on the anterior side of the A/P border (Figure 15.25).

Hh signaling via the receptor Patched (Ptc) is very complex, but we shall discuss it briefly because it illustrates yet another wrinkle in the regulation of signaling pathways. When Ptc is not engaged with its ligand Hh, it somehow modifies Smoothened (Smo) to make it inactive. When Hh does engage Ptc, Smo is phosphorylated (and possibly modified in other ways too), and both Ptc and Smo leave the membrane and move into the cytosol. In a way still not understood, the internalized, active form of Smo modifies a transcription factor called Cubitus Interruptus (Ci). Ci is a precursor that can be converted, by proteolysis, into either a repressive (Ci_{rep}) or active (Ci_{act}) form. These alternate forms can either repress or activate a battery of responsive genes, such as *dpp* and *ptc* itself. Use of the cellular internal-protein trafficking machinery as an important component of signal transduction may be unique, at least as far as our current understanding goes.

Dpp is thought to be a long-range morphogen, diffusing anteriorly and posteriorly from this signaling source, and organizing both anterior and posterior compartments. Dpp, of course, works through receptors (Thick Vein and Punt), stimulates signal transduction pathways, and affects the activity of target genes. Some of these target genes have been identified. Dpp may help organize a pattern by differentially

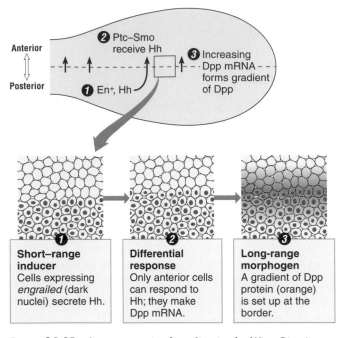

Figure 15.25   Anteroposterior Signaling in the Wing Disc A schematic dorsal view of the developing wing. The line through the middle separates the anterior and posterior compartments. Expression of *engrailed* posteriorly (En+) causes secretion of Hedgehog (Hh). which is received by anterior cells by means of Ptc and Smo. This induces *dpp* expression. and a gradient of Dpp is thereby formed. Details of these events are shown as diagrammatic enlargements of a cell cluster at the A/P border over time.

engaging genes with different response thresholds. Some transcription factor genes (for example, *spalt*) apparently require high levels of Dpp activity in order to be turned on, while others (such as *optomotor blind*) have a much lower threshold. Genes requiring lower concentrations of morphogen for activation may be expressed at a much greater distance from the source of morphogen than genes needing relatively higher morphogen concentrations.

Of course, there is much still to be learned about how these signaling molecules work to produce the final morphology of the adult wing; this includes details about the interactions between various signaling components, and the chain of command to the downstream genes that results in the final phenotype.

Patterning of the Dorsoventral (D/V) Compartments of the Wing Is Also Governed by Cellular Interactions

The wing not only has a front and back, but a top and bottom. The dorsal side is regulated by the homeobox gene *apterous*.

Cells expressing *apterous* become dorsal, while non-*apterous*-expressing cells become ventral. A sharp border of *apterous* expression marks the D/V compartmental boundary. D/V signaling is possibly somewhat more complicated than A/P signaling. We shall sketch it here because the signaling molecules are some of the same ones already encountered in other situations, and which we shall encounter again when examining vertebrate development. What is important, then, is not only the details of this story but also the fact that the same players recur in so many different contexts.

Figure 15.26 outlines what is known about the cellular D/V interactions in the developing wing. Apterous has at least two distinctive actions: first, it gives dorsal cells a compartmental address so that the dorsal and nondorsal (and thus ventral) regions are now distinct; second, Apterous initiates a signaling cascade between these two compartments. Let us examine this second function.

Apterous stimulates dorsal cells to express Serrate, a mebrane-bound ligand for the receptor Notch and for another secreted protein, Fringe; together, Fringe and Notch form a complex. Fringe is a glycosyl transferase, an enzyme that glycosylates a hydroxyl-linked fucose in the extracellular domain of Notch. In so doing, Fringe modifies the sensitivity of the Notch receptor to its ligands: Fringe reduces the sensi-

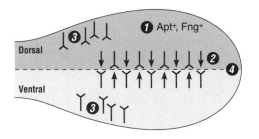

❶ Dorsal cells express *apterous* (*apt*) and *fringe* (*fng*).

❷ Apterous stimulates *serrate* expression dorsally. Fringe reduces sensitivity of Notch to Serrate and increases sensitivity of Notch to Delta.

❸ Notch receptors distant from the D/V border are not engaged by the ligand.

❹ Localized expression of *fng* leads to a band of activation of the Notch pathway, resulting in a band of *wingless* expression at the D/V border.

Figure 15.26   Dorsoventral Signaling in the Wing Disc This schematic of the developing wing. divided into dorsal and ventral compartments. illustrates interactions occurring between cells across the two compartments. Arrows indicate ligands. and notched lines indicate receptors.

tivity of Notch to Serrate, whereas it increases the sensitivity of Notch to Delta, the other ligand for Notch. The result is that dorsal, Fringe-secreting cells, which are upregulating their production of Serrate due to Apterous, at the same time have decreasing sensitivity to Serrate. This response probably reduces the potential effects of Serrate on the dorsal cells that are synthesizing it. However, ventral cells are still sensitive to Serrate. The ventral cells are thought to use Delta for signaling to dorsal cells. Some details of this pathway are still hazy, but the principle is clear: Fringe modifies (probably by glycosylation) the response of Notch to different ligands. While the receptor Notch is ubiquitous, its activation seems to be restricted to the zone of interaction between dorsal and ventral compartments.

Notice we have stated here that a ligand (for instance, Delta) can "activate" a receptor (such as Notch). What this means is that ligand binding changes the conformation of the receptor, which in turn leads to activation of second-messenger pathways. In Chapter 2, we discussed how ligand binding can change a G protein receptor from a GDP-binding to a GTP-binding molecule, with important secondary consequences. Another example, this time from Chapter 12, is the phosphorylation of RTK receptors after they have been activated by a ligand.

The activation of Notch is complex, and an ongoing subject of investigation. Ligand interactions produce several proteolytic cleavages (Figure 15.27). The extracellular domain is trimmed, probably by action of a protease called Kuzbanian. Then a membrane-bound protease complex cuts Notch again,

thereby releasing an active, intracellular form of Notch. The enzyme complex that carries out this intramembrane proteolysis is a huge, multiprotein assemblage. (One of its components is related to Presenilin, a protease found in human brain tissue that is associated with the onset of Alzheimer's disease.) The active intracellular Notch then migrates to the nucleus. There it interacts with some transcriptional cofactors, called CSL cofactors (including one called Suppressor of Hairless, or Su(H), that otherwise repress transcription of Notch target genes. Working together, intracellular Notch and the CSL cofactors stimulate transcription of a family of genes, among them *hairy* and *enhancer of split*.

A consequence of activating Notch at the D/V border is that *wingless* transcription is activated; the expression domain of *wingless* straddles this D/V border. It is noteworthy that Notch activates a developmental cascade in this instance, rather than acting as a repressor of transcription. Notch probably has different functions because it interacts with so many different partners, including Wg, Disheveled (Dsh), Numb, Fringe, and Ras. As is evident from its name, *wingless* is essential for wing development; you should recall that *wg* first acts in the embryo as a pair-rule gene defining body segments. This gene and its homologs in other phyla define a large group of signaling molecules, the Wnt family, which are active in many situations. Rather a lot is known about how Wnt signals are transduced; since you will encounter them in the next chapter, we shall briefly describe what is known about them here.

Wnt needs processing, at least in *Drosophila*, before it can act as a ligand. The membrane protein Porcupine (Porc) is

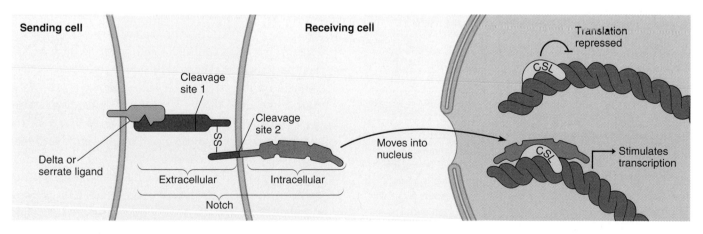

Figure 15.27 Notch Signal Transduction In its inactive state. Notch is a dimer composed of a transmembrane portion (having a significant intracellular segment) bound through a disulfide bridge to an extracellular portion. Signal transduction begins when the ligand on the sending cell interacts with the extracellular portion of Notch. After ligand binding. the extracellular portion of Notch is cleaved at site 1 and then at site 2. The now released intracellular portion of Notch moves into the nucleus. where it interacts with CSL cofactors that are repressing the target genes. The complex of cofactors and intranuclear Notch activates transcription.

necessary for the processing or secretion of Wnt. (This is inferred from the observation that without Porc function, Wnt stays associated with its cell of origin.) Wingless signaling also requires heparin sulfate proteoglycans in the extracellular matrix where signaling occurs. The active Wnt ligand is sensed by a family of receptors called Frizzled. (There are two *frizzled* genes in *Drosophila*.) Frizzled inhibits a system involved in the destruction of a vital component (β-catenin) of its signal transduction pathway. When this component is destroyed or rendered ineffective, as presumably is the case in the absence of Wnt signaling, then genes responsive to the signal are not turned on. But when Wnt stimulates Frizzled, the system that disarms β-catenin is turned off, allowing β-catenin to participate in turning on Wnt-responsive genes (Figure 15.28).

And what does Wnt (Wg) signaling via β-catenin do? It results in turning on other genes that code for transcription factors (T-cell transcription factor, or TCF, for example), which in turn organize subsequent development. Wg may be acting as a morphogen in the embryo, diffusing from its origin to induce transcription of *distalless* and *vestigial* in nearby cells sensitive to Wg. Dpp, which you recall is a morphogen involved in A/P organization, may act as an antagonist to Wg signaling, which

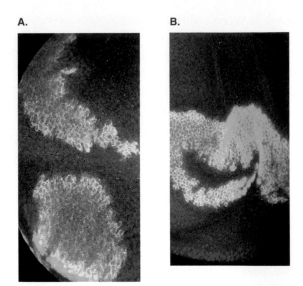

Figure 15.29 The Local Response to Free and Tethered Forms of Wingless The effect of restricting Wg to its cell of origin. **(A)** Cells are producing Wg, which is free to diffuse; the response of the gene *vg*, indicated by an antibody *(red)* against the Vg protein, occurs up to 10 or more cell diameters away from the Wg-secreting cell *(green)*. **(B)** By means of genetic engineering, Wg is tethered to the cell membrane *(green cells)*. Note that only those cells adjacent to the Wg-producing cells are responsive, as indicated by expression of Vg *(red)*.

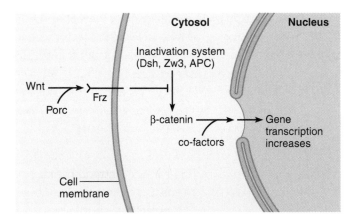

Figure 15.28 The Wnt (Wingless) Signaling Pathway The various steps of the Wnt signaling pathway are shown diagrammatically. Wnt needs the assistance of Porcupine (Porc) for secretion. The receptor Frizzled (Frz) receives the ligand Wnt. This signal inhibits a system (comprised of Dsh, Zw3, and APC) that ordinarily destroys β-catenin; as a result, β-catenin is allowed to survive. Together with partners, β-catenin can then enter the nucleus to influence the transcription of target genes. The actual pathway may be more complex and less linear than shown here; ongoing research is likely to modify this simple model to some extent. For *Drosophila*, the Wnt family molecule involved is Wingless, but a similar pathway using *wg* homologs is found in many animal groups.

could impose another layer of complexity on all these interactions. (There is evidence that Wg and Dpp are antagonists in the development of the leg disc as well.) The protein encoded by a recently discovered gene, *wingful (wf)*, is also an extracellular antagonist of Wg. We should add that Wg as a signal may operate in some instances through a β-catenin-independent pathway; for example, it helps organize the polarity of bristle patterns in the wing epithelium and bristle sensory organs without utilizing the canonical β-catenin second-messenger pathway, as we discussed earlier in the chapter.

Most of the ideas about pathways specifying parts of the wing derive from experiments in which a reporter gene for one of the participating molecules has been introduced into the embryo, and then the activity of the reporter in normal and abnormal locations was assessed. For instance, by use of a genetic trick, a small clone of *wg*-expressing cells can be created at a distance from the D/V border. Bristle patterns and neuroblast positions characteristic of the wing margin are thus re-created as a little island far from the edge of the wing blade. And transcription of *distalless* and *vestigial*, targets of Wg action, can be seen as many as 10 cell diameters away from the Wg-secreting clone (Figure 15.29A).

Now suppose *wg* is mutated (through molecular biological engineering) to produce Wg protein that cannot be secreted but remains tethered to the surface of its cell of origin. Then the modified gene can be injected into the embryo to establish a transgenic embryo. When the activity of the mutated transgene is monitored (by measuring the upregulation in expression of *distalless* and *vestigial*), the action of the modified Wg protein is now visibly limited to cells that can touch the clone of Wg-secreting cells (Figure 15.29B). From such experiments, it is clear that the action of Wingless as a ligand depends on how far it can diffuse.

How Does the Developing Embryo Establish Local Patterns of Organ Formation?

The examples we have just discussed could be replayed in other contexts many times. For instance, the development of the leg discs to form legs—which involves formation of the D/V, A/V, and proximodistal axes—utilizes many of the same signaling systems; there are differences, of course, but the tactics have a familiar ring. The compound disc that gives rise to the eye and antenna, which has been studied in considerable detail, has been found to involve many of the same molecules. Over and over again researchers encounter signaling systems involving three of the molecular families discussed in this chapter—Wnt, TGF-β (of which Dpp and the BMPs are family members), and Hedgehog—though sometimes disguised with unfamiliar names.

We shall encounter these signaling systems again in the next chapter, which explores gene regulation in vertebrates. Also parts of the vertebrate story are gradients of morphogens (or postulated morphogens) and responses of interacting cells depending on varying morphogen levels. We shall see that borders and edges are created from combinations of asymmetric cell divisions and complex intercellular signaling. Nature sometimes seems to calls on any and every trick available from the book of regulatory mechanisms in order to create an embryo.

KEY CONCEPTS

1. Cellular interactions in the embryo that are essential for determination and differentiation are based on regulating the machinery for sending and receiving chemical signals.

2. Nonequivalent cells result from asymmetric cell divisions, which in turn are based on polarization of the cytoskeleton and cell surface. Cytoskeletal asymmetry is either inherited from the antecedent cell or imposed by the local microenvironment. Cells use a complex biochemical apparatus to translate the cytoskeletal asymmetry into a nonequivalent partitioning of macromolecules during division.

3. Once different cell populations are established, one cell can signal other nearby cells *not* to follow the fate adopted by the former; this lateral inhibition (for example, in Delta and Serrate signaling via Notch in *Drosophila*) thus reinforces earlier developmental distinctions.

4. Development of the body plan in early *Drosophila* development is a stepwise process utilizing a cascade of transcription factors that influence each other through their genes. Concentration differences of different transcription factors establish gradients. Differential affinity of transcription factors for their target sites provides a basis for differential effects of one gene product upon another gene. This provides discontinuities in gene activity, resulting in (1) domains with borders and (2) activation of different effector genes in different parts of the body plan, downstream in the developmental process. Affinities of transcription factors for control-region DNA also underlie the action of homeotic selector genes.

5. Regulated cell signaling is also used as a tactic once early body regions have been delimited. Concentration gradients of ligands can act as morphogens. Differences in affinities of ligands for their cognate receptors can result in different cells interpreting the same morphogen in different ways.

6. Similar cohorts of signaling and receiving molecules are used in many different developmental situations in *Drosophila*. Some examples are (a) Delta and Serrate activating the receptor Notch (modified by Fringe); (b) Dpp (or other TGF-β members) activating their cognate receptors, termed Thick Vein or Punt in *Drosophila*; (c) Hedgehog activating the Patched–Smoothened receptor complex; and (d) Wingless activating the receptor Frizzled. Many other animal embryos use homologs of these molecules.

STUDY QUESTIONS

1. Suppose we could produce an embryo of *C. elegans* that had no maternal nor zygotic source of Mom5 protein. What would be the consequence of the absence of this protein?

2. The receptor Notch is involved in lateral inhibition in which the cell with Notch receives signals telling it *not* to adopt the same fate as the cell sending the signal. How far do you think this lateral inhibition could spread? Can you devise experiments to test your proposal?

3. In a *Drosophila* embryo bearing a null mutation of the gene *eve*, the embryo forms seven normal odd-numbered segments, but the seven even-numbered segments are all very defective. Why is this so?

4. Below is the spatial order of expression of the genes *otd* and *ems* in the three anterior (ant.) head segments; also shown are the principal sites of expression of the eight homeotic genes of the HOM complex expressed in the 14 segments of the posterior part of the head (H1 through H3), the thorax (T1 through T3), and the abdomen (A1 through A8): Suppose the entire bithorax complex of genes has been rendered null by a deletion; in other words, there is no Ubx, AbdA, or AbdB function. What would be the identity of the different segments? Explain your answers.

otd	*ems*	*lab*	*scr*	*antp*	*ubx*	*abdA*					*abdB*		
ant	ant	ant	H	H	H	T1	T2	T3	A1	A2	A3	A4	A5	A6	A7	A8

SELECTED REFERENCES

Anderson, K. 1995. One signal, two body axes. *Science* 269:489–490.

A brief review of Gurken signaling.

Anderson, K. V. 1998. Pinning down positional information: Dorsal-ventral polarity in the *Drosophila* embryo. *Cell* 95:439–442.

A paper on recent developments in the control of dorsoventral polarity in the fly, by a leader in the field.

Bray, S. 2000. Notch. *Curr. Biol.* 10:R433–R436.

A brief, incisive review of what Notch does and how it does it.

Gonzalez-Reyes, A., Elliott, H., and St. Johnston, D. 1995. Polarization of both major body axes in *Drosophila* by Gurken-Torpedo signalling. *Nature* 375:654–658.

A research paper on how follicle–oocyte interactions polarize the oocyte.

Gonzalez-Reyes, A., and St. Johnston, D., 1998. Patterning of the follicle cell epithelium along the anterior-posterior axis during *Drosophila* oogenesis. *Development* 125:2837–2846.

Research on the patterning of the follicle cells.

Greenwald, I. 1998. Lin-12/Notch signaling: Lessons from worms and flies. *Genes Dev.* 12: 1751–1762.

A detailed review of Notch function in several different developmental situa-tions encountered in both *C. elegans* and *Drosophila*.

Hawkins, N., and Garriga, G. 1998. Asymmetric cell division: From A to Z. *Genes Dev.* 12:3625–3638.

A review of the Numb–Inscuteable pathway.

Ingham, P. W., and McMahon, A. P. 2001. Hedgehog signaling in animal development: Paradigms and principles. *Genes Dev.* 15:3059–3087.

A thorough review of all aspects of Hedgehog signaling and the transduction of the signal.

Jan, Y. N., and Jan, L. Y. 1998. Asymmetric cell division. *Nature* 392:775–778.

Another review of the role of Numb, by one of the principal laboratories in the field.

Kosman, D., Ip, Y. T., Levine, M., and Arora, K. 1991. Establishment of the mesoderm-neuroectoderm boundary in the *Drosophila* embryo. *Science* 254:118–122.

A research paper showing how the Dorsal gradient is interpreted by the genes *twist* and *snail* to create a sharp border.

Lohmann, I., and McGinnis, W. 2002. Hox genes: It's all a matter of context. *Curr. Biol.* 12:R514–R516.

A brief review of new findings on how *ubx* and other homeotic genes act on target genes.

Misra, S., Hecht, P., Maeda, R., and Anderson, K. V. 1998. Positive and negative regulation of Easter, a member of the serine protease family that controls dorsal-ventral patterning in the *Drosophila* embryo. *Development* 125: 1261–1267.

A research paper exploring the role of proteases in regulating dorsoventral patterning.

Nusslein-Volhard, C., Frohnhofer, H. G., and Lehmann, R. 1987. Determination of antero-posterior polarity in *Drosophila*. *Science* 238: 1675–1681.

A historical review of how maternal effect genes control axis formation, by members of the laboratory that did the work.

Qi, H., Rand, M. K. D., Wu, X., Sestand, N., Weiyi, W., Rakic, P., Xu, T., and Artavanis-Tsakonas, S. 1999. Processing of the Notch ligand Delta by the metalloprotease Kuzbanian. *Science* 283:91–94.

A research paper showing how a protease cleaves Notch to allow signaling.

Roth, S. 1998. *Drosophila* development: The secrets of delayed induction. *Curr. Biol.* 8: R906–R910.

A review of how many different genes work together to establish the dorsoventral axis.

Roth, S., Neuman-Silberberg, S., Barcelo, G., and Schupbach, T. 1995. *Cornichon* and the EGF receptor signaling process are necessary for both anterio-posterior and dorsal-ventral pattern formation in *Drosophila*. *Cell* 81:967–978.

A seminal research paper on the basis of axial patterning.

Sen, J., Goltz, J. S., Stevens, L., and Stein, D. 1998. Spatially restricted expression of *pipe* in the *Drosophila* egg chamber defines embryonic dorsal-ventral polarity. *Cell* 95:471–481.

A research paper describing the breakthrough in understanding what localizes the activation of Dorsal.

St. Johnston, D., 1995. The intracellular localization of messenger RNAs. *Cell* 81:161–170.

A wide-ranging review of how embryonic cells localize mRNA.

Technau, G. M., 1987. A single cell approach to problems of cell lineage and commitment during embryogenesis of *Drosophila melanogaster*. *Development* 100:1–12.

A summary of the embryological work on cell lineages in the fly.

Vincent, J.-P., and Briscoe, J. 2001. Morphogens. *Curr. Biol.* 11:R581–R584.

A lovely exposition of what morphogens are and how they work.

CHAPTER

16

DEVELOPMENTAL REGULATORY NETWORKS II: VERTEBRATES

Recent discoveries of the importance of cascades of transcription factor genes and of genes encoding members of signaling pathways have revolutionized our understanding of development. In addition to providing new insights into development, these discoveries, derived from studying *Drosophila*, have also given researchers a new set of experimental approaches. Different organisms present different experimental challenges and opportunities. For example, the gene-knockout approach in mice allows one to create a null mutation in order to find out whether a gene is essential for development. The large frog egg is amenable to the microinjection of potentially active molecules, thereby converting the developing embryo into a kind of test tube for this unusual kind of biochemistry.

CHAPTER PREVIEW

1. Amphibian embryos utilize many signaling systems found in *Drosophila*.

2. Amphibian embryos employ localized mRNA to establish the Nieuwkoop center.

3. The Nieuwkoop center establishes the Spemann organizer, where gastrulation is initiated.

4. The Spemann organizer secretes several proteins that antagonize vegetal signals, thereby allowing dorsal mesoderm and the neural tube to form.

5. Vertebrates also use homeotic (HOX) genes to organize the body plan and limb development.

In this chapter, now that we are armed with our experience of *Drosophila* and other invertebrates such as *Caenorhabditis elegans* and the sea urchin, we can reexamine more closely the molecular basis for the formation of the vertebrate body plan and for some aspects of organogenesis. Are the same genes and cascades involved here as in *Drosophila*? Is the strategy of nature similar in these different creatures? Genetic manipulations provide a powerful tool; hence, the mouse embryo has been important for research on vertebrates. Another vertebrate, the zebrafish *(Danio rerio)*, is also being used in intensive efforts to identify genes and regulatory circuits important in vertebrate development (see Box 16.1).

We begin our reexamination in this chapter with amphibian embryos; the very large body of embryological data and the ease of manipulating these creatures experimentally have led to important insights in vertebrate development. We focus on the regulatory circuits that set up germ layers and the body plan in *Xenopus*. We then broaden our discussion to amniotes, dwelling especially on the homeotic selector genes. And finally, we turn to one example of terminal differentiation, the limb, an instance where complex regulatory circuits have been especially well studied.

SIGNALING AND DEVELOPMENT

A Frog Is Not a Fly

If the same tactics used in *Drosophila* were to be found in *Xenopus*, we would expect to find localized Bicoid, Hunchback, and Nanos proteins influencing transcription during early development. But this is not the case. You may recall (from Chapter 4) that the unfertilized amphibian egg is symmetric, having no anteroposterior or dorsoventral polarity. It is rotation of the cortex relative to the internal cytoplasmic core (the rotation being entrained by the point of sperm entrance) that breaks the symmetry and initiates events that set up the axes of the body plan. You may also recall that in *Drosophila*, intracellular organelle movements during oogenesis aid in setting up the dorsoventral polarity.

The cortical rotation in *Xenopus* is essential for setting up body axes and for developing the dorsal axial structures, which are the hallmarks of the vertebrate body plan. Figure 16.1A shows an embryo in which cortical rotation has been prevented by irradiation of the vegetal pole with UV light. The result is a hyperventralized embryo. Inspection of its internal anatomy reveals some differentiation of ventral mesodermal structures, but that is all. For comparison, Figure 16.1B

shows a hyperdorsalized embryo produced by treatment with D_2O; D_2O stabilizes microtubules, which otherwise are very dynamic.

When vegetal cytoplasm is transferred from a normal embryo to the equatorial region of such a UV-irradiated zygote, axial development can be restored. It is even possible to induce a fairly complete secondary embryo by injecting vegetal cortical cytoplasm into prospective ventral cells of a normal embryo (Figure 16.1C). Obviously, some factor or factors must be present in the cortical cytoplasm of the vegetal pole of the zygote to establish a center of some kind, resulting in subsequent dorsal axial development. It is presumed that these factors are rearranged somehow by cortical rotation. A currently popular hypothesis is that Wnt signaling molecules, or components of the Wnt pathway, may be distributed by rotation to the prospective dorsal side.

A finding consistent with (though not proving) this hypothesis is the finding that β-catenin, which as we saw in Chapter 15 is part of the Wnt signaling pathway, becomes enriched on the prospective dorsal side after egg rotation. If GSK3β and its partners do not destroy β-catenin, then the latter can act as a transcriptional activator; you will recall from Chapter 15 that Wnt signaling inhibits GSK3β so that β-catenin survives and can enter the nucleus. (In Figure 15.28, the GSK3β molecule is labeled Zw3, the name of this ortholog in *Drosophila*.) A group of proteins inhibiting the machinery responsible for the destabilization of β-catenin (GSK3β and its partners) has recently been identified in *Xenopus*. It is noteworthy that UV irradiation not only prevents the loss of GSK3β that occurs in normal development; it also stops dorsoanterior development (see Figure 16.1A). Taken together, these findings suggest that manipulation of the Wnt–β-catenin–GSK3β pathway is crucial for dorsoanterior development.

Many Signaling Molecules and Transcription Factor Domains Are Found in Virtually All Animals

It is time to be explicit about the nearly ubiquitous presence of many of the molecules we discussed in *Drosophila*. Some transcription factors found in *Drosophila*, such as Bicoid, are not found in other phyla. But many, many of the transcription factors and signaling molecules found in the fruit fly do have close relatives in most metazoans. These transcription factors may not be identical, of course, but functionally important parts of the molecules (such as the homeobox motif in homeobox genes) are often well conserved. Furthermore, some creatures may have many similar but nonidentical members of a given family. The number of family members may be different

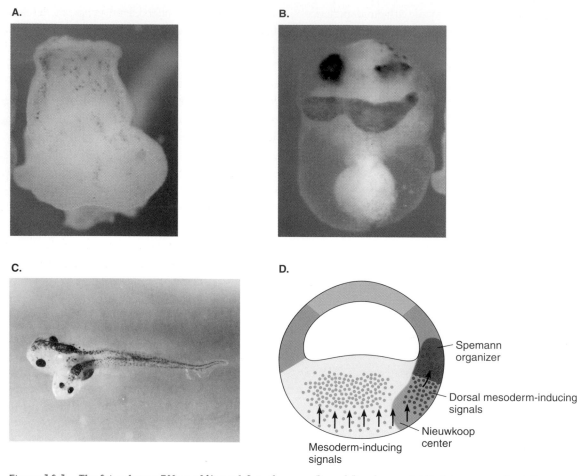

Figure 16.1   The Stimulatory Effect of Vegetal Cytoplasm on Dorsal Development **(A)** A hyperventralized *Xenopus* embryo in which only ventral mesoderm and some endoderm have developed. There are no dorsal axial structures, only epidermis, blood, and a reduced gut. This embryo was produced by irradiating the vegetal pole of a zygote prior to cortical rotation. **(B)** For comparison, a hyperdorsalized embryo, produced by treatment with D_2O (a microtubule-stabilizing agent), is also shown. This embryo possesses a circumferential band of eye pigment, a large cement gland, and a very large heart. **(C)** When cortical cytoplasm from the vegetal pole of a fertilized egg was injected into the prospective ventral vegetal blastomeres of a normal 16-cell-stage embryo, an almost complete secondary embryo formed, due to the inductive action of this vegetal cytoplasm. **(D)** The *Xenopus* blastula resulting from cortical rotation, shown diagrammatically; note the approximate positions of the Nieuwkoop center, the Spemann organizer, and the regions from which the different germ layers arise.

in different organisms. For example, *Drosophila* has only a few known members of the TGF-β family, while the mouse and the human have at least a dozen members of this family.

THE NIEUWKOOP CENTER REVISITED

The Nieuwkoop Center Is a "Dorsalizing" Center

The Nieuwkoop center, which we discussed in Chapter 4, is a group of cells in the vegetal hemisphere of *Xenopus* embryos that induces, or at least assists in, the formation of the Spemann organizer and dorsal-type mesoderm. Without dorsal mesoderm and the Spemann organizer, there is no dorsal axial development—no notochord, somites, neural plate, or pharyngeal endoderm. There is a lack of exact agreement between workers in the field as to precisely where the center is located and how primary its role is. This is not surprising, for definitions of its position and activity are all operational; that is, they depend on the nature of the defining experiment. Although any portion of the vegetal hemisphere of the blastula can, as was originally shown by Nieuwkoop, induce some kind of mesoderm from an animal cap, it is the *vegetal cells from the prospective dorsal side* that induce dorsal mesoderm. The

Spemann organizer, which also arises from subequatorial cells on the prospective dorsal side of the vegetal hemisphere, may actually arise from a dorsal portion of what was formerly the Nieuwkoop center. This scenario is shown in Figure 16.1D.

This matter of definitions is not just semantic hocus-pocus. There is, as we shall come to see, a very complex traffic of signaling between cells during the period when the germ layers are specified and gastrulation occurs. Knowing exactly which cell is doing what, where a given cell is located, and where it moves are essential for evaluating the experimental evidence.

Several different kinds of molecules are known to act as dorsalizing factors; among them are several members of the Wnt signaling family, especially Xwnt8 and Xwnt8b; two members of the TGF-β signaling family (activin and Vg1); and a homeobox-containing transcription factor, Siamois. Soon after the midblastula transition, *siamois* is expressed in cells of the Nieuwkoop center. If any of the just-mentioned molecules is injected into the ventral portion of a developing embryo, a duplicate axis can occur ventrally. And each molecule can also rescue developing embryos from ventralization imposed by UV irradiation (Figure 16.2). Are all of these molecules in-volved in the signaling necessary for dorsal development? Are they part of the same pathway? Another complication is that, except for Siamois, each of the molecules mentioned earlier in this paragraph is expressed at other times and in other places; undoubtedly, they have other functions in these different contexts. Nor are their actions all identical; activin and Vg1 both can induce mesoderm in animal caps, but Xwnt8 cannot induce mesoderm by itself in an animal cap assay.

Thus, the interrelationships among these various signaling factors, transcription factors, and their actual roles in the embryo are complicated, and have been difficult to dissect. Genetic analyses of what is upstream and what is down-stream of one molecule or another in a pathway are not feasible. What has been done is a combination of clever biochemistry and molecular biology, and educated guesses based on what has been found in *Drosophila* and *C. elegans*. We can sketch the current ideas, and will do so to some extent, but you can be certain that the details of the picture will be modified as time passes. New genes involved in dorsalization and other pat-terning activities are being discovered almost daily. Interpre-tations of various experiments are frequently incomplete or at odds with one another.

A.

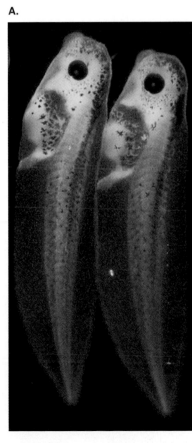

B.

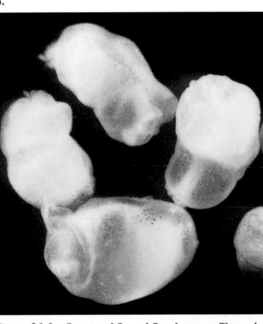

C.

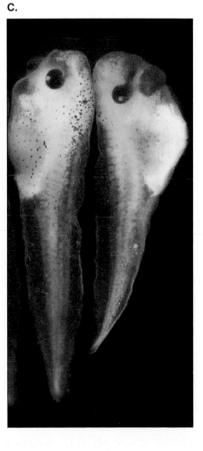

Figure 16.2   Rescue of Dorsal Development Through Microinjection of Wnt mRNA Photographs of embryos from a typical assay for dorsal development. **(A)** Normal embryos. **(B)** Embryos in which UV irradiation has elimi-nated dorsal development. **(C)** A UV-irradiated embryo in which dorsal development has been "rescued" by injecting Xwnt8 mRNA. This same mRNA has an opposite, ventraliz-ing activity later in development.

Box 16.1 Zebrafish

During the last several years, Nusslein-Volhard and her colleagues in Tübingen, Germany, decided to launch a massive study on a vertebrate in which the production of mutations and genetic analysis could be employed with relative ease. Although the mouse is used for this purpose with impressive results, its generation time of several months means results cannot be obtained quickly; consequently, the enterprise of making large numbers of mutations and scoring them is difficult in the mouse. So the German workers selected the zebrafish, *Danio rerio*, as their model organism. The zebrafish had been brought to the attention of developmental biologists by a dedicated group of scientists led by George Streisinger and Charles Kimmel, working at the University of Oregon.

There are several advantages to using the zebrafish for this kind of work. Large numbers of embryos can be obtained throughout the year, fertilization can be carried out in vitro, and the embryos are optically clear. Moreover, the generation time is short, and homozygous diploids can be obtained by parthenogenesis. Last but not least, important genes found in the zebrafish, because it is a vertebrate, are likely to be present in amniotes and could lead to useful study of diseases with genetic etiology in humans.

The zebrafish develops somewhat differently from other vertebrates we have considered. The accompanying figure depicts the sequence of its early development. After fertilization, most cytoplasm accumulates near the animal pole, and the yolk is concentrated vegetally. Only the animal cap of cytoplasm undergoes (meroblastic) cleavage, forming a blastodisc. Nuclear divisions occur at the periphery of the blastodisc to form a **yolk syncytial layer (YSL)**, composed of many nuclei in a single large yolk "cell" lying underneath the blastodisc. Meroblastic cleavage and formation of a syncytium at the edge of the blastodisc are also features of avian development. As in *Xenopus*, microtubule-mediated movements of ß-catenin occur on one side of the egg, which is destined to form the future dorsal side of the embryo.

A sagittal section through the embryo shows the yolk cell with its many syncytial nuclei, overlaid by cells of the **deep layer**, which in turn is covered by an **enveloping layer** (see part A of the figure). The enveloping layer is a specialized covering of the developing embryo not found in amphibians or amniotes. The deep-layer cells will form the embryo proper.

Cells around the edge of the blastodisc expand over the yolk cell (see figure, part B). This movement results mainly from autonomous expansion of the YSL, which pulls the enveloping layer over the yolk cell. Microtubule function is essential for this "pulling" action. As epiboly proceeds, many deep cells at the margin of the blastodisc undergo involution and ingression, thereby forming a new

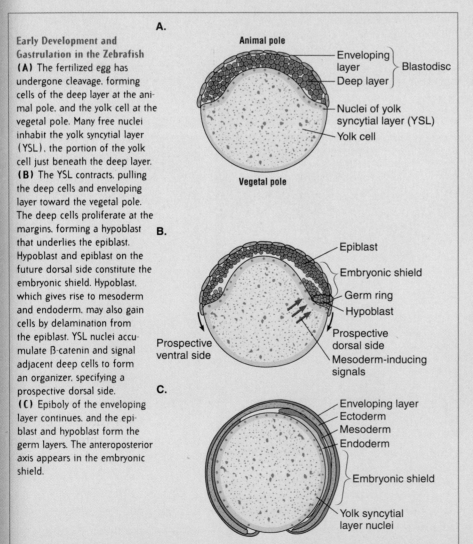

Early Development and Gastrulation in the Zebrafish
(A) The fertilized egg has undergone cleavage, forming cells of the deep layer at the animal pole, and the yolk cell at the vegetal pole. Many free nuclei inhabit the yolk syncytial layer (YSL), the portion of the yolk cell just beneath the deep layer. **(B)** The YSL contracts, pulling the deep cells and enveloping layer toward the vegetal pole. The deep cells proliferate at the margins, forming a hypoblast that underlies the epiblast. Hypoblast and epiblast on the future dorsal side constitute the embryonic shield. Hypoblast, which gives rise to mesoderm and endoderm, may also gain cells by delamination from the epiblast. YSL nuclei accumulate ß-catenin and signal adjacent deep cells to form an organizer, specifying a prospective dorsal side. **(C)** Epiboly of the enveloping layer continues, and the epiblast and hypoblast form the germ layers. The anteroposterior axis appears in the embryonic shield.

A.

Animal pole

Enveloping layer
Deep layer
} Blastodisc

Nuclei of yolk syncytial layer (YSL)

Yolk cell

Vegetal pole

B.

Epiblast

Embryonic shield

Germ ring

Hypoblast

Prospective ventral side

Prospective dorsal side

Mesoderm-inducing signals

C.

Enveloping layer
Ectoderm
Mesoderm
Endoderm

Embryonic shield

Yolk syncytial layer nuclei

layer of cells between the original deep layer and the yolk cell. A thickening of the margin of the blastodisc, called the **germ ring**, is composed of a superficial layer (next to the enveloping layer) called the epiblast, and an inner layer, the hypoblast. The portion of the germ ring and blastodisc on the future dorsal side, the side where β-catenin accumulates in nuclei of the YSL, thickens. The accumulation of β-catenin indicates that the Wnt pathway is probably involved in specification of the forming germ layers. The thickened germ ring and blastodisc on the prospective dorsal side are now called the **embryonic shield.**

In the shield, the hypoblast cells undergo convergent extension to form endoderm and dorsal mesoderm (see figure, part C). Involution and ingression of the germ ring and formation of the shield are the zebrafish version of gastrulation.

The YSL nuclei that accumulate β-catenin are thought to be equivalent to a Nieuwkoop center, and the embryonic shield acts as an organizer. Evidence for this comes from the fact that when the shield cells, including the deep cells of this region, are transplanted to the prospective ventral side of a blastodisc, a complete secondary embryo can be induced.

Cells in the midline of the epiblast of the shield coalesce to form a solid cellular rod, which later cavitates to form a neural tube. This mode of neural tube formation is also employed for neural tube formation in the tail bud of amphibians and amniotes.

The involvement of Wnt and β-catenin in the induction of the organizer is similar to what is found in other vertebrates, and neural induction involves homologs of BMP and Wnt and their antagonists (homologs of Dickkopf, a protein introduced later in the chapter, and Chordin). Thus, the underlying signaling pathways driving early zebrafish development are closely related to those in other vertebrates. On the other hand, many morphological aspects are different. In addition to the unique enveloping layer, formation of a blastodisc and syncytium is different from what is found in *Xenopus* but rather similar to what occurs in birds. The involvement of the YSL in epiboly is distinctive, and formation of the anterior neural tube is different from that in amphibians and amniotes. Though we shall not discuss organ formation in zebrafish as a separate topic, the basic outlines of most of organogenesis are similar to what was described in Chapters 6 and 7.

A community of researchers have already made substantial progress using the zebrafish to identify genes important in development. A large-scale screen for mutants has produced about 700 different mutations affecting embryonic development. We can expect that many new genes affecting both early development and organogenesis will be identified and studied in the zebrafish, and there is every reason to believe that the results may be applied to understanding amniote development.

In spite of the likelihood of revising our understanding of these events, there is good reason to be concerned with the details of how the body plan of the embryo is established. First, the sheer complexity of the signaling circuits and their interrelationships, even though we do not fully understand them, constitutes an important principle. Development is being driven by interlocking and overlapping, finely tuned networks of signaling and regulation of gene expression. Second, much of the apparatus used in these networks is similar in different kinds of embryos, and is employed in many different events during development. Development is driven by very conservative yet robust machinery.

The siamois *Gene Is a Reliable Indicator of Nieuwkoop Center Activity*

The expression of Siamois, a homeodomain-containing transcription factor, is a good indicator of Nieuwkoop center activity. It is possible to abolish *siamois* expression (hence, the Nieuwkoop center)—and indeed, any dorsal development—by interfering with maternal β-catenin. This loss-of-function type of experiment can be done in two ways: one is to inject into the zygote a so-called antisense oligonucleotide—an oligonucleotide possessing a nucleotide sequence exactly complementary to the nucleotides encoding amino acids (usually, but not always, near the N-terminus) in the β-catenin protein (see Box 7.2). This interferes with the synthesis of β-catenin from maternal mRNA early in development. Another way is to inject an mRNA that encodes C-cadherin (also called EP-cadherin; see Table 12.3) into the zygote, causing an overexpression of C-cadherin. The abundant C-cadherin binds to the adherens junction and causes sequestration of β-catenin in these junctions at the cell surface. Then β-catenin cannot fulfill its other signal-transducing functions, resulting in a loss of that particular β-catenin function.

In either case, interfering with β-catenin signaling interferes with dorsal development, stops *siamois* expression, and prevents the Nieuwkoop center from forming. If the mRNA for Siamois, activin, or Vg1 is injected into such a perturbed embryo, rescue of dorsal development will occur. This is interpreted to mean that activin, Vg1, and Siamois (as well as other

axis-inducing molecules) act downstream in the pathway from β-catenin. Only Wnt and other molecules upstream from β-catenin in its action pathway, such as GSK3, can induce *siamois* expression in ectopic locations. (For a review of the β-catenin pathway, see Figure 15.28.)

Figure 16.3 presents the result of these injection experiments diagrammatically. Notice that the signaling pathway is similar to the signaling pathway we encountered in development of the *Drosophila* wing. The actual Wnt or Wnt-like molecule that is the normal endogenous inducer of this pathway in *Xenopus* has still not been identified, but Xwnt8b is a possible candidate since it is maternally expressed. However, it has also been proposed that the Nieuwkoop center could arise solely from activation of the cytoplasmic components of the Wnt pathway in prospective dorsal cells, so there may not be a requirement for a Wnt ligand after all. Just how this

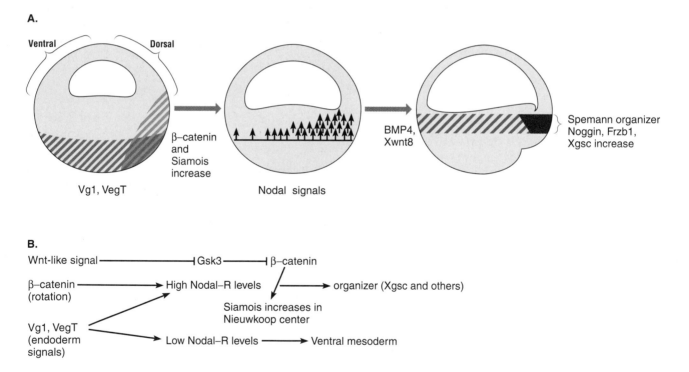

Figure 16.3   Wnt Signaling and Formation of the Nieuwkoop Center (A) Steps in the formation of a Nieuwkoop center. Fertilization causes a cortical rotation, which shifts some organelles. This results in stabilization of β-catenin on the prospective dorsal side; β-catenin then may enter the nucleus and collaborate with transcription factor(s), leading to activation of the *siamois* gene in the Nieuwkoop center. This in turn leads to activation of Spemann organizer-specific genes such as *xgoosecoid (xgsc)*. Also, maternal mRNA encoding the transcription factor vegT is necessary for endoderm formation, and for a synergy with Nodal-related factors (Nodal-R) needed for mesoderm formation. Nodal factors are more concentrated on the prospective dorsal side than on the prospective ventral side. **(B)** A schematic of some of the probable molecular interactions in this Wnt signaling pathway: unstimulated Wnt pathways lead to β-catenin destruction by the action of a protein kinase called GSK3. It is proposed that β-catenin accumulates on the prospective dorsal side because of cortical rotation. Also, some Wnt-like signal, or some other endogenous activation, leads to inhibition of GSK3, thereby freeing β-catenin to become active and eventually resulting in transcription of *siamois* and establishment of the Nieuwkoop center. VegT and possibly Vg1 stimulate endoderm formation, and result in high levels of Nodal-R signaling on the prospective dorsal side and lower levels of Nodal-R on the prospective ventral side.

"activation" of the Wnt pathway in dorsal cells occurs is not known, but since deletion of the Wnt receptor Frz7 reduces dorsal (but not ventral) mesoderm formation, it is indeed likely that Wnt is involved.

The Transcription Factor Gene vegT Also Plays a Role in Germ-Layer Specification

Let us stop for a moment and summarize. Cortical rotation causes stabilization of β-catenin on the prospective dorsal side. Maybe Wnt signaling activates or helps the β-catenin pathway to stimulate Siamois production, which then causes other genes to carry out Nieuwkoop center signaling; this in turn leads to dorsal mesoderm specification and formation of the organizer (see Figure 16.3A).

This sequence of events helps to explain mesoderm specification and eventually neural ectoderm development. But what about the role of endoderm? Are TGF-β family members involved? They certainly are in some way, because expression of vg1, xnr (1,2, or 4), or derriere can all induce mesoderm in animals caps. And there are other nagging questions. Nieuwkoop's experiments were based on transplantation that demonstrated vegetal cells *could* induce mesoderm, but do vegetal cells actually do this in normal development? There are also discordant experimental results. For example, Patrick Lemaire and John Gurdon, working in Cambridge, England, in 1994 were able to completely dissociate cells of the cleaving embryo from one another; yet some genes characteristic of the Spemann organizer, such as *xgoosecoid*, were turned on anyway, even in the absence of an organized tissue.

These hints of something more have now been realized in experiments resulting from a collaboration between Jian Zhang, David Houston, and Mary Lou King at the University of Miami, and Chris Wylie and Janet Heasman, then at the University of Minnesota (now at the University of Cincinnati). They worked on a gene called *vegT*, which encodes a transcription factor with a so-called T Box domain; *vegT* is expressed during oogenesis, and localized transcripts of it are present in the vegetal hemisphere of the egg. When the maternal VegT mRNA concentration was reduced through injection of an antisense oligonucleotide against VegT, the embryos that developed lacked endoderm and did not form a Nieuwkoop center. Furthermore, vegetal cells from such a VegT-depleted embryo could not induce mesoderm when combined with an animal cap taken from a normal embryo; that is, they failed the Nieuwkoop center "test." Ectodermal epithelium and ventral-type mesoderm formed instead of endoderm. Thus, in

addition to Wnt signaling and β-catenin activation of *siamois*, there is also a localization of maternal VegT mRNA and protein, which is essential for endoderm formation and for the mesoderm-inducing activity of prospective endoderm.

An improved second generation of antisense experiments resulted in almost complete elimination of maternal VegT, and no endoderm and almost all mesoderm failed to form in such embryos. Furthermore, expression of many signaling molecules important in mesoderm development (FGF; Xnr1, -2, -4; and Derriere) was eliminated. It seems that *vegT* is at or near the top of a hierarchy of genes leading to endoderm formation and of genes directing expression of TGF-β family members involved in mesoderm formation. Both the Wnt-β-catenin and the VegT-TGF-β cascades help to establish endoderm, mesoderm, and axial organization of the embryo (Figure 16.3B). The TGF-β family members that comprise the Nodal subfamily (Xnr1 -2, -3, and -4) are key mediators of the action of VegT. Injection of the mRNA of *nodal related* genes can reverse the effects of depletion of VegT, and other experiments show that Xnr is directly involved in mesoderm specification. There is some evidence that components of the TGF-β and Wnt pathways can physically interact, thereby affecting expression of an important downstream gene (*xtwin*) in the organizer.

Since VegT is a transcriptional activator, and transcription only starts after the midblastula transition, VegT could account for Nieuwkoop center activities. However, certain transplantation studies have indicated that some activity of the Nieuwkoop center for induction of dorsal mesoderm is present in the vegetal hemisphere as early as the 32-cell stage. Other experiments seem to contradict this conclusion. It is not clear how all this information will fit together. Perhaps there is an early Nieuwkoop center activity independent of VegT, and a second and stronger VegT-sponsored wave of activity. It seems clear that the main effects of VegT occur in responding cells after the midblastula transition.

THE SPEMANN ORGANIZER REVISITED

The Spemann Organizer Arises as a Consequence of Nieuwkoop Center Activity

Though the activity of the Nieuwkoop center is crucial, its role is transient. Experimental evidence suggests that the center only begins strong signaling activity shortly after the midblastula transition, and some of its action occurs in the absence of concurrent protein synthesis. As of this writing, it is not known exactly what causes the prospective dorsal marginal zone to become the Spemann organizer, which is the center of axis formation and gastrulation. Current ideas invoke another factor or factors—perhaps other Wnt family members, perhaps antagonists of TGF-β family members—as helping to define the location of the Spemann organizer on the prospective dorsal side. Nodal related factors and bFGF are known to act just before gastrulation and to induce the formation of bottle cells, which indicates the onset of gastrulation. Nieuwkoop center cells are vegetal; some of them may physically overlap with the prospective Spemann organizer.

Adjacent to the Nieuwkoop center is an area composed of surface ectoderm, subjacent prospective mesoderm, and deeper mesoderm; it is the initial site of both gastrulation and the Spemann organizer. Cells in this area will secrete a number of signaling molecules that reinforce the dorsal mesodermal character of this population. The organizer cells will also secrete ligands involved in neuralizing the ectoderm to form a nervous system. As in the case of the Nieuwkoop center, we shall encounter myriad factors involved in Spemann organizer activity—factors that interact with one another to produce a robust, complex, signaling apparatus.

The Spemann Organizer Has Distinctive Gene Expression and Secretes Many Ligands

The Spemann organizer is not a homogeneous population of cells. Recently, a careful fate map has been made, in which the borders of the cells inhabiting the organizer are defined and the distributions of two ligands and two transcription factors are shown (Figure 16.4). This map is topographically complex. There are ligands present in the organizer, such as Noggin and Chordin, that have strong neuralizing influences as well as dorsalizing influences on mesoderm. And there are others, such as activin and bFGF, that primarily influence mesoderm. Furthermore, the organizer is a changing population of cells, and the action of these cells continues throughout gastrulation and even into the neurula stages. Add to this the changing competence (that is, the changing functional state of receptors) of cell populations receiving organizer signals, and you can appreciate the potential complexity of the Spemann organizer.

The Spemann Organizer Is a Neuralizing Center

The organizer may be considered to have two kinds of activities: neuralization, and dorsalization of the mesoderm. The latter may be a consequence of the synergy of TGF-β and

β-catenin-initiated events started after the midblastula transition. As we discussed in Chapter 6, the discovery of the powerful neuralizing factors Noggin and Chordin (there may others as well, such as Follistatin) and an understanding of how they work has radically altered our view of how the organizer causes neuralization. Noggin and Chordin physically interact with and antagonize BMPs, which are members of the TGF-β family. This antagonism of BMP activity removes the impetus for ectoderm to adopt an epidermal fate, allowing the dorsal ectoderm to follow its "default" neural pathway. What is missing here is detailed knowledge of how this so-called default neural differentiation pathway originates in the first place. The notochordal tissue is active in dorsoventral patterning of the neural tube, by means of active signaling with the ligand Sonic Hedgehog, as discussed in Chapter 6.

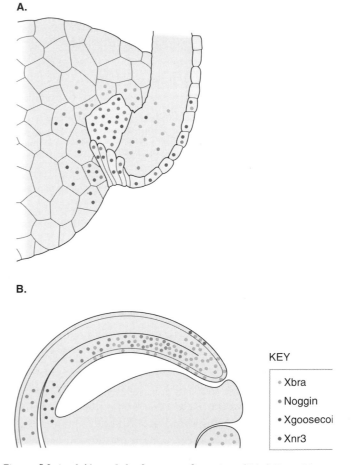

A.

B.

KEY

- Xbra
- Noggin
- Xgoosecoi
- Xnr3

Figure 16.4 A Map of the Spemann Organizer (A) Cells at the site of formation of the dorsal lip of the organizer, showing the distribution of different organizer-specific mRNAs. **(B)** After involution has occurred and the archenteron is established, the different organizer regions are visibly organized into different regions of the neurula.

A. Two factors

B. Secondary posteriorization

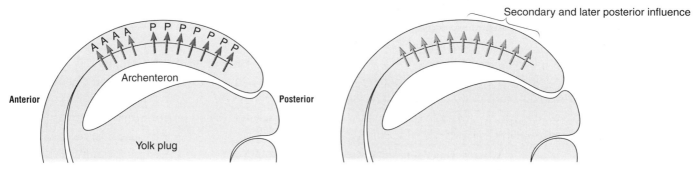

Figure 16.5   Two Models of How Regional Inducing Properties of the Organizer Operate Sagittal sections through the late gastrula of *Xenopus* are shown diagrammatically. **(A)** The two-factor model posits that separate anteriorizing (A) and posteriorizing (P) factors are released from different regions of the notochordal cells originating from the Spemann organizer. **(B)** The two-step model emphasizes the presence of only one neuralizing signal along the entire notochord, followed by a secondary posteriorizing signal coming later on.

There may be other subtle interactions as well. When animal cap tissue is exposed to Noggin, neural tissues develop without any mesoderm formation. The organization of this neural response may be assessed by looking at the expression of marker genes that are characteristic of dorsal parts of the brain, or of ventral parts. (It has been noticed that expression of these markers is neither random nor intermingled, but rather occurs in distinct areas.) Exposure to different levels of Noggin leads to differential expression of neural markers, indicating Noggin may be a morphogen. On the other hand, there is evidence that Noggin-treated cells interact with one another to attain some dorsoventral neural organization.

The Spemann Organizer Provides Anteroposterior Neural Patterning

Since the early days of analyzing the Spemann organizer, researchers have appreciated that it plays a role in anteroposterior patterning. Explants of the early dorsal lip together with competent ectoderm resulted in brainlike structures, while organizer tissue from the midgastrula stage combined with ectoderm could produce hindbrain and spinal cord. Recent tissue recombination experiments using different parts of the organizer from very early gastrulae showed that it is regionally organized to induce anterior or posterior neural structures. Two theories about how anteroposterior organization might be produced have contended over the years: one is that the organizer produces at least two substances (today we might say at least two different *mixtures* of substances), one favoring anterior, the other posterior neural development. The other theory is that the organizer produces

an all-anterior-type neural organization, and a later posteriorizing signal, delivered regionally, stimulates formation of posterior neural tissues. These are called the two-factor and two-step models, respectively (Figure 16.5). Both theories could be partially true, of course; nothing requires that they be mutually exclusive.

There are candidates for posteriorizing molecules. Retinoic acid (RA) is found to be more concentrated in posterior parts of the embryo than in anterior regions; when RA is combined with Noggin treatment of animal caps, posterior neural markers as well as anterior ones are expressed. Similarly, simultaneous treatment with Noggin and bFGF can encourage posterior neural development, and a ligand of the Wnt family has been implicated in posteriorization by bFGF. Other experiments have shown that production of Xwnt8 by posterior paraxial mesoderm contributes to posterior neural development. Further experiments are needed to sort out whether these or other candidate molecules are actually involved in the intact *Xenopus* embryo.

Another possibility, which also may only be part of the picture, is that regional influences from "outside" the organizer tissues—for example, laterally or ventrally—could help pattern neural development. Two genes have recently been discovered, *dickkopf* (German for "big head") and *cerberus* (Figure 16.6), that can induce ectopic heads, as well as duplicated livers and hearts, when injected into *Xenopus* embryos. These genes encode novel secreted protein factors, and both have been shown to be powerful antagonists of Wnt signaling. In fact, simultaneous inhibition of both the Wnt and BMP signaling pathways leads to induction of a secondary axis including a head. Cerberus has been shown to directly antagonize BMP,

A.

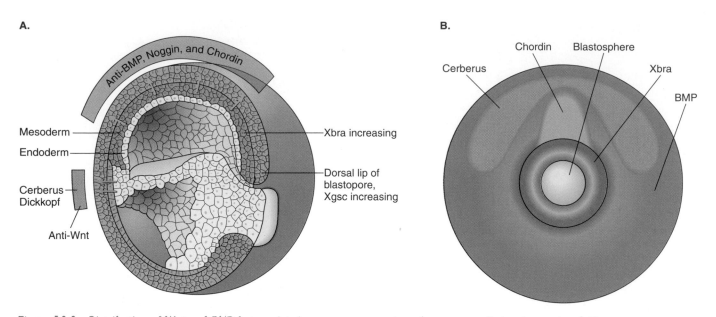

B.

**Figure 16.6 Distribution of Wnt and BMP Antagonists in
Xenopus (A)** A cutaway view of the *Xenopus* gastrula. The BMP antagonists Noggin and Chordin (and Follistatin) are present in the dorsal
midline. Invaginated mesoderm is expressing *xbra*, and the organizer is
expressing *xgoosecoid*. Wnt antagonists Cerberus and Dickkopf are
released from anterior cells that will soon give rise to endoderm and
mesoderm (sometimes called endomesoderm). Their action is necessary for head formation. **(B)** A diagram of the posterodorsal aspect of
the *Xenopus* gastrula with the overlying ectoderm stripped away, showing the broad anterior distribution of Cerberus. Expression of the
marker gene for mesoderm, *xbra*, is shown as an involuting torus of
cells around the blastopore.

Nodal, and Wnt signaling by binding to these ligands and preventing their interaction with the cognate receptors. Insulin-like
growth factors and their receptor have also been discovered in
Xenopus, and overexpression of these growth factors (brought
about by microinjection of the cognate mRNAs) increases the
amount of head tissue at the expense of the trunk. The organization of neural tissues evoked by the Spemann organizer will
likely be shown to depend on a complex set of interactions
involving substances released by different regions within the
organizer, as well as synergistic and antagonistic influences,
such as Wnt, Nodal, and BMP signaling, emanating from
outside the organizer itself.

The Spemann Organizer Also Dorsalizes Mesoderm

Spemann and Mangold's experiments demonstrated that the
transplanted organizer induced both neural and dorsal mesodermal tissue. The organizer itself forms notochord (the dorsalmost mesoderm) and some pharyngeal endoderm. Factors
that dorsalize and stabilize mesoderm are released from the
organizer. It is likely that a complex mixture of factors is
being released from the organizer—some counteracting Wnt
and TGF-β (BMP) signals, which are "ventralizing," others
acting as morphogens to help pattern mesoderm.

What makes understanding these relationships difficult is
that we know a little bit about a lot of things. The transcription factors and signaling-component molecules encoded by
over 30 genes have been found to be enriched in early mesoderm. We shall discuss a few that illustrate possible mechanisms for the patterning of mesoderm. We know that Noggin
is a powerful neural inducer and works by antagonizing signaling by BMP2 and BMP4. Noggin also stimulates explants
of ventral mesoderm, which otherwise would form only
blood, to differentiate more dorsally as muscle (Figure 16.7)

Several other TGF-β-like factors, members 1 through 4 of
the Xnr group, are present in the organizer and induce muscle formation in animal cap explants. (Xnr stands for "nodal-
related genes found in *Xenopus*"; Nodal was first cloned from
mice.) When Xnr1 and Noggin are injected together into the
ventral side of a normal embryo, a complete secondary axis
can be produced (Figure 16.8). Similarly, when the same combination is injected into a UV-arrested *Xenopus* embryo, there
is complete rescue of normal development. Noggin, working
as a BMP antagonist, together with one or a number of other
TGF-β molecules, thus apparently helps to organize and
pattern mesoderm.

Another BMP antagonist emanating from the organizer is
Chordin; when Chordin binds to BMP, the latter is prevented

A.

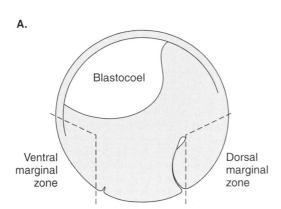

B.

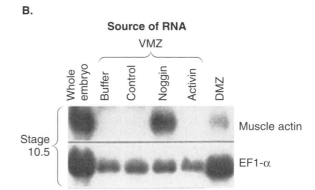

Figure 16.7 The Dorsalization of Mesoderm by Noggin Prospective ventral mesoderm was isolated from a *Xenopus* gastrula and exposed to Noggin. **(A)** An early gastrula (stage 10.5), shown here diagrammatically in side view with the collapsing blastocoel at top left, was dissected to isolate either ventral or dorsal marginal zone tissues *(dashed lines)*. Each of these tissues was then cultured in vitro in a medium either containing Noggin protein or not. **(B)** *Upper strip:* the presence of muscle tissue in these cultures was assessed by isolating RNA and assaying for actin mRNA. This assay is done by subjecting RNA extracted from the cell to electrophoresis, then transferring the RNA to a nitrocellulose membrane, and hybridizing the membrane to a radioactive DNA probe specific for actin mRNA. When RNA separated by electrophoresis is transferred to a membrane, it is termed a Northern blot. *Lower strip:* a probe for the mRNA of EF1-α, a ubiquitous enzyme needed for protein synthesis, was also used, in order to ensure that the RNA used for the analysis had not been degraded. Various additions were made to the culture medium for the ventral marginal zone (VMZ), whereas the dorsal marginal zone (DMZ) was cultured without any additions; note that ventral marginal tissue does not produce muscle actin unless cultured with Noggin.

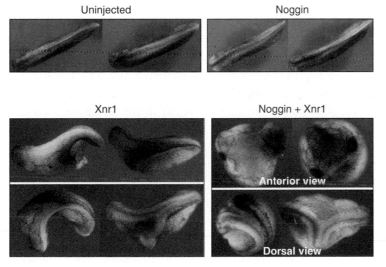

Figure 16.8 A Secondary Embryo Produced by Noggin and Nodal Acting Together A single ventral blastomere at the four-cell stage was injected with either Noggin mRNA *(top right)*, Nodal related (Xnr1) mRNA *(bottom left)*, or both *(bottom right)*, and then allowed to undergo gastrulation and neural tube formation. The combination of Noggin and Xnr1 produced a complete secondary axis.

from interacting with its receptor. An elaborate mechanism regulates the amount of this antagonism, thereby establishing a finely tuned gradient of BMP. Though the details are still being worked out, the inactive Chordin–BMP complex can be digested by the protease Xolloid, which releases some active BMP fragments from the inactive complex. Xolloid is regulated to some extent by another protein, called Twisted Gastrulation (Tsg), though the mechanism by which it does so is still under investigation. The complexity of these inter-

actions gives adequate opportunity, at least in theory, for different kinds of BMP activity gradients to form.

There are probably several other factors emanating from the Spemann organizer. For example, Lunatic Fringe is related to the extracellular portion of the *Drosophila* protein Fringe; Lunatic Fringe has been isolated from *Xenopus* and shown to induce mesoderm in animal caps. Another molecule found in the *Xenopus* organizer, Frizb, is a secreted factor related to the extracellular domain of Frizzled in *Drosophila*,

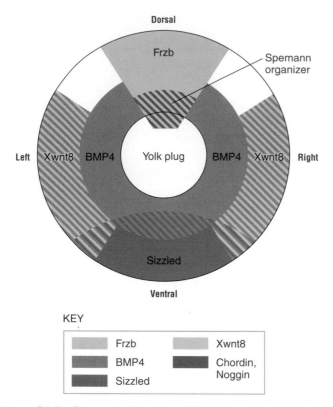

Figure 16.9   Expression Domains of Wnt Antagonists Frizb and Sizzled A diagram showing the approximate expression domains for various factors active in the dorsoventral organization of mesoderm. The view is of the gastrula seen from the vegetal pole. with the Spemann organizer region toward the top and the yolk plug in the middle. The strong ventralizing molecules. Xwnt8 and BMP4. are represented as nonoverlapping concentric rings. but their regions actually do overlap with each other. The domain of Frizb overlaps with the domains of Chordin and Noggin in the organizer. and the ventrally located Sizzled overlaps with both Xwnt8 and BMP4.

which we discussed in Chapter 15. Frizb is a powerful antagonist of Wnt signaling and is present in the organizer as a crescent next to a region of Xwnt8 activity just outside the organizer; presumably, Frzb prevents Xwnt8 from ventralizing cells in this area. Another Wnt antagonist, Sizzled, has recently been characterized; it borders Xwnt8 on the ventral side, thereby likely preventing Xwnt8 from acting ventrally (Figure 16.9).

Antagonisms Between Ventralizing and Dorsalizing Factors Pattern the Mesoderm

The developing mesoderm is subject to powerful influences other than those emanating from the organizer. Current think-

ing is focusing predominantly on the BMPs and Xwnt8 as ventralizing influences, while activin is thought to be important in evoking mesodermal differentiation. How can two ligands, such as activin and the BMPs—all members of the same TGF-β family—bring about such different responses?

Recent discoveries about the molecular basis for the transduction of TGF-β signals indicate that activin and the BMPs are probably transduced by parallel but different members of a complex signal apparatus. In brief, when a TGF-β family member interacts with its receptor—a dimer of two serine-threonine kinase proteins, designated types I and II—the type II molecule phosphorylates its type I partner. This activated type II–type I complex then combines with and phosphorylates a cytosolic protein (Smad in *Xenopus*, Mad in *Drosophila*, and Sma in *C. elegans*). The Smad family has many members, and they can dimerize. The dimers may be active or inactive; when active, they enter the nucleus, where they participate in regulating transcription. It has been shown that activin and its close relatives (which use either Smad2 or Smad3) and BMP family members (which use either Smad1 or Smad5) utilize different Smad transducers; presumably, this is why they may have distinct effects. The pathway-specific Smad molecules must then interact with a so-called co-Smad, which in vertebrates is designated Smad4. Interestingly, both β-catenin and TGF-β can interact with Smad4, providing a molecular node where these two signaling pathways can interact (Figure 16.10).

Because the BMPs are a numerous class of ligands, it has taken a considerable effort to determine which ones are active, and what roles they play in vertebrate development. Apparently, both BMP2 and BMP4 are powerful ventralizing agents in *Xenopus*, leading to epidermal differentiation of ectoderm, and stimulating formation in mesoderm of ventral derivatives such as blood islands and body-wall mesenchyme. The Wnt family is also large; many different *wnt* genes have been cloned from *Xenopus*. The gene *xwnt8* is known to be very active, but there may be other family members playing a role as well. Xwnt8 shows quite different activities when assayed at different stages, for example, at midcleavage versus after gastrulation. Recall that Xwnt8 helps to stimulate formation of a Nieuwkoop center when injected into the early embryo, but when overexpressed during gastrulation, Xwnt8, like BMP2 and BMP4, ventralizes mesoderm differentiation. While *bmp2* and *bmp4* seem to be expressed throughout the ventrolateral mesoderm, expression of *xwnt8* later on is more restricted, hemmed in by the antagonism of Sizzled ventrally and Frizb dorsally. A current idea is that signaling by BMP2 and BMP4 is the dominant influence in establishing ventrolateral mesodermal

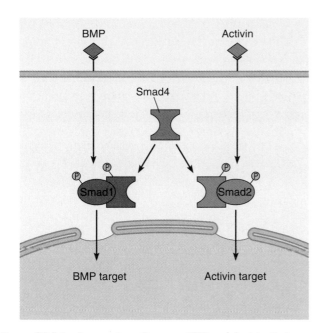

Figure 16.10 Interactions Between BMP and Activin Pathways, as Modulated by Smad A model of how different TGF-β family members may activate parallel, partially overlapping pathways of intracellular signal transduction through the Smad transducers. In this mechanism, BMP activates a Smad1-Smad4 combination, while activin activates a Smad2-Smad4 combination. It is believed that Smad1 and Smad2 are specific for different ligands, while Smad4 is a partner regulating the strength of signaling. When both activin and BMP activate the same cell, the two parallel pathways may compete for Smad4, which would obviously affect the strength and/or nature of the response induced by the ligands.

MORPHOGENS

Morphogens Are Involved in Positional Information

There is another facet to ligand-induced developmental changes that we have mentioned but not explored deeply, and that is the effect of concentration on ligand action. As we saw in Chapter 3, when a given substance has different developmental consequences at different concentrations, it is called a *morphogen.* A powerful theoretical discussion of how morphogens might work was provided by Lewis Wolpert, working in London, who presented the problem clearly in graphical form. Suppose a concentration gradient of a substance is spread over a volume of competent responding tissue in an embryo. Wolpert pointed out that if the cells could somehow "interpret" the concentration of the substance, the gradient might then lead to different developmental consequences in different regions (Figure 16.11A). This theory of **positional information** has two key components: a gradient of ligand concentrations *and* the ability of cells to differentially respond to different concentrations. With the advent of modern cell biology, we can see how prescient Wolpert's idea was. We now know ligands and their receptors can discriminate concentrations by a number of different methods, and different receptors can be linked to different intracellular pathways.

BMPs and Activin Are Morphogens

differentiation; Xwnt8 then cooperatively refines this BMP influence to prevent notochord differentiation (except where Frizb antagonizes it) and to promote myogenic differentiation. Precisely how these two signaling pathways interact in the cell and what their essential downstream gene targets are have yet to be worked out.

Even though our understanding of how different signaling pathways interact is incomplete, and even though new discoveries will alter the details, a few important lessons nevertheless stand out. Important organizing centers such as the Nieuwkoop center and the Spemann organizer operate by carefully *orchestrated* release of many different ligands, which can antagonize or reinforce each other and possibly establish feedback loops. The receptor populations responding to the ligands are also complex and changing. Making a creature is a big deal, so maybe it is not surprising that the biochemical regulation is so complex. Table 16.1 indicates the principal activities of the factors we have discussed so far.

Let us look at some examples of how ligands that are morphogens may be employed for patterning mesoderm. A ligand important for mesoderm differentiation is activin. Jeremy Green and Jim Smith, working in London, first showed that if different concentrations of activin were applied to explanted animal caps, different kinds of mesodermal differentiation would occur. And rather small differences in activin concentration, even twofold, had very different effects. The higher the activin concentration used, the more dorsal the character of the mesodermal derivatives that formed in the animal cap explant.

This analysis has now been extended by Steven Dyson and John Gurdon using radioactively labeled activin (Figure 16.12). They manipulated the absolute numbers of type II activin receptors by injecting the mRNA encoding this receptor into responding cells. Dissociated blastula cells bound activin with high affinity and considerable stability. As the absolute number of receptors occupied by activin increased,

TABLE 16.1 FACTORS ASSOCIATED WITH AXIAL DEVELOPMENT AND GERM-LAYER FORMATION IN *XENOPUS*

Center of Influence	Name	Principal Activity
Maternal localization	VegT	Needed for Nieuwkoop center, endoderm, and mesoderm
	β–catenin	Essential for Nieuwkoop center and Spemann organizer
Nieuwkoop center	Xwnt8	Dorsalizes Does not induce mesoderm in cap
	Vgl	Can induce secondary embryos
	Activin	Can induce secondary embryos
	Siamois	Can induce secondary embryos Needed for Spemann organizer formation
	GSK3	Can induce Siamois
	Xnrl, –2, –3, –4	Induced by VegT Can induce mesoderm
	Derriere	Induced by VegT Can induce mesoderm
Spemann organizer	Xgoosecoid	Transcription factor Activates convergent extension Dorsalizes mesoderm
	Noggin	Antagonizes ventralizing BMPs
	Chordin	Acts in concert with Noggin
	Activin	Dorsalizes mesoderm
	bFGF	Similar to activin May favor posterior mesoderm
	Xbra	Important for mesoderm differentiation
	Xnrl, –2, –3, –4	Important for mesoderm differentiation
Head organizer	Dickkopf	Wnt pathway antagonist Induces ectopic head structures
	Cerberus	Similar to Dickkopf
Other	BMPs	Essential for epidermal and ventral development
	Lunatic Fringe	Can induce mesoderm
	Frizb	Wnt antagonist
	Sizzled	Another Wnt antagonist
	Sonic Hedgehog	Helps pattern neural tube and dorsal mesoderm

A. Activin

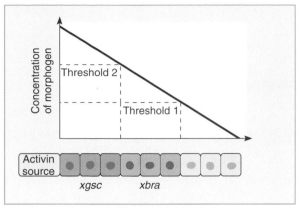

B. BMP4

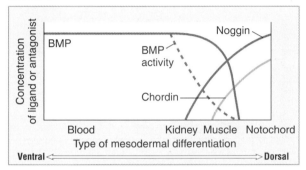

Figure 16.11 Morphogen Gradients (A) A theoretical model of how a single substance, a morphogen, can activate different responses at different concentrations. In this example, a morphogen such as activin is produced by a single source, after which it diffuses away. The concentration of the morphogen decreases as its distance from the source increases, as shown in this graph. Suppose the enhancer-promoter region for *xbra* has a strong affinity for activin, so that when the concentration for threshold 1 is reached, the *xbra* gene may be turned on. But the promoter for *xgoosecoid (xgsc)* has a lower affinity for the activin, and hence is not turned on until the concentration of activin reaches threshold 2. Thus, *xbra* (a mesodermal response gene) is upregulated at some distance from the activin source, while *gsc* (a Spemann organizer-specific gene) is turned on only close to the source. A possible alternative model involves a so-called relay mechanism, in which cells close to the activin source are stimulated to activate *gsc*, and also may then send a different signal to adjoining cells to activate *xbra*. **(B)** A gradient of BMP4, produced by interactions with antagonists, as shown graphically. BMP4 is present everywhere except in the extreme dorsal mesoderm, where absence of BMP4 allows differentiation of the notochord. The dorsal mesoderm produces Noggin and Chordin, which are antagonists of BMP4 and thus reduce its activity, creating a BMP4 gradient. This gradient results in differentiation: When BMP4 activity is reduced extensively, the muscle genes, such as *myf5*, can be activated. Higher levels of BMP4 allow kidney differentiation. And where there is no attenuation of BMP4 whatsoever, blood formation can occur. The interaction of ligand and antagonist may be much more complex than shown here, involving other components affecting the stability of the ligand itself.

transcription of different marker genes changed; when about 100 receptors per cell were occupied, the pan-mesodermal gene *xbra* was transcribed; when approximately 300 receptors per cell were occupied, the Spemann organizer gene *xgoosecoid*, was transcribed (Figure 16.12). This analysis, together with experiments showing that activin can, indeed, diffuse at low concentrations across many cell diameters, indicates that activin is probably acting as a morphogen.

Another probable morphogen is BMP. If pregastrula cells of *Xenopus* are exposed to different concentrations of BMP, different types of differentiation will take place. One way of doing this experiment is to inject different concentrations of the mRNA for BMP4 into the very early embryo (the four-cell stage was used), then isolate the dorsal marginal zone cells at gastrula stage and culture them. The experiment has to be done this way so that influences of other signaling centers will not override the effects of exogenously added BMP. Low doses of BMP inhibit notochord formation, allowing muscle, kidney, and blood to form. Higher BMP concentrations allow only kidney and blood to form, and still higher levels allow only blood and mesenchyme. In other experiments, BMP

mRNA and a lineage tracer were co-injected into one cell of a 32-cell-stage embryo. The cell overproducing BMP4 could be identified by the lineage tracer, but the extent of its influence could be seen up to about 10 cell diameters away. The influence of BMP4 is counteracted by Noggin, and various experiments in which both BMP4 and Noggin mRNAs were co-injected into embryos showed clearly that different concentrations of Noggin could actually produce a gradient of BMP4 activity: the higher the Noggin concentration, the lower the BMP4 activity (Figure 16.11B).

The formation of morphogen gradients can be very subtle and specific. Recent work using *Drosophila* has shown that the concentration of active Dpp, a BMP family member, is regulated by two antagonists called Sog (Short Gastrulation) and Tsg, as well as by a protease (Tolloid) that can clip Sog and help release active Dpp. Even if active Dpp is being released by Tolloid, it quickly becomes rebound by excess Sog; only at some distance from high Sog levels can active Dpp molecules persist. Thus, rather than producing a smooth gradient of active Dpp, the involvement of Tsg, Sog, and Tolloid provides a way for the developing fruit fly embryo to have

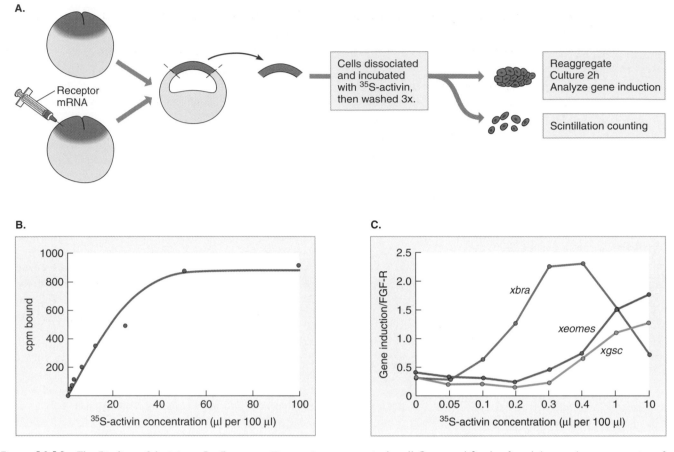

Figure 16.12 The Binding of Activin to Its Receptors The number of activin receptors that are occupied by activin were measured and correlated with the degree of activation of different genes. **(A)** Some *Xenopus* zygotes were injected with different amounts of mRNA for activin receptors. Animal caps from blastulae of either untreated or microinjected zygotes were dissociated to single cells, which were then incubated with radioactive activin. **(B)** After the cells were washed, it was possible to measure the number of activin molecules stably bound to a single cell. Dyson and Gurdon found that, as the concentration of activin was increased, the amount of bound activin increased. There are about 500 receptors per animal cap cell. **(C)** The activation of different genes—*xbra, xeomesodermin (xeomes)*, and *xgsc*—could be measured by looking at the level of expression with the number of activin receptors actually occupied. The level of induction is normalized in this graph to a molecule (FGF-R) that does not change when activin concentrations are manipulated.

steep local concentrations of active Dpp. This behavior is relevant to our discussion of vertebrate morphogens because these same players have been found in *Xenopus*. Chordin is the homolog of Sog, BMP4 the homolog of Dpp, Tolloid is present, and a homolog of Twisted Gastrulation has recently been found in the frog. The potential importance of these *Xenopus* homologs in the regulation of BMP activity was discussed in the section on the dorsalization of mesoderm by the Spemann organizer. Even though this line of research is very recent, it is safe to conclude that BMP activity gradients in *Xenopus* are likely regulated in a way similar to that found in *Drosophila*.

Left-Right Body Patterning Also Involves Signaling Pathways

We shall leave our discussion of the inductive centers in *Xenopus* in order to briefly mention another aspect of the body plan we have thus far ignored, namely bilateral asymmetry: the left and right sides of the vertebrate body are not identical. In humans, for example, the apex of the heart points to the left, the aorta loops to the right, the inferior vena cava is on the left side of the spinal column, the right and left sides of the lung have different numbers of lobes, and the intestine runs from right to left. In one out of approximately 20,000 people, this left–right asymmetry is reversed, fortunately with no ill effect whatsoever.

A number of genes have recently been shown to play a role in establishing this asymmetry in several different vertebrates. One of these is *nodal*, which (as noted earlier in the chapter) encodes a TGF-β-type ligand. In postgastrula embryos of the frog, the chick, and the mouse, *nodal* is expressed on the left side but not the right (Figure 16.13). Researchers have recently found that cells in the organizer of all the major vertebrate groups possess cilia, and these cilia beat in such a way that an anticlockwise flow of fluid is created in the region of the organizer. And the flow is powerful enough to sweep particles containing the Nodal ligand predominantly to the left side of the embryo. This surprising result suggested that if the fluid containing Nodal could be made to flow in the reverse (clockwise) direction, that left–right asymmetry could be abolished or at least attenuated, and that is indeed what happens.

Investigators are busy identifying the signaling pathways upstream and downstream of *nodal*. While asymmetric *nodal* expression seems to be well conserved, other genes upstream of *nodal* are not always the same in different vertebrates, so earlier parts of the pathway may not be so highly conserved. In *Xenopus*, the entire left–right asymmetry can be completely reversed by microinjecting a mature form of Vg1 into certain vegetal blastomeres on the right side. This finding implies that left–right asymmetry is embedded in the earliest signaling events establishing the basic body plan.

AMNIOTE HOX GENES

Amniote Embryos Use Similar, but Not Identical, Regulatory Networks

We have concentrated on *Xenopus* as an example of the complexity of regulatory circuits in vertebrate development because so many of the important experiments have been done in the frog. Chick, zebrafish, and mouse embryos, and especially the use of mouse genetics, have also been of great importance. The same signaling circuits keep showing up over

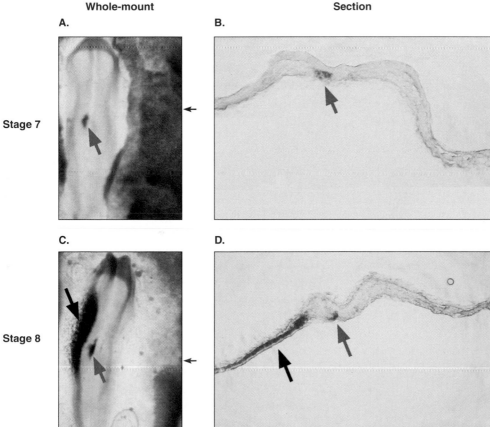

Whole-mount   **Section**

A.   **B.**

Stage 7

C.   **D.**

Stage 8

Figure 16.13 The Asymmetric Expression of *nodal* In this series of experiments, the *nodal* gene of chickens was cloned and used to prepare a probe labeled with a tagged nucleotide that can be detected with an antibody against the tag. Chick embryos of different early stages were fixed, and the probe was hybridized either to (A,C) the entire embryo (whole mount) or to (B,D) a histological section cut approximately perpendicular to the primitive streak. Images were then obtained for localization of the probe. **(A,B)** At stage 7 (one somite present), *nodal* is expressed in a patch just to the left of the notochord *(arrow)*. **(C,D)** At stage 8 (four somites), there is a small patch of expression next to the midline, and a much broader region of expression in the blastoderm on the left side but not on the right side *(arrows)*.

and over again, as do TGF-β family members and receptors, Sonic Hedgehog and its relatives, and Wnt pathway members. It would be misleading, however, to give the impression that the details are the same in all vertebrate classes. They definitely are not.

The BMP antagonist encoded by *noggin* in *Xenopus* has been cloned in the mouse, and a null mutation of the gene examined in a knockout experiment. Though in amniotes there is no dorsal lip of the blastopore—no "Spemann organizer"—you will recall that Hensen's node fulfills an organizer function, and the node does express the mouse homolog of *noggin*. Yet the embryo of a *noggin* null mutant mouse still forms a notochord and a neural tube! In such mice, the neural tube develops dorsal elements—a BMP-responsive event—but the Sonic Hedgehog–mediated ventral development of the neural tube is missing in the *noggin* mutant mouse. Evidently, Noggin is essential for ventral neural development in the mouse. There are also defects in somite formation in the *noggin* mutant, so Sonic Hedgehog and Noggin are probably synergists in ventral neural and paraxial mesoderm development.

One interpretation of these findings is that Noggin is important in organizer function in *Xenopus* but not in the mouse. It may be more complicated than that, however. Even in *Xenopus*, *noggin* is expressed in the notochord and ventral neural tube after induction, an example of the same gene being expressed in different contexts. There are other organizer molecules that are believed to be important in *Xenopus* and perhaps elsewhere. For example, Chordin is another BMP antagonist and Follistatin is an activin antagonist, and both of these are also expressed in the Spemann organizer, as well as in Hensen's node in amniotes. Perhaps Noggin is not in and of itself essential but simply one of several players; it may take multiple null mutants of Follistatin, Chordin, Noggin, and possibly other molecules in order to fully dissect these relationships.

In *Xenopus*, activin and Vg1—both TGF-β family members—are potent inducers of mesoderm. In the mouse, however, none of the TGF-β members studied in loss-of-function mutations seems to play a general role in mesoderm formation. Mutants of BMP2 in the mouse typically display defects in heart development. So, while TGF-β members as a class may be crucial in the formation of germ layers, gastrulation, and early differentiation, the precise roles of different homologs differ in amniotes and amphibians.

Recent evidence has shown rather convincingly that the mouse embryo has a separate anterior inducing-organizing center located in the anterior hypoblast layer, and even more recent evidence indicates that amphibians, too, may have an "anterior" organizing center located in the endoderm. As more homologs of different genes encoding transcription factors and signaling molecules important in body plan formation are tested functionally, the extent to which we can generalize the details across the different vertebrate classes will become clearer. We shall revisit this question of the conservation of developmental mechanisms in the next chapter.

Homeotic Selector Genes Are Present in Vertebrates as Well as Drosophila

As soon as the first homeobox (HOM) motif was discovered in *Drosophila*, molecular biological and genetic methods were rapidly applied to finding the homologs, if any, in vertebrates. And found they were. Figure 16.14 presents a comparison of the homeobox gene clusters found in *Drosophila* and the mouse. The homeobox motifs characteristic of the *Drosophila* genes are sufficiently distinct that genes cloned in the mouse could be assigned to one or another members of the *Drosophila* family; thus, mouse HOX gene *b6* is more similar to *antennapedia* than to other HOM genes. Astonishingly, representatives of all the HOM genes have been found in the mouse; moreover, they are present in the same linear order on the chromosome as in *Drosophila*. The mouse has four groups of linked genes, presumably the result of wholesale duplications during evolution of the entire HOM region. And there are additional genes located "downstream" on the chromosome from the *abdB* representatives. Thought to be modifications of *abdB*-type HOM genes, they are expressed in the postanal tail regions, a chordate invention.

The HOM–HOX relationships illustrate several kinds of gene correspondence that can exist between groups. Genes closely related to one another in different organisms and thought to be derived from one common ancestral gene during evolution are called **homologs,** a concept we have been using throughout the book. Homologs believed to have a one-to-one relationship are termed **orthologs.** When gene duplication has occurred, creating a larger family of related genes, the additional family members are considered **paralogs** of one another. Hence, the mouse contains orthologs of the *Drosophila* HOM genes, and wholesale duplications of the entire cluster created paralogs. For example, in the mouse, the HOX9 paralog group is comprised of *a9, b9, c9,* and *d9*. Each of these is an ortholog of the *abdB* gene of *Drosophila*. The *a10, a11,* and *a13* genes are more distantly related members of the *abdB* gene family, and would be considered homologs (but not orthologs) of *abdB*.

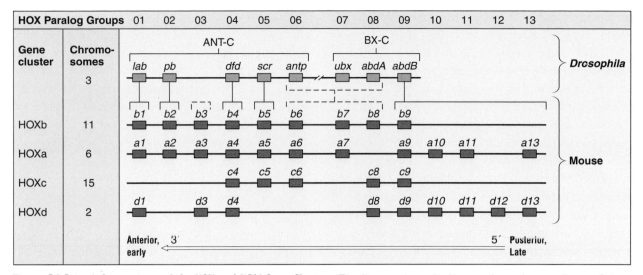

HOX Paralog Groups		01	02	03	04	05	06	07	08	09	10	11	12	13	

Figure 16.14 A Comparison of the HOX and HOM Gene Clusters The diagram shows the linear order and nomenclature of the clustered homeobox genes in the fly and mouse. Abbreviations for individual *Drosophila* genes *(left to right)*: *lab, labial; pb, proboscipedia; dfd, deformed; scr, sex combs reduced; antp, antennapedia; ubx, ultrabithorax; abdA, abdominal A; abdB, abdominal B.* The orthologous genes in the mouse are found in four linear clusters—a, b, c, and d (formerly called 1, 2, 3, and 4)—and each HOX gene is identified by its cluster letter and a number instead of being individually named.

The anterior borders of expression of these linked homeobox genes, called HOX clusters in the mouse and other vertebrates, become located more and more posteriorly the farther away from the most anteriorly expressed labial homologs one goes. Figure 16.15 presents the results of some whole-mount in situ hybridization studies on mouse embryos, showing the expression domains of genes in one of the four HOX clusters. Even though the anterior borders of expression of the different members of a cluster are distinct, there is considerable overlap in expression domains. Furthermore, at any given time during development, the precise domains of expression of a given gene may be changing; that is, their expressions are dynamic in both time *and* space.

There are homeobox-containing genes outside the HOM or HOX clusters that are expressed anterior to the *labial* orthologs. In *Drosophila,* the genes *empty spiracle (ems)* and *orthodenticle (otd),* as you may recall, are expressed in the anterior head regions; mutations of *ems* and *otd* cause serious defects in head development, and thus are thought to be important for patterning anterior parts of the body. Homologs of these two genes (*otd* is abbreviated *otx* in vertebrates) are also expressed in the head of mice, where they probably play a similar role.

The mouse is not unique in having HOX clusters; all other metazoans so far examined also have some version of these clusters; we shall discuss this finding further in the next

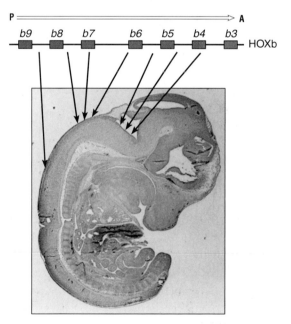

Figure 16.15 Expression Domains of Some HOX Genes in the Mouse A sagittal section of a 12.5-day mouse embryo. Serial sections were cut from this embryo, and the sections were hybridized with different probes of the HOXb gene cluster. The anterior-most borders of expression of the different members of this cluster are indicated by the arrows pointing to the neural tube of the embryo.

chapter, where we consider development and evolution. The promoter and regulatory regions of various HOX genes can be linked, using a reporter gene, to the time and place of expression. Figure 16.16 shows the expression domain of the HOX gene *b2*, as reported by β-galactosidase expression.

HOX Genes Function as Selector Genes

The sheer number of HOX genes in vertebrates and the inability to carry out genetic manipulations have made analysis of HOX gene function challenging. Fortunately, the ability to delete gene function in the mouse makes analysis possible by observing loss-of-function mutations, which give clear evidence for homeotic transformation. Figure 16.17 shows an example in which the HOX gene *b4* was knocked out, altering the axial skeleton; cervical vertebra C2 now has the morphology of C1, indicating an anterior transformation, just as might be expected from comparable work with *Drosophila*. Many HOX genes have been examined this way.

Loss-of-function mutations often result in transformations to anterior structures; for example, loss of any one of the HOX genes *c8*, *b4*, *a2*, or *d3* results in anterior transformations. Sometimes loss-of-function mutations generates a posterior transformation (as with *a5* or *a11*)—just as occurs in loss-of-function mutations in *dfd* and *scr* in *Drosophila*. So far, so good.

The interpretation of HOX gene knockouts is not altogether simple, however, and we would be mistaken to conclude that HOX genes function exactly the same in vertebrates as HOM genes do in *Drosophila*. Often the loss of function of a homeotic mouse gene does not show a severe or dramatic phenotype; sometimes there is no detectable homeotic transformation, simply an abnormal morphology or loss of a structure. These results may be due to the fact that vertebrates have overlapping domains of expression as well as putative synergism and antagonism between different HOX members. So, while HOX genes are clearly expressed in anteroposterior regional domains and are clearly implicated in control of regional identities, it is not clear that they function in precisely the same way in all

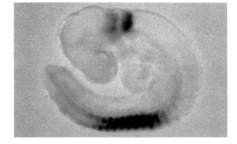

Figure 16.16   HOX Gene Expression in a Reporter Construct The bacterial β-galactosidase structural gene *(lacZ)* was linked to the enhancer-promoter sequence of HOX *b2*. and this reporter construct was injected into the male pronucleus of a fertilized mouse egg. The DNA was integrated at random into the chromosomes and transmitted to all the cells of the embryo. The mouse embryo was then stained at somite stage 13 (9.5 days postcoitum) for β-galactosidase activity. Enzyme activity is visible only where one would expect the HOX *b2* gene to be expressed. namely. rhombomeres r3 and r5 (dark bands on left and right. respectively) in the hindbrain and in dorsal root ganglia. posterior.

A.

B.

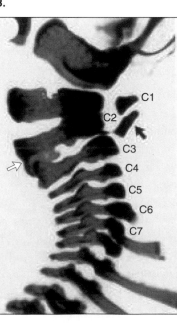

Figure 16.17   A Homeotic Transformation in the Mouse The results of a knockout experiment in which HOX *b4* was knocked out. In these photographs. ventral is to the right. **(A)** The spinal column of an untreated mouse (the control) is shown. with the seven cervical vertebrae identified. **(B)** In the vertebrae of the knockout mouse. C3 now shows some of the characteristics of C2. and C2 has a ventral tubercle *(solid arrow)* and a partial duplication of the neural arch *(open arrow)*. which are characteristic of C1.

creatures. Nonetheless, it is amazing that HOM and HOX genes seem to be generally involved in such a similar way in creatures as disparate as flies and mice.

What regulates the expression of HOX genes in vertebrates, and what are the targets of HOX gene expression? The much more difficult enterprise of answering these questions is under way, and as we have seen in *Drosophila*, it is still a work-in-progress. For example, the mouse has a homolog, called *bmi1*, of one of the polycomb gene group of *Drosophila*. The polycomb gene group in *Drosophila* is known to regulate (by repression) some genes of the HOM cluster. If the mouse *bmi1* gene is rendered null by a knockout procedure, the result is some posterior skeletal transformations in the mouse. This implies that *bmi1* in the mouse plays a role similar to its polycomb homologs in *Drosophila*. Another question involves the overlap in expression of HOX genes, which has not been closely analyzed at the cellular level, even though the regional expression of different HOX genes is known. Is the expression of two overlapping HOX genes a salt-and-pepper phenomenon—one gene turned on in one cell (or region), another gene in another cell—or can two HOX genes be expressed simultaneously in the same cell? The answer to this question will be important in constructing a detailed model of HOX gene action.

SIGNALS IN LIMB DEVELOPMENT

HOX Genes and Signaling Pathways Play Important Roles in Limb Development

Much of the analysis of HOX genes has so far centered on the axial skeleton, neural crest derivatives, the central nervous system, and the limbs. A flood of information is now coming in about the expression patterns of various HOX genes in many different organ systems, and on the roles these various HOX genes play as determined by loss-of-function and gain-of-function experiments. It would require an entire book to record and analyze all these findings, and at the end of the day it would be out of date.

We can, however, gain some insight into how the homeotic selector genes, as well as signaling pathways, fit into downstream events of organogenesis by visiting once again the development of the vertebrate limb (see Chapters 7 and 13). The vertebrate limb is an instance for which we have information, though incomplete, on how selector genes and regulatory circuits may operate.

Limb Bud Placement Is Probably Regulated by Several Factors

The limb buds form in the lateral plate (including both mesoderm and ectoderm). Cells that originate from outside the lateral plate also contribute to the developing limb; limb muscle cells originate from the somites, and the nerves and blood vessels of the developing limb arise, respectively, from ectoderm and mesoderm situated outside the limb bud proper (see Chapter 7). The position along the anteroposterior axis of the embryo of the limb fields has been investigated through a large number of microsurgical transplantations and interventions. Interactions between intermediate mesoderm and the somites have been implicated. If a barrier is placed between intermediate and lateral somatopleuric mesoderm, the limb bud will fail to form. Hence, there must be some influences passing from medial to lateral tissues to stimulate initial formation of the limb bud.

There are also regional, dynamically changing patterns of HOX gene expression along the anteroposterior axis (especially the HOX group of *a9* through *a13* and their paralogs, *d9* through *d13*) that are thought to play a role in positioning the limb fields. If the area of the early embryo known to give rise to a limb bud is transplanted to another location along the body axis, the transplanted tissue will still form a limb. Implantation of a bead coated with FGF (a number of different FGF molecules will do) can evoke an ectopic supernumerary limb that is perfectly formed but reversed in polarity. Hence, there is reason to believe that whatever system sets up the early limb field, that system originates in the intermediate mesoderm and involves some member(s) of the FGF family (FGF2 and FGF8 are current candidates). When a bead of FGF 2 is implanted into the flank, transcription of *snR*, the chick homolog of *snail* (a gene we met in *Drosophila* that is involved in mesoderm formation), increases dramatically within one hour, followed by transcription of *tbx*, a group of genes with the T box domain (soon to be discussed); *fgf10* transcription follows at 17 hours after bead implantation, and Shh can be detected by 24 hours.

We do not understand what drives the proposed early FGF production in the prospective limb locations. The combined action of several HOX genes may be involved. A detailed understanding of how the limb fields are positioned along this axis remains elusive. The lateral mesoderm synthesizes retinoic acid (RA) throughout the flank prior to limb bud formation, and RA affects signaling pathways in the developing limb, as we shall soon discuss. The character of the limb bud formed, whether anterior or posterior, seems to be governed by the action of genes for two T box–containing transcription factors: *tbx4* (expressed in the hindlimbs) and *tbx5* (in the forelimbs). Ectopic expression of either of these genes in concert with FGF activity results in formation of either fore- or hindlimbs, depending on which T box gene is expressed. Recent experiments indicate that the action of *tbx4* is regulated by a gene encoding another transcription factor, PitX1. Malcolm Logan and Cliff Tabin, working at Harvard University, have been able to obtain partial transformations of wings to legs, accompanied by Tbx4 accumulation, by expressing *pitX1* in the wing. And Toshihiko Ogura and his colleagues, working in Nara, Japan, were able to transform legs to wings and vice versa, by manipulating expression of *tbx4* and *tbx5*.

On the other hand, the positioning of the limb along the dorsoventral axis is somewhat better understood, and molecules that play a role in defining the three axes of the limb—anteroposterior (A/P), dorsoventral (D/V), and proximodistal (P/D)—are now known. Furthermore, the role of HOX genes as selector genes acting upon an axially defined limb bud is beginning to be understood. A very great aid to this investigation has been progress made in understanding the development of the wing in *Drosophila* (see Chapter 15). We turn now to examine how the signaling centers for determining the limb axes in amniotes are developed, and how HOX genes may then operate.

Dorsoventral Organization Is Mediated by a D/V Compartmental Boundary

As we saw earlier (Chapters 7 and 13), dorsoventral organization depends on the positioning of the apical ectodermal ridge (AER), which controls the position of outgrowth, and on the pattern of overall differentiation of the limb tissues. A homolog of the *Drosophila* gene *engrailed*, called *en1* in amniotes, is expressed in ventral ectoderm but not dorsally in either mesoderm or ectoderm. There is evidence that the lateral somatopleuric mesoderm adjacent to prospective ventral ectoderm is needed for this *en1* expression. This delineation of expression domains into *+en1* and *−en1* occurs before development of the limb buds proper, and demarcates a boundary between prospective dorsal and ventral ectoderm.

In prospective dorsal ectoderm, where *en1* is not expressed, there is expression of a cell membrane protein homologous to *fringe*, called *radical fringe* (*rfng*). Dorsal ectoderm also requires unknown signals from the nearby somite. The *en1* gene antagonizes *rfng* expression; hence, there are *rfng*-expressing tissues dorsally and *en1*-expressing tissues ventrally, with a border between them (Figure 16.18). The AER forms at this border. When the area where *rfng* is expressed is altered experimen-

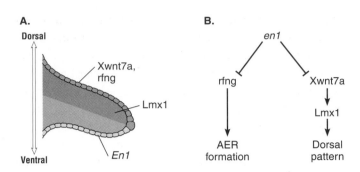

Figure 16.18   Dorsoventral Patterning of the Limb (A) This diagrammatic frontal section through the early limb bud shows the domains of expression of *engrailed1* (*en1*) in the ventral ectoderm and *lmx1* in the dorsal mesoderm. Both *en* and *lmx1* encode homeodomain-containing transcription factors. **(B)** A simplified flow diagram shows the functional relationships between these molecules: *en1* expression in ventral ectoderm represses both *radical fringe* (*rfng*) and *xwnt7a* in the ventral portion of the limb bud. But since *en1* is not expressed dorsally, both *xwnt7a* and *rfng* (which are secreted ligands) can be expressed there, resulting in formation of the AER and appropriately patterned dorsal mesoderm.

tally, the position of the AER changes to correspond to the new border between *rfng* and *en1*. Thus, the position of the AER is set by these opposing ectodermal signals. Yet the AER will not form unless underlying mesodermal factors (possibly FGF10) act upon this prepatterned AER position. So ectodermal signals set the position of the future limb, while mesodermal signals enable its development.

Not only does *en1* play a role in D/V positioning of the AER; it is also involved in D/V patterning of the limb as a whole. Dorsal ectoderm expresses the *wnt* homolog *xwnt7a* as well as *rfng*. Xwnt7a signals to the underlying mesoderm to produce the transcription factor Lmx1, which is known to be necessary for forming dorsal-pattern elements in the limb. Lmx1 is absent ventrally because En1 suppresses it. Perhaps ventral development of the limb is a default route of differentiation possible only in the absence of Lmx1 (Figure 16.18).

Anteroposterior Patterning Is Controlled by Sonic Hedgehog

The zone of polarizing activity (ZPA) was mentioned in Chapter 7 as the area controlling anteroposterior patterning of the limb. John Saunders and his colleagues discovered many years ago that when a block of mesoderm near the posterior margin of the limb bud is transplanted to the anterior margin, a near perfect mirror-image duplication of the limb can occur (Figure 16.19). Expression of the growth factor Sonic Hedgehog (Shh), a vertebrate homolog of *hedgehog*, is congruent with the known limits of the ZPA. Implantation of beads coated with Shh can create ectopic limbs, just as does transplantation of the ZPA. Furthermore, the nature of the supernumerary limb may be dictated to some extent by the dose of Shh administered; low doses tend to produce limbs with mainly anterior digits, while higher doses induce both anterior and posterior digits. Hence, there is reason to suspect that Shh is acting as a morphogen; the debate over *how* Shh acts centers on whether it acts directly by diffusing from the ZPA to set up a concentration gradient, or indirectly through some relay mechanism. There is some evidence that Shh stimulates BMP2 production, and BMP2 has weak anteroposterior polarizing activity in the limb bud. Could BMP2 be the relay morphogen?

Shh also stimulates the proliferation of mesenchyme, essential for limb outgrowth, and there is a complex interplay between the AER and the ZPA. A prominent idea in the field is that there is a positive control circuit between the AER and ZPA: Shh from the ZPA stimulates the AER to secrete FGF (possibly FGF4), which in turn stimulates more Shh production. However, a mouse possessing a null mutation of FGF4 has nearly normal limbs, so several FGF factors other than FGF4 may play a role. In fact, an FGF4/FGF8 double null mutant shows limbs that are either severely malformed or absent.

Experiments that have imposed either over- or underexpression of BMPs on the developing limb have produced confusing and conflicting results. In an attempt to understand the role of BMPs, researchers have investigated yet another secreted growth factor, a relative of Noggin called Gremlin, a BMP antagonist that has been shown to play an important role. Gremlin, which is present in cells adjacent to the Shh-secreting cells of the ZPA, provides some fine-tuning of BMP levels, which is important for production of FGF from the AER. Both the ZPA and the AER are essential for the integrity of limb outgrowth; both play essential yet distinct roles in patterning.

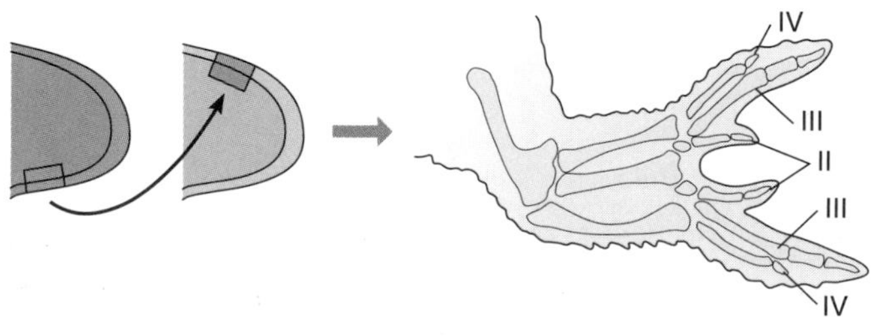

Figure 16.19   Transplantation of the ZPA in the Chicken
(A) When a block of ZPA tissue from the posterior side of the limb bud is transplanted to an anterior portion of another limb bud, a duplication occurs in the transplant recipient, as evidenced by **(B)** the skeleton of the wing of the newly hatched chick; note that duplicated digits IV, III, and II have formed in mirror image to the normal digits.

What sets up the localization of Shh to the posterior mesoderm (ZPA region)? The *dHAND* gene, which encodes the dHAND transcription factor, has been known to be important for development of the heart and of the face. Recent studies in mice, chicks, and fish have shown that dHAND is also required for development of the limbs and fins. The gene is expressed in lateral plate mesoderm before the limb buds appear. As the limb buds emerge, dHAND becomes restricted to posterior regions of the forelimb and hindlimb and also to some flank mesoderm between the limb buds. The synthesis and secretion of Shh in the ZPA is dependent on the prior expression of dHAND.

HOX gene *b8* is normally expressed in the region where the ZPA forms, and ectopic expression of HOX *b8* can result in the appearance of an ectopic ZPA. Only the HOX *b8*-expressing cells at the posterior distal margin of the mesoderm form a ZPA, however, so it may be that signals from the AER are necessary as well.

Retinoic acid (RA) can induce expression of *dHAND* when applied to the anterior part of the limb bud, and RA also can also induce the HOX gene *b8*, while inhibition of RA synthesis blocks its expression. (See Figure 16.20.) Indeed, general suppression of RA synthesis early in development blocks formation of the limb buds. Beware, however, of overinterpreting "negative" results acquired from loss-of-function mutants or from experiments using inhibitors.

P/D Patterning Requires Ectoderm and Mesoderm

The AER is essential for limb outgrowth, and once it is established, it produces FGF2, -4, and -8. Applying a bead of FGF to an AER-denuded limb will restore outgrowth that results in a properly patterned limb. How do the cells formed during P/D outgrowth acquire their patterned information? The tissues of the developing limb are exposed, as we have discussed, to influences of the ectoderm (for dorsoventrality) and of the ZPA (for anteroposteriority), but we still do not know how the information on P/D differentiation is set up. Once early cartilage-forming cells have been left behind in a proximal position (as distal outgrowth continues), how do they know to form a humerus? Conversely, how do cartilage-forming cells arising late in limb outgrowth (and hence occupying a more distal position) know to form digits?

A venerable answer to these questions is the proposition that cells proliferating underneath the AER constitute a "progress zone," a kind of stem cell population that moves distally as the limb extends (Figure 16.21). Hence, the progeny of progress zone cells that leave the zone early are exposed to AER influences for a shorter time than those leaving the zone later. These

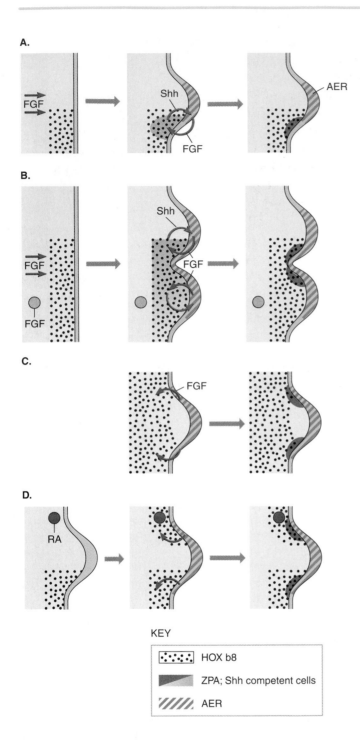

KEY

HOX b8

ZPA; Shh competent cells

AER

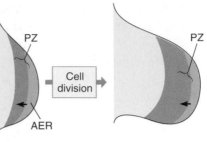

PZ cells receive FGF signals from AER specifying P/D fate

Only half the cells are still in PZ

AER respecifies cells still in PZ

Figure 16.21 The Progress Zone Model of Limb Differentiation
The progress zone (PZ), which is subjacent to the ectodermal AER, is colored red. *Left:* While cells are in the PZ, they acquire positional information from interactions with the AER. *Middle:* As proliferation continues, some cells leave the PZ but remain in a more proximal location (closer to the body). Other cells remain in the PZ, where they are exposed to further influences of the AER. *Right:* Cells remaining longer in the PZ acquire additional positional information (*dark red*), which presumably specifies a more distal (away from the body) program of differentiation.

Figure 16.20 Experimental Inductions of Limb Buds with FGF, HOX *b8*, and Retinoic Acid (A) Signaling in the normal limb bud. The HOX *b8* gene is expressed in some mesenchymal cells (*stippled*). These cells secrete FGF and activate production of Sonic Hedgehog (Shh) in competent cells (*light red*), which sets up a positive feedback loop between the ZPA (*red*) and the newly formed AER (*blue cross hatch*). **(B)** A bead coated with FGF is implanted into the anterior limb bud. This induces ectopic proliferation of mesenchyme, inducing Shh and setting up an additional positive feedback loop between another ZPA and an ectopic AER. **(C)** In this experiment, HOX *b8* expression is established across the entire limb bud in a transgenic mouse. HOX *b8*-expressing cells are adjacent to the AER only at anterior and posterior locations of the limb bud, and only there can feedback loops be established. So a mirror-image-symmetric limb develops. **(D)** A bead coated with retinoic acid (RA) is implanted in the anterior limb bud, producing a mirror-image duplication of digits in the wing. It is supposed this happens because RA induces expression of HOX *b8*, thus leading to the same kind of situation seen in part C.

AER "influences" could, of course, consist either of changing cocktails of different FGF molecules and other factors, or simply of quantitative differences in different FGF levels, or of a combination of the two. The proposal that residence time in the progress zone is crucial to developmental P/D differentiation has been influential on workers in the field, but experimental support is lacking. The original experiment involving AER removal is now known to induce considerable cell death in the subjacent mesoderm, so the mechanism by which the AER influences P/D patterning is still not clear.

Discovery of two related homeobox genes, *meis1* and *meis2*, has provided new insights into proximodistal organization. These genes play a role in defining the identity of proximal limb elements, and show localized expression in proximal limb elements. Overexpression of these genes leads to truncation of distal limb elements. Recent work by Nadia Mercader and her colleagues in Madrid have shown that RA not only activates transcription of *meis1* and *meis2* but is also necessary for maintaining their expression. RA, as we have stated previously, is present early and throughout lateral plate mesoderm, and becomes restricted only after appearance of the AER. Mercader and her colleagues have shown that it is FGF from the AER that limits the influence of RA; moreover, they propose this antagonism of RA by FGF is what results in localization of Meis1 and -2 and consequent proximal development.

Clearly, there is an interdependence of mesodermal and ectodermal signaling, and an interdependence of FGF and Shh signaling coming from the different centers. Removing the AER results in loss of Shh expression in the ZPA. There also is an interaction between dorsal ectoderm, which expresses *xwnt7a*, and the ZPA. Mice bearing a knockout of the *xwnt7a* gene show reduced Shh expression and lack some posterior-pattern elements of the limb. Discovery of new players, such as dHAND and the *meis* genes, is beginning to fill in some of the voids, and it may not be too long before we have a fairly thorough understanding of the pathways involved in formation of the limb buds and all three axes of the limb.

HOX Genes May Regulate Limb Differentiation

A number of different HOX genes have now been rendered null in mice using knockout procedures, and while some of these unfortunate animals have severe malformations of the limbs, the majority of them have much more subtle defects. Many HOX genes are expressed in the limb areas, and the pattern of expression is dynamic, changing so rapidly that it is hard to relate patterns of expression to specific developmental roles. As mentioned previously, there is some correlation between the expression domains of HOX paralogs *a9* though *a13*, *d9* though *d13*, and the limb element formed. Mesoderm forming the humerus expresses *d9* and *d10*; the domain of the prospective radius and ulna express a nested array of HOX genes *d9* though *d13*; and prospective wrist and digit express *a13* and *d13*. A primary role of HOX genes, based on examination of HOX knockouts and gain-of-function experiments, seems to be to regulate the rate and timing of cartilage (and hence, eventually, of bone) differentiation.

HOX genes may also function as true selector genes. Whether a hind- or forelimb differentiates, for example, may depend on HOX gene activity as well as the T box genes. The HOX9 paralog group is expressed in a distinct dynamic pattern in wings and hindlimbs. In the experiment carried out by Martin Cohn, Ketan Patel, and their colleagues in London, and shown in Figure 16.22, ectopic limbs were induced in the flank by

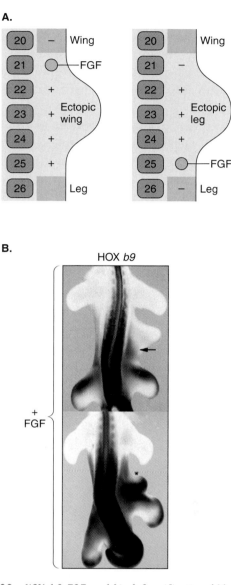

Figure 16.22 HOX *b9*, FGF, and Limb Specification (A) An experiment in which a bead coated with FGF was implanted in the flank of an early chick embryo. Implantation opposite somite 21 *(left)* produced an ectopic wing, while implantation opposite somite 25 *(right)* produced an ectopic leg. Marking the region with a lipophilic dye created a long-lasting tag for cells in the area. Cells which are indicated by + in the schematics contributed to the ectopic limb bud. As can be seen from the cells from the regions opposite somites 22-24 can contribute to either a wing or leg bud, depending on the position of implantation of the FGF-containing bead. **(B)** Whole-mount hybridization was used to detect expression of the HOX *b9* gene after such an FGF bead implantation was carried out. *Top:* By 57 hours after implantation of the bead opposite somite 24, the anterior boundary of *b9* was at the junction between the ectopic limb bud and the flank *(arrow)*. *Bottom:* By 72 hours after implantation opposite somite 24, the ectopic limb had taken on a hindlimblike pattern of expression.

implanting beads soaked in FGF2. Depending on its exact site of placement, the FGF2 could induce a supernumerary hindlimb or wing. Cell-marking experiments have shown that the same population of flank cells can form either anterior or posterior limbs. The patterns of HOX9 expression observed in these induced supernumerary limbs were appropriate to whether the induced outgrowth was anterior (winglike) or posterior (hindlimblike); in other words, the HOX9 expression pattern induced by FGF2 corresponded to the type of limb being formed. These results suggest (without proving) that HOX gene expression is involved in the choice of whether anterior or posterior limb elements are formed. Even though HOX gene expression changed in flank cells as a result of FGF bead implantation, there were no changes in adjacent vertebrae, nor in expression of HOX9 paralogs in neighboring neural tube or paraxial mesoderm tissues. The selection of the type of limb by HOX9 paralogs in these experiments was specific.

We do not know any of the targets of the HOX genes in limb development (nor, indeed, in other situations in vertebrate development as well). We do know enough, however, to realize that the regulation of complicated intercellular signaling networks orchestrates the pattern and differentiation of the developing embryo. And even though the details of gene regulatory circuits in the limb and other parts of the body are currently incomplete, it is clear that many of the same (or related) players that we met in *Drosophila* and *C. elegans* are also playing important and similar roles in *Xenopus*, the chick, and the mouse. The similarity of pathways used to regulate development in different organisms was surprising to workers in the field, and this has invigorated the study of the relationship between development and evolution. We shall examine that relationship in the next chapter.

KEY CONCEPTS

1. Intracellular "currents," which probably utilize cytoskeletal elements such as microtubules, may be used by the egg and zygote to properly position maternal determinants.

2. Complexity and overlapping functions of the different components are hallmarks of developmental regulatory circuits.

3. We often encounter the same or similar molecular players in embryos of very different organisms. We may also encounter the same or similar molecular players within the same embryo in different developmental contexts, for example, Shh in the notochord and floor plate and in the ZPA.

4. Signaling centers—such as the Nieuwkoop center, the Spemann organizer, and the apical ectodermal ridge—use complicated molecular "cocktails" as signals. These cocktails can be composed of both synergists and antagonists, and the mixtures may have several disparate activities, such as neuralization, dorsalization, or both.

5. Different but similar ligands may produce different responses by utilizing intracellular second-messenger pathways that are distinct or partially distinct (as with the Smad family of homologs).

6. Morphogens have different developmental consequences depending on their concentration; in the case of activin, it is the absolute number of occupied receptors that produces different responses.

7. HOX genes are organized similarly in the genome in very different organisms, and the sequence of expression domains along the anteroposterior axis is more or less similar in different organisms.

8. Borders between expression domains of some crucial signaling molecules may specify the position of subsequent signaling centers; for example, the boundary between dorsal and ventral ectoderm sets the position of the AER in the developing limb.

STUDY QUESTIONS

1. Is cortical rotation necessary for the formation of the Nieuwkoop center?

2. Can you predict the effect of injecting an antisense oligonucleotide against Noggin mRNA into a *Xenopus* embryo?

3. Speculate on how a morphogen, for example, activin, could bring about different kinds of differentiation depending on the concentration to which responding cells are exposed.

4. In amniote embryos, the neural tube and somites form sequentially from anterior to posterior end. What might be responsible for this?

5. Predict the outcome of an experiment in which the expression of the gene *radical fringe* is reduced by a loss-of-function mutation. What would be the effect on the wing bud?

SELECTED REFERENCES

Agius, E., Oelgeschlager, M., Wessely, O., Kemp, C., and DeRobertis, E. 2000. Endodermal Nodal-related signals and mesoderm induction in *Xenopus*. *Development* 127:1173–1183.

A research paper that probes the details of how VegT, Vgl, members of the Xnr family, and β-catenin work together to induce mesoderm.

Cohn, M. J. 2000. Giving limbs a hand. *Nature* 406:953–954.

A summary of the evidence for the involvement of dHAND in formation of the ZPA.

Davis, A. P., Witte, D. P. Hsieh-Li, H. M., Potter, S., and Capecchi, M. R. 1995. Absence of radius and ulna in mice lacking *hoxa 11* and *hoxd-11*. *Nature* 375:791–795.

Analysis of knockouts that serve as the basis for how HOX genes might pattern limb elements.

Dudley, A. T., and Tabin, C. J. 2000. Constructive antagonism in limb development. *Curr. Opin. Genet. Dev.* 10:387–392.

A review of the new finding on the role of BMPs in limb patterning.

Dyson, S., and Gurdon, J. B. 1998. The interpretation of position in a morphogen gradient as revealed by occupancy of activin receptors. *Cell* 93:557–568.

A research paper on the mechanism by which morphogens activate cellular responses.

Harland, R. 2000. Neural induction. *Curr. Opin. Genet. Dev.* 10:357–362.

An incisive review of new complications for understanding the role of the Spemann organizer in neural induction.

Harland, R. 2001. A twist on embryonic signalling. *Nature* 4110:423–424.

A summary of experiments carried out by several groups on the role of antagonists and proteases in forming BMP activity gradients.

Harland, R., and Gerhart, J. 1997. Formation and function of Spemann's organizer. *Annu. Rev. Cell Dev. Biol.* 13:611–667.

An in-depth and up-to-date review of the Spemann organizer

Heasman, J., Kofron, M., and Wylie, C. 2000. β-Catenin signaling activity dissected in the early *Xenopus* embryo: A novel antisense approach. *Dev. Biol.* 222:124–134.

How a new kind of antisense reagent is used to eliminate β-catenin signaling and thereby reveal its function.

Johnson, R. L., and Tabin, C. J. 1997. Molecular models for vertebrate limb development. *Cell* 90:979–990.

A detailed treatment of the role of the various signaling molecules and transcription factors implicated in limb organization.

Joseph, E. M., and Melton, D. A. 1998. Mutant Vgl ligands disrupt endoderm and mesoderm formation in *Xenopus* embryos. *Development* 125:2677–2685.

Experiments in which a mutated Vgl molecule is used to block endogenous Vgl signaling (dominant negative) leading to the conclusion that Vgl is indeed necessary for induction of dorsal mesoderm.

Kimelman, D., and Griffin, K. J. P. 1998. Mesoderm induction: A postmodern view. *Cell* 94:419–421.

A review of the new information on the role of VegT.

King, T., and Brown, N. A. 1999. Embryonic asymmetry: The left side gets all the best genes. *Curr. Biol.* 9:R18–R22.

A thorough discussion of the interrelationships of the many genes involved in left–right asymmetry and how they may interact with each other.

Kodjabachian, L., and Lemaire, P. 1998. Embryonic induction: Is the Nieuwkoop center a useful concept? *Curr. Biol.* 8:R918–R921.

An interesting discussion of the generality of the Nieuwkoop center.

Kofron, M., Demel, T., Xanthos, J., Lohr, J., Sun, B., Sive, H., Osada, S., Wright, C., Wylie, C., and Heasman, J. 1999. Mesoderm induction in *Xenopus* is a zygotic event regulated by maternal VegT via TGFβ growth factors. *Development* 126:5759–5770.

A research paper detailing the mechanism by which VegT induces mesoderm.

Lane, M. C., and Smith, W. C. 1999. The origins of primitive blood in *Xenopus*: Implications for axial patterning. *Development* 126:423–434.

A demonstration that the old fate maps for *Xenopus* were probably off by about 90°.

Mercader, N., Leonardo, E., Piedra, M. W., Martinez, C., Ros, M. A., and Torres, M. 2000. Opposing RA and FGF signals control proximodistal vertebrate limb development through regulation of Meis genes. *Development* 127:3961–3970.

Research paper showing the importance of RA for activation of *meis* genes and proximal limb development.

Moon, R. T., and Kimelman, D. 1998. From cortical rotation to organizer gene expression: Toward a molecular explanation of axis specification in *Xenopus*. *BioEssays* 20:536–545.

An excellent review, especially emphasizing the role of Wnt family members, on how the germ layers form in *Xenopus*.

Oelgeschlager, M., Larrain, J., Geissert, D., and DeRobertis, E. 2000. The evolutionarily conserved BMP-binding protein Twisted gastrulation promotes BMP signaling. *Nature* 405:757–763.

Description of a new factor that helps regulate the BMP signaling activity.

Pennisi, E. A. 1998. How a growth control path takes a wrong turn to cancer. *Science* 281:1438–1440.

An interesting discussion of how the Wnt pathway works in embryos and how its malfunction can result in cancer.

Piccolo, S., Agius, E., Leyns, L., Battacharyya, S., Grunz, H., Bouwmeester, T., and DeRobertis, E. M. 1999. The head inducer Cerberus is a multifunctional antagonist of Nodal, BMP, and Wnt signals. *Nature* 397:707–710.

A research paper analyzing how the Cerberus antagonist works.

Stennard, F. 1998. *Xenopus* differentiation: VegT gets specific. *Curr. Biol.* 8:R928–R930.

An analysis of the work on the *vegT* gene and how it might be involved in the formation of endoderm and the Nieuwkoop center.

Stern, C. D. 2002. Fluid flow and broken symmetry. *Nature* 418:29–30.

A brief analysis of recent papers showing that the direction of ciliary beating is a mechanism for generating left–right asymmetry in vertebrates.

Vogel, G. 1999. New findings reveal how legs take wing. *Science* 283:1615–1616.

A brief summary of the role of *tbx* genes in determining limb identity.

Wylie C., Kofron, M., Payne, C., Anderson, R., Hosobuchi, M., Joseph, E., and Heasman, J. 1996. Maternal β-catenin establishes a dorsal signal in early *Xenopus* embryos. *Development* 122:2987–2996.

The research paper that showed the importance of the Wnt–β-catenin pathway in axial organization of the amphibian embryo.

Zhang, J., Houston, D. W., King, M. L., Payne, C., Wylie, C., and Heasman, J. 1998. The role of maternal VegT in establishing the primary germ layers in *Xenopus* embryos. *Cell* 94:515–524.

A research paper on the newly discovered maternal transcription factor that plays an important role in germ-layer formation.

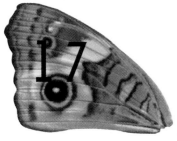
DEVELOPMENT AND EVOLUTION

DEVELOPMENT AND EVOLUTION

Are There Laws of Development?

We have stressed the importance of the basic strategies that underlie the development of organisms, and the similarities of molecular tactics that bring about development. But if there is so much similarity of method, so much unity of plan, how does the astonishing diversity of living creatures come about? Some biologists contend that there really are no fundamental tenets that apply to organismal development, and that each species has its own history, its own unique set of devices for assembling itself.

CHAPTER PREVIEW

1. Many genes that are active in development (for example, genes of the HOM and HOX complexes) are well conserved.

2. Entire signaling pathways, such as the Wnt pathway, are used in many embryos and in many different developmental events.

3. Differences in how conserved pathways are used show how different phenotypes can arise.

4. Within a particular phylum, embryos may differ in kinds of eggs, cleavage, and organ formation yet share a common postgastrula phylotypic stage.

5. Intercellular signaling can be modified to provide considerable diversity.

6. Larval stages are thought to provide a platform for evolutionary change.

Our point of view is somewhat different—that while there may be no short set of "Newton's laws" for development, there are indeed unifying and underlying principles and mechanisms. We have emphasized them in the preceding chapters. What we have seen is that the basic tactics of development are extraordinarily flexible and robust. They underwent considerable modification during the evolution that led to extant creatures. We assert that the cell and molecular biology sustaining development is so flexible and robust that the appearance of very diverse creatures during the last 500 million years is not so surprising after all. But let us remind you: this is a point of view, a hypothesis. When making inquiries into how creatures came to be the way they are, decisive experimental verification may well prove difficult or even impossible.

There Is a Close Link Between the Study of Development and the Study of Evolution

The middle of the 19th century saw incredible ferment in the study of biology. All kinds of creatures, from the curious and exotic to the familiar, were studied intensively, including their anatomy, life history, and embryology. In 1828, the German biologist Karl von Baer recognized the similarity of various vertebrate embryos at the neurula stages. This idea was developed further by Ernst Haeckel in the 1860s; he proposed that vertebrate embryos develop by a succession of anatomical stages similar to different vertebrate classes, resulting in the famous dictum that "ontogeny recapitulates phylogeny." If taken literally, this captivating idea is utter nonsense, but it did nonetheless attempt to relate development and evolution. After Darwin's theory of evolution by natural selection broke forth upon the biological world in 1859, it became obvious that the actual development of a creature was of great importance in evolution. The idea that relationships between different animals can be elucidated, at least in part, by study of their embryology is truly useful.

After all, the information encoded in the DNA for the traits of an organism can only come into being during the actual development of the creature. The embryo is the genotype in the process of being translated into the adult phenotype. Of course, you now know that it is not just the DNA, but the information in the DNA interacting with its environment, that underpins development and selective gene expression. The genome without an egg cannot construct an embryo. As well as producing the adult itself, development of the organism is what provides the source of heritable variation and invention that is the material basis upon which natural selection can operate. No wonder the subjects of evolution and development are closely linked.

Molecular Biology and Genetics Have Invigorated the Relationship Between Development and Evolution

The ability to clone and sequence DNA adds a set of powerful tools for study of the relationships between creatures. It is beyond the scope of this text to document how salutary has been the addition of DNA sequence comparison to morphological comparison in the study of phylogeny. However, we can note that inferences of relationship based on careful use of DNA sequence comparison, coupled with traditional comparative anatomy and life history analysis, have greatly increased our knowledge about the relationships between extant creatures, and provided a solid basis for hypotheses about their possible descent in evolutionary time.

Two discoveries of modern molecular genetics have also been of great importance. The first is that genes not only encode proteins; by one means or another, many genes also regulate the activities of other genes. The second is that proteins involved in regulating gene expression are linked in regulatory pathways in a fashion analogous to the linkage of enzymes in metabolic pathways. These realizations all started, of course, with the work of François Jacob and Jacques Monod on the *lac* operon of *E. coli*. The product of the *I* gene is a protein that can interact with specific small molecules in the environment and then directly regulate transcription of the target gene, which encodes β-galactosidase. Since then, as we have outlined in this book, many signal transduction pathways and modular enhancers have been discovered. A number of basic elements of gene regulation appear in unicellular organisms, but the linkage of the elements into pathways of signal transduction and gene regulation occurs predominantly in multicellular organisms. It is during the development of multicellular plants, animals, and fungi that the spectacular employ of regulatory pathways becomes so manifest. This discovery of regulatory circuits has not only revolutionized developmental biology; it has also breathed new life into ways to think about how changes in development could come about and how developmental changes might result in new phenotypes.

CONSERVATION OF GENES AND NETWORKS

Many Genes Important in Development Are Conserved

Our concern in this chapter is to focus on how modifications in the regulation of development can ultimately lead to morphological and physiological variation. What is conserved in animal development, and what is not? What has been conserved but extensively modified during evolution? One thing should be readily apparent from the preceding chapters: there are enough similarities in some regulatory mechanisms that we can assume many regulatory genes, and the regulatory circuits for signal transduction and gene control, have a very ancient origin. As mentioned in Chapter 6, the genes *pax6* (expressed in vertebrate eyes) and *eyeless* (in insects) are structurally similar; they are also functionally equivalent, insofar as the vertebrate gene has been shown experimentally that it functions in eye development in *Drosophila* (Figure 17.1), and the *Drosophila* gene can function in experimental circumstances in eye development in vertebrates. Consequently, the ancestral gene giving rise to *eyeless* and *pax6* must have existed prior to the divergence of arthropods and chordates. Figure 17.2 shows explicitly the profound similarities among *eyeless* and its homologs in different animals. There are enough other known genes possessing sufficient structural and/or functional similarity to one another to conclude that their progenitor genes likely had an ancient origin predating the arthropod–chordate divergence.

Some scientists have termed the collection of signals, signal transduction pathways, and gene regulatory mechanisms known to operate in most animals the "tool kit" of development. Just what has to be in the minimal tool kit? In this chapter, we describe examples of how changes in the tools and in their use produce variations and novelty, the raw material for selection; certain basic, necessary tools provide an extraordinarily robust yet flexible platform for developmental diversity, as we shall see.

There is also much we have to neglect in this chapter, which is not intended for study of evolutionary biology per se. We shall not dwell on hypotheses about how evolution may have occurred using this or that particular developmental mechanism. We shall keep our sights rather narrowly focused on development itself, albeit in an evolutionary context.

Useful Motifs Are Conserved

All living organisms must have a supply of nitrogen, reduced carbon, water, and electron acceptors such as oxygen. Consequently, a series of enzymes are linked in order to sustain a metabolism of compounds based on these elements. For example, some amino acid stretches are particularly well adapted for binding ATP. A small number of ATP-binding motifs show up again and again in different proteins. These motifs may have arisen independently to serve the same purpose—examples of so-called convergent evolution—or they may descend, with modification, from some ancestral, canonical ATP-binding motif. Certain versions of ATP-binding motifs in proteins are very likely descended from a common ancestral version. Figure 17.3 shows how similar the motifs are in ATP-binding proteins from vastly different organisms. We now know that nucleotide stretches may be duplicated and transposed from one site to another within the genome, and hence opportunities abound for useful motifs, such as the ATP-binding site, to appear in different modules in a number of

Figure 17.1   The Ability of Vertebrate Gene *pax 6* to Form Ectopic Eyes in *Drosophila* The *pax6* gene from a mouse was expressed in the fruit fly, providing a small ectopic eye on the antenna. Note the presence of ommatidia, which are the individual facets characteristic of the compound eyes of insects. Clearly, *pax6* is a functional ortholog of *eyeless*. The ability of *eyeless* to form ectopic eyes in *Drosophila* is shown in Figure 6.17.

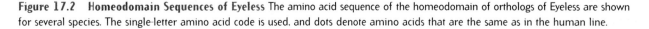

Human	SHSGVNQLGGVFVNGRPLPDSTRQKIVELAHSGARPCDISRILQVSNGCVSKILGRYYETGSIRP
Mouse	• •
Quail	• •
Zebrafish	• •
Sea urchin	• •
Drosophila	• •
Nematode	• • T • • • • • • • • G • • • • • A • • R • D • • K C • C • • • • S • T • •

Figure 17.2   Homeodomain Sequences of Eyeless The amino acid sequence of the homeodomain of orthologs of Eyeless are shown for several species. The single-letter amino acid code is used, and dots denote amino acids that are the same as in the human line.

A.

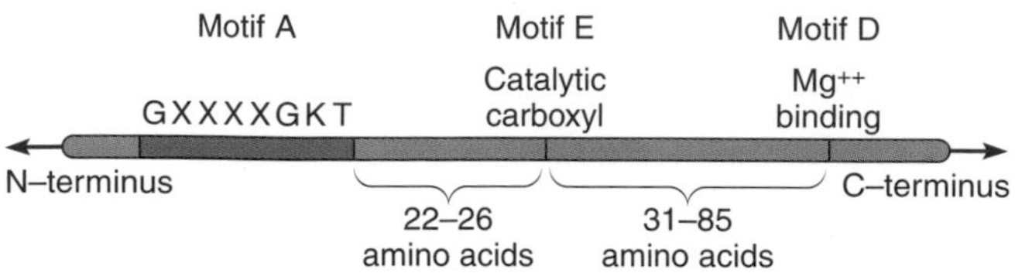

Motif A   Motif E   Motif D

            Catalytic Mg^{++}

GXXXXGKT   carboxyl   binding

N–terminus   C–terminus

          22–26   31–85

     amino acids   amino acids

B.

Protein	Function	Motif A	Catalytic carboxyl
Consensus		GXXXXGK$^{T}_{S}$	E
E. coli Rec A	DNA recombination	GPESSGKT	E
Bovine ATPase	Release PO^4	GGAGVGKT	E
Yeast CDC 48	Cell cycle control	GPPGTGKT	E
NifH	Nitrogen fixation	GKGGIGKS	E

Figure 17.3   The Canonical ATP Binding Motif (A) A schematic showing the arrangement of various amino acid motifs in the ATP binding site: these motifs are designated A (nucleotide binding), E (a glutamic acid in the active site), and D (an aspartic acid necessary for Mg^{2+} binding). Single-letter abbreviations are used to identify amino acids, and X indicates positions where the amino acid varies across species. **(B)** A brief list of some varied proteins that bind ATP, together with the sequence of its binding motif, called motif A.

different proteins. The exon/intron organization of eukaryotic genomes also favors such shuffling.

The duplication and dispersal through the genome of nucleotide sequences encoding particularly well adapted protein motifs may go well beyond short amino acid stretches binding such molecules as ATP, retinoic acid, or steroids. We might also expect terminal products of differentiation that are crucial for important intracellular functions, or for the function of a particular organ or tissue, to show up again and again. And so they do. The similarities, among all animals, of myosin, actin, or tubulin molecules, for example, are easy to demonstrate; these molecules show a high degree of conservation. The same is true for the collagens, hemoglobin, and other structural proteins essential for organ functioning. Not only are they clearly related in different animals; these molecules are also ancient, often present in unicellular creatures, thus predating the appearance of multicellularity in evolution.

A rather striking example of conservation of proteins serving important functions in terminal differentiation is found in the voltage-gated L-type Ca^{2+} channels. These channels have been implicated in the transfer of massive amounts of calcium from extracellular fluids into the cells involved in the deposition of vertebrate bone (osteoblasts) and teeth (dentino-

blasts). The search for a similar gene in coral polyps was successful. The genes from rat and coral encoded a protein with a striking similarity in overall molecular anatomy and possessed extensive conservation of amino acid sequences. Corals (coelenterates) and vertebrates are about as far apart on any taxonomic scheme as it is possible to be; moreover, the form of calcium in the two groups differs, coral being composed of calcium carbonates versus the calcium phosphates in bone. Yet extant organisms requiring massive amounts of calcium for biomineralization evidently rely on a gene derived from some common ancestor.

Entire Signaling Pathways Are Conserved

Not so obvious, but falling in the same category as genes for collagen, hemoglobin, and Ca^{2+} channels, are the members of various transcription factor and cofactor families, as well as ligands and receptors involved in signaling. The 60–amino acid homeodomain motif is an obvious example. We could add to the list ligands such as Wnt and receptors such as Notch and Toll. What is truly astonishing, however, and not foreseen 20 years ago, is the conservation of *entire regulatory pathways*. We have had occasion to examine many of these pathways in this book. The pathways can be very similar in very different organisms, and they may be utilized over and over again in different developmental situations in a given organism, a testimony to their robust and flexible character. Table 17.1 lists some of the examples we have studied in this book—from *Drosophila*, *Xenopus*, the mouse, and other metazoans—that fall into this category.

We learned in Chapters 15 and 16 that all of these pathways are used multiple times in both *Drosophila* and vertebrates, and that the individual components of each pathway are similar in different organisms and sometimes functionally exchangeable. Many other animals possess these same pathways, and some have been exceptionally well studied in *C. elegans*, as well as in vertebrates. There are two homologs of *Delta* and *Notch* in nematodes, while in vertebrates there is a large family of such homologs. As different taxa diversified over the course of evolution, the number and variety of different signaling and response teams also had the opportunity to diversify. And some of these signaling pathways may work together in certain situations. Hh and Dpp signaling interact in development of the *Drosophila* eye. Hh, Dpp, and Notch/Delta signaling are used in segment and appendage development. In vertebrates, there are at least three Hh family members, at least 11 Wnt family members, and over 20 Dpp (BMP) family members. Hh, Wnt, and BMP are used in development of limbs in vertebrates and of appendages in arthropods, as well as in many other roles in development that we have not discussed.

Newly discovered parallels between signaling pathways in well-studied embryos of different phyla are becoming commonplace. For example, we mentioned in Chapter 13 the role of members of the fibroblast growth factor family (FGFs) in the branching of epithelial ducts, especially the lung. You may recall that FGF10 is expressed in mesenchyme adjacent to duct tips, where it interacts with its receptor tyrosine kinase (RTK), stimulating proliferation and consequent bud outgrowth and branching. An FGF homolog has been found in *Drosophila* (*branchless*), and it, too, interacts with its RTK (*breathless*) to stimulate branching of the tracheoles of the insect respiratory system. In the case of tracheole branching in insects, the FGF homolog (*branchless*) induces branching without proliferation, acting to stimulate migration of specific lead cells and induce changes in cellular shape necessary for terminal differentiation of the tracheole.

Until recently, it has been puzzling just how FGF stimulation of epithelial branching can be restricted; why do neighboring cells in the vertebrate lung not also become stimulated to proliferate; similarly, why do cells next to the specific lead cells of the developing insect tracheole not migrate and change shape? In *Drosophila*, a gene *sprouty* has been characterized whose exact function is to restrict FGF action. Cells stimulated by FGF respond by secreting Sprouty, which then acts on neighboring cells to interfere with their ability to respond to the FGF signal (Figure 17.4A). In null mutants of *sprouty*, the tracheole branching starts to occur in neighboring cells, and proper branching of tracheoles degenerates into a fuzzy terminus. Not surprisingly, a *sprouty* homolog (Sprouty2) was recently found in vertebrates, where it, too, apparently acts to restrict the spread of FGF signaling in lung duct proliferation and branching (Figure 17.4B). So we see that a complex system for the regulation of branching in airways has been conserved between arthropods and chordates, even though the cellular basis of how branching occurs differs.

But before we proceed too far with our generalizations, a note of caution is in order. Lest we assume that everything is the same in arthropods, vertebrates, and other taxa, let us state clearly that there are also important differences, some of which we shall consider later in the chapter.

The HOM/HOX Complex Illustrates the Partial Conservation of Selector Genes

We have already discussed in Chapter 15 how the genes of the HOM complex in *Drosophila* act as selector genes, specifying the nature of the terminal differentiation that occurs in various

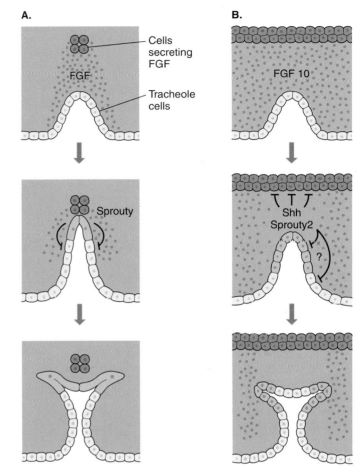

Figure 17.4   FGF and Branching in Respiratory Organs (A) The diagram shows how *sprouty* limits the response of the *branchless* receptor to *breathless*. In *Drosophila*. FGF-secreting cells stimulate branch formation by developing tracheole cells *(top)*. The responding cells *(orange)* are stimulated to secrete Sprouty. which acts on adjacent cells of the forming tracheole to inhibit their response to FGF *(middle)*. This lateral inhibition results in further branching *(bottom)*. **(B)** In the analogous situation in the developing mouse lung, FGF10 stimulates outgrowth of a lung bud *(top)*. Cells at the bud tip secrete Sprouty2. which in turn inhibits the response to FGF10 by neighboring cells; tip cells also secrete Shh. which inhibits FGF10 secretion next to the tip *(middle)*. resulting in branching *(bottom)*.

segments. And we mentioned in Chapter 16 that vertebrates possess relatives of these genes that are grouped into four HOX gene clusters. In fact, homeobox genes related to the canonical *Drosophila* HOM complex are found in almost all animals, and the importance of HOM-related genes in regional differentiation of the body plan has been clearly established (Box 17.1) The HOM/HOX gene families are one of the favorite subjects of investigation of modern evolutionary biology.

Thus far, all arthropods examined possess orthologs of the HOM family members. So the genes are present—are they doing the same thing in all arthropods? The answer is, "Well,

TABLE 17.1 EXAMPLES OF COMMONLY USED SIGNALING PATHWAYS

Ligand[a]	Receptor/Downstream Mediators
FGF	Tyrosine kinase/Ras, G protein
TGF-β/Dpp	Serine-threonine kinase/Smad family
Wg/Wnt	Frizzled/β-catenin
Hedgehog/Shh	Patched/protein kinase A
Delta/Serrate	Notch/Su(H)[b], E(split)[b]
Steroid hormone	Cytosolic receptor/dimer with R X R

[a] The slash in the left column indicates a homolog or a functionally similar ligand.

[b] Su(H) and E(split) are abbreviations for Suppression of Hairy and Enhancer of Split, respectively; both are transcription factors that respond to Notch in this signaling pathway.

kind of." The HOM cluster genes are themselves subject to generalized regulation by upstream genes (such as activators like *trithorax* and repressors like *polycomb*) and HOM genes in turn regulate a surprisingly diverse and large number of downstream genes (see Chapter 15). HOM complex genes are thus "in the middle." When we look at the details of HOM gene activity in different embryos, we observe that there are occasional changes in how they are regulated or what their downstream targets are. In other words, the relationships between HOM genes and their upstream and downstream partners may change, which may have important consequences. Let us look at a specific example.

The Gene ubx Helps Govern the Formation of Butterfly Wings

All insects possess forewings and hindwings arising from body segments T2 and T3; the two wing sets are never identical. In *Drosophila*, for instance, the hindwings on T3 are extensively modified to form the halteres, which function as balancing organs. As we discussed in Chapter 15, differentiation of the T3 segment is regulated by the HOM gene *ubx*. In the absence of *ubx* function, both T2 and T3 develop forewings. Since T2 forms wings in the absence of homeotic gene function, wing formation constitutes a kind of ground state. Other BX-C genes suppress wing formation in the other body segments.

Butterflies, on the other hand, develop true (but nonidentical) wings on both T2 and T3. And *ubx* is expressed in the T3 segment of butterflies. How does *ubx* function in T3 so that a haltere forms in Diptera but a true hindwing forms in Lepidoptera? At least in part, this is due to changes in the target genes regulated by *ubx* and by the regulation of *ubx* itself.

A recent collaboration led by Sean Carroll at the University of Wisconsin and Frederick Nijhout at Duke University, working with the butterfly *Precis coenia*, isolated a dominant mutant *(hindsight)* in which patches of ventral hindwing have become identical to corresponding areas in the forewing (Figure 17.5). Expression of *ubx* occurs throughout the hindwing of wild-type *P. coenia*, but expression of this gene in *hindsight* mutants is missing in some places in the ventral hindwing. The patches of missing *ubx* function display the homeotic transformation of hindwing (T3) to the T2-type forewing differentiation. It may be that *hindsight* is a novel gene present only in Lepidoptera; it apparently regulates *ubx* expression.

Further analysis of mutants in *P. coenia* showed that *ubx* regulates *distalless (dll)*, which in turns regulates the production of eyespots characteristic of lepidopteran wings (Figure 17.6). *Drosophila* wings have no eyespot, but they do express *dll*. Apparently, *ubx* regulation of *dll* and other downstream genes needed for eyespot formation must have evolved in butterflies after Lepidoptera and Diptera became distinct. Thus, an ancient regulatory gene *(ubx)* governs unique target-gene pathways in Lepidoptera. This means *ubx* has acquired a function in lepidopterans that differs from its function in dipterans. Changes in *ubx* function do not mean changes in the homeobox motif itself, for much of the specificity of homeo-

A.

B.

Forewing

Hindwing

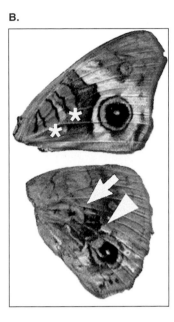

Figure 17.5 The Wings of the Butterfly *P. coenia* (A) The ventral aspect of forewings *(top)* and hindwings *(bottom)* are shown for wild-type specimens. **(B)** Ventral wing pigmentation patterns of the *hindsight* mutant show homeotic changes, especially in the hindwing *(bottom)*. The region denoted by the arrow shows pattern elements *(asterisks)* normally found in the forewings. The arrowhead indicates a region of the hindwing in which the scales have a morphology similar to that of the forewing.

domain proteins is known to be due to amino acid sequences outside the homeodomain proper. For example, the *ubx* gene from a member of Onychophora (a sister phylum of the arthropods, with a simpler body plan) can substitute for many, but not all, of the functions of *ubx* in *Drosophila* when inserted into a null *ubx* mutant. The differences between the versions of *ubx* in onychophorans and fruit flies lie outside the well-conserved homeobox.

One set of functions of *ubx* in *Drosophila* is to repress expression of certain downstream genes in T3, thereby leading to haltere development rather than wing formation. For example *wingless* (*wg*) is expressed at the dorsoventral boundary of developing forewings in both *Drosophila* and *P. coenia*. In *Drosophila*, *ubx* represses *wg* expression at the posterior margin of the developing haltere in T3, but in *P. coenia, ubx* does not repress *wg* in hindwing development of T3. Once again, the relationship between *ubx* and its downstream target genes has been modified. This exemplifies a situation found again and again in comparisons of genes from different creatures, whether within the same phylum or across different phyla: many genes, serving both regulatory and structural roles, are conserved, yet they have been obviously modified during evolution. These modifications encompass both physical changes—alterations of the amino acid sequence of the encoded protein—but also changes in their functional role in regulatory networks. One of the major contributions of developmental biology to the study of evolutionary mechanisms has been to underline the importance of *descent with modification* during evolution.

A.

B.

C.

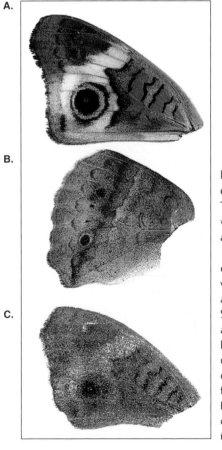

Figure 17.6 Eyespots on *P. coenia* Wings The ventral surface of a wild type **(A)** forewing and **(B)** hindwing. **(C)** Cells of the developing butterfly wing were transformed with a viral vector (the Sindbis virus) carrying a *Drosophila ubx* gene. Expression of *ubx* in the developing forewing caused a homeotic transformation to a more hindwing-type morphology, including changes in eyespots.

Box 17.1 The HOX Cluster

The accompanying diagram sets forth the known members of the HOX cluster in members of several different animal phyla, including coelenterates (*Hydra*), nematodes (*C. elegans*), annelids, molluscs, arthropods (*Drosophila*), cephalochordates (*Amphioxus*), and vertebrates (mice).

There is some evidence that differences between closely related genes in a cluster are primarily due to sequences lying outside the homeobox proper. The logical outcome of this view is that the homeobox sequence is the workhorse, but surrounding sequences provide specificity. However, the true situation is probably much more complicated than that. For instance, recent work by Yuanxiang Zhao and Steve Potter, working at Children's Hospital in Cincinnati, shows clearly that small differences in the homeobox sequence itself can also profoundly affect the expression of HOX genes. Zhao and Potter were able to swap the sequence encoding the homeobox domain in mouse HOX genes *a10* for *a11* (and these two homeobox sequences are very similar). Nonetheless, even though the axial skeleton was normal in mice containing the swapped homeobox, the skeleton of the limb and the formation of kidneys and female reproductive organs were underdeveloped and abnormal. So the actual function of Hox genes undoubtedly depends both on the homeobox sequence *and* its surrounds.

The number of known genes and paralogous groups continues to change as more creatures are investigated. An additional paralog group in *Amphioxus* and added genes in the coelenterates have been reported.

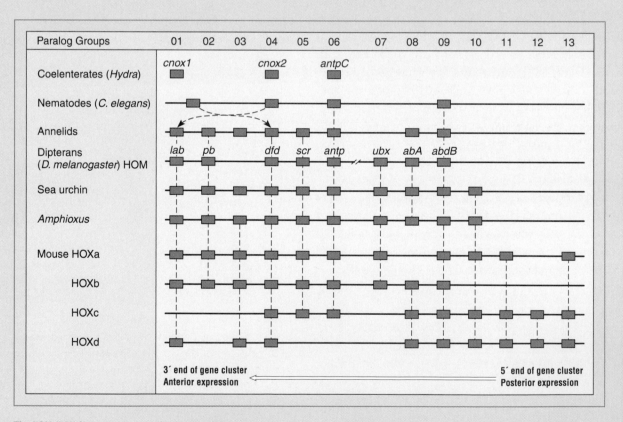

The HOM/HOX Cluster in Various Organisms Horizontal lines indicate known linkages, with double slashes to show breaks. Vertical dashed lines indicate probable homologies. Genes expressed more anteriorly are shown to the left, and their linear order on the chromosome corresponds with the order of the spatial expression in the embryo. Note that expression of the genes in the nematode complex does not correspond to the canonical order.

Vertebrates Have Altered Patterns of HOX Gene Expression

It is not only *Drosophila* and other invertebrates that show alterations in HOM gene expression. Recall from Chapter 16 that the basic organization of homeotic genes is maintained in vertebrates but with wholesale duplication of the set. You may wish to refer to Box 17.1 to compare the organization of HOM/HOX clusters in different creatures. HOM/HOX genes function as selector genes and specify domains of certain kinds of differentiation. You will recall that in amniotic vertebrates, the four sets of HOX sequences have overlapping functional domains, and that all three germ layers can be loci of HOX gene expression. HOX genes ensure that particular kinds of differentiation characteristic of different anteroposterior axial levels develop appropriately.

Does this mean that HOX genes specify such particulars as which rhombomere is correct for becoming cerebellum and which somite number for forelimb development? Apparently not. When HOX expression domains of the mouse and chick are compared, for example, HOX *c6* has an anterior expression limit at somites 12 to 13 in the mouse, but its anterior limit is somites 19 to 20 in the chick. Interestingly, one of these somite locations is the site of formation of the first thoracic vertebra in the chick, the other is the site for the same development in the mouse. The precise anterior limit of the HOX expression domain has shifted relative to the somites when one compares these two classes. *Descent with modification may involve changes in regulation of the HOX gene with attendant alterations in time or domain of expression, or changes in the target genes of HOX.* Because there are four groups of HOX genes in most vertebrates, one cannot assume that the details of expression of different genes within a given paralog group are necessarily identical. In fact, many examples show clearly that this is not so.

Limblessness in Snakes Involves Changes in HOX Gene Expression

Snakes may have enormous numbers of somites, and correspondingly huge numbers of vertebrae. In the python, for example, there are hundreds of similar vertebrae, forelimbs are completely absent, and hindlimbs are vestigial, consisting of a pelvic girdle and truncated femur. Marvin Cohn and Cheryl Tickle, working in London, have shown that most of these numerous vertebrae are all of the thoracic type, which bear ribs. They propose that much of the increase in number of vertebrae is due to expanded domains of expression of HOX genes specifying thoracic vertebrae. They examined the distribution of two HOX proteins known to be associated with development of thoracic vertebrae in tetrapods (*c6* and *c8*). They also looked at the expression of *b5*, whose anterior limit of expression is the first cervical vertebra in tetrapods.

In tetrapod vertebrates, all three of these HOX genes have as their anterior limit of expression the position of the forelimb, and they are known to be involved in specification of the forelimb and shoulder (Figure 17.7A,C). In python embryos, all three of these genes are expressed from the anterior-most somite all the way posterior to the cloaca (Figure 17.7B,C).

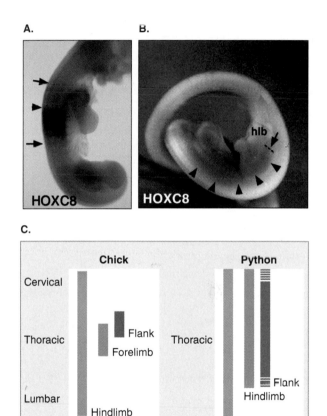

Figure 17.7 Expression of HOX Genes in Chick and Python Embryos (A) Antibody staining of HOX *c8* in the chick embryo, where it is restricted to the thoracic region. The anterior end of the embryo is up. The arrows mark the limits of expression, and the arrowhead shows the boundary between weak and strong expression. **(B)** In the python, the domain of HOX *c8* expression extends posteriorly to the level of the hind limb bud (hlb). The anterior end of the embryo is in the lower part of the image. The dashed line and arrow indicate the sharp posterior border of expression, and the arrowheads denote the extent of expression more anteriorly. **(C)** A comparison of the expression domains of the HOX genes *c6* (*red*), *c8* (*blue*), and *b5* (*green*) in the chick and python. The broken line at anterior and posterior extremes of *c6* in the python (*red*) indicates uncertainty about its exact limits of expression.

The entire area where the vertebral column will form anterior to the cloaca exhibits patterns of HOX gene expression consistent with thoracic identity; the expansion of expression domain of these HOX genes—at the expense of cervical vertebrae (that is, there is no neck) probably underlies the expansion of the vertebral column in snakes. There is no anterior boundary of expression for *c6*, *c8*, and *b5* in the somites, which could be why the anterior limb bud does not form in snakes.

The posterior limb bud does form in the python, however, exactly at the level of the cloaca where *c8* expression ends posteriorly. In the snake embryos studied by Cohn and Tickle, however, no apical ectodermal ridge (AER) could be detected. Nor could they detect any gene expression associated with AER function (such as FGF2), even though FGF2 and other genes were expressed elsewhere in mesodermal organs, such as kidneys and scale buds. Likewise, there was no Sonic Hedgehog expression in the python's hind limb-bud mesenchyme, even though there is Shh expression in the neural tube of the snake. Cohn and Tickle have transplanted python hind limb-bud mesenchyme into chick embryo limb buds, showing that python mesenchyme is capable of producing Shh when stimulated by chick AER. They speculate that the expansion in the domains of HOX gene expression in the mesoderm of the python might be responsible for aberrant development of the AER in the hindlimb ectoderm, which then leads to failure of hindlimb development. It seems clear that changes in expression domains of HOX genes may lead to radical changes in body organization.

THE PHYLOTYPIC STAGE

A Phylotypic Stage Exists in the Postgastrula Embryos of Many Animal Phyla

We have been asking whether there are similarities in the development of very different creatures. When we consider expression of certain regulatory genes, such as the HOX family, genes that are clearly homologous may exhibit modifications in timing and place of their expression and in their relationship to target genes. In the 19th century, Karl von Baer, Ernst Haeckel, and others noticed another striking similarity. In vertebrates and arthropods, even though the adult members of different classes look very different, there is a stage of development, occurring somewhat after gastrulation, during which

the embryos look surprisingly like one another. Haeckel's drawings of vertebrates have been criticized in recent years for their idealized "artistic license." More accurate representations of vertebrates do show, however, that postgastrula embryos share phyletic characters and display more similarity than do later stages. This time of morphologically similar appearance is called the **phylotypic stage.**

Examples of some of the major classes of arthropods are shown in Figure 17.8. While all arthropods have a chitinous exoskeleton, articulated body segments, and jointed appendages, there are a vast diversity in morphologies among the million or so different species (Figure 17.8A). Differences include number of segments, the way segments are grouped, and details in kinds of appendages to mention only a few characters showing great diversity. Yet in all arthropod classes, the segmented germ band stage, which arises after gastrulation and neurulation and contains 10^3 to 10^4 different cells per embryo, is similar across classes, displaying a conserved body plan at that stage (Figure 17.8B).

In Chordates, the Phylotypic Stage Is the Pharyngula

It is after vertebrates and other chordates have completed gastrulation and neurulation that the four defining chordate features (notochord, dorsal hollow nerve cord, gill slits, and postanal tail) first become visible. The famous drawings by Haeckel are so stylized that his claims for similarity of early vertebrate embryos are no longer well regarded. Figure 17.9 is a more accurate representation, and it shows lateral views of the anterior portion of several different vertebrate embryos. As with arthropods, all three germ layers have now formed, the gut has extended from prospective mouth to anus, and the typical chordate dorsal axis (with notochord, nerve cord, and somites) is evident.

Clearly, these embryos are not identical, but there are similarities. The meaning and implications of the phylotypic stage have been softened. Thus, the phylotypic stage is now considered to be a postgastrula stage when embryos are *more* similar to one another than they are in later development. This is also the stage of expression of regionally distinct, anteriorly expressed HOX genes such as *emx* and *otx*, and others. These similar embryos gives rise to creatures as diverse as fish and fowl, snakes and squirrels. Not only can postphylotypic development take very different pathways, giving rise to very different creatures; the roads to formation of the phylotypic stage, called the *pharyngula*, also differ across vertebrate classes, as you will remember from our discussion of amphib-

A. Adults

| Myriapoda | Insecta | Arachnida | Crustacea |
| (millipede: *Pauropus*) | (beetle: *Tenebrio*) | (funnel web spider:*Agelena*) | (crayfish: *Astacus*) |

B. Embryos

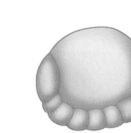

Figure 17.8   Phylotypic Stages of Various Arthropods (A) Adult phenotypes of representative animals from four different classes: myriapods. insects. spiders. and crustaceans. **(B)** What their postgastrula (germ band stage) embryos look like.

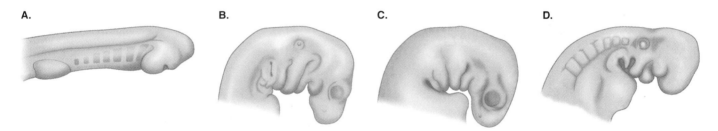

A.   **B.**   **C.**   **D.**

Figure 17.9   Vertebrate Embryos at the Phylotypic Stage Drawings of the head and cervical region of some selected vertebrate embryos after the completion of neurulation and tail bud formation: **(A)** sea lamprey. **(B)** European terrapin. **(C)** chicken. and **(D)** domestic cat.

ian and amniote development. In fact, we may generalize and state that development both preceding and following the phylotypic stage is extraordinarily diverse. John Gerhart and Marc Kirschner, in their recent book *Cells, Embryos, and Evolution*, have called the phylotypic stage a "robust platform" from which the different creatures within a given phylum may develop. Now we shall briefly examine the stages leading up to and following the chordate pharyngula, in order to see what is the role of the phylotypic stage in the development of intraphyletic diversity.

Development up to the Pharyngula Stage Takes Place in Different Ways

While the postneurula stage of the frog, chick, and mouse appear morphologically similar and do indeed possess a conserved basic vertebrate body plan, each of them gets to this stage very differently. The amphibian egg is polarized in the animal-vegetal axis before fertilization and is cylindrically symmetric around this axis. The sperm aster then imposes a cortical rotation that breaks this symmetry and

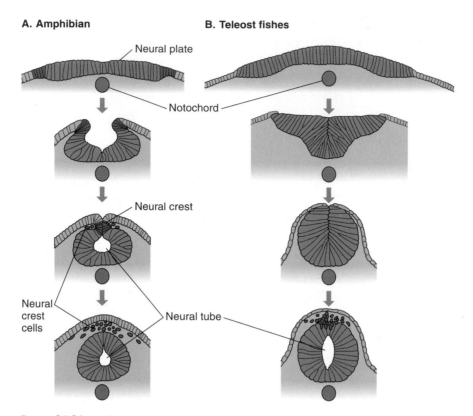

Figure 17.10   A Comparison of Neural Tube Formation in Frogs and Fishes Diagrams comparing neural tube formation in frogs and fishes. **(A)** In anurans, the neural tube develops from a sheet rolling up to form a tube, as described in Chapter 6. **(B)** In teleost fishes, a solid cord develops that subsequently forms an internal cavity.

results in a dorsoventral axis. Successive inductions by the Nieuwkoop center and the Spemann organizer then organize the pharyngula stage.

The avian egg, on the other hand, is organized in a completely different way from the amphibian. These large, yolky eggs possessing scant cytoplasm enable reproduction on land, and they are probably the result of a very long evolutionary process. Animal-vegetal polarity develops during oogenesis in birds, just as in amphibians, but the direction of body axes is difficult to predict, even when cleavage is quite far advanced and tens of thousands of cells have formed. As we discussed in Chapter 5, some bias for the positioning of the primitive streak occurs under the influence of gravity as the fertilized egg passes down the oviduct, but this is not an especially strong influence and can be overcome by self-organizing properties of the blastoderm itself. Only during multicellular stages after egg laying does a definitive organi-zation resulting in formation of the primitive streak actually arise. Of the approximately 60,000 cells present when the egg is laid, only about 500 will actually construct the pharyngula; the remainder will be devoted to forming the yolk sac, amnion, and chorion. The germ layers are generated by a suite of morphogenetic movements very different from those that occur during amphibian gastrulation. A gradual and protracted series of inductive interactions between germ layers and between the migratory Hensen's node and other tissues help to form the pharyngula. Large portions of the early blastoderm are quite capable of self-organization to form an axis if surgically isolated.

Finally, there is the still different situation in the mouse egg. The adaptation of placentation renders the large yolk deposits in the egg nonessential. (There is still a yolk sac in mammals, albeit empty, an illustration of descent with modification, or making do with what you've got.) There is little or no asymmetry in the early mouse egg, which is true of most mammalian eggs. There may, however, be an axial bias in the undisturbed mouse egg. Early cleavage generates totipotent blastomeres. The compaction and generation of differences between "outside" and "inside" cells, occurring at the fourth and fifth cell divisions, is crucial for development of the inner cell mass, which gives rise to the pharyngula proper. Once again, these early events are very different from what occurs in birds and amphibians.

The different cell biological tactics of prepharyngula development found in different classes is not unique to chordates, but merely an example of the diversity found throughout many, perhaps most, animal phyla. Looking more deeply at the diverse tactics found in chordates, we may see that all of them result in a spatial organization that allows the successive inductions to take place and then results in the anteroposterior and dorsoventral organization characteristic of chordates. All the chordates embrace an early induction of endoderm and mesoderm, an organizer-type induction during gastrulation, and inductions by the notochord after gastrulation. The phylotypic pharyngula is flexible and robust precisely because there are a number of ways to generate it, and because it serves as a flexible platform for postphylotypic development.

Development After the Pharyngula Stage Generates Great Diversity

The chordate pharyngula provides a platform for generating an astonishing diversity of chordate creatures. The cellular tactics can play upon the archetypal body plan, much as a musician plays an instrument, thereby producing many wondrous forms from a few "notes." Even the morphogenesis of the pharyngula itself may proceed differently in different vertebrates. For example, in our primary examples of frogs, chicks, and mice, the anterior neural tube is generated by a folding of surface cells (Figure 17.10A; see also Chapter 6), and portions of the posterior neural tube are formed by cavitation of a solid rod of cells. But in teleost fishes, as well as in species of the class Agnatha (lampreys and hagfishes), there is no invagination to form the neural tube. Rather, the dorsal midline first forms a solid neural keel; this region then forms the solid, rod-shaped anlage of the neural tube, which then forms a tube throughout its entire length by cavitation (Figure 17.10B).

DIVERSIFICATION OF SIGNALING SYSTEMS

Vertebrate Limbs Provide an Example of Postphylotypic Diversity

Many features are used to classify species into the various vertebrate classes. These include integument (scales, feather, hairs), skeletal composition (cartilage, bone), physiological function (gills, lungs), to name a few. Comparisons of DNA sequences in creatures of different vertebrate classes has mainly confirmed the classical taxonomic groupings. The kinds of limbs and the skeletal elements in them comprise a morphological character historically favored by taxonomists and evolutionary biologists for discerning vertebrate relationships. The development of limbs is also a subject that has been well explored by developmental biologists (see Chapters 7 and 16). For instance, the presence of flippers, hooves, or wings can be diagnostic of different orders of vertebrates; in this chapter, we have already discussed limblessness in snakes, attributing it to HOX gene differences affecting downstream target induction systems.

The establishment of limb-bud domains is probably due, at least in part, to the action of HOX genes, as discussed previously. Once selected, the limb buds all employ reciprocal signaling between the AER and underlying mesoderm, and between the zone of polarizing activity (ZPA) and its targets. The actual details of numbers and kinds of long bones and digits being formed does not depend on this signaling, however. When the ZPA is transplanted, or other surgical interventions are imposed, or mutants are isolated and studied, the duplicated portions or the parts persisting when others are absent still conform to the basic skeletal elements characteristic of the species that contributes the mesenchyme. The arrangement of bones in the forelimb of various vertebrates is shown in Figure 17.11. The consequences of mirror-image duplication brought about by ZPA transplantation or mutation is shown in Figure 17.12. From these figures, it is obvious that when ectopic digits are induced by manipulations, only appropriate, species-specific skeletal elements are being duplicated. Chick and mouse limb buds always produce chicklike and mouselike digits, respectively, even when the digits are duplicated.

The pattern of bones formed is due to the pattern of precursor cartilaginous elements formed in the mesenchyme.

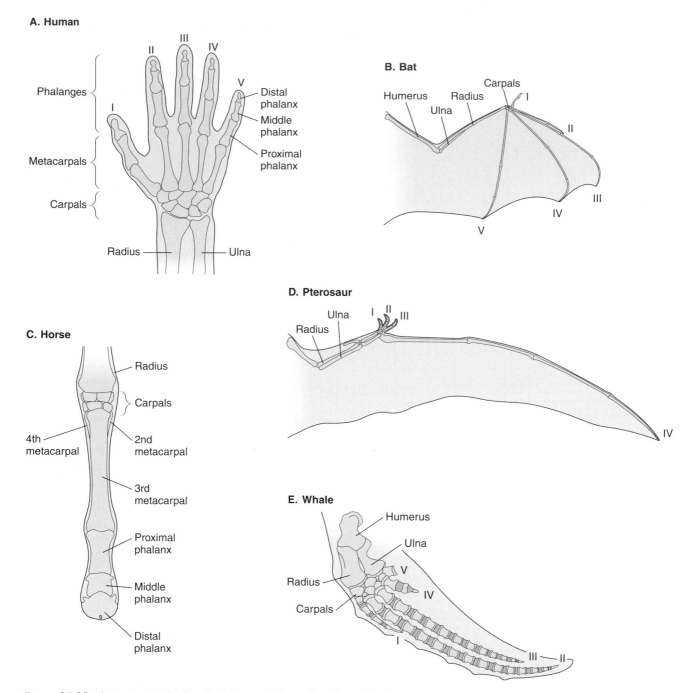

Figure 17.11   Variation in the Forelimb Bones of Some Vertebrates The bones of the human forelimb and examples of variations encountered in different vertebrates.

Though our knowledge is far from complete, we do know that genes in the HOXa and HOXd groups have substantial effects on the exact placement and form of these cartilaginous elements. Thus, current ideas about the basis of species-specific limb-bud differentiation invoke different combinations of HOX selector genes acting to provide the micropattern of compartments needed for skeletal growth and differentiation. The HOX gene targets are probably ligands of the BMP and FGF families, which in turn are capable of affecting cell proliferation, adhesion, aggregation, motility, and the like. Local morphogenetic tactics then give rise to the variety of limb skeletal patterns. There is much to learn even in this well-studied example, but the general way in which at least some of the diversity of limb types comes about can now be appreciated. Strategic selector gene combinations regulate ligand–receptor circuits, which organize morphogenesis and terminal differentiation.

A. Chick

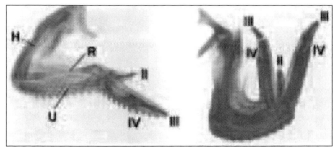

Control   ZPA graft

B. Mouse

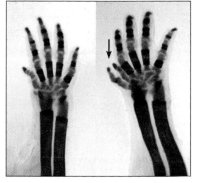

alx4⁺/⁺   *alx4*⁻/⁻

Figure 17.12 ZPA-Induced Duplication of Digits (A) The results of transplanting the ZPA from its normal location to the anterior portion of a limb bud of a chick embryo at stages 19 to 23. The controls shows the typical II-III-IV pattern from anterior to posterior, while the transplant shows a mirror-image duplication in which the digits are strictly winglike. **(B)** For comparison, partial duplication in a mouse forelimb is shown, in which a mutation in the *alx4* gene causes an ectopic ZPA to form. The duplicated digit is nevertheless characteristic of a mouse digit I *(arrow)*.

Signaling Systems May Adopt New Roles in Different Animal Groups

We have stressed the similarity of signaling systems used in different creatures. We should not forget, however, that these systems are used not only in different organisms but also for many different purposes at different times and places in the developing embryo of a given organism. For instance, Notch signaling in *Drosophila* is involved in development of wings, bristles, gut, muscle, and ommatidia of the eye. Notch is a communication system used in many different contexts. Notch signaling is also employed in vertebrates as well, in T-lymphocytes, somites, and the nervous system, for example, all of which are instances unrelated to the signaling contexts in *Drosophila*.

Signaling systems and transcription factor circuits apparently can be coopted to subserve novel functions in development. Chris Lowe and Greg Wray, working at the State University of New York at Stony Brook, examined the expression of some well-known homeobox-containing regulatory genes in echinoderms. For example, the gene *orthodenticle (otd)* is involved in the specification of anterior head structures, both in arthropods and chordates. When Lowe and Wray looked at *otd* in brittle stars and sea urchins, they found it expressed in the podia (which function in locomotion, feeding, and sensory reception) of the water vascular system, an organ system unique to the echinoderm phylum. This gene and the two other genes examined, *engrailed* (Figure 17.13) and *distalless (dll)*, were found to be expressed in various structures having a fivefold radial symmetry, which again is a unique echinoderm feature. The gene *dll* is also expressed in the podia of sea urchins and sea stars, but was not in brittle stars. Hence, these unique functions of Dll are not found in all echinoderm classes.

Signaling Systems May Themselves Be Modifiable

Not only is the role of a particular system modifiable, but the details of the operation of the system itself may change. A fascinating example is found in recent work looking at the expression of the *toll* gene in different insect orders. Recall that *toll* encodes a membrane receptor involved in setting up the dorsoventral pattern in *Drosophila*. Toll is ubiquitously

A.   **B.**

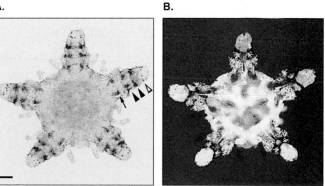

Figure 17.13 Expression of *engrailed* in an Echinoderm Juvenile (A) A juvenile of the brittle star *Amphipholis squamata*, as photographed in bright-field illumination. Ectodermal stripes of *en* expression are present *(solid arrowheads)*, and a few nuclei have begun to produce Engrailed protein where the boundary of the next arm segment will appear *(open arrowhead)*. The arrow denotes En on neurons in a more proximal (hence, developmentally older) segment of the arm. **(B)** The same specimen photographed under polarized light, which highlights the calcified skeletal elements.

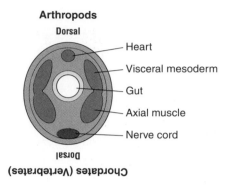

Figure 17.14   Contrasting the Dorsoventral Organization of Chordates and Arthropods This stylized diagram shows the location of heart, visceral mesoderm, gut, axial muscle, and central nervous system. What is dorsal in an arthropod is ventral in a chordate, and vice versa.

present in the cell membrane of the *Drosophila* egg; it interacts with the locally produced ligand in the perivitelline space, thereby establishing spatial specificity. However, in the beetle *Tribolium*, Toll mRNA is present in a strong gradient, very present on the ventral side but nearly absent dorsally. Evidently, the same signaling components are being used in these beetles and flies but are deployed differently.

The origin of some signaling molecules and their linkage into functional systems may predate the divergence of plants and animals. For example, some domains of Toll must have a very ancient origin, for there is considerable similarity, and hence implied homology, between cytoplasmic portions of Toll and the N-terminal portions of receptor proteins involved in disease resistance in *Arabidopsis* and other higher plants. Another class of plant genes conferring disease resistance possesses serine-threonine kinase domains similar to Pelle, another *Drosophila* protein utilized in the Toll–Dorsal pathway; and the plant gene *clavatal* (see Chapter 11) also shares homology with kinase domains in *pelle*. It is interesting to note that the receptor Toll of *Drosophila* also shows substantial similarity to the receptor for interleukin-1, found in cells of the mammalian immune system; moreover, Toll and other members of the Toll–Dorsal pathway are involved in responses to bacterial infection in *Drosophila*.

A final example of alterations of signaling systems involves the differences between arthropods and chordates with regard to patterning of the dorsoventral axis during development. Chordates exemplify strong dorsal development, with a notochord, dorsal hollow nerve cord, and somites arranged on the dorsal side. In stark contrast, arthropod nerve cords and mesoderm formation occur ventrally. You will recall that dorsal neural development in chordates requires antagonism

of the BMP ligands by the secreted "antiligands" Chordin and Noggin. A homolog of *chordin*, called *short gastrulation (sog)*, has been found in *Drosophila*. Furthermore, the homolog of the genes encoding BMPs of vertebrates is *dpp* in *Drosophila* (discussed in Chapter 15). Chip Ferguson, Eddie DeRobertis, and their colleagues have shown that these molecules are interchangeable between the species. Sog can function like Chordin in frogs, and Chordin like Sog in flies. Evidently there is a conservation of molecules in the signaling system, and a conservation of their roles in axis specification, but the whole organization is turned upside down with respect to dorsoventral polarity (in other words, the side closer to earth) in one phylum compared with another (Figure 17.14). One might argue, then, that the context in which a signaling system operates can have tremendous effects on the outcome, and so changes of context also provide a source of novelty.

We should add, however, that the hypothesis that a difference between chordates and arthropods involves an inversion of the dorsoventral axis is a subject of ongoing investigation. While many genes, such as *sog* and *chordin*, do fit the hypothesis, some markers of neural differentiation in the hemichordates (a little-investigated deuterostome phylum possessing an epidermal nerve net rather than a centralized nervous system) do not fit so neatly in to this scheme.

LARVAE AND EVOLUTION

Many Animals Develop Indirectly from Within a Larva

We have purposefully neglected descriptions of the development of a very large number of animals. Eric Davidson and his colleagues at Caltech have argued that the developmental modes of long germ-band insects (which they call type 3 development) and vertebrates (type 2) are distinct from the mode by which the preponderance of animals develop (which they term type 1). In type 1 development, utilized by most of the invertebrate phyla (except long germ-band insects), cleavage patterns are invariant, transcription is activated early in cleavage, and signaling that results in specification of various lineages occurs before gastrulation. The embryos of a minority of these species can develop directly to generate juvenile forms displaying the rudiments of the definitive adult body plan. In most instances, however, the embryos form a ciliated swimming larva. Often, such larvae bear little or no resemblance to the adult, and the juvenile develops from a small population of precursor cells embedded within the larva. The larva is a "life support system" for the developing juvenile, which becomes independent only as a result of metamorphosis.

When we discussed metamorphosis in Chapter 8, our examples did not focus on this version of "indirect development," in which the large parts of the adult organism are built from what Kevin Peterson, Andrew Cameron, and Eric Davidson have called *set-aside* cells. Metamorphosis involves a radical and rapid transformation from a free-living juvenile or precursor form to the sexually mature adult. Although set-aside cells are a conspicuous element of metamorphosis of marine invertebrate larvae, there are many other instances of metamorphosis that do not involve set-aside cells. For example, amphibian metamorphosis is a remodeling process that does not involve set-aside cells. Insect metamorphosis does involve many groups of set-aside cells—the imaginal discs—but does not represent the most typical instance of this mode, nor does it involve a fundamental reorganization of the basic body plan.

A comparison of the young adult sea urchin with its larva, called the pluteus, reveals distinct differences in morphology and body plan. Figure 17.15A shows the larva after three days of development. Note that it bears no recognizable similarity to the adult sea urchin. Figure 17.15B depicts a section through this larva at a later stage, when the set-aside cells attached to the gut are gradually forming the juvenile urchin (Figure 17.15C).

Most terrestrial vertebrates show direct development, while marine invertebrates utilize indirect development and set-aside cells. The variety of these larvae is astonishing, as are the morphological differences between larva and adult. The concept of a phylotypic stage is not easily applied to the numerous members of these phyla, though in all instances examined, HOX gene clusters are present and their expression seems involved in realization of the body plan of the adult.

It can be argued that the existence of set-aside cells is actually a very primitive condition, and that living chordates and some arthropods have lost this mode, at least to some extent. Another point of view is that larval development and metamorphosis have evolved independently in several taxa, so that creatures possessing larvae with set-aside cells are not ancestral to the majority of extant animals. We shall have to leave the debate on this question to evolutionary biologists, but insofar as juveniles and adults develop within a distinct larva that eschews the phylotypic body plan, the role of the set-aside cells in development of different creatures is grist for our mill. And the role of set-aside cells in evolution has to be reckoned with.

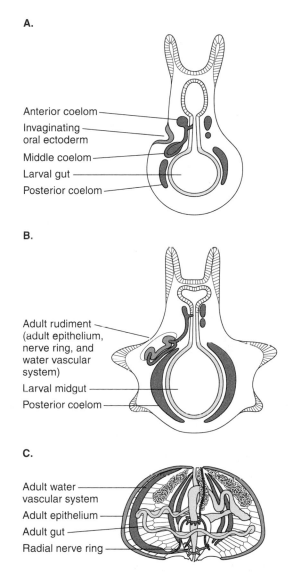

A.

Anterior coelom

Invaginating oral ectoderm

Middle coelom

Larval gut

Posterior coelom

B.

Adult rudiment (adult epithelium, nerve ring, and water vascular system)

Larval midgut

Posterior coelom

C.

Adult water vascular system

Adult epithelium

Adult gut

Radial nerve ring

Figure 17.15   A Pluteus Larva of a Sea Urchin Showing the Juvenile Rudiment (A) An early pluteus larva is shown in frontal view, with the oral aspect toward the top. Various portions of the early coeloms, which will contribute to the juvenile urchin after metamorphosis (anterior, middle, and posterior coelom), have formed by evagination from the embryonic gut. **(B)** A somewhat later larva is now in the process of forming an adult rudiment. **(C)** A cross section through an adult urchin is shown, with the anal opening at the top and the mouth at the bottom.

HOX Genes in Sea Urchins Are Expressed in Set-Aside Cells

Homologs of the HOX gene cluster have now been cloned from the sea urchin (see Box 17.1). The genes of this cluster are arranged in the same prototypical linear order as their homologs in dipterans and mammals. Since the sea urchin displays the maximal form of indirect development, we may ask, How are the HOX genes expressed in larva and juvenile? Recent evidence, using whole-mount in situ hybridization to detect the spatial expression patterns of these genes, shows that most members of this gene cluster are *not* expressed in the embryo, but clearly the HOX genes are expressed in the developing juvenile rudiment (in other words, in descendants of the set-aside cells). An implication of this finding is that the purpose of HOX gene clusters in the sea urchin may be to specify adult features along the body axis. If so, it follows that the timing of expression of HOX genes will be found to differ in direct- and indirect-developing creatures. So far, no metazoans have been found to develop without expression of HOX gene clusters.

The Linkage Between Direct and Indirect Development May Not Be So Complex

Of the approximate 860 known species of echinoids (sea urchins and sand dollars), most show maximal indirect development, that is, a larva with set-aside cells. The pluteus larva, a larval type at least 200 million years old, is the result of embryonic development, which secondarily undergoes metamorphosis. Direct development without a feeding pluteus larva, and hence without consequent metamorphosis, does crop up, however, in about a fifth of the species in six different orders, separated by geography and evolutionary history. It is clear from a careful examination of phylogenetic relationships of echinoderms that indirect development is the ancestral condition among echinoderms and that loss or truncation of metamorphosis has occurred independently on several occasions.

Two closely related species of sea urchins of the genus *Heliocidaris*, found in the waters off Australia, have been studied by Rudolf Raff and his colleagues. *H. tuberculata* shows typical indirect development, while *H. erythrogramma* has abandoned this mode (Figure 17.16A,B). The latter has large yolky eggs, and the embryo does not form a functional gut. The juvenile urchin develops directly during embryonic development. These two species are estimated to have diverged relatively recently, approximately 10 million years ago. This is hardly enough time for massive changes to occur in the genome, yet the changes in the embryonic phase of development are substantial. The direct-developing *H. erythrogramma* has about 2,000 mesenchymal cells that ingress, rather than the 32 found in *H. tuberculata*. Mesenchyme in *H. tuberculata* forms a larval skeleton, but the large number of mesenchymal cells in *H. erythrogramma* skip larval skeleton formation and begin to

construct adult calcified structures, such as test plates and spines, only a couple of days after fertilization.

One gene expressed by mesenchymal cells undergoing biomineralization is *msp130*, a cell surface protein important for forming skeletal elements. In *H. erythrogramma*, this gene is not expressed until adult skeletal elements are being formed, and then only in the small subset of the 2,000 mesenchymal cells directly involved in biomineralization. In contrast, all mesenchymal cells in *H. tuberculata* express *msp130*, and do so just before ingression of the 32 cells into the blastocoel. Raff and his colleagues isolated the promoter region of the *msp130* gene from the direct-developing *H. tuberculata* and attached it to a β-galactosidase reporter. They then introduced this construct into embryos

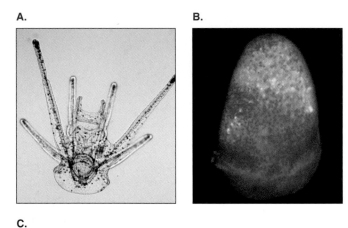

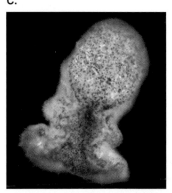

Figure 17.16   A Comparison of the Postgastrula Forms of Two Species of *Heliocidaris* (A) *H. tuberculata* is a typical indirect-developing sea urchin that forms a pluteus larva. A 17-day-old larva with arms (ar), mouth (m), hindgut (hg), and rudiment of the adult (r). **(B)** A lateral view of a three-day-old *H. erythrogramma* larva, which has no mouth, gut, or skeletal arms; an incomplete ciliary band (cb) is present. **(C)** The hybrid developing from an *H. tuberculata* sperm and an *H. erythrogramma* egg looks like neither parent, though it now has several paternal characteristics and undergoes metamorphosis. This dorsal view shows a lobed body and complete ciliary band; a mouth and arms are present on the ventral side.

of the indirect-developing *H. erythrogramma*, and found that the *msp130* promoter from *H. tuberculata* was able to direct gene expression at a time and place appropriate for the indirect development. In other words, the promoter responded differently, and appropriately, in the presence of different sets of regulatory factors in the direct- and indirect-developing species.

While not definitively proven, it is difficult to avoid the conclusion that the substantial changes seen when *H. erythrogramma* and *H. tuberculata* are compared may be due principally to changes in gene regulation, rather than changes in structural genes themselves. And just possibly the number of such changes needed for a dramatic change in development might be smaller than we might have originally guessed. This kind of information bears on theories that consider the origins of groups such as the vertebrates, which do not use set-aside cells.

Changes in Relative Timing of Developmental Processes Can Create Substantial Differences in Organisms

The change in timing of expression of *msp130* is an example of changes in the timing of a developmental process. It is obvious that changes in the *msp130* expression is one of very many changes in timing of gene expression that must occur when the indirect developing ancestor is compared with its direct-developing descendants. When changes in the timing of a developmental process (relative to other processes, or relative to the timing of that process in some ancestor) occurs, it is termed *heterochrony*. In Chapter 8, we discussed one of the more well known examples of heterochrony in connection with amphibian metamorphosis. You will recall that some species of salamanders can become sexually mature without undergoing a full-scaled metamorphosis to terrestrial existence. This condition of neoteny in such salamanders involves several changes in the timing of production of and responses to hormones that cause sexual maturation. The molecular mechanisms that underlie the timing of entire processes are not well understood, though we now know they are governed by many different signaling pathways.

NOVELTY

Some Radical Changes in Body Plan May Not Be So Complex

A second example of profound morphological change, directed by what may be relatively simple changes in the genes of regulatory circuits, comes from comparisons of closely related species of ascidians, members of our own phylum, Chordata. Ascidians are also members of the subphylum Urochordata; they do not possess set-aside cells as such, but do undergo metamorphosis. As we so often encounter in biology, they do not fit neatly into a niche; it is the postgastrula embryo that displays chordate characteristics (dorsal nerve cord, notochord, and postanal tail), while the adult is difficult to recognize as a chordate, having adopted a sessile, filter-feeding existence on the benthos.

The typical ascidian tadpole has a distinct tail with a notochord and somites, but some closely related species in the same genus, which are tailless, have been studied by Bill Jeffery and Billie Swalla, working at the Roscoff Marine Station in France. *Mogula oculata* has the conventional tail, while *M. occulta* is tailless (Figure 17.17A,C). Only about 20 of the ascidian species are tailless, and experts on their phylogeny conclude that the tailed condition is the ancestral one; in other words, taillessness represents the loss or modification of characters during relatively recent evolutionary times.

A few differences in gene expression have been found between the two *Mogula* species, among them a tyrosine kinase (Cymric), a leucine zipper transcription factor (Lynx), and a zinc finger transcription factor dubbed Manx. Expression of *manx* is restricted to the primordium, which generates the tail and is downregulated in the tailless species (*M. occulta*). When hybrids develop by combining sperm from the tailed species and eggs from the tailless species, there is a partial restoration of tail development (Figure 17.17B) and an upregulation of *manx*. This gene has recently been shown to be linked to another gene, *bobcat*, which is also needed for tail development. Jeffery and Swalla propose that the negative regulation of this *manx–bobcat* locus is a crucial part of the transition to taillessness.

While there is much we do not know, this example and the preceding one (*Heliocidaris*) illustrate the likelihood that changes in development resulting in major morphological changes may involve only a relatively small number of changes in the crucial regulatory genes.

Apparent Similarities May Mask Regulatory Differences

The exchange of regulatory genes between different animals provides powerful evidence for the conservation of regulatory circuits. The equivalent gene function of *eyeless* in vertebrates and insects is one example we have mentioned; there are many more. We should be aware, however, than there are many instances in which subtle, or even not so subtle changes have

A.

B.

C.

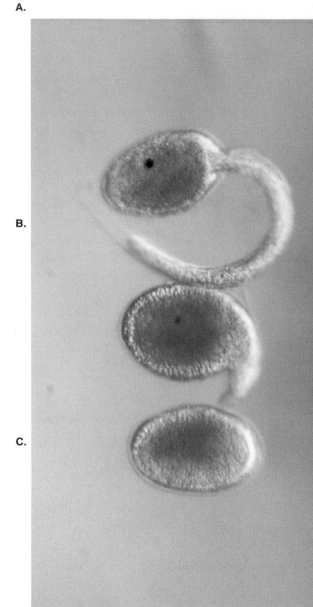

Figure 17.17   The Larvae of *Mogula* Species (A) A hatched larva of a typical tailed ascidian. *M. oculata.* **(B)** A hybrid between *M. occulata* and **(C)** a tailless species. *M. occulta.*

taken place in regulatory circuits. For instance, Noggin is an important signaling molecule emanating from the organizer of amphibians (see Chapter 16). As we saw in Chapter 16, the homolog of Noggin from the zebrafish *(Danio rerio)* has been isolated. Even though the zebrafish version of Noggin has dorsalizing properties, the gene *noggin* is *not* expressed in the zebrafish organizer. Nor is Sonic Hedgehog, required for the induction of floor plate cells, produced in the neural tube of zebrafish.

While there are many other examples of the inexact nature of apparent conservation of regulatory genes (recall differences in *Toll* regulation in insects), we shall mention here only a final example: *slug* gene expression in mice. Slug is the homolog of Snail in *Drosophila*, an important transcription factor needed in the development of mesoderm in flies; Slug is also important in neural crest development in fish, frogs, and chick. When *slug* is knocked out in mice, however, neural crest migration and development are not impaired. Hence, the biological function of Slug may not be precisely the same in all vertebrates. We should point out that vertebrates have undergone considerable genomic duplication early in their evolutionary history, at least judging from the HOX complex, so that overlap and partial functional redundancy complicate interpretations of gene knockout experiments.

The Neural Crest Is a Vertebrate Invention

Sometimes developmental novelties arise that are not obviously related to some progenitor or to some relative in other phyla. This is the case for the neural crest of vertebrates, discussed in the previous paragraph. The neural crest has no obvious relatives in other phyla. The invertebrate chordates, urochordates (ascidians), and cephalochordates (lancelets) are sessile or slow swimmers; they are also filter feeders. These invertebrate chordates lack neural crest cells entirely. But the lowliest of the vertebrates, the ancient group of jawless fishes (Agnatha, comprised of hagfishes and lamprey eels) do possess a neural crest. Here, as in all other vertebrates, the crest cells arise from the ectodermal lips of the closing neural plate and become a migratory population with an extraordinarily diverse set of destinations and kinds of terminal differentiation.

Earlier (Chapter 6), we outlined the development of the neural crest. There has been considerable speculation by evolutionary biologists on its origins. The fact that anterior neural crest cells populating the head region form muscles and bone, often thought of as mesodermal derivatives, supports the idea that in a distant ancestral group there may have been some transformation of ectoderm to mesoderm at the epidermal-neural plate border, giving rise to this maverick population of cells. Some mesodermal genes, such as the aforementioned *slug*, are expressed in neural crest cells in many vertebrates.

Neural crest cells are often found in structures derived from more than one germ layer, and are often engaged in intercellular signaling involved in tissue interactions. Most of the cephalic structures with crest contributions (such as teeth, various cartilages and bones of the skull, and various derivatives of the pharyngeal arches) utilize extensive tissue interactions in their development. A striking instance of differences and conservation in structures containing cephalic neural crest has been revealed by interspecific transplantation of ectoderm of the ventral facial region. Newts (order Urodela) have balancers on either side of and ventral to the mouth; these structures are formed from head ectoderm and ventral head mesoderm, the latter originating from the neural crest. Frogs (order Anura) possess adhesive organs, called cement glands, in the same location; cement glands are also formed from head ectoderm and neural crest–derived mesenchyme (Figure 17.18A) If a square of ectoderm from the ventral facial region of a newt is transplanted to the ventral head region of a frog, it will heal to form structures characteristic of a newt in the equivalent location, in other words, balancers (Figure 17.18B).

The frog neural crest mesenchyme thus has retained the ability to induce the newt ectoderm, but it cannot instruct the ectoderm what to form. The newt ectoderm follows its species-specific program of differentiation, but this requires a signal from neural crest mesenchyme. The tissue responding to the crest has changed, but the crest has maintained and conserved its inducing role. Urodeles and frogs diverged some 200 million years ago, so this signaling property of neural crest must be very ancient.

Developmental Strategies Generate Organismal Novelty

It has been our purpose to underline the importance of development when considering mechanisms of evolutionary change. We have highlighted the extraordinary flexibility of developmental mechanisms, and the virtual certainty that changes in expression of regulatory genes afford a way to understand how novelty might arise, thereby providing the raw material for selection of diversity. Since development is the way the reproducing adult organism is formed, changes in evolution necessarily require changes in development.

We stated at the beginning of this chapter that many hypotheses about precisely how development produces evolutionary change may be difficult to confirm. Experiments on developing organisms may, however, enable us to test many of these ideas. Raff and his colleagues recently created sea urchin hybrids by bringing together the large yolky eggs of a direct-developing species of *Heliocidaris* and the sperm of an indirect-developing species of the same genus (Figure 17.16C). The hybrid formed a larva in which many genes and characteristics of the paternal, indirect-developing species

were expressed. This is not the usual outcome in echinoderm hybrids; usually the maternal species dominates. But in these *Heliocidaris* hybrids, the larvae resembled neither parent and bore some resemblance to a dipleurula larva, a form thought to be related to larvae of ancestral echinoderms.

Could probing the mechanism of the changes seen in interspecific hybrids enable one to glimpse the crucial changes leading to development of related but different kinds of larvae? Could experiments in which assemblages of regulatory genes from one creature are substituted for those of another by DNA transformation procedures allow one to reproduce some putative evolutionary changes? In our judgment, important new insights will likely be gained, perhaps in the not too distant future, by experiments that probe the mechanisms by which novelty of form and function arise during development.

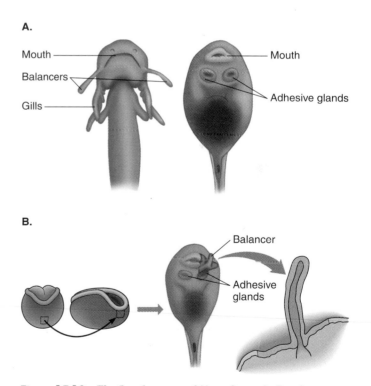

Figure 17.18   The Development of Newt Gastrula Ectoderm on the Cranioventral Aspect of an Anuran Embryo (A) Ventral views of the head of a newt and a frog, the former with balancers and gills, the latter with adhesive glands. **(B)** The experiment described in the text, shown diagrammatically Ectoderm is removed from a newt neurula and transplanted to the facial region of a frog neurula. The transplanted newt tissue forms a newtlike balancer composed of frog neural crest cells and newt ectoderm.

KEY CONCEPTS

1. Many apparently novel traits of different organisms are the result of genes present in a shared ancestor that become modified during evolution.

2. Many differences between animals can arise during development through changes in regulatory genes. These can include (a) changes in the upstream genes that control a particular regulatory gene, (b) changes in the downstream target of a particular regulatory gene, and (c) changes in the domain of expression of a particular regulatory gene.

3. The phylotypic postgastrula stage characteristic of a given phylum provides a flexible and robust platform for generating species diversity.

STUDY QUESTIONS

1. Define the vertebrate phylotypic stage, and compare the ways birds and mammals attain that stage.

2. Both insectivorous bats and insects possess wings for flight. Some common regulatory circuits might be involved in the development of these two structures. Suggest experimental approaches to evaluate that hypothesis.

3. Cephalochordate lancelets do not possess a neural crest. Speculate on how peripheral nerves might arise in such a creature. Can you propose an experiment to test your idea?

SELECTED REFERENCES

Baker, B., Zambryski, P., Staskawicz, B., and Dinesh-Kumar, S. P. 1997. Signaling in plant-microbe interactions. *Science* 276:726–733.

An account of similarities and differences in protein motifs used in signaling in plants and animals.

Burke, A. C., Nelson, C. E., Morgan, B. A., and Tabin, C. 1995. Hox genes and the evolution of vertebrate axial morphology. *Development* 121:333–346.

A review of the role of HOX genes in limb development.

Carroll, S. B., Grenier, J. K., and Weatherbee, S. D. 2001. *From DNA to diversity*. Blackwell Science, Malden, Mass.

An entire book on "evo-devo."

Cohn, M. J., and Tickle, C. 1999. Developmental basis of limblessness and axial patterning in snakes. *Nature* 399:474–479.

A research article describing HOX gene expression in snakes.

French, V. 2001. Genes, stripes and segments in 'Hoppers'. *Curr. Biol.* 11:R910–R913.

Differences and similarities in the role of homologous genes in early development of insects and grasshoppers.

Gerhart, J., and Kirschner, M. 1997. *Cells, embryos and evolution.* Blackwell Science, Malden, Mass.

A recent influential book about the impact of cellular and molecular biology on evolutionary studies.

Hall, B. K., and Wake, M. H., eds. 1999. *The evolution of larval forms.* Academic Press, San Diego.

A collection of articles on the role of larvae in evolution.

Holland, P., and Garcia-Fernandez, J. 1996. Hox genes and chordate evolution. *Dev. Biol.* 173:382–395.

The title says it all.

Holley, S. A., Jackson, P. D., Sasai, Y., Lu, B., DeRobertis, E., Hoffmann, F. M., and Ferguson, E. L. 1995. A conserved system for dorsal-ventral patterning in insects and vertebrates involving *sog* and *chordin. Nature* 376:249–253.

A research paper bearing on the hypothesis of axis inversion in the comparison of arthropods and vertebrates.

Lowe, C. J., and Wray, G. A. 1997. Radical alterations in the roles of homeobox genes during echinoderm evolution. *Nature* 389: 718–721.

A research paper examining the expression of some homeobox genes in echinoderms.

Lowe, C. J., Wu, M., Salic, A., Evans, L., Lander, E., Stange-Thomson, N., Gruber, C. E., Gerhart, J., and Kirschner, M. 2003. Conserved expression map of anteroposterior neural patterning genes between hemichordates and chordates: implications for early deuterstome evolution and chordate origins. *Cell*, in press.

New research bearing on the hypothesis of axis inverson between chordates and arthropods.

Maxton-Kuchenmeister, J., Handel, K., Schmidt-Ott, U., Roth, S., and Jackle, H. 1999. Toll homolog expression in the beetle *Tribolium* suggests a different mode of dorso-ventral patterning than in *Drosophila* embryos. *Mech. Dev.* 83:107–114.

A research paper on differences in *toll* expression in flies and beetles.

Peterson, K. J., Cameron, R. A., and Davidson, E. H. 1997. Set aside cells in maximal indirect development: Evolutionary and developmental significance. *BioEssays* 19: 623–631.

A statement of the set-aside cell hypothesis and the important role of indirect development in evolution.

Placzek, M., and Skaer, H. 1999. Airway patterning: A paradigm for restricted signalling. *Curr. Biol.* 9:R506–R510.

A review of *branchless, breathless, sprouty,* and their vertebrate homologs.

Raff, R. A. 1996. *The shape of life.* University of Chicago Press, Chicago.

An influential book on "evo-devo."

Raff, E. C., Popodi, E. M., Fly, B. J., Turner, F. R., Villinski, J. T., and Raff, R. A. 1999. A novel ontogenetic pathway in hybrid embryos between species with different modes of development. *Development* 126:1937–1945.

A research paper on the nature of the hybrid produced by crossing direct- and indirect-developing sea urchin species.

Richardson, M. K., Hanken, J., Gooneratne, M., Pieau, C., Raynaud, A., Selwood, L, and Wright, G. M. 1997. There is no highly conserved embryonic stage in the vertebrates: Implications for current theories of evolution and development. *Anat. Embryol.* 196:91–106.

The careful reexamination of Haeckel and von Baer's proposals about phylotypic stages of vertebrates.

Swalla, B. J., Just, M. A., Pederson, E. L., and Jeffery, W. R. 1999. A multigene locus containing the *Manx* and *bobcat* genes is required for development of chordate features in the ascidian tadpole larva. *Development* 126:1643–1653.

A research paper on the origin of taillessness in ascidians.

Weatherbee, S. D., Nijhout, H. K. F., Grunert, L. W., Halder, G., Galant, R., Selegue, J., and Carroll, S. 1999. Ultrabithorax functions in butterfly wings and the evolution of insect wing patterns. *Curr. Biol.* 9:109–115.

Comparisons of the homeobox gene *ubx* in flies and butterflies.

Zoccola, D., Tambutte, E., Senegas-Balas, F., Michiels, J.-F., Failla, J.-P., Jaubert, J., and Allemand, D. 1999. Cloning of a calcium channel α1 subunit from the reef-building coral, *Stylophora pistillata. Gene* 227:157–167.

A research paper on the cloning of a calcium channel gene in coral, and its similarity to some vertebrate genes.

STUDY ANSWERS

Chapter 1

1. Cultured cells are probably a better source because the conditions provided by tissue culture stimulated entry of cells into the cell cycle. Hence, the transplanted nuclei could more easily synchronize with the very rapid cell division cycles of the cleaving egg.

2. The mRNA, since is transcribed from the sense strand, has an antisense sequence; an antisense probe cannot form a complementary duplex with an antisense mRNA.

3. The key is to somehow mark the differentiated cells that give rise to new tissues. If it were possible to mark a differentiated bone cell, for example, with some kind of permanent heritable marker, then one could detect whether the progeny of this bone cell formed another kind of tissue during regeneration. Such marking techniques will be discussed at various places in the book.

3. There must obviously be some change in the cytoskeleton, which involves microtubules, microfilaments, and intermediate filaments. In this instance, it is the formation of actin microfilaments that is responsible. How can you distinguish between microtubules and microfilaments? (Hint: use inhibitors of microtubule or microfilament assembly.)

4. It should be possible to treat eggs with ammonia without fertilizing them, and then measure the amount of cytoplasmic adenylation. The result should then be compared to the adenylation in fertilized eggs. One problem in interpreting the results might be that the ammonia acts as a parthenogenetic agent. How could you evaluate whether the ammonia has side effects and causes a more or less complete fertilization reaction, thus making the link between pH and polyadenylation difficult to evaluate? (Hint: measure calcium transients.)

Chapter 2

1. The presence of polar bodies ought to be a clue. If any exist, then meiosis I has been completed. If there are no polar bodies, you could try to fertilize the egg and examine how many polar bodies, if any, form after fertilization.

2. Since eggs can only survive using internal reserves as sources of energy, there must be ways to regulate metabolism so that it remains low until fertilization or development actually begins. Furthermore, the egg must be protected against environmental insults, like dehydration, temperature changes, and UV light. We would therefore expect to see special membranes deposited around such eggs.

Chapter 3

1. Since ring canal function is important in oogenesis, it would be a maternal effect mutation, having effects only when the mother is homozygous for the recessive gene. Ring canal function is essential for delivering materials from the nurse cells to the oocyte; therefore, all events for which nurse cells are responsible, such as ribosome delivery or delivery of Bicoid and Nanos proteins, would be compromised. Oocyte development would be grossly abnormal, and no viable embryos would form.

2. Laser ablation in cycle 4 might not have much effect. Remember that the patterning of the cellular blastoderm is

very much the result of interactions between maternally localized factors. It might be that the laser action did not impair diffusion of substances like Bicoid, Nanos, and Hunchback, since there are no cells at cycle 4 and hence the determination of cellular blastoderm nuclei might be rather normal. Of course, the devil is in the details. If the laser irradiation caused considerable damage to the cortical cytoplasm, then cell formation and morphogenesis might be impaired and produce abnormalities in the wing that formed.

3. A greater number of genes would probably result in greater synthesis of Bicoid mRNA by nurse cells, and a higher level of Bicoid mRNA would be tethered in the anterior portion of the oocyte. Since diffusion is concentration dependent, Bicoid would diffuse more posteriorly than normal and attain higher levels during cleavage. This would result in accentuated development of anterior structures. Anterior structures might also extend more posteriorly, at the expense of thorax development.

4. Tube and Pelle work in the dorsoventral pathway, participating in the reactions that allow Cactus to dissociate from Dorsal. If Tube or Pelle were absent or reduced, Cactus likely would not dissociate from Dorsal, thereby preventing Dorsal from entering nuclei. A dorsalized phenotype (one lacking in ventral structures) would result.

Chapter 4

1. Since eggs and early zygotes contain everything needed for rapid cell division, it is likely they contain all the materials needed for chromatin formation and assembly into condensed chromosomes. This surplus store of such materials is unlikely to exist in most "regular" cells.

2. There are probably a variety of ways to analyze this, but it is likely that procedures inhibiting one of the events could be used to see whether it inhibited the others. For example, we could inhibit transcription by exposing the blastula to a drug that arrests transcription, and see whether motility has been affected.

But this procedure might not be decisive. Suppose a common regulatory switch is "upstream" of all three events. Then, inhibiting one function might not reveal the link that all three have to some regulator. The only route might be to genetically analyze an organism in which it is practical to find large numbers of mutants and suppressors of mutants.

3. Surely gastrulation would be initiated, since the bottle cells form below the equator (see Figure 4.13). But convergent extension of the ectoderm, epiboly, and the presence of a substrate for motility of involuted mesoderm would be absent, as well as most of the prospective mesoderm. The gas-

trulation process would encounter enormous impediments and probably produce monsters.

Chapter 5

1. *Drosophila*: the ventral furrow forms mesoderm, ingressing cells form neuroblasts, and anterior and posterior invaginations form gut. Xenopus: invaginating cells form endoderm and some dorsal mesoderm. Chick: ingressing cells form all the mesoderm and endoderm. (These comments apply strictly to surface cells.)

2. Rotation of the entire hypoblast would involve movement of a putative anterior center, and thus we would expect to see changes in placement of the head structures. This has not been observed.

3. If it were possible to obtain a chimera between wild-type ES cells (which form only embryonic epiblast) and BMP4 homozygous mutant embryos (which would form all of the extraembryonic ectoderm), and if the embryo formed no primary germ cells, then these results would be consistent with the necessity for a BMP4 signal from the extraembryonic ectoderm. This experiment was recently performed by Kirstie Lawson et al. (cited in Selected References at the end of the chapter).

4. The usual junctions between epithelial cells would be expected to form during compaction. Thus, cells located central to peripheral cells—that is, the ICM—would be isolated by tight junctions from the ionic environment surrounding the trophectoderm of the embryo.

Chapter 6

1. You would be hard pressed to find any kind of differentiation in vertebrates in which tissue interactions do not play a part. It once was argued that the epidermis forms autonomously, without input from any other tissues, but we now know that the influence of BMPs from cells of the vegetal hemisphere is essential for epidermal development, and there are later contributions from the dermal mesenchyme as well. Contemporary ideas about neural development posit that it is the absence of ligand influence that is important for neural development , but we have no satisfactory explanation for how ectoderm acquires this bias in the first place.

2. The Spemann organizer has a special status for several reasons. First, its discovery played an important role in the history of developmental biology, for reasons explained in the text. Second, the extent of its influence in organizing the axis of the animal—the hallmark of the vertebrate body plan—far exceeds that of other known tissue interactions. Third, the organizer tissue itself undergoes convergent extension to elongate the anteroposterior axis of the embryo (see Chapters

12 and 13), and thereby deploys the prospective notochord in exactly the right place to induce the central nervous system.

3. One can study the influence of crest cell age on their migration by simply looking at the destination of marked cells. As a control, some early-emerging crest cells are taken from an embryo in which all cells are identified by a lineage marker, such as a vital stain or a species-specific marker, and then transplanted to another embryo of the same age; this allows the researcher to follow the migratory pathway of the cells. Then a *heterochronic transplantation* is done, in which embryos of different ages are used for donor and host. It should be possible to see whether "young" neural crest cells placed in an "older" embryo migrate in the same pathway as they do in a host of the same age.

4. The usual approach to the problem of determining whether a particular growth factor is involved in development is to design ways of demonstrating both a *loss of function* and a *gain of function* of the suspected factor. In this instance, a loss of FGF8 function might be studied in mouse embryos in which the gene for FGF8 had been knocked out by homologous recombination. A negative result would not, however, necessarily mean that FGF8 was not involved, since the active agent might be comprised of a cocktail of factors, and lack of FGF8 might not produce a recognizable phenotype.

Another approach would be the gain-of-function experiment of locally exposing developing neural tissue to additional FGF8, perhaps by implanting a small bead soaked in FGF8. This approach has actually been carried out (see the 1999 paper by Salvador Martinez et al. listed in the Selected References at the end of Chapter 6).

Finally, it is essential to determine whether the expression of FGF8 during normal development is consistent with its proposed role; is FGF present at the right time and correct place?

Chapter 7

1. Somites dictate the segmental arrangement of the spinal cord ganglia. The ventral neural tube influences the differentiation of the sclerotome into cartilage, and the dorsal neural tube influences the development of striated muscle from the myotome. (See Figure 7.5.)

2. As shown in Figure 7.7, Shh helps pattern the sclerotome, and together with Wnt helps the myotome to form skeletal muscle.

3. This question has two parts. First, what determines the direction of elongation? One experiment to ascertain the answer might be to explant intermediate mesoderm containing an actively extending pronephric duct and place either anterior or posterior intermediate mesoderm next to the posterior aspect of the duct. This setup might demonstrate whether anterior tissue has a chemorepellent effect. Can you think of other strategies? Could the intermediate mesoderm be transplanted in a reversed anteroposterior orientation.

The second part of the question is, What is the actual mechanism of the elongation? The text mentioned both proliferation of the tip and recruitment of adjacent cells. Excising the tip of the pronephric duct might help distinguish the two mechanisms.

4. Using the quail–chick cell-marking system, this should be a straightforward experiment. Transplant some quail yolk sac to a chick embryo, allow the embryo to hatch, and then look at the nuclear morphology of the red blood cells in the adult. (Birds have nucleated red blood cells.) Any quail-type erythrocytes that are observed must have originated from yolk sac. This result would not show, however, that *all* erythrocytes arise from yolk sac precursors, only that *some* of them do.

Chapter 8

1. Both ecdysone and T_3 are positive regulators of metamorphosis in their respective settings. Ecdysone stimulates molting and enhances the formation of pupal and/or adult characteristics. Likewise, T_3 is the hormone in amphibian larvae that drives metamorphosis toward adult characteristics. Both ecdysone and T_3 have antagonistic hormones that help regulate the progression of metamorphic changes.

2. The relative amounts of hormones, the precise timing of their release, and the tissue distribution of receptors are thought to determine the character of a molt. At the simplest level, the ratio of juvenile hormone to ecdysone determines whether the molt will enhance development of more differentiated characteristics.

3. Some experimental data would help to explain this difference. For instance, is muscle really the target of T_3 in those different tissues? Or is another tissue primarily responding? Muscle degeneration in the tail is a rather late event, requiring high hormone levels. Perhaps apoptosis is induced by the release of ligands favoring apoptosis from another tissue, such as dermis, and muscle is only secondarily affected. Likewise, are limb muscles responding directly? At the basis of the different responses must be some difference in the distribution of receptor, the kind of receptor (or receptor partner), the second-messenger pathway utilized by the activated receptor, or some combination of these. Can you design an experiment to find out?

Chapter 9

1. You might make two meristems.
2. Look for mutants that do not show apical dominance.

3. Plants could regulate certain cell types for the environment; for example, the layer that faces the sun would have an increase in photosynthetic cells.
4. False
5. A leaf is an organ that originates from the flank of the shoot meristem and is dorsoventrally organized.
6. A meristem is a group of organized cells that do not have a determined fate.

Chapter 10

1. The unfertilized flower includes sporophyte tissue and the central cell, which are both diploid, and haploid embryo sac cells such as the egg or synergids. In the fertilized flower, the endosperm is triploid.
2. The sequence from outer to inner whorl will be carpel, stamen, stamen, carpel.
3. Cold and daylength.
4. The seed would be exactly like the mother with no chance for recombination of unfavorable traits.

Chapter 11

1. Isolate mutants in which the spacing differs from that in the wild type.
2. See if a hormone mutant has altered light responses and vice versa.
3. Adjacent plant cells are connected by membrane-lined pores, called plasmodesmata. Small molecules freely move through the plasmodesmata.
4. Proteins have the capacity to increase the size of plasmodesmata pores.
5. Plants have photoreceptors that absorb different wavelengths of light.
6. The phytochrome response is induced by red light and reversed by far-red light.
7. GA-deficient mutants are short and dark green.

Chapter 12

1. If you could devise a way to stop or slow mitosis during gastrulation, it should illuminate the role of mitosis. There are mitotic inhibitors that could be used, but the chorion around the embryo makes delivery of these inhibitors very difficult. A better experiment would be to exploit a mutant that regulates mitotic rates. Several such mutations are known, among them *string*. Of course, if mitosis were severely compromised, then development would cease, so it would be necessary to put the mutant under control of an inducible promoter, such as a "heat shock" promoter, which responds only when the temperature is elevated.

2. Cell marking is the key to following the cells. Refer to the methods discussed in Chapter 2 for marking cells, and judge which marking methods might be stable and nontoxic under the experimental conditions.
3. Since collagen type IV is nonfibrillar and found in the basal lamina, this knockout in the homozygous condition would likely compromise the structure and function of the basal lamina. Havoc would probably ensue, and such embryos would not be viable. Any structure possessing an epithelial-mesenchymal boundary might be compromised. The list would be very long. For example, any migration along a basal lamina, as occurs among certain cells during gastrulation, might be affected. Morphogenesis of epithelial buds lying on a basal lamina, for example in the lung and pancreas, would be affected.

Chapter 13

1. (a) Two examples of ingression in amniote development are epiblast cells moving from the surface through the primitive streak, and neural crest precursors leaving the surface epithelium to become migratory mesenchyme. (b) A knockout of known metalloprotease genes in mouse embryos might affect either of these two processes. But a negative result would have little meaning, since one or more unknown metalloproteases might actually be involved; you cannot knock out an unidentified gene.
2. Haptotaxis is the theory that adhesive strength between a migrating cell, or cell group, and the substrate plays a role in guiding that migration.

This can be tested in vitro by culturing the growing mesonephric duct on substrates with different adhesivity for those cells. Positively charged molecules, such as polylysine, or different basal lamina components, such as laminin or type IV collagen, might be used in the experiments. Also, one could inject enzymes that might influence the adhesivity of the substrate in the embryo on which the mesonephric duct migrates, and see whether these injections influenced the growth of the duct toward the posterior.

Neither of these experiments would be decisive, however, because the results would be subject to many alternate interpretations. (Can you think of any of them?) Would having appropriate mutations be decisive?
3. Pancreatic epithelium has a requirement for mesenchymal factor(s) that is not tissue specific; any mesenchyme can apparently supply it.
4. Look at the time and place of expression of each of these FGF ligands by using specific antibodies against them. Also, use antibodies against the receptors of these ligands to determine when and where they appear. Only ligands and recep-

tors that are present at the time outgrowth starts could be involved. Check the text for the ones involved.

Chapter 14

1. Since incorporation of 5-azaC into the DNA would prevent or impair the methylation, regulatory mechanisms that depend on changes in methylation status of the DNA would be compromised. Presumably, the DNA of the tissue culture cells would be much less methylated after incorporation of 5-azaC. Perhaps the tissue culture cells do not normally form muscle because crucial genes are methylated and therefore are inactive. Following this reasoning, the 5-azaC incorporated into such genes, by preventing methylation, would allow them to become active, thereby leading to muscle differentiation.

2. Certainly havoc would occur. The nature of the effects would depend on several factors. One is the strength of the mutation—that is, whether the mutation completely eliminates function or only hobbles it. Another is whether the mutation is recessive or dominant. A third factor is that some polymerase molecules might be passed on to the zygote from the nurse cells of heterozygous mothers. Eventually, however, newly synthesized RNA polymerase III would be absent. Since processing of almost all transcripts requires some of the small RNAs, since ribosome structure and function depend on the 5S precursor RNA, and since translation requires transfer RNA—all of which are processed by RNA polymerase III—the null mutations would be lethal; there would be no development.

3. Modules A and B in endo16 work positively with the basal promoter to promote expression, while E and F negatively regulate this expression so that it does not trespass into the ectoderm. However, the absence of modules C and D in the reporter construct means that the negative regulation that ordinarily prevents endo16 expression in mesenchyme would be absent. Thus, we would probably observe reporter expression in skeleton-forming mesenchymal cells, as well as in the gut.

4. If Nanos is not localized, then local repression of Hunchback mRNA translation will be ineffective. Hunchback activity in the posterior region of the embryo would favor development of anterior structures there. Certainly, posterior development would be seriously deranged, and would possess some anterior character.

Chapter 15

1. Since Mom5 is the receptor for the Wnt-like signal Mom2, there would be no Wnt signal to downregulate Pop1. This would result in high Pop1 levels everywhere in the EMS cell; division of this cell would then produce only anterior-type progeny, in other words, only MS cells. There would be no gut in the larvae developing from such zygotes.

2. Since both Notch and its ligands (Delta and Serrate) are membrane-bound, the signaling could only be exerted on cells touching the Delta or Serrate ligand. However, if a bucket-brigade kind of relay mechanism operates (see Figure 15.7A), the receiving cell could then possibly relay the initial signal by a Notch pathway, or by an independent pathway, to other surrounding cells.

Experimental tests to ascertain this would be very difficult to set up. If the influence of Notch extends to only one cell, then nearby clones of cells that express the same phenotype as the primary Notch cell should be observed in either normal or experimental situations. Creation of mosaic patches of cells by somatic recombination might be informative.

3. From Figure 15.13, it is clear that eve is expressed only in regions that develop into even-numbered segments.

4. The gene antp is usually expressed all the way back in the embryo from its normal anterior border in T1/T2, but its action is repressed by the BX-C genes. In their absence, Antp is free to determine the segment identity for all segments from T2 to the posterior tip of the embryo as T2.

Chapter 16

1. In a strict sense, cortical rotation is necessary for Nieuwkoop center formation. If rotation is prevented by any means, dorsal mesoderm and a Spemann organizer fail to develop, leading to a monstrously "ventralized" embryo. Recall, however, that Nieuwkoop showed that cells from anywhere in the vegetal hemisphere could induce some kind of mesoderm from an animal cap, the nature of the mesoderm depending on the source of vegetal cells. Evidently, mesoderm induction is broadly distributed in the vegetal hemisphere, and a ventralized embryo will still form mesenchyme and blood cells characteristic of ventral mesoderm. What rotation apparently does is produce a specific area (the Nieuwkoop center) that can induce dorsal mesoderm.

2. Injecting an antisense oligonucleotide against Noggin mRNA into a Xenopus embryo would cause a loss of Noggin function. Noggin is necessary to counteract the ventralizing influences of BMP2 and BMP4. So a loss of Noggin influence would lead to epidermal development instead of neural development in the axial regions of the dorsal part of the embryo. The exact result would probably depend heavily on just how low one is able to "drive" the Noggin mRNA concentrations.

3. Such speculation would practically amount to a short course in receptor biology! One could propose that there are two classes (or more) of receptors with different affinities for the ligand, so that high and low concentrations activate different classes of receptors. Or the number of receptors occupied might lead to clustering of receptors in the plasma membrane, which could lead to localized interactions with certain intracellular constituents. Or activation of different numbers of similar receptors might lead to different levels of phosphorylation of a second-messenger component, and monophosphorylated and multiphosphorylated versions of this component might have very different biochemical activities. And so on.

4. While we do not know, the sequential anteroposterior formation of somites might be due to the timing of expression of regionally specific HOX genes. Or, since the organizer (Hensen's node) moves from anterior to posterior end, it is likely due to a timed delivery of ligands as the signaling center moves posteriorly.

5. The gene *radical fringe* (*rfng*) antagonizes *engrailed1* (*en1*), which is necessary to set the position of the AER. If *rfng* were completely lacking, there would be no border, no AER would develop, and hence no limb would develop. On the other hand, if *rfng* were simply reduced in concentration, perhaps the *en1–rfng* border would shift dorsally; the outcome would depend on the exact concentrations needed to form an adequate "border."

Chapter 17

1. The phylotypic stage in vertebrates is the tail-bud stage. Birds undergo meroblastic cleavage of yolky eggs, and gastrulation by ingression of surface cells. Axis formation occurs by regression of Hensen's node. On the other hand, mice do not have yolky eggs, and cleavage is complete. Gastrulation in the mouse, however, is similar to that in birds.

2. It is known that members of important regulatory circuits play a role in limb development in vertebrates and wing development in insects. Bat wing development could thus be examined for expression of *wnt*, *engrailed*, *hedgehog*, and other such genes, to see whether the bat wing is utilizing molecules similar to those employed in other vertebrate limbs.

3. This is really a question of fact, and so would require careful examination of living and fixed lancelet embryos to diagnose the origins of various nerves. Or this could be the perfect opportunity to go to the library and read up on what is known about the origin of peripheral nerves in *Amphioxus*.

PHOTO CREDITS

Chapter 1: 1.1 A © Montes De Oca; 1.1 B-C, 1.4 Gould, James L., and William T. Keeton, *Biological Science*, 1996, Sixth Edit. W. W. Norton.

Chapter 2: 2.3 B © Jerome Wexler; 2.11 A,B Courtesy of Laurinda Jaffe; 2.12 A-C Vodicka M. A. and Gerhart J. C. (1995). *Development* 121, 3505-3518.

Chapter 3: 3.4 Pepling M. E. and Spradling A. C. (2001). *Dev. Biol.* 234, 339-351; 3.5 B Ray R. P. and Schupbach T (1996). *Genes & Development* 10, 1711-1723; 3.7 Courtesy of A. P. Mahowald.

Chapter 4: 4.1 Dumont (1972). *J. Morphol.* 136,153; 4.2 Gall J. G., Stephenson E. C., Erba H. P., et al (1981). *Chromosoma* 84, 159-171. 4.5 Courtesy of M. Tegner; 4.6 Cha, B-J and Gard, D. L. (1999). *Dev. Biol.* 205,275-286; 4.7 B Vincent et al. (1986). *Dev. Biol.* 113,484-500; 4.7 C top Ellinson (1997). *Dev. Biol.* 128,183, 4.7 C bottom Rowning et al (1988) *PNAS* 94,118-9; 4.9 B Courtesy of D. M. Green.

Chapter 5: 5.6 A-C, 5.7 H. L. Hamilton, *Lillie's Development of the Chick*, plate 2, p. 23, 1952, 3rd Edit. Holt, Rinehart and Winston.

Chapter 6: 6.12 A, B Levi-Montalcini (1964). *Science* 143,105-110; 6.15 A-C Courtesy of Robert Hilfer; 6.17 A,B Halder, Callaerts, And Gehring (1995). *Science* 267,1788-1792.

Chapter 7: 7.15D © Dwight Kuhn 7.20A H. L. Hamilton, *Lillie's Development of the Chick*, plate 6, p. 88, 1952, 3rd Edit. Holt, Rinehart and Winston; b7.2D,F,G, and H Courtesy of H. Nishida.

Chapter 8: 8.4A,B Lam, Hall, Bender and Thummel (1999). *Dev. Biol.* 212, 208; 8.7 Courtesy of Jack Bostrack; 8.8 Ashburner (1974). CSH *Symp Quant Biol* 38,655; 8.13A-C Courtesy of Alejandro Sanchez Alvarado.

Chapter 9: 9.5A-B Courtesy of David Jackson, diagrams adapted from Paul Green; 9.6A-B Courtesy of David Jackson; 9.6C-D Courtesy of Mark Running; 9.8B Courtesy of D. Reinhardt; 9.10 A Courtesy of E. Sussex; 9.10B Courtesy of R. Iverson; 9.10C-E Courtesy of Andrew Hudsen and Marja Timmermans; 9.12 Courtesy of Phil Benfey;

9.13A Casimiro et al (2001). *The Plant Cell Online* 56, 766; 9.13B Courtesy of R. Iverson.

Chapter 10: 10.3A-B Courtesy of Joe Colasanti; 10.4A-E Courtesy of P. Yanofsky; 10.5A-B Courtesy of L. Weigel; 10.9 A Courtesy of Sheila McCormick; 10.13A-B Courtesy of E. Goldberg; 10.14A Courtesy of D. Tasaka (cucl/cuc2 mutant), Courtesy of J. Barton (*Arabidopsis pinhead* mutation), all others Sarah Hake; 10.14B Adapted from Bowman, *Trends in Plant Sciences*; 10.14C Sarah Hake.

Chapter 11: 11.1B Courtesy of J. Schiefelbein; 11.2A-B David G. Oppenheimer, *Current Opinion in Biology*, 1998; 11.3TL and TR DB Sztmanski, RA Jilk, SM Pollock, and MD Marks, Development, 1998; 11.3BL and BR Masucci JD, Rerie WG, Foreman DR, Zhang M, Galway ME, Marks MD, Schiefelbein JW, *Development*, 1996; 11.4A Courtesy of Y. Mizukami; 11.4B-E J. Larkin, M.D. Marks, J. Nadeau, F. D. Sack, *Plant Cell*, 1997; 11.6A A Schnittger A, Folkers U, Schwab B, Jurgens G, Hulskamp M, *Plant Cell*, 1999; 11.7L, TR, and BR Courtesy of Maureen McCann; 11.8 Lucas, W. J. B. Ding and C. Van Der Schoot, *New Phytol*, 1993; 11.10A L, C, and R Courtesy of Allen Sessions; 11.11B, B insert and C Keiji Nakajima, Giovanni Sena, Tal Nawy & Philip N. Benfey, *Nature*, 2001; 11.12 Courtesy David Jackson; 11.13A-D Cleary, A. L. and Smith, L. G. *The Plant Cell*, 1998.

Chapter 12: 12.14E Steinberg, 1962; 12.14F Steinberg, 1962; 12.18A Winklbauer in Keller et al., 1990; 12.18B Leptin and Grunewald, 1990; 12.18C Niswander and Martin, 1993; 12.18D Strome and Wood, 1983; 12.19 Courtesy of Matt Kofron.

Chapter 14: 14.2 F Gould, James L., and William T. Keeton, *Biological Science*, 1996, Sixth Edit. W. W. Norton; 14.13A-B Thomsen and Melton, 1993; B14.1A(b) and B(b) Armone et al. *Development*, 1997.

Chapter 15: 15.12A-D Frasch, M., Hoey, T., Rushlow, C., Doyle, H., and Levine, M. *EMBO Journal*, 1987; 15.16A-D Fujioka et al. *Development*, 1999; 15.18 A-C Lehman, *Development*, 1988; 15.21C Ed Lewis; 15.22C Tom Kaufman; 15.23A-B Levine M; 15.29A-B Zecca, Basler and Struhl,

Cell 87, 1996; b15.2A-B Wang, D., G-W., Kirchhamer, C. V., Britten, R. J., and Davidson, E. H., Development, 1995.

Chapter 16: 16.1A Gerhart, 1989; 16.1B Fujisue, Kabayakawa, and Yamana, 1993; 16.1C Gilbert, Scott. *Developmental Biology*. Sixth Edition, Figure 10.12, Sinauer Associates, Inc., Sunderland, Massacusetts, 2000; 16.2A-C Smith and Harland, 1991; 16.7B Smith, Knecht, Wu and Harland, 1993; 16.8 Lustig et al., 1996; 16.13A-D Levin et al., 1995; 16.15 Graham, Papalopulu and Krumlauf, 1989; 16.16 Sham et al., 1993; 16.17A-B Ramirez-Solis, R., Zheng, H., Whiting, J., Krumlauf, R., and Bradley, A., 1993; 16.22B Cohn et al., 1997.

Chapter 17: 17.1 Halder, G., Callaerts, P., and Gehring, W. J., 1995; 17.5A-B Weatherbee et al., 1999; 17.6A-C Lewis et al., 1999; 17.7A-B Cohn and Tickle, 1999; 17.12A-B Riddle, Johnson, Laufer and Tabin, 1993; 17.13A-B Lowe and Wray, 1997; 17.16A-C EC Raff et al., 1999; 17.17 Swalla and Jeffery, 1990.

NAME INDEX

Entries in **boldfaced** type indicate bibliographic citations. Information found in illustrations, tables, and boxes is indexed in *italic* type.

SUBJECT INDEX

Boldface type indicates where terms are defined in the text, while *italic* type identifies information found in boxes, illustrations, and tables. Parentheses indicate equivalencies, while brackets identify a category or group of which the entry is a member. In most cases, genes and their associated products (mRNA and/or protein) have been indexed in the same entry.